T5-CAO-692

# ORDER AND CHAOS
# IN STELLAR AND PLANETARY SYSTEMS

*COVER ILLUSTRATION:*

An orbit in Hénon-Heiles potential. For more information see Figure 4, paper by Agekian, this volume, page 3.

# ASTRONOMICAL SOCIETY OF THE PACIFIC CONFERENCE SERIES

A SERIES OF BOOKS ON RECENT DEVELOPMENTS IN ASTRONOMY AND ASTROPHYSICS

**Volume 316**



A listing of all other ASP Conference Series and IAU volumes published by the ASP may be found at the back of this volume

ASTRONOMICAL SOCIETY OF THE PACIFIC
CONFERENCE SERIES

**Volume 316**

# ORDER AND CHAOS IN STELLAR AND PLANETARY SYSTEMS

Proceedings of a meeting held in
Saint Petersburg, Russia
17–24 August 2003

Edited by

**Gene G. Byrd**
*University of Alabama, Tuscaloosa, Alabama, USA*

**Konstantin V. Kholshevnikov**
*Saint Petersburg State University, Saint Petersburg, Russia*

**Aleksandr A. Mylläri**
*University of Turku, Turku, Finland*

**Igor' I. Nikiforov**
*Saint Petersburg State University, Saint Petersburg, Russia*

and

**Victor V. Orlov**
*Saint Petersburg State University, Saint Petersburg, Russia*

SAN FRANCISCO

ASTRONOMICAL SOCIETY OF THE PACIFIC

390 Ashton Avenue
San Francisco, California, 94112-1722, USA
Phone: 415-337-1100
Fax:415-337-5205
E-mail: service@astrosociety.org
Web Site: www.astrosociety.org


ASP Conference Series - First Edition



ISBN: 1-58381-172-9

Library of Congress Cataloging in Publication Data
Main entry under title
Card Number: 2004109610

Printed in United States of America by Sheridan Books, Ann Arbor, Michigan

# Dedication

The meeting was dedicated to:

The 90th birthday of
Professor Tatheos Artem'evich Agekian

and

The 70th birthday of
Professor Vadim Anatol'evich Antonov

# Contents

## Part V. Stucture and Kinematics of the Milky Way Galaxy



## Part VI. Outside the Milky Way

# Foreword

I regard being able to work on the organizing committee for this meeting and help arrange for the proceedings' publication as repayment for the important influence of Russian science in my childhood as well as an opportunity to bring important astronomical results to the world.

When I am asked how I became an astronomer, I reply, "I am a child of Sputnik." Sputnik helped change the educational environment of my childhood to make becoming a "pure" or "basic" scientist possible. Although our family's farm in Texas was in an isolated area, attitudes there reflected a larger perception that scholarly education and science were deeply suspect. I remember that my intelligent and dedicated sixth grade teacher would not show a film strip about the planned United States Vanguard satellite because it was "against the Bible." Not long after, even in our isolated part of Texas, past the muddy roads, and the thick stands of trees, the shocking beeping sound of the Russian satellite Sputnik, was played instead of the country and rock music of the local radio stations. For a while at least, the people of the United States realized that fear of education and exploration could be hazardous to the importance and very existence of our nation.

Happily, the shock of Sputnik helped open many doors so that I became the first in my family to graduate from college and one of those fortunate enough to work exploring the universe and, even better, explaining its wonders to others from kindergartners through graduate students. Happily, also, at least between Russia and America, the threat of nuclear destruction has receded from those terrifying days of the 1950's and 1960's. Even in the Texas of my childhood, things have changed. Many years later, while I was in graduate school, when my mother and I visited my now retired sixth grade teacher, she brought out a carefully preserved diagram of the solar system I had drawn for a parent-teacher open house so many years before. I think she had changed her mind.

As my education progressed, I began to appreciate those who built the scientific foundations of Russian space and astronomical achievements such as Konstantin Tsiolkovsky among many others as part of a continuing national commitment to astronomy and space exploration. Characteristically, the United States, after a post-Sputnik Apollo program race to reach the moon, abandoned the effort with no astronauts having been on the moon for many years. Most recently, in response to a second Space Shuttle disaster, we see a plan to prematurely end the Hubble Space Telescope's operation. Another result is that Russian rockets are the principal means of supply and support for the International Space Station.

During my recent scientific visits to Russia, I have been impressed by efforts under difficult circumstances by my Russian hosts to maintain the institutional foundations of Russian scientific achievements. Aside from strengthening the institutional foundations, this meeting and the papers presented in it extend the dynamical foundations of astronomy in the tradition of our two meeting honorees, Professors Tatheos Artem'evich Agekian and Vadim Anatol'evich Antonov on the occasion of their 90th and 70th birthdays, respectively.

Finally, I would like to thank the members of the scientific and local organizing committees as well as the co-editors of these proceedings for all their hard work in initiating, planning, and carrying out a well-planned and interesting meeting. A supplement to NSF grant AST 02-06177 made this book's publication possible. Other support was provided by RFBR (Grant 03-02-26071) and Leading Scientific School (Grant Nsh-1078.2003.02).

Gene Byrd

# Preface

In astronomy, numerous difficulties appear first at the stage of observations. Those are overcome by the hard step-by-step labors of observers and equipment developers. However, theoretical understanding of the results is also far from a straightforward elementary interpretation. It looks like Nature resists our efforts to put its phenomena in the framework of theorists' ideas. The last (but not least) role belongs to one of eternal problems of all thinking people—its essence is in discriminating the important from the insignificant. For example, radiation pressure and tidal interactions which are secondary factors to those on the Earth's surface may become quite important in space exploration. Therefore, it is essential for the theorist not only to create an idea, but also imagine its consequences, e.g., how taking into account additional factor(s) will change the general picture. Moreover, one should learn to envision a continuous evolutionary process since not everything existing in a static state could appear by itself without an external agent.

The same difficulties concern those in the more specialized area of the dynamics of stellar systems. We have very good foundations for which we should be thankful to our great predecessors. However (fortunately for us) there are a number of new problems, as well as long-standing issues that have not been resolved due to lack of observational data or the improper calculation technique, e.g., many theoretical methods that turned out to be useful in the dynamics of gravitating systems were invented in plasma physics since this area was more thoroughly developed due to its applications. The process works in the opposite direction as well—Professor K. F. Ogorodnikov was pleased by the fact that his classic book "Dynamics of Stellar Systems" was popular among plasma physicists. As a rule, individual theorists and research groups have different skills and experience in using methods of calculations and observational facilities—hence there is (some) unavoidable subjectivity of the conclusions. However, we hope that this subjectivity is of a temporary nature, since gradually points of view are converging, mistakes are revealed, unsuccessful theories are thrown away, and constructions, which contradict facts, are forgotten. In order to promote these processes, an exchange of ideas between researchers is required. This meeting is a good example of such an exchange. We hope that these Conference Proceedings will introduce a wide community of astronomers to modern ideas and trends in astronomical dynamics.

The place for the conference is not accidental. The Saint Petersburg school of stellar dynamics has a long-standing history starting from works of Professor Kyrill Ogorodnikov. Many "disciples" of the Ogorodnikov School took part in the organization and work of the conference. In particular, we would like to mention two of those disciples, namely Professors Tatheos Agekian (who was the first disciple of Professor Ogorodnikov) and Vadim Antonov who started studies in stellar dynamics twenty years later. Both are world-famous scientists and supervise future generations of disciples keeping the traditions of the School alive. The conference was devoted to their 90th and 70th birthdays.

The conference functioned in a businesslike but friendly atmosphere, and had public "resonances" at the University and in the city of Saint Petersburg. Some contributions were followed by lively discussions. The topics of contributions cover a wide range of topics. Some of them are rather specialized (e.g., the solar neighborhood, stellar orbits in galactic models, the few-body problem, etc.). Other contributions have a generalized character (e.g., Antonov's and Fridman's reviews). The astronomical areas of the Conference were quite wide, ranging from planetary systems to galactic dynamics. As a whole, this formed a combination of small-size and large-scale approaches to difficult dynamical problems.

G. Byrd, K. Kholshevnikov, A. Mylläri, I. Nikiforov, and V. Orlov

The Scientific Organizing Committee:
G. Byrd (USA, co-chair), K. Kholshevnikov (Russia, co-chair), L. Ossipkov (Russia, deputy chair), L. Sokolov (Russia, secretary), E. Athanassoula (France), V. Brumberg (Russia), A. Chernin (Russia), L. Ksanfomaliti (Russia), J. Makino (Japan), S. Nuritdinov (Uzbekistan), D. Pfenniger (Switzerland), S. White (Germany).

The Local Organizing Committee:
A. Mylläri (co-chair), V. Orlov (co-chair), A. Rubinov (secretary), B. Eskin, Yu. Gnedin, A. Kasaurov, E. Kazakevich, S. Kutuzov, A. Minz, I. Nikiforov, N. Petrov, N. Pitjev, D. Sidorin, L. Sokolov, P. Tarakanov, V. Troyan.

## Participants

T. A. Agekian, Sobolev Astronomical Institute, Saint Petersburg State University, Universitetskij pr. 28, Staryj Peterhof, Saint Petersburg 198504, Russia ⟨ vor@astro.spbu.ru ⟩

V. A. Antonov, Central Astronomical Observatory at Pulkovo, Pulkovskoe sh. 65/1, Saint Petersburg 196140, Russia ⟨ antonov@gao.spb.ru ⟩

S. A. Astakhov, Department of Chemistry & Biochemistry, Utah State University, Logan, Utah 84322-0300, USA ⟨ astakhov@newmail.ru ⟩

A. V. Bagrov, Institute of Astronomy of Russian Academy of Science, Pyatnitskaya ul. 48, Moscow, 119017, Russia ⟨ abagrov@inasan.ru ⟩

Y. V. Barkin, Sternberg Astronomical Institute, Moscow State University, Universitetskij pr. 13, 119992, Moscow, Russia ⟨ yuri.barkin@ua.es ⟩

A. A. Bekov, Fesenkov Astrophysical Institute, Almaty, 480020, Kazakhstan ⟨ bekov@mail.ru ⟩

V. V. Bobylev, Central Astronomical Observatory at Pulkovo, Pulkovskoe sh. 65/1, Saint Petersburg 196140, Russia ⟨ vbobylev@gao.spb.ru ⟩

G. G. Byrd, Department of Physics and Astronomy, University of Alabama, Tuscaloosa, 35487-0324, USA ⟨ byrd@possum.astr.ua.edu ⟩

A. D. Chernin, Sternberg Astronomical Institute, Moscow State University, Universitetskij pr. 13, 119992, Moscow, Russia ⟨ chernin@sai.msu.ru ⟩

I. Ciufolini, Dipartamento d'ingeneria dell'Innovazione, Universita di Lecce, Via per Arnesano, 73100, Italy ⟨ ciufoli@sigfrido.ifsi.rm.cnr.it ⟩

M. Croustalloudi-Flerianou, Department of Mechanics, Faculty of Applied Sciences, National Technical University of Athens, 5 Heroes of Polytechnion Ave, 157 73, Athens, Greece ⟨ markr@mail.ntua.gr ⟩

R. Cubarsi, Dept. Matematica Aplicada IV, Universitat Politecnica de Catalunya, Jordi Girona, 1-3, 08034 Barcelona, Spain ⟨ santiago@alcobe.net ⟩

V. V. Dikarev, MPI für Kernphysik, Postfach 103980, 69117 Heidelberg, Germany ⟨ dikarev@galileo.mpi-hd.mpg.de ⟩

Y. N. Efremov, Sternberg Astronomical Institute, Moscow State University, Universitetskij pr. 13, 119992, Moscow, Russia ⟨ efremov@sai.msu.ru ⟩

B. Famaey, Institut d'Astronomie et d'Astrophysique, Université Libre de Bruxelles, CP 226, Boulevard du Triomphe, B-1050 Bruxelles, Belgium ⟨ bfamaey@astro.ulb.ac.be ⟩

S. Franck, Potsdam Institute for Climate Impact Research, P.O. Box 601203, 14412 Potsdam, Germany ⟨franck@pik-potsdam.de⟩

T. Freeman, Bevill State Community College, 2631 Temple Ave. North, Fayette, AL 35555, USA ⟨tfreeman@bscc.edu⟩

A. M. Fridman, Institute of Astronomy of Russian Academy of Science, Pyatnitskaya ul. 48, Moscow, 119017, Russia ⟨fxela@online.ru⟩

M. L. Gozha, Rostov State Pedagogical University, Dneprovsky, 116, Rostov-on-Don, 344065, Russia ⟨galina@iubip.ru⟩

G. A. Gontcharov, Central Astronomical Observatory at Pulkovo, Pulkovskoe sh. 65/1, Saint Petersburg 196140, Russia ⟨georgegontcharov@yahoo.com⟩

D. L. Gorshanov, Central Astronomical Observatory at Pulkovo, Pulkovskoe sh. 65/1, Saint Petersburg 196140, Russia ⟨shakht@gao.spb.ru⟩

E. Griv, Ben-Gurion University of the Negev, P.O. Box 653, Beer-Sheva 84105, Israel ⟨griv@bgumail.bgu.ac.il⟩

T.-Y. Huang, Astronomy Department, Nanjing University, Nanjing, 210093, China ⟨tyhuang@nju.edu.cn⟩

T. Kalvouridis, Department of Mechanics, Faculty of Applied Sciences, National Technical University of Athens, 5 Heroes of Polytechnion Ave, 157 73, Athens, Greece ⟨telkal@central.ntua.gr⟩

E. E. Kazakevich, Sobolev Astronomical Institute, Saint Petersburg State University, Universitetskij pr. 28, Staryj Peterhof, Saint Petersburg 198504, Russia ⟨elen@ek3286.spb.edu⟩

A. A. Kholopov, Syktyvkar State University, Oktjabrsky pr. 55, Syktyvkar, 167001, Russia ⟨polsm@syktsu.ru⟩

K. V. Kholshevnikov, Sobolev Astronomical Institute, Saint Petersburg State University, Universitetskij pr. 28, Staryj Peterhof, Saint Petersburg 198504, Russia ⟨kvk@astro.spbu.ru⟩

A. A. Kisselev, Central Astronomical Observatory at Pulkovo, Pulkovskoe sh. 65/1, Saint Petersburg 196140, Russia ⟨lrom@gao.spb.ru⟩

B. P. Kondratyev, Mathematical Faculty, Udmurt State University, Universitetskaya 1, Izhevsk, 426034, Russia ⟨kond@uni.udm.ru⟩

T. S. Kozhanov, Kazakh National Agrarian University, ave. Abai 8, Almaty, 480100, Republic of Kazakhstan ⟨koblandy@mail.ru⟩

A. Kovačević, Dept. of Astronomy, Faculty of Mathematics, University of Belgrade, Studentski trg. 16, Belgrade, P.O. Box 550, Serbia and Montenegro ⟨andjelka@poincare.matf.bg.ac.yu⟩

S. A. Kutuzov, Department of Cosmic Technologies and Applied Astrodynamics, Saint Petersburg State University, Universitetskij pr. 35, Staryj Peterhof, Saint Petersburg 198504, Russia ⟨Kutuzov@apmath.spbu.ru⟩

E. D. Kuznetsov, Astronomical Observatory, Urals State University, Lenin pr. 51, Ekaterinburg, 620083, Russia ⟨Eduard.Kuznetsov@usu.ru⟩

Y. V. LESKOV, Astronomical Observatory, Urals State University, Lenin pr. 51, Ekaterinburg, 620083, Russia ⟨ Vladimir.Danilov@usu.ru ⟩

A. V. LOKTIN, Astronomical Observatory, Urals State University, Lenin pr. 51, Ekaterinburg, 620083, Russia ⟨ Alexhander.Loktin@usu.ru ⟩

V. V. MAKAROV, US Naval Observatory, 3450 Massachusetts Ave NW, Washington D.C. 20392-5420 USA ⟨ makarov@usno.navy.mil ⟩

G. A. MALASIDZE, Tbilisi State University, University St. 2, Tbilisi 0143, Georgia ⟨ tmdzinarishvili@yahoo.com ⟩

M. A. MARDANOVA, Department of Cosmic Technologies and Applied Astrodynamics, Saint Petersburg State University, Universitetskij pr. 35, Staryj Peterhof, Saint Petersburg 198504, Russia ⟨ mardanova@hotmail.com ⟩

A. I. MARTYNOVA, Forest Technical Academy, Institutskij per. 5, Saint Petersburg 194018, Russia ⟨ vor@astro.spbu.ru ⟩

A. V. MELNIKOV, Central Astronomical Observatory at Pulkovo, Pulkovskoe sh. 65/1, Saint Petersburg 196140, Russia ⟨ melnikov@gao.spb.ru ⟩

A. A. MINZ, Sobolev Astronomical Institute, Saint Petersburg State University, Universitetskij pr. 28, Staryj Peterhof, Saint Petersburg 198504, Russia ⟨ minzastro@yandex.ru ⟩

A. A. MYLLÄRI, Department of Information Technology, University of Turku, Lemminkaisenkatu 14-18 A, 20520, Turku, Finland ⟨ alemio@utu.fi ⟩

I. I. NIKIFOROV, Sobolev Astronomical Institute, Saint Petersburg State University, Universitetskij pr. 28, Staryj Peterhof, Saint Petersburg 198504, Russia ⟨ nii@astro.spbu.ru ⟩

S. NINKOVIĆ, Astronomical Observatory of Belgrade, Volgina 7, 11160 Beograd-74, Serbia and Montenegro ⟨ sninkovic@aob.bg.ac.yu ⟩

S. N. NURITDINOV, Astronomy Department, Physical Faculty, National University of Uzbekistan, Vuzgorodok, Tashkent, 700174, Uzbekistan ⟨ nurit@astrin.uzsci.net ⟩

V. V. ORLOV, Sobolev Astronomical Institute, Saint Petersburg State University, Universitetskij pr. 28, Staryj Peterhof, Saint Petersburg 198504, Russia ⟨ vor@astro.spbu.ru ⟩

L. P. OSSIPKOV, Department of Cosmic Technologies and Applied Astrodynamics, Saint Petersburg State University, Universitetskij pr. 35, Staryj Peterhof, Saint Petersburg 198504, Russia ⟨ leo@dyna.astro.spbu.ru ⟩

N. A. PETROV, Sobolev Astronomical Institute, Saint Petersburg State University, Universitetskij pr. 28, Staryj Peterhof, Saint Petersburg 198504, Russia ⟨ petrov@astro.spbu.ru ⟩

A. V. PETROVA, Sobolev Astronomical Institute, Saint Petersburg State University, Universitetskij pr. 28, Staryj Peterhof, Saint Petersburg 198504, Russia ⟨ an@astro.spbu.ru ⟩

N. P. PITJEV, Sobolev Astronomical Institute, Saint Petersburg State University, Universitetskij pr. 28, Staryj Peterhof, Saint Petersburg 198504, Russia ⟨ ai@astro.spbu.ru ⟩

S. M. POLESHCHIKOV, Syktyvkar Forest Institute, Lenin st. 35, Syktyvkar, 167000, Russia ⟨ polsm@syktsu.ru ⟩

E. N. POLYAKHOVA, Sobolev Astronomical Institute, Saint Petersburg State University, Universitetskij pr. 28, Staryj Peterhof, Saint Petersburg 198504, Russia ⟨ pol@astro.spbu.ru ⟩

R. R. RAFIKOV, Institute for Advanced Study, Einstein Drive, Princeton, NJ, 08540, USA ⟨ rrr@ias.edu ⟩

N. V. RASPOPOVA, Department of Cosmic Technologies and Applied Astrodynamics, Saint Petersburg State University, Universitetskij pr. 35, Staryj Peterhof, Saint Petersburg 198504, Russia ⟨ natalya_rasp@mail.ru ⟩

A. S. RASTORGUEV, Sternberg Astronomical Institute, Moscow State University, Universitetskij pr. 13, 119992, Moscow, Russia ⟨ rastor@sai.msu.ru ⟩

L. G. ROMANENKO, Central Astronomical Observatory at Pulkovo, Pulkovskoe sh. 65/1, Saint Petersburg 196140, Russia ⟨ lrom@gao.spb.ru ⟩

A. V. RUBINOV, Sobolev Astronomical Institute, Saint Petersburg State University, Universitetskij pr. 28, Staryj Peterhof, Saint Petersburg 198504, Russia ⟨ rav@astro.spbu.ru ⟩

M. SAITO, National Astronomical Observatory of Japan, 2-21-1, Osawa, Mitaka-shi, Tokyo, 181-8588, Japan ⟨ saitohms@cc.nao.ac.jp ⟩

N. A. SHAKHT, Central Astronomical Observatory at Pulkovo, Pulkovskoe sh. 65/1, Saint Petersburg 196140, Russia ⟨ shakht@gao.spb.ru ⟩

M. E. SHARINA, Special Astrophysical Observatory of Russian Academy of Sciences, N. Arkhyz, KChR, 369167, Russia ⟨ sme@sao.ru ⟩

I. I. SHEVCHENKO, Central Astronomical Observatory at Pulkovo, Pulkovskoe sh. 65/1, Saint Petersburg 196140, Russia ⟨ iis@gao.spb.ru ⟩

A. SHUKSTO, Sobolev Astronomical Institute, Saint Petersburg State University, Universitetskij pr. 28, Staryj Peterhof, Saint Petersburg 198504, Russia ⟨ simply@pisem.net ⟩

V. V. SIDORENKO, Keldysh Institute of Applied Mathematics of Russian Academy of Sciences, Miusskaya sq. 4, Moscow 125047, Russia ⟨ sidorenk@spp.keldysh.ru ⟩

L. L. SOKOLOV, Sobolev Astronomical Institute, Saint Petersburg State University, Universitetskij pr. 28, Staryj Peterhof, Saint Petersburg 198504, Russia ⟨ lsok@astro.spbu.ru ⟩

V. G. SURDIN, Sternberg Astronomical Institute, Moscow State University, Universitetskij pr. 13, 119992, Moscow, Russia ⟨ surdin@sai.msu.ru ⟩

E. I. TIMOSHKOVA, Central Astronomical Observatory at Pulkovo, Pulkovskoe sh. 65/1, Saint Petersburg 196140, Russia ⟨ elenatim@gao.spb.ru ⟩

M. VALTONEN, Tuorla Observatory, University of Turku, Väisäläntie 20, 21500 Piikkiö, Finland ⟨ mvaltonen2001@yahoo.com ⟩

S. Y. Vernov, Skobeltsyn Institute of Nuclear Physics, Moscow State University, Vorob'evy Gory, Moscow, 119992, Russia ⟨svernov@theory.sinp.msu.ru⟩

E. Y. Vilkoviskij, Fesenkov Astrophysical Institute, Almaty 480020, Kazakhstan ⟨vilk@afi.south-capital.kz⟩

V. V. Vityazev, Sobolev Astronomical Institute, Saint Petersburg State University, Universitetskij pr. 28, Staryj Peterhof, Saint Petersburg 198504, Russia ⟨vityazev@venvi.usr.pu.ru⟩

A. A. V'yuga, Sobolev Astronomical Institute, Saint Petersburg State University, Universitetskij pr. 28, Staryj Peterhof, Saint Petersburg 198504, Russia ⟨vegalira@rambler.ru⟩

M. V. Zabolotskikh, Sternberg Astronomical Institute, Moscow State University, Universitetskij pr. 13, 119992, Moscow, Russia ⟨zabolot@sai.msu.ru⟩

O. A. Zheleznyak, Uman' State Pedagogical University, Salova, 2, Uman', Cherkasska reg., 258900, Ukraine ⟨uman-astro@yahoo.com⟩

Zheng Jia-Qing, Tuorla Observatory, University of Turku, Väisäläntie 20, 21500 Piikkiö, Finland ⟨zheng@oj287.astro.utu.fi⟩

Zhou Liyong, Tuorla Observatory, University of Turku, Väisäläntie 20, 21500 Piikkiö, Finland ⟨liyzho@utu.fi⟩

# Part I

# General Properties of Orbits

*Order and Chaos in Stellar and Planetary Systems*
*ASP Conference Series, Vol. 316, 2004*
*G. Byrd, K. Kholshevnikov, A. Mylläri, I. Nikiforov, and V. Orlov, eds.*

# Motion in the Rotationally Symmetric Potential Field

T. Agekian

*Sobolev Astronomical Institute, St. Petersburg State University, Universitetskij pr. 28, St. Petersburg 198504, Russia*

**Abstract.** New equations of analytical mechanics are found. Those supplement the known equations of classical analytical mechanics. A new analytical mechanics is formed.

## 1. Formulation of the Problem

The rotationally symmetric potential is the most widespread form of potential in nature. The gravitational fields of the Earth, other planets and their satellites, stars, globular clusters, elliptical and spiral galaxies are exactly or almost exactly rotationally symmetric. There are grounds for believing that rotationally symmetric force fields are also widespread in the microworld.

In a rotationally symmetric potential, the infinite trajectory of a mass point fills a torus, all of those meridional cross sections are identical. To determine the laws of particle motion in a rotationally symmetric potential, it will suffice to study the features of its motion in the meridional (co-moving) plane that contains the particle and axis of the rotational symmetry.

This paper is devoted to properties of the field formed by the winds of infinite trajectory in meridional plane. We introduce a method to study this field, using new equations following from classic equations of motion.

## 2. Basic Equation

Let us denote the angle between the tangent to a trajectory in the meridional plane and the abscissa axis in this plane by $f$ (see Figure 1).

Based on some obvious properties of the third integral of motion and using the Boltzmann equation, we derived the following equation of the trajectory in meridional plane:

$$\frac{\partial f}{\partial R}\cos f + \frac{\partial f}{\partial z}\sin f + \frac{1}{2(U+I)}\left(\frac{\partial U}{\partial R}\sin f - \frac{\partial U}{\partial z}\cos f\right) = 0, \tag{1}$$

where $U(R, z)$ is the potential in meridional plane defined by the equality

$$U(R, z) = \Phi(R, z) - \frac{J^2}{2R^2},$$

here $\Phi(R, z)$ is the rotationally symmetric potential, $J$ is the area integral, $I$ is the energy integral.

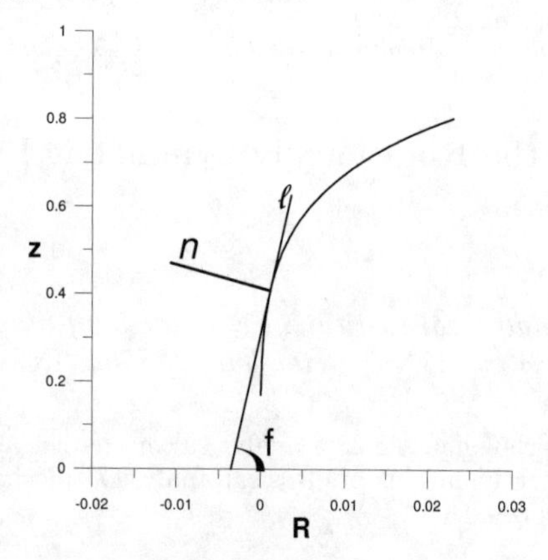


Figure 1. Direction field in meridional plane.

Given $U(R,z)$, $I$, and the angle $f$ at some point $(R,z)$ in the meridional plane, equation (1) defines the orbit that is filled with the winds of the infinite particle trajectory. In this case, the value of the third integral of motion is fixed (if it exists), although it remains unknown.

Figures 2 and 4 show results of our numerical integrations of equation (1) over long time intervals for a Contopoulos potential

$$U(R,z) = -\frac{1}{2}(R^2 + \frac{2}{3}z^2) + \frac{1}{3}Rz^2,$$

$$I = 0.366,$$

and for the Hénon–Heiles potential

$$U(R,z) = -\frac{1}{2}(R^2 + z^2) - Rz^2 + \frac{1}{3}R^3,$$

$$I = 0.051.$$

The angle $f$ was specified at some initial point $(R_0, z_0)$.

In each point of the orbit a region where two conjugate trajectory winds with the angles $f$ and $\alpha$ intersect, the equation

$$\frac{\partial \alpha}{\partial R}\cos\alpha + \frac{\partial \alpha}{\partial z}\sin\alpha + \frac{1}{2(U+I)}\left(\frac{\partial U}{\partial R}\sin\alpha - \frac{\partial U}{\partial z}\cos\alpha\right) = 0 \qquad (2)$$

defines the same orbit as equation (1).

In each point of the regions where four or more winds intersect (there is one such region in Figure 2 and three such regions in Figure 4), the winds each time break down into pairs of conjugated winds. We can always consider the pairs of conjugate equations (1) and (2).

When differentiated with respect to $f$ and $\alpha$ directions, the equations (1) and (2) preserve the orbit and the integrals of motion.

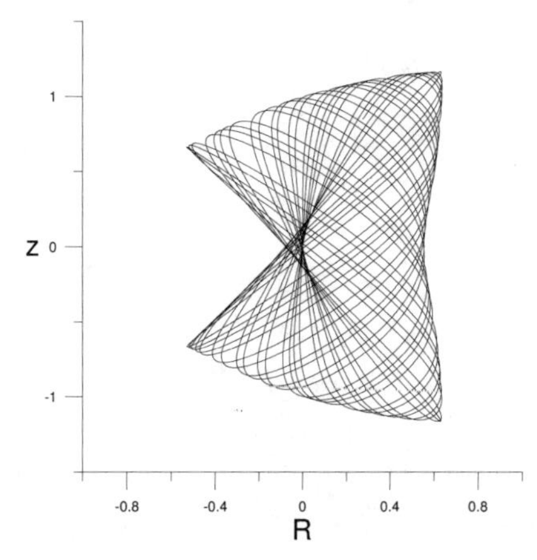


Figure 2. A trajectory in the Contopoulos potential. The coordinates of start point are $R_0 = 0$, $z_0 = 0$. The initial angle $f$ is $\pi/2$.

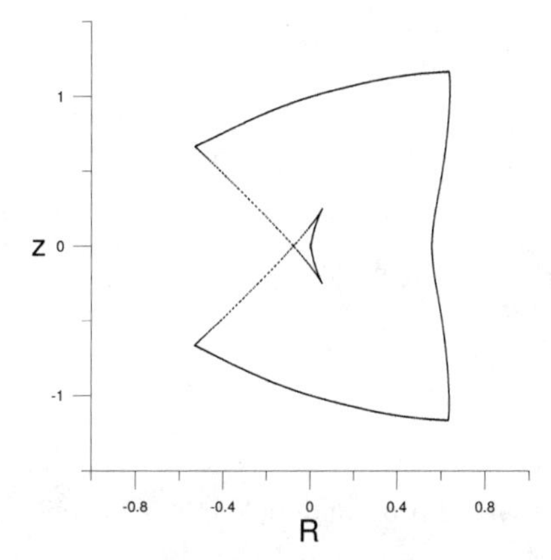


Figure 3. Contours of orbit and folds in the Contopoulos potential.

## 3. An Equation Valid on the Contours of the Orbit and the Folds of the Direction Field

The derivative of the angle $f$ along the normal to trajectory $\frac{\partial f}{\partial n}$ (see Figure 1) has a peculiar property. Agekian & V'yuga (1973) and Saginashvili (1975) considered the problem in the fields of quasi-Newtonian and quasi-Hooke potentials, respectively. They have shown that $\frac{\partial f}{\partial n} = \pm\infty$ on the contours of the orbit and the folds of the direction fields. This property is valid for an arbitrary rotationally symmetric potential.

Previously we derived the following equation for $\frac{\partial f}{\partial n}$:

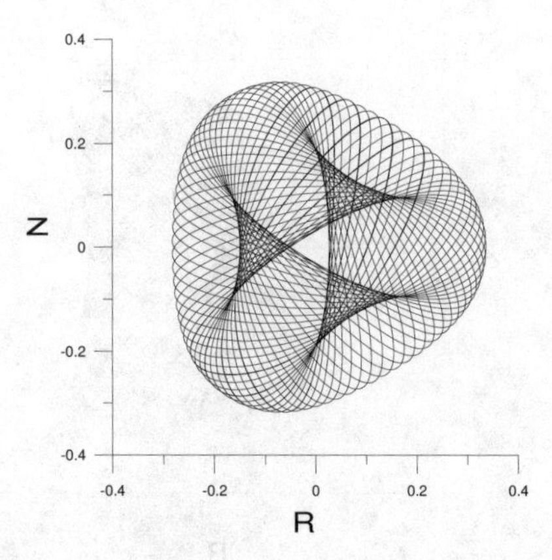


Figure 4. A trajectory in the Hénon–Heiles potential. The coordinates of start point are $R_0 = -0.15$, $z_0 = 0$. The initial angle $f$ is $\pi/2$.

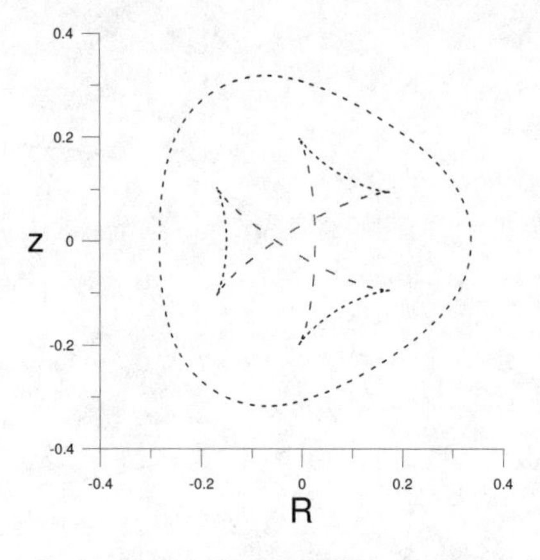


Figure 5. Contours of orbit and folds in the Hénon–Heiles potential.

$$\frac{\partial}{\partial l}\frac{\partial f}{\partial n} + \left(\frac{\partial f}{\partial n}\right)^2 + \frac{1}{2(U+I)}\left(\frac{\partial U}{\partial R}\cos f + \frac{\partial U}{\partial z} sin f\right)\frac{\partial f}{\partial n}$$
$$+\frac{3}{4(U+I)^2}\left(-\frac{\partial U}{\partial R}\sin f + \frac{\partial U}{\partial z}\cos f\right)^2$$
$$+\frac{1}{2(U+I)}\left(-\frac{\partial^2 U}{\partial R^2}\sin^2 f + 2\frac{\partial^2 U}{\partial R \partial z}\sin f\cos f - \frac{\partial^2 U}{\partial z^2}\cos^2 f\right) = 0. \quad (3)$$

In general, this equation is approximate, because it was derived by using differentiation along the normal to trajectory.

However, Agekian & Yakimov (1976) showed that, if one numerically integrates equations (1) and (3) simultaneously along the trajectory and marks the points at which $\frac{\partial f}{\partial n} = \pm\infty$, then the contours of the orbit and the folds can be accurately outlined. In this case, there is no need to plot the trajectory winds;

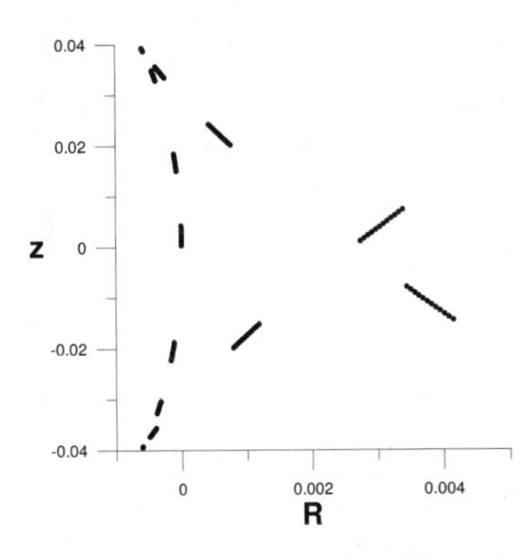


Figure 6. A fold of the third order in the Contopoulos potential.

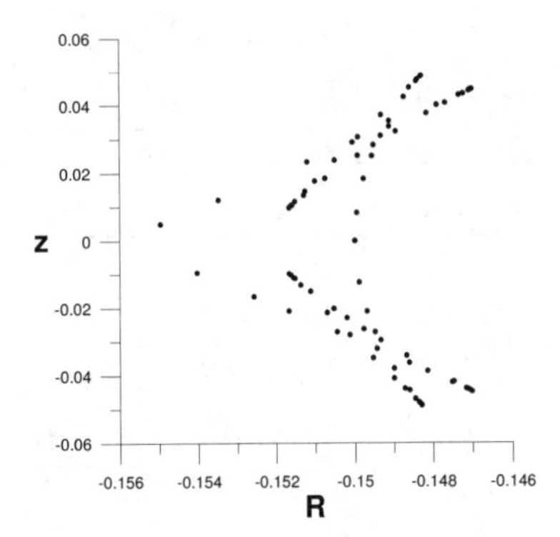


Figure 7. A fold of the third order in the Hénon–Heiles potential.

the pattern of the contours of the orbit and the folds is well defined in agreement with what is obtained by numerical integration of equation (1). This is shown in Figures 3 and 5.

Many authors (Pit'ev 1980, 1981; Agekian et al. 1992; Myllari & Orlov 1993; Dubinina 2001; Myllari et al. 2002) confirmed the accuracy of this method by numerical integration. Agekian & Orlov (2002) showed analytically that this method is accurate—differentiating along the normal to trajectory does not change the orbit and the integrals of motion.

The formation of the folds of the higher order at the middles of the lower order folds or in vicinities of their angular points is possible (see Figures 6 and 7).

## 4. Additional Equations

Let us denote

$$h(R, z) = U(R, z) + I;$$

$$a_1 = \frac{\partial^2 f}{\partial z^2} - \frac{\partial f}{\partial R}\frac{\partial f}{\partial z} + \frac{1}{2h}\frac{\partial f}{\partial z}\frac{\partial U}{\partial z} + \frac{1}{2h}\frac{\partial^2 U}{\partial R \partial z} - \frac{1}{2h^2}\frac{\partial U}{\partial R}\frac{\partial U}{\partial z},$$

$$a_2 = \frac{\partial^2 f}{\partial R \partial z} - \left(\frac{\partial f}{\partial R}\right)^2 + \frac{1}{2h}\frac{\partial f}{\partial R}\frac{\partial U}{\partial z} + \frac{1}{2h}\frac{\partial^2 U}{\partial R^2} - \frac{1}{2h^2}\left(\frac{\partial U}{\partial R}\right)^2,$$

$$a_3 = \frac{\partial^2 f}{\partial R \partial z} + \left(\frac{\partial f}{\partial z}\right)^2 + \frac{1}{2h}\frac{\partial f}{\partial z}\frac{\partial U}{\partial R} - \frac{1}{2h}\frac{\partial^2 U}{\partial z^2} + \frac{1}{2h^2}\left(\frac{\partial U}{\partial z}\right)^2,$$

$$a_4 = \frac{\partial^2 f}{\partial R^2} + \frac{\partial f}{\partial R}\frac{\partial f}{\partial z} + \frac{1}{2h}\frac{\partial f}{\partial R}\frac{\partial U}{\partial R} - \frac{1}{2h}\frac{\partial^2 U}{\partial R \partial z} + \frac{1}{2h^2}\frac{\partial U}{\partial R}\frac{\partial U}{\partial z},$$

$$a_5 = \frac{\partial^2 f}{\partial z^2} + \frac{1}{2h}\frac{\partial f}{\partial R}\frac{\partial U}{\partial R} + \frac{1}{2h}\frac{\partial^2 U}{\partial R \partial z} - \frac{3}{4h^2}\frac{\partial U}{\partial R}\frac{\partial U}{\partial z},$$

$$a_6 = 2\frac{\partial^2 f}{\partial R \partial z} - \frac{1}{2h}\frac{\partial f}{\partial z}\frac{\partial U}{\partial R} - \frac{1}{2h}\frac{\partial f}{\partial R}\frac{\partial U}{\partial z} + \frac{1}{2h}\frac{\partial^2 U}{\partial R^2} - \frac{1}{2h}\frac{\partial^2 U}{\partial z^2}$$
$$- \frac{3}{4h^2}\left(\frac{\partial U}{\partial R}\right)^2 + \frac{3}{4h^2}\left(\frac{\partial U}{\partial z}\right)^2,$$

$$a_7 = \frac{\partial^2 f}{\partial R^2} + \frac{1}{2h}\frac{\partial f}{\partial z}\frac{\partial U}{\partial z} - \frac{1}{2h}\frac{\partial^2 U}{\partial R \partial z} + \frac{3}{4h^2}\frac{\partial U}{\partial R}\frac{\partial U}{\partial z};$$

$$b_1 = \frac{\partial^2 \alpha}{\partial z^2} - \frac{\partial \alpha}{\partial R}\frac{\partial \alpha}{\partial z} + \frac{1}{2h}\frac{\partial \alpha}{\partial z}\frac{\partial U}{\partial z} + \frac{1}{2h}\frac{\partial^2 U}{\partial R \partial z} - \frac{1}{2h^2}\frac{\partial U}{\partial R}\frac{\partial U}{\partial z},$$

$$b_2 = \frac{\partial^2 \alpha}{\partial R \partial z} + \left(\frac{\partial \alpha}{\partial z}\right)^2 + \frac{1}{2h}\frac{\partial \alpha}{\partial z}\frac{\partial U}{\partial R} - \frac{1}{2h}\frac{\partial^2 U}{\partial z^2} + \frac{1}{2h^2}\left(\frac{\partial U}{\partial z}\right)^2,$$

$$b_3 = \frac{\partial^2 \alpha}{\partial R \partial z} - \left(\frac{\partial \alpha}{\partial R}\right)^2 + \frac{1}{2h}\frac{\partial \alpha}{\partial R}\frac{\partial U}{\partial z} + \frac{1}{2h}\frac{\partial^2 U}{\partial R^2} - \frac{1}{2h^2}\left(\frac{\partial U}{\partial R}\right)^2,$$

$$b_4 = \frac{\partial^2 \alpha}{\partial R^2} + \frac{\partial \alpha}{\partial R}\frac{\partial \alpha}{\partial z} + \frac{1}{2h}\frac{\partial \alpha}{\partial R}\frac{\partial U}{\partial R} - \frac{1}{2h}\frac{\partial^2 U}{\partial R \partial z} + \frac{1}{2h^2}\frac{\partial U}{\partial R}\frac{\partial U}{\partial z},$$

$$b_5 = \frac{\partial^2 \alpha}{\partial z^2} + \frac{1}{2h}\frac{\partial \alpha}{\partial R}\frac{\partial U}{\partial R} + \frac{1}{2h}\frac{\partial^2 U}{\partial R \partial z} - \frac{3}{4h^2}\frac{\partial U}{\partial R}\frac{\partial U}{\partial z},$$

$$b_6 = 2\frac{\partial^2 \alpha}{\partial R \partial z} - \frac{1}{2h}\frac{\partial \alpha}{\partial z}\frac{\partial U}{\partial R} - \frac{1}{2h}\frac{\partial \alpha}{\partial R}\frac{\partial U}{\partial z} + \frac{1}{2h}\frac{\partial^2 U}{\partial R^2} - \frac{1}{2h}\frac{\partial^2 U}{\partial z^2}$$
$$- \frac{3}{4h^2}\left(\frac{\partial U}{\partial R}\right)^2 + \frac{3}{4h^2}\left(\frac{\partial U}{\partial z}\right)^2,$$

$$b_7 = \frac{\partial^2 \alpha}{\partial R^2} + \frac{1}{2h}\frac{\partial \alpha}{\partial z}\frac{\partial U}{\partial z} - \frac{1}{2h}\frac{\partial^2 U}{\partial R \partial z} + \frac{3}{4h^2}\frac{\partial U}{\partial R}\frac{\partial U}{\partial z}.$$

Let us differentiate equations (1) and (2) along the directions of $f$ and $\alpha$:

$$\begin{aligned} a_1 \tan f \tan \alpha + a_2 \tan f + a_3 \tan \alpha + a_4 &= 0, \\ b_1 \tan f \tan \alpha + b_2 \tan f + b_3 \tan \alpha + b_4 &= 0, \\ a_5 \tan^2 f + a_6 \tan f + a_7 &= 0, \\ b_5 \tan^2 \alpha + b_6 \tan \alpha + b_7 &= 0. \end{aligned} \tag{4}$$

Eliminating $\tan f$ and $\tan\alpha$ from equation (4), we obtain

$$\begin{aligned}(-a_1a_5b_4 + a_1a_6b_2 + a_2a_5b_3 - a_2a_6b_1 - a_3a_5b_2 + a_4a_5b_1)\\ \times(-a_1a_7b_4 + a_2a_7b_3 + a_3a_6b_4 - a_3a_7b_2 - a_4a_6b_3 + a_4a_7b_1)\\ +(a_1a_7b_2 - a_2a_7b_1 - a_3a_5b_4 + a_4a_5b_3)^2 &= 0,\end{aligned} \tag{5}$$

$$\begin{aligned}(-b_1b_5a_4 + b_1b_6a_3 - b_2b_5a_3 - b_3b_6a_1 + b_3b_5a_2 + b_4b_5a_1)\\ \times(-b_1b_7a_4 + b_2b_6a_4 - b_2b_7a_3 + b_3b_7a_2 - b_4b_6a_2 + b_4b_7a_1)\\ +(b_1b_7a_3 - b_2b_5a_4 - b_3b_7a_1 + b_4b_5a_2)^2 &= 0.\end{aligned} \tag{6}$$

Equations (5) and (6) contain derivatives up to the second order with respect to $f$ and $\alpha$ only. The system of equations and equalities that defines the motion in the field of a rotationally symmetric potential may be presented in the following form:

(I) In the region of possible motions in the meridional plane, $f$ and $\alpha$ are the angles of conjugate trajectories. Equations (1), (2), (5), and (6) are valid.

(II) On the contours of the orbit and the folds of the direction field, $f = \alpha$, $\frac{\partial f}{\partial n} = \pm\infty$, and equation (3) is valid.

(III) When equations (1) and (3) are numerically integrated simultaneously, the points, at which $\frac{\partial f}{\partial n} = \pm\infty$, accurately outline the contours of the orbit and the folds of the direction field.

## References

Agekian, T. A., & Orlov, V. V. 2002, Ast.Lett., 28, 63

Agekian, T. A., & V'yuga, A. A. 1973, Vestnik Leningr.Univ.Ser.1, 7, 128

Agekian, T. A., & Yakimov, S. P. 1976, Vestnik Leningr.Univ.Ser.1, 13, 177

Agekian, T. A., Myllari, A. A., & Orlov, V.V. 1992, Soviet AZh, 69, 469

Dubinina, L. 2001, in Stellar Dynamics: from Classic to Modern, ed. L. P. Ossipkov & I. I. Nikiforov (St. Petersburg: St. Petersburg Univ. Press), 198

Myllari, A. A., & Orlov, V. V. 1993, Vestnik SPb Univ.Ser.1, 15, 120

Myllari, A. A., & Orlov, V. V., Pit'ev, N. P. 2002, Vestnik SPb Univ.Ser.1, 17, 105

Pit'ev, N. P. 1980, Vestnik Leningr.Univ.Ser.1, 19, 98

Pit'ev, N. P. 1981, Soviet AZh, 58, 528

Saginashvili, M. G. 1975, Bull.Abastuman.Obs., 46, 125

*Order and Chaos in Stellar and Planetary Systems*
*ASP Conference Series, Vol. 316, 2004*
*G. Byrd, K. Kholshevnikov, A. Mylläri, I. Nikiforov, and V. Orlov, eds.*

# Individual and Statistical Aspects of Star Motions

Vadim A. Antonov

*Central Astronomical Observatory at Pulkovo, Pulkovskoe sh. 65/1, St. Petersburg 196140, Russia*

## 1. Introduction

A star system cannot exist without star motions. Their vanishing means the mutation of the system into a quasar or a black hole. The star motions can be treated in different ways. An individual investigation is possible. Such an investigation starts from astrometric observations and continues by construction and the analysis of galactic orbits. We can consider the star motions as a mass phenomenon too. Both points of view are valid but the statistical one outweighs the other. The reason is evident. We can construct the galactic orbit of the Sun but we find it difficult to apply it. What does such an orbit say about the past history of the Solar System and the Earth itself? Or does it say nothing? For the present the answers are rather hazy (e.g., Marochnik & Mukhin 1983). On the contrary, the statistical conception of the Galaxy and other stellar systems are developed more or less successfully. Nevertheless, the individual and the statistical descriptions of the star motions interlace often in the theory and we shall mention this circumstance later.

An essential place belongs not only to the nature of bodies but also their relation to the human civilization. Why is the individual approach to the Solar System bodies predominant? The situation is that any such body (a planet, a satellite, an asteroid, a comet, a meteor flow too) is of importance either itself or as an indicator for the gravitational field along the body trajectory. On the contrary, the statistical methods open with difficulty the way into dynamics of the Solar System, e.g., in connection with Oort's cloud (Antonov & Todrija 1984) or to the rings of the giant planets (Antonov & Baranov 1996a,b). On the other hand, we have a continuous flow of quickly written theses on the motion of either body of the Solar System.

It should be mentioned that the statistical point of view in atomic physics suppresses sharply the individual one in the analogous way. The individual orbits of particles are seldom calculated (in accelerators, etc.). As a rule, the observations of individual atoms are the destiny of a fantastic Maxwell's demon. On the contrary, we are ourselves microscopic demons with respect to the celestial bodies. However, we observe the motion of the Solar System bodies with our own eyes and extrapolate the motion of stars in our imagination only. The discrepancy of the space-time scales adds the different colours of the theories to their distinctions for an abstract impartial observer.

## 2. Stability of Orbits and Stability of a System

Generally speaking, these conceptions are different. However, the situations arise when the stability of orbits predetermines the corresponding properties of the system as a whole. The Lindblad instability of circular orbits is a typical example. It can be treated as "edge" instability because it is revealed at the periphery of a strongly flattened body. Lindblad yet substantiated that the self-gravitation amplifies this instability: the segments coming off draw out the interior ones. The confirmations are found when analyzing the oscillation spectra of the Maclaurin liquid spheroid (Bryan 1888) and the plane kinetic model (Kalnajs 1972). By the way, the edge instability is a not strongly pronounced type, it interflows imperceptibly with other types of the instability.

The sense of the edge instability consists of the fact that the displacement outside of particles with nearly circular orbits decreases the kinetic energy provided the angular momentum is conserved. Even if the interior stars give part of the momentum to the exterior ones, this can give an advantage in the sense of energetics at the expense of the compression of that part of the system which loses the momentum.

A necessary condition of the edge instability is a sharp enough isolation of the instable orbits in the phase space. On the contrary, if the instability of the orbits takes place on the background of the smoothed-out distribution of stars, this circumstance does not involve the instability of the system: the orbital instability means simply that certain stars occupy interruptedly the place of other stars. This fact has no consequences due to indistinguishability of stars.

For instance, sometimes a steady state with a dense and slightly asymmetric core arises after the collapse of a rare system in numerical experiments (Polyachenko 1992). The state is steady in spite of the fact that in the analogous cases the analysis of the trajectories passing near the cusp reveals their strong instability.

## 3. Regular and Irregular Orbits

A modest name "Fatou problem" conceals a numerous complex of not entirely guessed properties of the motion of a particle in an arbitrary potential field. Of course, there are the numerous well-known classical integrable cases (Wintner 1941). The discussions in 1950–1960 evolved according to the principle: either all the motions are similar to the classical examples or all outside the classical schemes are absolutely disorderly. There was as a sensation the co-existence of both alternatives, for the same field often. First, the pioneers of the numerical experimentation (G. Contopoulos, P. O. Lindblad, A. Ollongren) discovered the unexpected regular motions outside the classical predictions. Second, Kolmogoroff–Arnold–Moser investigations spread the construction of conditionally periodic orbits to include rather general cases. Note that the domains of the phase space, where the convergence of the corresponding series has been proved, are rather small.

Let me make some comments.

I. A possible way for constructing the conditionally periodic orbits passes through the periodic ones of higher order. For this purpose I proposed a modification of the variational method in terms of the plane mappings which evidently allows the existence of such trajectories in rather wide regions under the minimal conditions of smoothness only (Antonov 1974). The analogous proofs have been somewhat later submitted by other authors. The passage itself from the periodic orbits to the conditionally periodic ones is automatically fulfilled under our circumstances (Antonov 1982).

II. Such conditionally periodic trajectories are obtained with the rotation number $\nu$ overlaying a certain interval without gaps, in contrast to the Kolmogoroff–Arnold theory which does not ensure the continuity of changing $\nu$. However, the Kolmogoroff–Arnold tori themselves are unbroken, they create the isolating barrages in the phase space at two degrees of freedom. Our tori are not entirely isolating, they leave slots for penetrating, possibly, the alien trajectories. New such tori are sometimes named "cantori" due to their similarity to the well-known Cantor manifolds.

III. Arnold explained to us that the description of a trajectory as a torus in the phase space is natural and uniform from the mathematical point of view. However, various variants arise when the phase torus projects in the configuration space. Namely, a simple "box" can be transformed and complicated by "folds" (Agekian & Yakimov 1976; Agekian, Mylläri, & Orlov 1992; Contopoulos & Moutsoulos 1965; Hasan & Norman 1990). The latter is very similar to caustics in optics (Arnold 1990).

IV. Earlier Poincaré (1892–1899) noted that the instability of periodic orbits reveals, in particular, the splitting at a small disturbances of parameters into $n = 2, 3, \ldots$ near-by "parallel" branches with a common period $\sim nT$ and this process can be infinitely continued. The numerical experiments of Contopoulos and his school confirm this fact very well. There is a certain similarity between the complication of periodic trajectories in dynamical systems and that of periodic regimes in hydrodynamic flows before appearing the turbulence (Arnold 1974; Belyaev & Yavorskaya 1980; L'vov & Predtechensky 1980).

V. Do periodic and conditionally periodic trajectories fill all the available phase volume in non-integrable models? Not to all appearances. A finite measure in the phase space, permitted for the really chaotic motions, remains. In the same way there is a true turbulence in hydrodynamics. Both domains are not accessible for rigorous mathematical proofs. More correctly, we could prove the chaotic character of the trajectories if the curvature of the configuration space in Jacobi's treatment were negative everywhere (see papers of Sinai et al.). Unfortunately, the curvature in stellar dynamics appears either with alternating signs or positive everywhere, so that prepared mathematical results are not applicable.

VI. Another approach makes sense when "integrability" means either the determination of trajectories in terms of known functions or at least in quadratures. Excepting the classical cases, such a success is rather rare. Our investigation (Antonov & Timoshkova 1993) can serve as an example, where we have found all the possible trajectories pictured as plane curves of the second order (including straight lines) at the motion of a particle in the field with the well-known potential of Contopoulos–Hénon:

$$U = -\,{}^{1}\!/_{2}\left(x^2 + \lambda y^2\right) + \alpha x^3 + x y^2 .$$

Their stability has been studied as well (Antonov & Timoshkova 1996).

VII. Instead of individual trajectories we can find invariant manifolds standing as isolating barriers in the phase space. Our publications (Antonov 1981; Antonov & Shamshiev 1993, 1994a,b) are examples. They give not true but "local" or "articular" integrals of motion.

## 4. Statistical Approach

At one time there were debates (Finlay-Freundlich & Kurth 1954) whether the statistical mechanics of stellar system is possible. It seems now that there are no reasons for discussions. In stellar dynamics the statistical approach with its two basic features is widely used. First, we disengage ourselves from individual properties of "particles": Vega, Mira, Algol, etc. are united in a general conception "stars". Second, we disengage ourselves from the concrete actual situation and describe the state of the system by means of probabilities. A probability itself is a profound initial conception only partly explained through as the space or temporal frequency. The spatial frequency (density) is used in dense stellar systems, the temporal frequency is used when investigating peripheral parts—crowns (e.g., Agekian & Baranov 1969).

The above enumerated characteristics are common features of the statistical approach to any physical system. However, there are principal objective distinctions. Particles interact either with the common field only or with each other. In the latter case the stellar dynamics talks about irregular forces, physicists simply use the term "collisions". They should be taken into account in singular, extremely dense systems (e.g., Antonov 1985; Zeldovich & Poduretz 1965).

As a rule, stellar systems are among collections in which individual particles interact with the common field only. There are other examples of the similar collections: the rare laboratory plasma, often the interplanetary and interstellar plasma, billiards, artificial atomic "traps", aggregates of small bodies of the Solar System. In the absence of individual energetic losses all the above systems are characterized by the conservation of the density $f$ in Boltzmann's phase space. This is the well known Liouville's theorem. The latter is directly applicable to stable steady systems and certain unstable ones. The stable steady systems maintain their structure if unhindered by the physical evolution of stars and the external disturbances.

On the contrary, there are cases when Liouville's theorem is valid only formally, not essentially. This phenomenon is called "the phase mixing": the function $f$ becomes the fast variable and therefore pliable with respect to insignificant effects of irregular forces and of various random influences. These inevitable disturbances smooth out the function $f$ in a irreversible way. In other words, neighbour cells of the phase space mix with each other and acquire a certain average phase density. Earlier we gave the classification of the kinds of phase mixing (Antonov, Nuritdinov, & Ossipkov 1973). The phase mixing can be considered as a displacement of the representing point in the auxiliary space of different possible models of stellar systems with the given global invariants: the total mass $M$, the energy $E$ and the momentum $K$. The restriction on this displacement is following (Antonov 1985; Lynden-Bell 1967; Tremaine, Hénon,

& Lynden-Bell 1980): any functional

$$S_C \equiv \iint C(f)\, d\vec{r}\, d\vec{v}$$

over all the phase volume, occupied by the system, can be only increased if $C(f)$ is a convex function, i.e.,

$$C''(f) \leq 0$$

everywhere. Actually, it is enough to take the functions $C$ of the special type

$$C(f|f_0) = 0, \quad \text{if } f \leq f_0, \qquad \text{or} \qquad C(f|f_0) = f - f_0, \quad \text{if } f \geq f_0,$$

with an arbitrary $f_0 \equiv$ const, since our general function $C$ is represented by a superposition of these special ones. The quantities

$$S(f_0) = \iint C(f|f_0)\, d\vec{r}\, d\vec{v}$$

can be considered as "coordinates" in the space of models and the drift is steady in the same direction. Do the evolutionary tracks come together finally in the only point (at the fixed values $M$, $E$, and $K$)? When analyzing the preliminary problem, the positive answer comes to mind. However, this conclusion is erroneous. Actually, the representing point comes across the zone of the stability and the mixing stops. These final states are different: the "universal" statistical law of Lynden-Bell has very coarse and limited character. So much the better, since the plurality of galactic states makes stellar dynamics much more attractive for the intellect of the theorists, in contrast to the cheerless monotony of the thermodynamically steady states.

In collisional systems the functional $S$ increases obligatorily with a definition only of the choice

$$C(f) = -f \ln f$$

and the system with the global invariants converges to the only state with max $S$. This is $H$-theorem of Boltzmann. Nevertheless, the gravitation has its specific character, namely: the thermodynamically steady state is transient only (see, for instance, Antonov 1985; Lynden-Bell 1967). The evolution continues up to the disintegration of the system or its catastrophic collapse.

## 5. Why the Irreversibility?

Any motions in the $N$-body problem are reversible. Nevertheless, both the mixing and the influence of irregular forces act at any time in the same direction. The hidden motive consists of the action of the Boltzmann principle of "the molecular chaos" (so it was called by Boltzmann, but it is possible, probably, to think over a more appropriate name in more general situations). Indeed, small disturbances can mix neighbor stellar populations in the phase space but not to separate again. All the more, two closely approaching stars "do not know" about each other. "The molecular chaos" as applied to physical systems, including the stellar ones, means always two things:

I. Two sufficiently distant objects are independent of each other.

II. The approaching from far away objects were mutually independent (but acquire the dependence after the impact).

These two points permeate through physics from the microscopic scales to the galactic ones. Point II indicates that although any motion can be reversed, however, nature as a whole has no invariance with respect to time: the latter is violated by probabilistic laws. The numerical experiments are also subject to the action of the principle of "the molecular chaos". Therefore they are capable of representing the evolution of real stellar systems.

Academician Vok attributed the irreversibility in nature to the influence of quantum effects in the motion of molecules. However, the star motions are beyond all measures from quantum borders but the irreversibility works very well. It seems to me, rather, *vice versa*, and the careful study of the irreversibility in the cosmic scales must throw light on that in the microworld.

## 6. Difficult Mathematical Problems of the Theory of Phase Mixing

I. Stable mixed states can be sometimes discerned, without the linearization of equations, taking into account the impossibility of the systems to mix further. Proof of the stability is more convincing and general but certain obstacles must be overcome. In particular, we considered (Antonov 1990) the polytropic model of a stellar cluster that undergoes disturbances of an unbounded amplitude. It has been shown that if the disturbances either satisfy the Liouville theorem or induce mixing, they demand energetic expenditures. Consequently, such disturbances do not spontaneously develop and the system is stable. It would be of interest to expend these reasonings to include different models.

II. We are able only very approximately to prognosticate the final state on which the evolution of the initially non-equilibrium model will stop. *Vice versa*, there is rather an indefinite problem to guess the origin of peculiarities on the picture of the phase density $f$ for the steady state stellar system. In any case, these peculiarities are more effectively destroyed at more intensive mixing, so that finally the galaxies keep mainly relics of not strong but moderate former non-steady processes; this fact could appear to be paradoxical at first sight.

III. All this does not obligatorily result in the steady state. Sometimes the evolution is delayed on regular global oscillations if the mechanism of quasi-diffusion does not work because of the absence of the stellar populations on the suitable resonant orbits (see, for instance, Smith & Miller 1990).

IV. The true diffusion is well described by related differential equations. On the contrary, the mixing is not represented in terms of the differential operations. There is the phase density before the mixing and that after mixing but it is difficult to define more accurately where and when the transmutation occurs. We could not work out general and usable equations of the mixing.

V. Sometimes there are qualitative indications showing the continuation of the mixing in the sequel. Keeping in mind the rare and homogeneous $y$- and $z$-directions plasma, the specialists established that a sharp boundary in the phase space $(x, v_x)$, if it turns up and down (i.e., the boundary value $v_x$ is not a single-valued function of $x$), cannot be straightened again and the mixing will occur inevitably. An analogous situation must take place in stellar dynamics as well. In any case, examples of self-maintained one-dimensional waves (Baranov 1991) manage without the loops at boundaries.

VI. The phase mixing is similar to that of an incompressible fluid. This analogy is well known in physics (e.g., Balescu 1975). A single liquid element is subject to the progressive stretching when the intensive mixing act. More exactly, when decomposing the process into separate random steps a separate result can consist of either in rounding off or subsequent stretching but the latter outweighs statistically. This is a variant of the large number law transformed from number sums to matrix products, such a variant has been studied by a number of mathematicians (Furstenberg, Tutubalin et al.). I improved their theories by departing from the condition of smoothness of the probability distribution. The results of my paper (Antonov 1984)—they themselves are more general in other terms as well—lead to the following conclusion. Let a circle of the unit radius is subject to the multistage linear transformation with matrices $M_1$, $M_2$, $M_3, \ldots$ Each matrix $M_n$ is a result of random choice from the fixed set of the matrices $A_1$, $A_2$, ..., $A_s$ ($\det A_i = 1$, $i = 1, 2, \ldots, s$) with the corresponding probabilities $p_i$, where $p_i \geq 0$, $p_1 + p_2 + \ldots + p_s = 1$, $p_i$ does not depend on $n$. With certain exceptions the elliptic image of the initial circle is persistently flattened under $n \to \infty$, namely:

$$P\{\lim_{n\to\infty} \frac{b_n}{a_n} = 0\} = 1$$

if $a_n$ and $b_n$ are semi-major and -minor axes of the ellipse. The exceptions are among the cases when the set of matrices $A_i$, $1 \leq i \leq s$ has a common determined invariant object: an ellipse, an axis or a cross of two axes.

I detected also that all this can be transferred to include the spatial case: a sphere is progressively stretched to a needle. The possible exceptions are the presence of a common invariant object: an ellipsoid, an axis, a plane or a cross consisting of tree axes. It is probable that very similar theorems must be revealed in the 4- or 6-dimensional phase space.

VII. The weak mixing (quasi-diffusion) appears often as an interaction between a regular global oscillation and a collateral stream of stars. As a rule, the oscillation is the wave running the center of the axisymmetric system with the angular velocity $\Omega$. A response of this process arises with the same angular velocity $\Omega$. This response is either a condensation of stars along the orbit with the angular velocity $\Omega_0 \sim \Omega$, or a orbit with $\Omega_0 = \Omega \pm m\kappa$, where $m$ is an integer, $\kappa$ is the epicyclic frequency (Lindblad's resonances). The stars of the near zone involved in the wave process with the velocity $\Omega + \varepsilon$ give back energy to the basic oscillation while those with the velocity $\Omega - \varepsilon$ take away energy. If the former phenomenon predominates, the oscillation is amplified, and *vice versa.* This result depends on subtle relations between gradients of $f$ in the phase space.

However, this widespread reasoning has a deficiency since it implies that the basic oscillation is close to the usual gravitational-sound ones having the positive energy $\Delta E$ over that of the steady state. However, if $\Delta E < 0$ for some modes of oscillations, this conclusion changes to the opposite one.

It is rather difficult to define the sign of $\Delta E$ since the calculation must be carried out with the accuracy up to the terms $\sim\varepsilon^2$ ($\varepsilon$ is the wave amplitude). Many publications give incorrect calculations. Let me submit a simplified method.

We shall introduce a fictitious lag of the gravitational field with respect to the original mass distribution for a time interval $\delta$. Thus, instead of the initial gravitational acceleration grad $U(\vec{r}, t)$ a new acceleration acts:

$$\operatorname{grad} U(t-\delta) \approx \operatorname{grad}\left(U - \delta\frac{\partial U}{\partial t}\right).$$

Let us introduce corresponding alterations into Boltzmann's equation:

$$\frac{\partial f}{\partial t} = -\vec{v}\operatorname{grad} f - \operatorname{grad}\left(U - \delta\frac{\partial U}{\partial t}\right)\operatorname{grad}_{\vec{v}} f.$$

The energy of the total system is defined as the following integral

$$E = m\int\!\!\int\left(\frac{\vec{v}^2}{2} - \frac{U}{2}\right) f\, d\vec{r}\, d\vec{v}$$

($m$ is the mass of a star) and its evolution is determined by the only superfluous term

$$\frac{\partial E}{\partial t} = m\int\!\!\int\left(\frac{\vec{v}^2}{2} - U\right)\frac{\partial f}{\partial t}\, d\vec{r}\, d\vec{v} = \delta\, m\int\!\!\int\frac{\vec{v}^2}{2}\operatorname{grad}\frac{\partial U}{\partial t}\operatorname{grad}_{\vec{v}} f\, d\vec{r}\, d\vec{v}.$$

Further we integrate by parts in the velocity space for a bounded system

$$\frac{\partial E}{\partial t} = -\delta\, m\int\!\!\int\vec{v} f\operatorname{grad}\frac{\partial U}{\partial t}\, d\vec{r}\, d\vec{v} = -\delta\int\vec{j}\operatorname{grad}\frac{\partial U}{\partial t}\, d\vec{r}$$

($\vec{j}$ is mass flux). It is necessary yet to integrate by parts in the configuration space

$$\frac{\partial E}{\partial t} = \delta\int\frac{\partial U}{\partial t}\operatorname{div} j\, d\vec{r} = -\delta\int\frac{\partial U}{\partial t}\frac{\partial \rho}{\partial t}\, d\vec{r} = -\frac{\delta}{4\pi}\int\left(\frac{\partial}{\partial t}\operatorname{grad} U\right)^2 d\vec{r} \le 0$$

($\rho$ is the density).

We show only the decreasing the energy of the disturbance, since the steady state is indifferent to the delay of gravitation. The disturbance will swing at $\Delta E < 0$ and damp in the opposite case.

Let the distribution be proportional to $\exp(i\omega t)$. Then the fictitious lag of gravitation is in the linearized equations expressed by the coefficient

$$e^{-i\omega\delta} \approx 1 - i\omega\delta$$

attached to the potential of the disturbance.

Thus, we obtain the following rule: when calculating the dispersion equation we substitute the constant of gravitation $G$ in the point of the definition of the disturbance gravitation on $G(1 - i\omega\delta)$. If as a result of this substitution the real root of the dispersion equation acquires a small positive part, we have $\Delta E > 0$, and a negative part corresponds to $\Delta E < 0$.

This is valid for any collisionless gravitational system and, apparently, for any initially gaseous gravitating system as well.

Even at present, in the epoch of common computers, the analytic methods in stellar dynamics have not exhausted themselves. However, we should be able to combine various approaches, in particular, to combine the local point of view with the global one. In stellar dynamics such a concord appears to be more actual even when compared to the theory of plasma.

## References

Agekian, T. A., & Baranov, A. S. 1969, Astrofizika, 5, 305 ( in Russian)

Agekian, T. A., Mylläri, A. A., & Orlov, V. V. 1992, AZh, 69, 469 (in Russian)

Agekian, T. A., & Yakimov, S. P. 1976, Vestn.Leningr.Univ., 13, 177 (in Russian)

Antonov, V. A. 1974, Trudy Astron.Observ.Leningr.Univ., 30, 111 (in Russian)

Antonov, V. A. 1982, Vestn.Leningr.Univ., 13, 86 (in Russian)

Antonov, V. A. 1981, Vestn.Leningr.Univ., 19, 97 (in Russian)

Antonov, V. A. 1984, Vestn.Leningr.Univ., 7, 67 (in Russian)

Antonov, V. A. 1985, in IAU Symp. 113, Dynamics of Star Clusters, ed. J. Goodman, P. Hut (Dordrecht: Reidel), 525

Antonov, V. A. 1990, in Problems of Celestial Mechanics and Stellar Dynamics, ed. T. B. Omarov (Alma-Ata: Nauka), 66 (in Russian)

Antonov, V. A., & Baranov, A. S. 1996a, AZh, 73, 482 (in Russian)

Antonov, V. A., & Baranov, A. S. 1996b, AZh, 73, 490 (in Russian)

Antonov, V. A., Nuritdinov, S. N., & Ossipkov, L. P. 1973, in Dynamics of Galaxies and Star Clusters, ed. T. B. Omarov (Alma-Ata: Nauka), 55 (in Russian)

Antonov, V. A., & Shamshiev, F. T. 1993, CeMDA, 56, 451

Antonov, V. A., & Shamshiev, F. T. 1994a, CeMDA, 59, 209

Antonov, V. A., & Shamshiev, F. T. 1994b, AZh, 71, 486 (in Russian)

Antonov, V. A., & Timoshkova, E. I. 1993, AZh, 70, 265 (in Russian)

Antonov, V. A., & Timoshkova, E. I. 1996, AZh, 73, 953 (in Russian)

Antonov, V. A., & Todrija, Z. P. 1984, Pis'ma AZh, 10, 67 (in Russian)

Arnold, V. I. 1974, Proc.Sympos.Pure Math., 28

Arnold, V. I. 1990, Singularities of Caustics and Wave Fronts (Dordrecht: Kluwer)

Balescu, R. 1975, Equilibrium and Nonequilibrium Statistical Mechanics (New York: Wiley-Interscience)

Baranov, A. S. 1991, AZh, 68, 1160 (in Russian)

Belyaev, Yu. I., & Yavorskaya, I. M. 1980, in Non-Linear Waves, The Stochasticity and the Turbulence, ed. M. I. Rabinowich (Gorky: Inst. Appl. Physics), 78 (in Russian)

Bryan, G. H. 1888, Philos.Trans., 180, 187

Contopoulos, G., & Moutsoulas, M. 1965, AJ, 70, 817

Finlay-Freundlich, E., & Kurth, R. 1954, Naturwiss., 42, 167

Hasan, H., & Norman, C. 1990, ApJ, 361, 69

Kalnajs, A. J. 1972, ApJ, 175, 63

L'vov, V. S., & Predtechensky, A. A. 1980, in Non-Linear Waves. The Stochasticity and the Turbulence, ed. M. I. Rabinowich (Gorky: Inst. Appl. Physics), 57 (in Russian)

Lynden-Bell, D. 1967, MNRAS, 136, 101

Marochnik, L. S., & Mukhin, L. M. 1983, Priroda, 11 (in Russian)

Poincaré, H. 1892–1899, Les méthodes nouvelles de la méchanique céleste, t. 1–3 (Paris: Gautheir-Villars)

Polyachenko, V. L. 1992, AZh, 69, 934 (in Russian)

Smith, B. F., & Miller, R. H. 1990, Bull.Amer.Math.Soc., 22, 1318

Tremaine, S., Hénon, M., & Lynden-Bell, D. 1980, MNRAS, 219, 285

Wintner, A. 1941, The Analytical Foundations of Celestial Mechanics (Princeton: Princeton Univers. Press)

Zel'dovich, Ya. B., & Podurets, M. A. 1965, AZh, 42, 963 (in Russian)

This paper is based on the text of the Brouwer Lecture of Prof. V. A. Antonov submitted to DDA meeting held at Yosemite National Park, California, USA on April 2000. Prepared for publication by L. P. Ossipkov.

*Order and Chaos in Stellar and Planetary Systems*
*ASP Conference Series, Vol. 316, 2004*
*G. Byrd, K. Kholshevnikov, A. Mylläri, I. Nikiforov, and V. Orlov, eds.*

# Analytical Estimates of the Maximum Lyapunov Exponents in Problems of Celestial Mechanics

Ivan I. Shevchenko

*Central Astronomical Observatory at Pulkovo, Pulkovskoe sh. 65/1, St. Petersburg 196140, Russia*

**Abstract.** Dynamical chaos caused by the presence of spin-orbit and orbital resonances in the motion of bodies of the Solar system is studied. The "separatrix map" method of analytical estimation of the maximum Lyapunov characteristic exponents (Shevchenko 2000a, 2002), developed in the framework of a model of the nonlinear resonance as a perturbed pendulum, is considered. The accuracy of some basic relations is improved. Current applications of the method are reviewed and discussed. These applications comprise the problem of the chaotic rotational dynamics of planetary satellites, the problem of the chaotic orbital dynamics of asteroids, and the problem of the chaotic orbital dynamics of satellite systems.

## 1. Introduction

Evaluation of the Lyapunov characteristic exponents (LCEs) is one of the major tools in studies of the chaotic motion, in particular in problems of celestial mechanics. The LCEs characterize the rate of divergence of neighbouring trajectories in phase space. The inverse of the maximum LCE gives the approximate time of predictable dynamics.

The development of methods of evaluation of the maximum LCE and full LCE spectra by means of numerical integration of dynamical systems has a thirty-year history (see, e.g., Lichtenberg & Lieberman 1992). On the contrary, analytical estimation of the maximum LCE started to be developed only recently. In this paper, a new method of analytical estimation of the maximum LCE (Shevchenko 2000a, 2002) and its applications to several problems of celestial mechanics are described. Some new refined formulas of the method are presented.

## 2. The Separatrix Map Method

At the present moment, there exist two theoretical approaches to analytical estimation of the maximum LCE of the motion in the neighbourhood of the separatrices of nonlinear resonances in Hamiltonian dynamics. First of them (Holman & Murray 1996; Murray & Holman 1997) is based on a method by Chirikov (1979) of analytical estimation of the maximum LCE of the standard map. Holman & Murray (1996) have obtained estimates of the maximum LCE for the cases of moderate and strong overlapping of resonances. Murray & Holman (1997) offered one general approximating formula linking two limits of variation of the stochasticity parameter. However, even in case of the standard map, this formula exaggerates real values of the maximum LCE by the order of magnitude if the value of the stochasticity parameter is about one. The use of the standard map paradigm is justified in presence of multiple resonances of approximately equal amplitudes. Such situation, however, occurs seldom.

A new method proposed by Shevchenko (2000a, 2002) is based on the theory of separatrix maps. The nonlinear resonance is modelled by the Hamiltonian of a perturbed nonlinear pendulum. A key role in the method is played by the fundamental dependence of the maximum LCE of the separatrix map on the parameter $\lambda$, which represents the ratio of the frequency of perturbation to the frequency of small phase oscillations on the resonance. Applicability of the separatrix map for description of the motion near the separatrices of the perturbed nonlinear resonance in the full range of $\lambda$ (i.e., both at high and low relative frequencies of perturbation) was established by Shevchenko (2000b).

The perturbed nonlinear resonance is modelled by the Hamiltonian:

$$H = \frac{\mathcal{G}p^2}{2} - \mathcal{F}\cos\varphi + a\cos(\varphi - \tau) + b\cos(\varphi + \tau), \tag{1}$$

where $\tau = \Omega t + \tau_0$. The variable $\varphi$ is the resonance phase, $p$ is the conjugate momentum, $\tau$ is the phase angle of perturbation with the initial value $\tau_0$. The quantity $\Omega$ stands for the constant perturbation frequency; $\mathcal{F}$, $\mathcal{G}$, $a$, $b$ are constants.

The separatrix map in Chirikov's form (Chirikov 1979; Lichtenberg & Lieberman 1992) describes the motion in the vicinity of the separatrices of a nonlinear resonance subject to symmetric periodic perturbation. In the case of symmetric perturbation ($a = b$) the separatrix map has two parameters. The first one, $\lambda$, is the ratio of $\Omega$, the perturbation frequency, to $\omega_0 = (\mathcal{F}\mathcal{G})^{1/2}$, the frequency of small oscillations on the resonance. The second one (the perturbation amplitude parameter) is given by the formula

$$W = \frac{a}{\mathcal{F}}\lambda[A_2(\lambda) + A_2(-\lambda)] = 4\pi\frac{a}{\mathcal{F}}\lambda^2 \operatorname{csch}\frac{\pi\lambda}{2}, \tag{2}$$

where

$$A_2(\lambda) = 4\pi\lambda\frac{\exp\frac{\pi\lambda}{2}}{\sinh(\pi\lambda)} \tag{3}$$

is the Melnikov–Arnold integral, as defined by Chirikov (1979) and in more detail by Shevchenko (1998, 2000b).

When the perturbation is asymmetric ($a \neq b$), one has two perturbation amplitude parameters instead of sole $W$. These two quantities, $W^+$ and $W^-$, are the values of $W$ for the prograde and retrograde motions of the model pendulum, respectively:

$$\begin{aligned} W^+(\lambda, \eta) &= \varepsilon\lambda\left[A_2(\lambda) + \eta A_2(-\lambda)\right], \\ W^-(\lambda, \eta) &= \varepsilon\lambda\left[\eta A_2(\lambda) + A_2(-\lambda)\right], \end{aligned} \tag{4}$$

where $\varepsilon = a/\mathcal{F}$, $\eta = b/a$ (Shevchenko 1999).

Following Shevchenko (2000a, 2002), we take the maximum LCE of the symmetric separatrix map in the form $L_{\text{sx}} = L(\lambda)$ with the function

$$L(\lambda) \approx C_h \frac{2\lambda}{1 + 2\lambda}, \tag{5}$$

where $C_h \approx 0.8$ is a constant. This form corresponds to the case of the least perturbed border of the chaotic layer, which is considered to be the most probable case; a dependence on the second parameter of the map is thus eliminated.

If $a \neq b$, the average periods of the prograde and retrograde chaotic near-separatrix rotations and the average half-period of the chaotic near-separatrix libration of the model pendulum all differ from each other. We put the formula for the periods of rotation as adopted in (Shevchenko 2000a, 2002) in the form:

$$T^{\pm} = T(\lambda) + c^{\pm}, \tag{6}$$

where

$$T(\lambda) \approx \lambda - \lambda \ln \lambda, \quad c^{\pm} = \lambda \ln \frac{32}{|W^{\pm}|}. \tag{7}$$

The index "+" refers to the prograde motion, and "−" to the retrograde one. To obtain the symmetric case ($a = b$), the indices in Eqs. (6) and (7) should be simply omitted.

The average (over the whole layer) of the maximum LCE is a sum of weighted contributions of the layer components corresponding to the librations, prograde rotations and retrograde rotations of the model pendulum:

$$\langle L_t \rangle = \Omega \left( \xi_{\text{libr}} \frac{L_{\text{libr}}}{T_{\text{libr}}} + \xi^{+} \frac{L^{+}}{T^{+}} + \xi^{-} \frac{L^{-}}{T^{-}} \right), \tag{8}$$

where the weights $\xi_{\text{libr}}$, $\xi^{+}$ and $\xi^{-}$ are directly proportional to the total time which the trajectory spends in the corresponding component; $\xi_{\text{libr}} + \xi^{+} + \xi^{-} = 1$. They can be expressed through relative measures of the components. In Eq. (8), the LCE is counted per unit of time $t$ of the original Hamiltonian system (1). In case of symmetric perturbation, one just has $\langle L_t \rangle = \Omega L_{\text{sym}} / T_{\text{sym}}$, where $L_{\text{sym}} = L_{\text{libr}} = L^{+} = L^{-}$, $T_{\text{sym}} = T_{\text{libr}} = T^{+} = T^{-}$.

Analytical estimation of the average (8), in particular the weights, is a complicated problem. The weights depend on the system parameters, the asymmetry of perturbation among them. Shevchenko (2000a, 2002) put simply $\langle L_t \rangle = \frac{\Omega}{2} [(L^{+}/T^{+}) + (L^{-}/T^{-})]$. This approximation turned out to be in good agreement with numerical integrations of equations of motion in a number of applications described in the following Sections.

Further in this Section, we present refined basic relations $L(\lambda)$ and $T(\lambda)$ obtained by means of numerical experiments with the separatrix map for the Hamiltonian (1) with symmetric perturbation. A convenient form of the separatrix map, used, e.g., by Chirikov & Shepelyansky (1984) and Shevchenko (1998), is the following:

$$\begin{aligned} y_{i+1} &= y_i + \sin x_i, \\ x_{i+1} &= x_i - \lambda \ln | y_{i+1} | + c \pmod{2\pi}, \end{aligned} \tag{9}$$

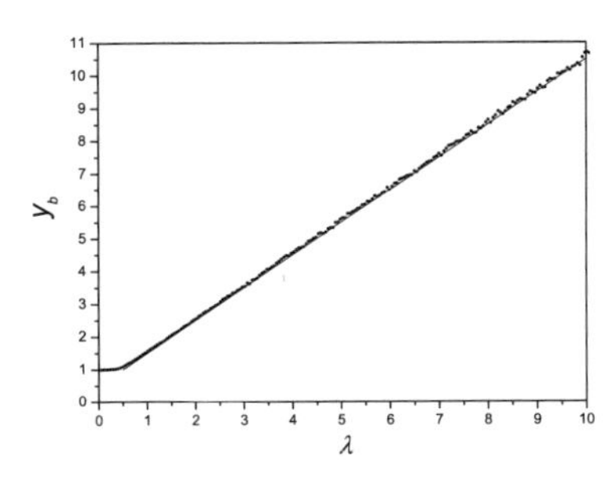


Figure 1. The dependence of the chaotic layer half-width $y_b$ on $\lambda$.

where $y = w/W$ (and $w$ is the relative deviation of the model pendulum energy from the separatrix energy value; see the definition in Chirikov 1979), $x = \tau + \pi$, $c = \lambda \ln 32/|W|$. The variables $x$ and $y$ are mapped asynchronously (Shevchenko 1998, 2000b): respectively when the model pendulum is at the lower and upper points of equilibrium.

The half-width $y_b$ of the main chaotic layer of the separatrix map, the maximum Lyapunov exponent $L$ of the motion in the layer, and the reduced period $T$ [corresponding to the first term in Eq. (6)] are presented as functions of $\lambda$ in Figures 1 and 2. They have been obtained by numerical experiments with the separatrix map. The value $y_b$ of $y$ at the border of the chaotic layer is determined as the maximum of $|y_i|$ obtained during computation of a single trajectory; the maximum LCE is calculated by the tangent map method (its description see, e.g., in Lichtenberg & Lieberman 1992); the reduced period $T$ is calculated as the increment $\Delta x_{i+1} = x_{i+1} - x_i - c$ averaged over $i$. The resolution (step) in $\lambda \in [0, 10]$ is equal to 0.05. At each step in $\lambda$, the values of $y_b$, $L$, and $T$ are computed for 100 values of $c$ equally spaced in the interval $[0, 2\pi)$. The number of iterations for each trajectory is $n_{\text{it}} = 10^7$. This is sufficient to saturate the computed values of $y_b$, $L$, and $T$. At each step in $\lambda$ we find the value of $c$ corresponding to the minimum $y_b$ (the case of the least perturbed border), and plot $y_b$, $L$, $T$ corresponding to this case.

The plot of thus obtained $y_b$ against $\lambda$ is shown in Figure 1. The observed dependence apparently follows the piecewise linear law

$$y_b(\lambda) = \begin{cases} 1 & \text{if } 0 \le \lambda \le \frac{1}{2}, \\ \lambda + \frac{1}{2} & \text{if } \lambda > \frac{1}{2}. \end{cases} \tag{10}$$

The linear function $y_b = \lambda + 1/2$ is drawn in Figure 1 at $\lambda > 1/2$. It practically coincides with the observed data. The sense of this relation is clear: it is the sum of the predicted border value of $y$ averaged over $\tau \in [0, 2\pi]$ [this value approximately equals $\lambda$ (Chirikov 1979)] and the the half-amplitude (approximately equal to 1/2) of the border line.

Figure 2a presents $L(\lambda)$ and its approximation by the rational function

$$L(\lambda) = \frac{b + c\lambda}{1 + a\lambda} \tag{11}$$

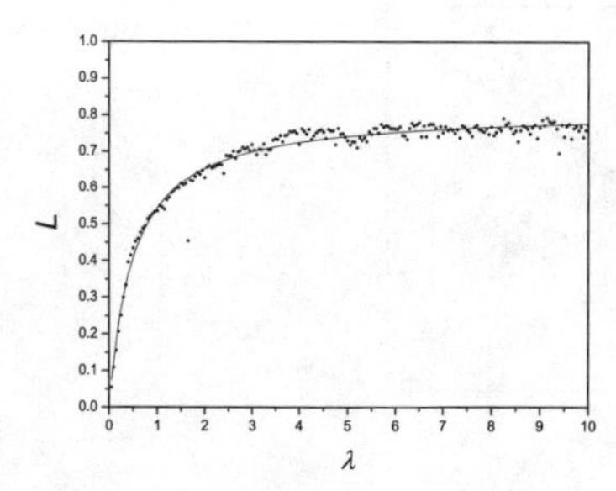

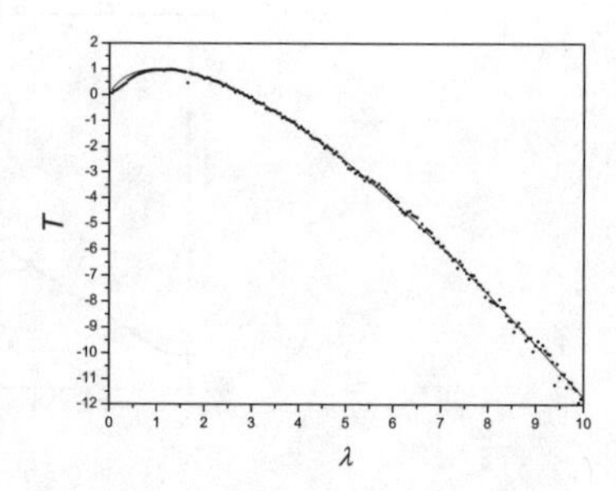


Figure 2. The dependences of the maximum Lyapunov exponent $L$ and the reduced period $T$ on $\lambda$ (dots) with their analytical approximations (solid curves).

with $a = 1.99 \pm 0.09$, $b = 0.05 \pm 0.01$, and $c = 1.61 \pm 0.07$. [Here $c$ is a different quantity than designated in Eq. (7).] The nonlinear fitting has been performed by the Levenberg–Marquardt method. The values of the parameters are given with their standard errors.

Chirikov's constant is given by the limit $L(\lambda \to \infty)$; so, according to the described numerical experiment, $C_h \approx 0.81$ in good agreement with our earlier inference $C_h \approx 0.8$ (Shevchenko 2000a, 2002).

From Eqs. (5), (10), and (11) it follows that $L(\lambda) \approx C_h \lambda / y_b$. The physical sense of this formula will be discussed elsewhere.

Figure 2b presents $T(\lambda)$ and its approximation by the function

$$T(\lambda) = a\lambda - b\,\lambda \ln \lambda \tag{12}$$

with $a = 0.993 \pm 0.009$, $b = 0.939 \pm 0.005$.

Comparison of Eqs. (11) and (12) to (5) and (7) testifies the good accuracy of simple formulas (5) and (7).

## 3. Chaotic Rotation of Satellites

A satellite is modelled as a triaxial rigid body in a fixed elliptic orbit. The ratios of the principal central moments of inertia, $A/C$ and $B/C$, and the orbital eccentricity, $e$, are the parameters of the model. For a number of natural planetary satellites, estimates of the maximum LCE of the chaotic rotational motion in the vicinity of the separatrices of the synchronous resonance were obtained by Shevchenko (2002) by the "separatrix map method" described in Section 2. The case of the planar chaotic rotation (in the orbital plane) was considered. The analytical maximum LCE estimates demonstrated good agreement with numerical ones obtained in numerical integrations of the equations of motion by the "shadow trajectory" method.

Shevchenko & Kouprianov (2002) computed the full Lyapunov spectra of the chaotic planar and spatial rotation of the satellites by the HQR-method of von Bremen, Udwadia, & Proskurowski (1997). The numerical LCEs were compared to analytical estimates calculated by the separatrix map method. Further evidence was given that the agreement of the numerical data with the separatrix map theory in the planar case is good. It was shown that the theory developed for the planar case is most probably still applicable in the case of spatial rotation, if the dynamical asymmetry of the satellite is sufficiently small ($A/C > 0.8$) and the orbital eccentricity is relatively large ($e > 0.02$).

Kouprianov & Shevchenko (2003) showed that the separatrix map method is in a particularly good correspondence with the results of numerical integrations of the chaotic spatial rotation of a prolate axisymmetric ($B/C = 1$) satellite. The dependences of the maximum LCE on the $A/C$ parameter were built for such a satellite at various values of $e$. They confirmed the applicability of the analytical estimation of the maximum LCE in the domain of parameters' values determined in Shevchenko & Kouprianov (2002).

The dependence of the Lyapunov spectra on the energy (the Jacobi integral) of the system for a triaxial satellite in a circular orbit was investigated in numerical experiments by Kouprianov & Shevchenko (2003). It was found that the dependence of the maximum LCE on the energy is linear when the energy is low, with one and the same slope (but various shifts) for the majority of our sets of the values of parameters. What is more, in the case of the prolate axisymmetric satellite, the dependence seems to be universal (the same) over the studied range of the energy in a broad range of values of the inertial parameters. Upper bounds on the values of the maximum LCE were obtained. The "energetic approach" provides a complementary method for analytical estimation of the LCEs; it might be useful when the energy is high.

## 4. Chaotic Asteroidal Dynamics

The mean motion resonances and secular resonances represent well-known examples of the orbital resonances. The mean motion ones represent commensurabilities of the periods of the motion along orbits. The secular ones are defined (see, e.g., Froeschlé & Scholl 1986, 1988) as the resonances between the rates of precession of the longitudes of pericenters or as the resonances between the rates of precession of the longitudes of nodes of the orbits of the perturbing and test bodies. An important role in the orbital dynamics of bodies of the Solar system, in particular asteroids, is played by the so-called three-body resonances (Nesvorný & Morbidelli 1998, 1999; Murray, Holman, & Potter 1998). In the case of a three-body resonance, the resonant phase is a combination of angular elements of the orbits of three bodies (a test one and two perturbing ones; e.g., an asteroid, Jupiter, and Saturn). Up to the present moment, large numerical material on estimates of LCEs of the chaotic motion conditioned by resonances of the indicated types in the Solar system has been accumulated in scientific literature. In overwhelming majority, these estimates were obtained by numerical integration of original dynamical systems. There exists a necessity of theoretical explanation of the observed variety of chaotic behaviour displayed in these estimates. Development of methods of theoretical estimation of LCEs in these problems requires, first of all, the identification of the guiding and perturbing resonances in each considered case.

Shevchenko, Kouprianov, & Melnikov (2003) studied dynamical chaos in the problem of the orbital dynamics of asteroids near the 3/1 mean-motion resonance with Jupiter. The maximum LCEs were determined for the trajectories on a representative plane of starting values. The trajectories were computed by the Wisdom (1983) map in the planar elliptic restricted three-body problem. Dependences of the maximum LCE on the problem parameters and on the initial data were analysed. Jupiter's orbital eccentricity and the initial value of the orbital eccentricity of an asteroid were varied. Dependences of the maximum LCE on Jupiter's orbital eccentricity for various initial values of asteroid's orbital eccentricity were built, as well as dependences of the maximum LCE on the initial value of asteroid's eccentricity for the minimum, maximum, and current values of Jupiter's orbital eccentricity. The inference was made that the domain of chaos in phase space of the problem consists of two components of different nature. The values of the maximum LCEs observed for one of the components were compared to the theoretical estimates obtained by the separatrix map method within the framework of a model of the resonance as a perturbed nonlinear pendulum. The perturbed pendulum approximate Hamiltonian of the problem, derived by Holman & Murray (1996) and Murray & Holman (1997), was used in the analytical treatment. The analytical maximum LCE estimates were found to be in agreement with the numerical ones.

Since the three-body resonances can be as well described by a model of the perturbed pendulum (Nesvorný & Morbidelli 1998, 1999; Murray et al. 1998), the maximum LCE estimates can be obtained for them by the same method as described in Section 2.

## 5. Chaotic Satellite Systems

Chaotic dynamics near the separatrices of the 3/1 resonance in the orbital motion of two Uranian satellites, Miranda and Umbriel, were considered by Shevchenko & Melnikov (2002). At present, these satellites are dynamically close to the 3/1 mean motion resonance, and they could reside in the chaotic layer near its separatrices during a stage of the long-term dynamical evolution of the system.

Shevchenko & Melnikov (2002) obtained analytical estimates of the maximum LCE by the separatrix map method. The expression by Malhotra & Dermott (1990) for the perturbed pendulum approximate Hamiltonian was used. The maximum LCEs were as well computed by means of direct numerical integration of the equations of motion. The analytical and numerical estimates were found to be in good agreement. This agreement allows one to conclude that the separatrix map method gives accurate enough estimates of the maximum LCE. Therefore, one can estimate the Lyapunov times without accomplishment of numerical integration of the equations of motion.

## 6. Conclusions

The separatrix map method of analytical estimation of the maximum Lyapunov exponents (Shevchenko 2000a, 2002) has shown its efficiency in a number of problems of celestial mechanics. The improved formulas, presented in this paper, are intended to increase the accuracy of the maximum LCE estimates in these and new applications.

**Acknowledgments.** This work was supported by the Russian Foundation for Basic Research under grant 03-02-17356.

## References

Chirikov, B. V. 1979, Phys.Rep., 52, 263

Chirikov, B. V., & Shepelyansky, D. L. 1984, Physica D, 13, 395

Froeschlé, Ch., & Scholl, H. 1986, A&A, 166, 326

Froeschlé, Ch., & Scholl, H. 1988, CeMDA, 43, 113

Holman, M. J., & Murray, N. W. 1996, AJ, 112, 1278

Kouprianov, V. V., & Shevchenko, I. I. 2003, A&A (in press)

Lichtenberg, A. J., & Lieberman, M. A. 1992, Regular and Chaotic Dynamics (New York: Springer)

Malhotra, R., & Dermott, S. F. 1990, Icarus, 85, 444

Murray, N., & Holman, M. 1997, AJ, 114, 1246

Murray, N., Holman, M., & Potter, M. 1998, AJ, 116, 2583

Nesvorný, D., & Morbidelli, A. 1998, AJ, 116, 3029

Nesvorný, D., & Morbidelli, A. 1999, CeMDA, 71, 243

Shevchenko, I. I. 1998, Phys.Scr., 57, 185

Shevchenko, I. I. 1999, CeMDA, 73, 259

Shevchenko, I. I. 2000a, Izvestia GAO, 214, 153

Shevchenko, I. I. 2000b, J.Exp.Theor.Phys., 91, 615 [2000, ZhETP, 118, 707]

Shevchenko, I. I. 2002, Cosmic Res., 40, 296 [2002, Kosmich.Issled., 40, 317]

Shevchenko, I. I., & Kouprianov, V. V. 2002, A&A, 394, 663

Shevchenko, I. I., & Melnikov, A. V. 2002, Izvestia GAO, 216, 371

Shevchenko, I. I., Kouprianov, V. V., & Melnikov, A. V. 2003, Solar System Res., 37, 74 [2003, Astronomicheskii Vestnik, 37, 80]

von Bremen, H. F., Udwadia, F. E., & Proskurowski, W. 1997, Physica D, 101, 1

Wisdom, J. 1983, Icarus, 56, 51

*Order and Chaos in Stellar and Planetary Systems*
*ASP Conference Series, Vol. 316, 2004*
*G. Byrd, K. Kholshevnikov, A. Mylläri, I. Nikiforov, and V. Orlov, eds.*

# The Painlevé Analysis and Construction of Solutions for the Generalized Hénon–Heiles System

E. I. Timoshkova

*Central Astronomical Observatory at Pulkovo, Pulkovskoe sh. 65/6, St. Petersburg 196140, Russia; e-mail: elenatim@gao.spb.ru*

S. Yu. Vernov

*Skobeltsyn Institute of Nuclear Physics, Moscow State University, Vorob'evy Gory, Moscow 119992, Russia; e-mail: svernov@theory.sinp.msu.ru*

**Abstract.** The generalized Hénon–Heiles system has been considered. In two nonintegrable cases new two-parameter solutions have been obtained in terms of elliptic functions. These solutions generalize the known one-parameter solutions. In these nonintegrable cases with the help of the Painlevé test three-parameter special solutions have been found as Laurent series, converging in some ring. One of parameters determines the singularity point location, other parameters determine coefficients of series. For some values of these parameters the Laurent series solutions coincide with the Laurent series of the elliptic solutions, obtained in this paper. The Painlevé test can be used not only to construct local solutions as the Laurent series, but also to find elliptic solutions.

## 1. The Painlevé Property and Integrability

When some mechanical problem is studied, time is assumed to be real, whereas the integrability of motion equations depends on the behavior of their solutions as functions of complex time. S. V. Kovalevskaya was the first, who proposed (Kowalevski 1889) to interpret time as a complex variable and to require that solutions of mechanical problems have to be single-valued functions meromorphic in the entire complex plane. This idea gave a remarkable result: S. V. Kovalevskaya discovered a new integrable case (nowadays known as the Kovalevskaya's case) for the motion of a heavy rigid body about a fixed point (Kowalevski 1889). The Kovalevskaya's result demonstrated that the analytic theory of differential equations can be fruitfully applied to mechanical and physical problems. The important stage of development of this theory was the Painlevé classification of ordinary differential equations (ODEs) with respect to the types of singularities of their solutions (Painlevé 1897).

Let us formulate the Painlevé property for ODEs. Solutions of a system of ODEs are regarded as analytic functions, may be with isolated singularity points. A singularity point of a solution is said *critical* (as opposed to *noncritical*) if the solution is multivalued (single-valued) in its neighborhood and *movable* if its location depends on initial conditions. The *general solution* of an ODE of order $N$ is the set of all solutions mentioned in the existence theorem of Cauchy, i.e., determined by the initial values. It depends on $N$ arbitrary independent constants. A *special solution* is any solution obtained from the general solution by giving values to the arbitrary constants. A *singular solution* is any solution which is not special, i.e., which does not belong to the general solution. A system of ODEs has the **Painlevé property** if its general solution has no movable critical singularity point (Painlevé 1897).

Investigations of many dynamical systems (Tabor 1989) show that systems with the Painlevé property are completely integrable (Liouville integrable). At the same time the integrability of an arbitrary system with the Painlevé property has yet to be proved. There is not an algorithm for construction of the additional integral by the Painlevé analysis. On the another hand there are many integrable systems without the Painlevé property. The Painlevé analysis can be connected with the normal form theory (Goriely 2001).

The **Painlevé test** is any algorithm, which checks some necessary conditions for a differential equation to have the Painlevé property. The original algorithm, developed by P. Painlevé, is known as the $\alpha$-method. The method of S. V. Kovalevskaya is not as general as the $\alpha$-method, but much more simple. The remarkable property of this test is that it can be checked in a finite number of steps. This test can only detect the occurrence of logarithmic and algebraic branch points. Up to the present there is no general finite algorithmic method to detect the occurrence of essential singularities. Different variants of the Painlevé test are compared in Conte (1999).

Developing the Kovalevskaya method (Kowalevski 1889) further, M. J. Ablowitz, A. Ramani and H. Segur constructed a new algorithm of the Painlevé test for ODEs (Ablowitz, Ramani, & Segur 1980). They also were the first to point out the connection between the nonlinear partial differential equations (PDEs), which are solvable by the inverse scattering transform method, and ODEs with the Painlevé property. Subsequently the Painlevé property for PDEs was defined and the corresponding Painlevé test (the WTC procedure) was constructed (Weiss, Tabor, & Carnevale 1983; Weiss 1983).

## 2. The Hénon–Heiles Hamiltonian

In the 1960s the models of the star motion in an axial-symmetric and time-independent potentials have been developed (Contopoulos 1963; Hénon, & Heiles 1964) to show either existence or absence of the third integral for some polynomial potentials. Due to the symmetry of the potential the considered system is equivalent to two-dimensional one. To clarify the question of the existence of the third integral Hénon and Heiles considered the behavior of numerically integrated trajectories (Hénon & Heiles 1964). Emphasizing that their choice does not proceed from experimental data, they have proposed the Hamiltonian

$$H = \frac{1}{2}\Big(x_t^2 + y_t^2 + x^2 + y^2\Big) + x^2 y - \frac{1}{3}y^3,$$

because on the one hand, it is analytically simple; this makes the numerical computations of trajectories easy; on the other hand, it is sufficiently complicated to give trajectories which are far from trivial.

The generalized Hénon–Heiles system is described by the Hamiltonian:

$$H = \frac{1}{2}\left(x_t^2 + y_t^2 + \lambda x^2 + y^2\right) + x^2 y - \frac{C}{3}\,y^3 \qquad (1)$$

and the corresponding system of the motion equations:

$$\begin{cases} x_{tt} = -\lambda x - 2xy, \\ y_{tt} = -y - x^2 + Cy^2, \end{cases} \qquad (2)$$

where $x_{tt} \equiv \frac{d^2x}{dt^2}$ and $y_{tt} \equiv \frac{d^2y}{dt^2}$, $\lambda$ and $C$ are numerical parameters. Due to the Painlevé analysis three integrable cases of (1) have been found:

(i) $C = -1$, $\lambda = 1$,
(ii) $C = -6$, $\lambda$ is an arbitrary number,
(iii) $C = -16$, $\lambda = 1/16$.

The general solutions in the analytic form are known only in integrable cases (Conte, Musette, & Verhoeven 2002; Conte, Musette, & Verhoeven 2003), in other cases not only four-, but even three-parameter exact solutions have yet to be found. Different special solutions for system (2) have been found in (Weiss 1984; Timoshkova 1991; Conte & Musette 1992; Antonov & Timoshkova 1993; Timoshkova 1999; Timoshkova 2001). The obtained solutions are the elliptic functions and, so, can be represented as the general solutions of some polynomial first order equations. The Ablowitz–Ramani–Segur algorithm of the Painlevé test appears very useful to find local solutions in the form of formal Laurent series. It have been proved (Melkonian 1999) that all local solutions for the generalized Hénon–Heiles system found as formal psi-series converge on some real time interval. Therefore, all solutions found as formal Laurent series are actual solutions. The knowledge of such solutions assists to find the elliptic solutions.

## 3. Special Solutions

To seek the global single-valued solutions we transform system (2) into the fourth order equation (Conte & Musette 1992; Timoshkova 1999):

$$y_{tttt} = (2C-8)y_{tt}y-(4\lambda+1)y_{tt}+2(C+1)y_t^2+\frac{20}{3}Cy^3+(4C\lambda-6)y^2-4\lambda y-4H, \qquad (3)$$

where $H$ is the energy of the system.

To find a special solution of eq. (3) one can assume that $y$ satisfies some more simple equation. For example, there exist solutions in terms of the Weierstrass elliptic functions, which satisfy the following equation

$$y_t^2 = \mathcal{A}y^3 + \mathcal{B}y^2 + \mathcal{C}y + \mathcal{D}, \qquad (4)$$

where $\mathcal{A}$, $\mathcal{B}$, $\mathcal{C}$ and $\mathcal{D}$ are some constants. $\mathcal{D}$ is proportional to energy $H$, therefore, these solutions are two-parameter ones. The following generalization of eq. (4)

$$y_t^2 = \tilde{\mathcal{A}}y^3 + \tilde{\mathcal{B}}y^{5/2} + \tilde{\mathcal{C}}y^2 + \tilde{\mathcal{D}}y^{3/2} + \tilde{\mathcal{E}}y + \tilde{\mathcal{G}} \qquad (5)$$

gives new one-parameter solutions in two nonintegrable cases: $C = -4/3$ or $C = -16/5$, $\lambda$ is an arbitrary number (Timoshkova 1999). Solutions with $\tilde{\mathcal{G}} \neq 0$ or $\tilde{\mathcal{E}} \neq 0$ are derived only at $\tilde{\mathcal{D}} = 0$, therefore, substitution $y = \varrho^2$ transforms eq. (4) into

$$\varrho_t^2 = \frac{1}{4}\Big(\tilde{\mathcal{A}}\varrho^4 + \tilde{\mathcal{B}}\varrho^3 + \tilde{\mathcal{C}}\varrho^2 + \tilde{\mathcal{D}}\varrho + \tilde{\mathcal{E}}\Big). \tag{6}$$

Let us consider the possibility to generalize these one-parameter elliptic solutions. Due to the Painlevé analysis it has been shown (Vernov 2003) that in the both above-mentioned nonintegrable cases there exist three-parameter local solutions in the form of the Laurent series. One of parameters determines the singularity point location, other parameters determine coefficients of the obtained Laurent series. For some values of these parameters the obtained Laurent series coincide with the Laurent series of the elliptic solutions found in (Timoshkova 1999). For each possible pair of values $C$ and $\lambda$ four different Laurent series solutions for (3) have been obtained. These solutions have nonzero residues. The sum of residues of an elliptic function in its parallelogram of periods has to be zero, hence, two local solutions with opposite residues correspond to one global elliptic solution. For $C = -4/3$ the three-parameter generalization of the two-parameter solution, which satisfies eq. (4), has been found as well. The residue of this three-parameter solution is equal to zero. Two-parameter generalizations of solutions of eq. (5) have been obtained only for some values of $\lambda$ (Timoshkova 2001). The Painlevé analysis shows that the similar solutions can exist for arbitrary value of $\lambda$. We seek solutions in the form $y(t) = \varrho(t)^2 + P_0$, where $P_0$ is an arbitrary constant. Substituting $y(t)$ in eq. (3) we obtain:

$$\begin{aligned} \varrho_{tttt}\varrho = {} & -4\varrho_{ttt}\varrho_t - 3\varrho_{tt}^2 + 2(C-4)\varrho_{tt}\varrho^3 + (2P_0(C-4) - 4\lambda - 1)\varrho_{tt}\varrho \\ & +2(3C-2)\varrho_t^2\varrho^2 + (2CP_0 - 4\lambda - 8P_0 - 1)\varrho_t^2 + \tfrac{10}{3}C\varrho^6 \\ & +(2C\lambda + 10CP_0 - 3)\varrho^4 + 2(2C\lambda P_0 + 5CP_0^2 - \lambda - 3P_0)\varrho^2 \\ & +\tfrac{10}{3}CP_0^3 + 2C\lambda P_0^2 - 3P_0^2 - 2\lambda P_0 - 2H. \end{aligned} \tag{7}$$

The function $\varrho(t)$ satisfies eq. (6), therefore, eq. (7) is equivalent to the following system:

$$\left\{\begin{array}{l} (3\tilde{\mathcal{A}} + 4)(-3\tilde{\mathcal{A}} + 2C) = 0, \\ \tilde{\mathcal{B}}(-21\tilde{\mathcal{A}} + 9C - 16) = 0, \\ 96\tilde{\mathcal{A}}CP_0 - 240\tilde{\mathcal{A}}\tilde{\mathcal{C}} - 192\tilde{\mathcal{A}}\lambda - 384\tilde{\mathcal{A}}P_0 - 48\tilde{\mathcal{A}} \\ \;-105\tilde{\mathcal{B}}^2 + 128\tilde{\mathcal{C}}C - 192\tilde{\mathcal{C}} + 128C\lambda + 640CP_0 - 192 = 0, \\ 40\tilde{\mathcal{B}}CP_0 - 90\tilde{\mathcal{A}}\tilde{\mathcal{D}} - 65\tilde{\mathcal{B}}\tilde{\mathcal{C}} - 80\tilde{\mathcal{B}}\lambda \\ \;-160\tilde{\mathcal{B}}P_0 - 20\tilde{\mathcal{B}} + 56C\tilde{\mathcal{D}} - 64\tilde{\mathcal{D}} = 0, \\ 16\tilde{\mathcal{C}}CP_0 - 36\tilde{\mathcal{A}}\tilde{\mathcal{E}} - 21\tilde{\mathcal{B}}\tilde{\mathcal{D}} - 8\tilde{\mathcal{C}}^2 - 32\tilde{\mathcal{C}}\lambda - 64\tilde{\mathcal{C}}P_0 - 8\tilde{\mathcal{C}} + 24C\tilde{\mathcal{E}} \\ \;+64C\lambda P_0 + 160CP_0^2 - 16\tilde{\mathcal{E}} - 32\lambda - 96P_0 = 0, \\ 10\tilde{\mathcal{B}}\tilde{\mathcal{E}} + (5\tilde{\mathcal{C}} + 8CP_0 - 16\lambda - 32P_0 - 4)\tilde{\mathcal{D}} = 0, \\ 384H = -48\tilde{\mathcal{C}}\tilde{\mathcal{E}} + 96C\tilde{\mathcal{E}}P_0 + 384C\lambda P_0^2 + 640CP_0^3 - 9\tilde{\mathcal{D}}^2 \\ \;-192\tilde{\mathcal{E}}\lambda - 384\tilde{\mathcal{E}}P_0 - 48\tilde{\mathcal{E}} - 384\lambda P_0 - 576P_0^2. \end{array}\right. \tag{8}$$

The system (8) has been solved by computer algebra software REDUCE. Solutions with $\tilde{\mathcal{G}} \neq 0$ or $\tilde{\mathcal{E}} \neq 0$ exist only for $C = -16/5$ and $C = -4/3$ and in integrable cases. These solutions can be presented in the following form

$$y(t-t_0)=\left[\frac{a\wp(t-t_0)+b}{c\wp(t-t_0)+d}\right]^2+P_0\qquad(ad-bc=1),$$

where $\wp(t-t_0)$ is the Weierstrass elliptic function, $a$, $b$, $c$ and $d$ are some constants. The parameter $P_0$ defines the energy of the system. There exist two different elliptic solutions for each possible pair of values of $C$ and $\lambda$. So we can conclude that all one-parameter solutions obtained in Timoshkova (1999) have two-parameter generalizations. Moreover, the Painlevé analysis shows that these solutions can have three-parameter generalizations.

Of course, solutions, which are single-valued in the neighborhood of one singularity point, can be multivalued in the neighborhood of another singularity point. So we can only assume that global three-parameter solutions are single-valued. If we assume this and moreover that these solutions are elliptic functions (or some degenerations of them), then we can seek them as solutions of some polynomial first order equations. The classical theorem, which was established by Briot and Bouquet, proves that if the general solution of the autonomous ODE is single-valued, then this solution is either an elliptic function, or a rational function of $e^{\gamma x}$, $\gamma$ being some constant, or a rational function of $x$. Note that the third case is a degeneracy of the second one, which in its turn is a degeneracy of the first one. Painlevé (1897) has proved that if the general solution of the autonomous ODE is single-valued, then the necessary form of this ODE is

$$\sum_{k=0}^{m}\sum_{j=0}^{2m-2k}h_{jk}\,y^j{y_t}^k=0,\qquad h_{0m}=1,\tag{9}$$

in which $m$ is a positive integer number and $h_{jk}$ are constants.

Recently a new method to find elliptic solutions has been proposed (Conte & Musette 2003). This method is based on the Painlevé test and uses the Laurent series expansion to find the analytic form of elliptic solutions. Rather than substitute eq. (9) into eq. (3), one can substitute the found Laurent series solutions of eq. (3) into eq. (9) and obtain a linear system in $h_{jk}$. This method is more powerful than the traditional method and allows in principal to find all elliptic solutions. We hope that the use of this method allows to find the three-parameter elliptic solutions.

## 4. Conclusions

Two-parameter elliptic solutions for the Hénon–Heiles system with $C=-16/5$ and $C=-4/3$ and an arbitrary $\lambda$ have been found. Two different solutions correspond to each pair of values $C$ and $\lambda$. Exact three-parameter solutions in nonintegrable cases of the Hénon–Heiles system are not known. With the help of the Painlevé test local three-parameter solutions as converging Laurent series have been found. These Laurent series generalize the Laurent series of the obtained two-parameter elliptic solutions. There are no obstacles to existence of three-parameter single-valued solutions, so, the probability to find exact three-parameter solutions, which generalize the obtained solutions, is high.

**Acknowledgments.** S.Yu.V. is grateful to F. Calogero, R. Conte, V. F. Edneral, A. K. Pogrebkov for valuable discussions. This work has been supported by Russian Federation President's Grants NSh-1685.2003.2 and NSh-1450.2003.2 and by the grant of the scientific Program "Universities of Russia".

## References

Ablowitz, M. J., Ramani, A., & Segur, H. 1980, J.Math.Phys., 21, 715; J.Math.Phys., 21, 1006

Antonov, V. A., & Timoshkova, E. I. 1993, AZh, 70, 265

Conte, R., ed. 1999, The Painlevé property, one century later, Proceedings of the Cargèse school (3–22 June, 1996, Cargèse), CRM series in mathematical physics (New York: Springer)

Conte, R., & Musette, M. 1992, J.Phys.A, 25, 5609

Conte, R., & Musette, M. 2003, Physica D, 181, 70; nlin.PS/0302051

Conte, R., Musette, M., & Verhoeven, C. 2002, J.Math.Phys., 43, 1906

Conte, R., Musette, M., & Verhoeven, C. 2003, TMF (Russ.J.Theor.Math. Phys.), 134, 148 (in Russian), 128 (in English); nlin.SI/0301011

Contopoulos, G. 1963, AJ, 68, 1; AJ, 68, 763

Goriely, A. 2001, Physica D, 152/153, 124

Hénon, M., & Heiles, C. 1964, AJ, 69, 73

Kowalevski, S. 1889, Acta Mathematica, 12, 177; 1890, Acta Mathematica, 14, 81 (in French); reprinted in Kovalevskaya, S. V. 1948, Scientific Works (Moscow: AS USSR Publ. House) (in Russian)

Melkonian, S. 1999, J.Nonlin.Math.Phys., 6, 139; math.DS/9904186

Painlevé, P. 1897, Leçons sur la théorie analytique des équations différentielles, profeesées à Stockholm (1895) sur l'invitation de S. M. le roi de Suède et de Norwège (Paris: Hermann); reprinted in CNRS 1973, Œuvres de Paul Painlevé, 1 (Paris); on-line version: The Cornell Library Historical Mathematics Monographs, http://historical.library.cornell.edu/

Tabor, M. 1989, Chaos and Integrability in Nonlinear Dynamics (New York: Wiles)

Timoshkova, E. I. 1991, AZh, 68, 1315

Timoshkova, E. I. 1999, AZh, 76, 470

Timoshkova, E. I. 2001, in Stellar Dynamics: from Classic to Modern, ed. L. P. Ossipkov & I. I. Nikiforov (St. Petersburg: St. Petersburg Univ. Press), 201

Vernov, S. Yu. 2003, TMF (Russ.J.TheorMath.Phys.), 135, 409 (in Russian), 792 (in English); math-ph/0209063; math-ph/0312048

Weiss, J. 1983, J.Math.Phys., 24, 1405

Weiss, J. 1984, Phys.Lett.A, 102, 329; Phys.Lett.A, 105, 387

Weiss, J., Tabor, M., & Carnevale, G. 1983, J.Math.Phys., 24, 522

*Order and Chaos in Stellar and Planetary Systems*
*ASP Conference Series, Vol. 316, 2004*
*G. Byrd, K. Kholshevnikov, A. Mylläri, I. Nikiforov, and V. Orlov, eds.*

# The Maximum Lyapunov Exponent of the Chaotic Motion in the Hénon–Heiles Problem

Alexander V. Melnikov and Ivan I. Shevchenko

*Central Astronomical Observatory at Pulkovo, Pulkovskoe sh. 65/1, St. Petersburg 196140, Russia*

**Abstract.** We explore the role of integration time in obtaining the correct dependence of the maximum Lyapunov exponent on the energy of the system in the Hénon–Heiles problem.

We consider the Hamiltonian of the problem by Hénon & Heiles (1964):

$$H = \frac{1}{2}\left(p_1^2 + p_2^2 + q_1^2 + q_2^2\right) + q_1^2 q_2 - \frac{1}{3}\, q_2^3, \tag{1}$$

where $q_1$ and $q_2$ are the coordinates, $p_1$ and $p_2$ are the conjugate momenta.

In our paper (Shevchenko & Melnikov 2003), the dependence of the value of the maximum Lyapunov characteristic exponent $L$ on the energy $E$ was obtained in the Hénon–Heiles problem by means of numerical integration of system (1). We concluded that, contrary to an earlier statement by Benettin, Galgani, & Strelcyn (1976), the observed dependence is not exponential, but is close to a power law with the power-law index $\alpha \approx 3.6$.

In this note, we explore in more detail the role of the time interval of numerical integration in obtaining the correct dependence of the maximum Lyapunov exponent on the energy of the system.

We calculate the maximum Lyapunov exponent on the time intervals $t = 10^4$, $10^5$, $10^6$, and $10^7$ for one and the same set of initial data. We use the "shadow trajectory" method (see, e. g., Lichtenberg & Lieberman 1992) and the HQR method (von Bremen, Udwadia, & Proskurowski 1997; Shevchenko & Kouprianov 2002). The equations of motion are integrated by an eighth-order Runge–Kutta method. For the details of the computing procedure see (Shevchenko & Melnikov 2003). The resulting dependences are plotted in Figures 1 and 2. In both Figures, the upper pair of plots corresponds to $t = 10^4$ and $10^5$, and the lower pair to $t = 10^6$ and $10^7$.

We approximate the upper bounds of the observed dependences by straight lines which, in the logarithmic coordinates, represent power laws. The upper bounds instead of averages are used to exclude the influence of the "stickiness effect", which manifests itself in downward spikes in the plots, and to exclude the effect of marginal resonances, which manifests itself in fine wave-like structure (in particular prominent at $t = 10^6$, $10^7$; see Figures 1c, 1d and 2c, 2d) of the dependences. On the stickiness effect in Hamiltonian dynamics and marginal resonances see (Shevchenko 1998a,b).

Computation by the HQR method produces the entire Lyapunov spectrum. The second element of the Lyapunov spectrum of the Hénon–Heiles system, the autonomous system with two degrees of freedom, is zero. While the integration

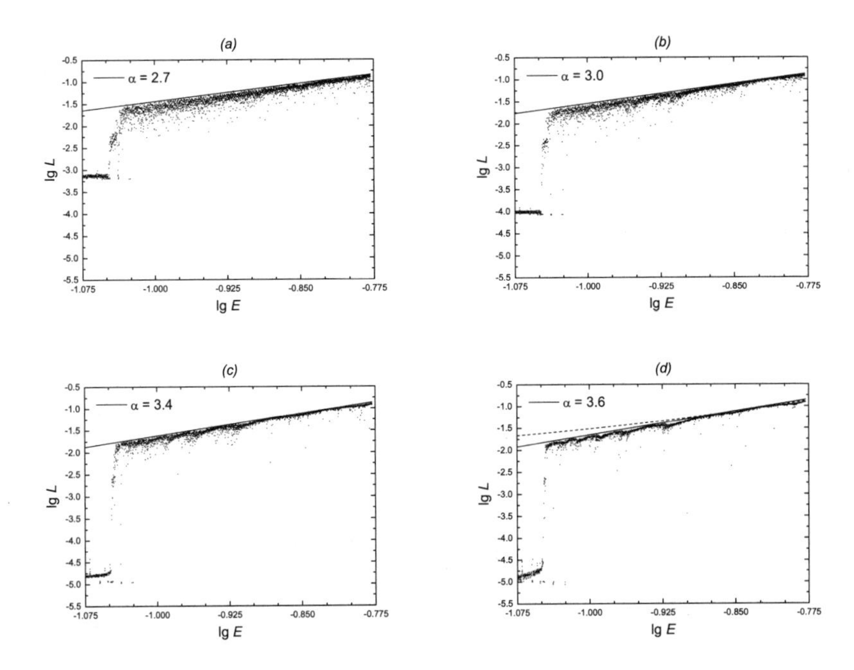


Figure 1. Energy dependence of the maximum Lyapunov exponent computed by the shadow trajectory method; $t = 10^4$, $10^5$, $10^6$, and $10^7$. The straight solid line is the power-law approximation.

time is increased, the curve of the energy dependence of the computed second element (the curve resembling a horizontal line in Figures 2a–2d) sinks as a whole tending to zero; whereas the curve of the energy dependence of the computed first element converges to a fixed power law. Shevchenko & Melnikov (2003) showed that $t = 10^7$ is sufficient to achieve stabilization that permits to determine $\alpha$ with the accuracy of two significant digits.

The indices of the approximating power laws are given in Figures 1a–1d and Figures 2a–2d for all used values of the integration time. One can see that though the upper bounds are well approximated by straight lines at any integration time, the indices increase with increasing the integration time, because the sinking of the dependence curve is greater at lower values of the energy. The time of saturation of the current Lyapunov exponents apparently increases with decreasing the energy.

Thus our numerical experiments demonstrate that although the correct form (a power law) of the energy dependence of the maximum Lyapunov exponent of the Hénon–Heiles system is evident at any integration time used in our study, the correct index of the power law can be obtained only at long enough integration times, due to the longer times of saturation of the current maximum Lyapunov exponent at low energies.

**Acknowledgments.** This work was supported by the Russian Foundation for Basic Research under grant 03-02-17356.

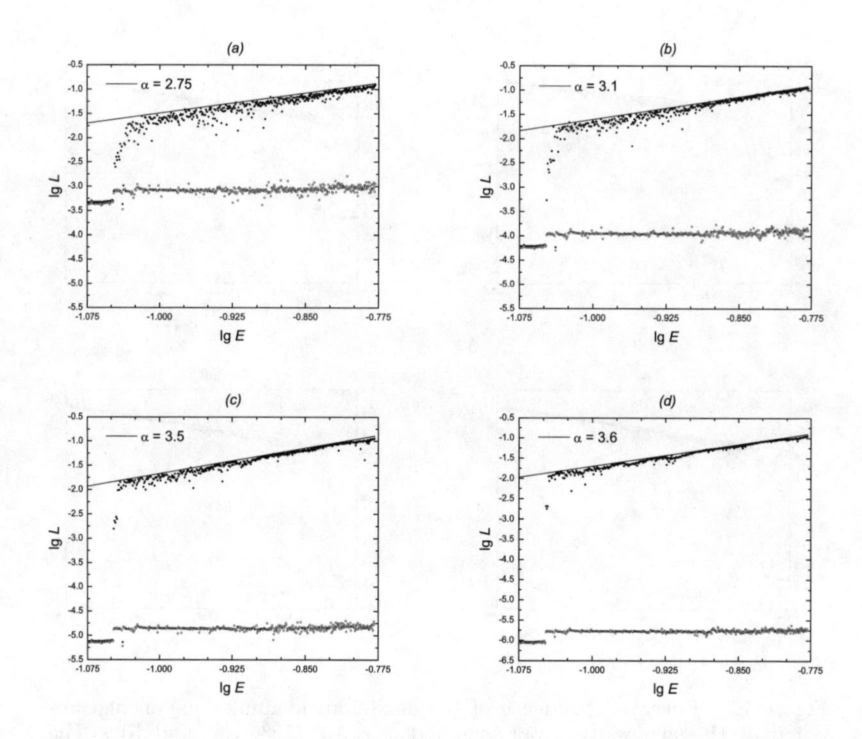


Figure 2. Energy dependence of the computed Lyapunov spectrum (by the HQR method) for the same times of integration.

## References

Benettin, G., Galgani, L., & Strelcyn, J.-M. 1976, Phys.Rev.A, 14, 2338

von Bremen, H. F., Udwadia, F. E., & Proskurowski, W. 1997, Physica D, 101, 1

Hénon, M., & Heiles, C. 1964, AJ, 69, 73

Lichtenberg, A. J., & Lieberman, M. A. 1992, Regular and Chaotic Dynamics (New York: Springer)

Shevchenko, I. I. 1998a, Phys.Lett.A, 241, 53

Shevchenko, I. I. 1998b, Phys.Scr., 57, 185

Shevchenko, I. I., & Kouprianov, V. V. 2002, A&A, 394, 663

Shevchenko, I. I., & Melnikov, A. V. 2003, JETP Lett., 77, 642 (Pis'ma ZhETP, 77, 772)

*Order and Chaos in Stellar and Planetary Systems*
*ASP Conference Series, Vol. 316, 2004*
*G. Byrd, K. Kholshevnikov, A. Mylläri, I. Nikiforov, and V. Orlov, eds.*

# Orbits in "Disk+Halo" Galaxy Model

S. A. Kutuzov

*Applied Math. & Control Processes Faculty, St. Petersburg State University, Universitetskij pr. 35, Staryj Peterhof, St. Petersburg 198504, Russia*

**Abstract.** Motion of a test particle in a gravitational field is considered. The steady-state rotationally symmetric model consists of a razor-thin disk and a halo. We extend the analytical method of constructing invariant curves by Tohline & Voyages (2001) to our model which is somewhat more general. The situation depends on a partition of model mass between the disk and the halo.

The periodic polar orbit obtained earlier is mentioned. Spatial trajectories have been calculated as well. Orbits are piecewise plane, so angular momentum changes discontinually and stochastically while crossing the disk. The distribution of angular momenta is considered in dependence on energy.

## 1. Introduction

In quite flattened Kuzmin's model (Kuzmin 1956) with Stäckel potential, distant stars "above" equatorial plane are attracted so as if whole system mass was concentrated at the point on a rotation axis "under" the plane and vice versa. This property is exactly fulfilled in the case of pure Kuzmin–Toomre disk (Kuzmin 1953; Toomre 1963). Tohline & Voyages (2001) have noticed that outside Kuzmin disk a motion is plane and Keplerian. Each of the Keplerian integrals of motion has its counterpart. The Laplace–Runge–Lenz vector is among them. A new form of the last integral was presented by Muñoz (2003).

Quite recently Lynden-Bell (2002) has obtained the fundamental result that an existence of the third integral of motion and an availability of two attraction centres are strictly connected. He has derived the integral in the case of two fixed centres and has generalized the method "to find the general form of the potential in which an exact third integral exists". But two attraction centres may also inhere in models of other kind where the third integral does not exist. We have such a situation in our model (Kutuzov & Ossipkov 1980) consisting of a razor-thin disk and a halo.

A problem of studying motion in bi-central field is of great interest. Energy as well as angular momentum vector component normal to the disk are constant in the course of motion. When crossing the disk angular momentum vector as well as Laplace vector change discontinuously. A trajectory leaps from one plane to the other, so we have leap-orbits.

Kuzmin & Malasidze (1990) have found three invariants which remain constant by crossing the disk and have derived the connection between Keplerian elements on two subsequent planes. Tohline & Voyages (2001) have suggested an analytical method to invesigate Poincaré sections. The periodic orbit in the meridional plane in our model was obtained earlier (Kutuzov 2001). Hunter (2002) has found in his comprehensive work that disk-crossing effect leads to a stochastic motion.

Here we study motion of a test-particle in the bi-central gravitational field of our model. Analytical approach as well as numerical computation are used. The meridional and spatial trajectories have been calculated. A dependence of orbit characteristics on the model parameters is discussed. The distribution of particle angular momenta is demonstrated in connection with its energy.

## 2. The Gravitational Field Model

We consider a special mass distribution model of stellar system which consists of infinitesimal thin "sharp" disk embedded in a halo. Such a model could be applied (of course in restricted sense) for Sombrero-like galaxies. Halo equidensities are lens-like surfaces formed by two families of concentric spheres with the fixed attraction centres on the rotation axis. The equidensities have a sharp edge on the disk plane and coincide with equipotentials of the system (Ossipkov & Kutuzov 1987).

We shall use dimensionless quantities. To get corresponding dimensional quantities one can use dimensional scale parameters $\hat{\Phi}$ , $\hat{r}$ which are potential and distance units respectively. The mass unit is $\hat{\mathcal{M}} = \hat{r}\,\hat{\Phi}/G$, where $G$ is the gravitational constant. The attractive centres are the points $C_\pm$ with cylindrical coordinates $R = 0,\ z = \pm\mu$. Denoting the origin of coordinate system as $O$ we define vector $\mathbf{a} = \overrightarrow{OC_+}$. The structure parameter $\mu \in [0,\ m]$ has meaning of the disk mass and the scale parameter $m$ is a total mass of the model. An arbitrary point $P(R, z)$ has the radius-vector $\mathbf{r}_\pm = \overrightarrow{OP}$ where the index corresponds to a sign of $z$. Introducing the radius-vector $\mathbf{s}_\mp = \overrightarrow{C_\mp P}$ we have the following connection between the vectors:

$$\mathbf{r}_\pm = \mp\mathbf{a} + \mathbf{s}_\mp, \qquad \mathbf{s}_\mp = \mathbf{r}_\pm \pm \mathbf{a}. \tag{1}$$

The distance of $P$ from the attraction centre $s_\mp = |C_\mp P|$ is simply expressed via cylindrical coordinates:

$$s_\mp^2 = R^2 + (z \pm \mu)^2\,, \qquad s_\mp \geq \mu. \tag{2}$$

We adopt the partial rational case of Kuzmin–Malasidze potential law (Kuzmin & Malasidze 1969):

$$\varphi(s) = \frac{m}{l+s}\,, \qquad m = l + \mu, \qquad l \in [0,\ m]. \tag{3}$$

From here the sign indices are omitted. The structure parameter $l$ is a characteristic length which coincides with the halo mass.

This potential governs motion outside the disk in any plane which passes through corresponding attraction centre. For a circular velocity $V(s)$ in the plane we have

$$V^2(s) = \frac{s}{m}\,\varphi^2(s). \tag{4}$$

It is a unimodal function with the maximum at $s = l$. We have obtained the equation of an envelope on $L^2E$ Lindblad diagram [see eqs. (10), (7)] in explicite form:

$$L_c^2 = \frac{(m+H)(3m-H)^3}{32m^2E}, \qquad H \equiv \sqrt{m^2 + 8lmE}. \tag{5}$$

## 3. Equations and Integrals of Motion

Let us consider general properties of motion. There are two kinds of motion: i) motion in the disk plane $z = 0$, ii) disk crossing motion. In the first case there is one attraction centre $O$ and the potential (3) takes the form:

$$\psi(R) = \frac{m}{l + \sqrt{\mu^2 + R^2}}. \tag{6}$$

We restrict ourselves to the second kind which is more complicated and seems to be more interesting. Owing to the steady state the energy integral

$$E = \varphi(s) - v^2/2 \tag{7}$$

exists. Here $v$ is a full velocity modulus. As our model possesses a rotational symmetry, the angular momentum integral

$$J_z = x\dot{y} - y\dot{x} = Rv_\theta \tag{8}$$

exists also. Here $x$, $y$, $z$ are Cartesian coordinates with origin at the point $O$ and $v_\theta$ is an azimuthal component of velocity. This integral is the component of the angular momentum vector with respect to the centre of the system $O$ which is normal to the disk,

$$\mathbf{J} = \mathbf{r} \times \mathbf{v}. \tag{9}$$

Another angular momentum vector with respect to the attraction centre is

$$\mathbf{L} = \mathbf{s} \times \mathbf{v} = \mathbf{J} \pm \mathbf{a} \times \mathbf{v}. \tag{10}$$

All its components are constant during motion along the fixed inclined plane outside of the disk between two subsequent crossings.

The Laplace–Runge–Lenz vector is very important (Muñoz 2003). For an arbitrary potential we could define it as

$$\mathbf{A} = \mathbf{L} \times \mathbf{v} + \varphi(s)\mathbf{s}. \tag{11}$$

Its derivative with respect to time $t$ for our potential (3) is

$$\frac{d\mathbf{A}}{dt} = \frac{l}{m}\,\varphi(s)^2\,\mathbf{v}. \tag{12}$$

The vector is constant in the Keplerian field only. Then $l = 0$, i.e., pure Kuzmin disk remains and the halo disappears. In general case $\mathbf{A}$ changes very slowly far from the centre. So its value there could be used for estimating orbit eccentricity $e$ since in the Keplerian case

$$A = me. \tag{13}$$

Taking into account that Laplace–Runge–Lenz vector is directed along the line of apsides we propose to call it **apsidian**.

The angular momentum vector with respect to the attraction centre defines the fixed inclined plane by the dot production

$$\mathbf{L} \cdot \mathbf{s} = 0. \tag{14}$$

We use polar coordinates $s$ and $\chi$ to describe motion along the plane. The angle $\chi$ is analogous to a true anomaly. Corresponding velocity components are

$$v_s = \dot{s}, \qquad v_\chi = s\dot{\chi}. \tag{15}$$

The momentum and energy integrals for this motion can be written in the form

$$L = s v_\chi = s^2 \dot{\chi}, \qquad E = \varphi(s) - \frac{L^2}{2s^2} - \frac{v_s^2}{2}. \tag{16}$$

Using these integrals we write differential equations of motion along the plane:

$$\frac{ds}{dt} = c\sqrt{2[\varphi(s) - E] - \frac{L^2}{s^2}}, \qquad \frac{d\chi}{dt} = \frac{L^2}{2s^2}, \qquad c = \mathrm{sign}(\mathbf{s} \cdot \mathbf{v}). \tag{17}$$

Here $c$ is a sign of radial velocity. These equations are integrable via elliptical integrals (Kuzmin & Malasidze 1969). Nevertheless it is more convenient to integrate them numerically.

## 4. Motion in a Meridional Plane

Here we extend analytical method by Tohline & Voyages (2001) to our model. We consider meridional (polar) orbits which lie in the $x = 0$ plane and intersect $z$ axis. We use section surfaces $z$, $v_z \equiv \dot{z}$, i.e., well known Poincaré sections. Since on $z$ axis $s = z \pm \mu$ and $v_s = v_z$ we obtain from eq. (16)

$$v_z^2 = 2\varphi(Z) - 2E - \frac{L^2}{Z^2}, \qquad Z = z \pm \mu. \tag{18}$$

In the last equation we have the same sign as for $z$. Since a picture is symmetrical with respect to the disk plane we restrict ourselves to the case $z \geq 0$.

That is an equation of the invariant curves' family with two parameters—$E$ and $L$. To find $Z$ when $v_z$ is given one has to solve a cubic equation. But inverse problem is elementary. The equation (18) permits to classify orbits. A character of invariant curve depend on pair of integral values, i.e., on the position of the point in Lindblad diagram. If $v_z \geq 0$ by passing through the system centre $O$ the trajectory is formed by independent oscillations in two mutually perpendicular directions and belongs to the class of box orbits. If a trajectory does not pass through the centre ($z > 0$) and $v_z = 0$ twice on $z$ axis, then it belongs to the class of loop orbits. Putting $z = v_z = 0$ into eq. (18) we obtain

$$E + \frac{L^2}{2\mu^2} = 1. \tag{19}$$

This equation describes a separating straight on $L^2E$ diagram. The situation depends on the partial weights of the disk and the halo in the model. In the case of pure spherical halo with $\mu = 0$ (two attraction centres merge) there are no box orbits—only loop ones remain.

There is an example of the periodic meridional orbit found by Kuzmin & Malasidze (1990) for very flattened Kuzmin (1956) model. It was rediscovered by Hunter (2002) for Kuzmin disk. Such orbit was also obtained by the author (Kutuzov 2001) for more general "disk+halo" model. The trajectory consists of a circle arc in the middle conjugated with two segments. Hunter (2002) added another example of periodic orbit for Kuzmin disk namely an ellipse. We have calculated trajectories starting in the neighbourhood of the edge point of periodic curve as well. The more start points deviate from the edge point of the periodic orbit, the more unstable trajectories become.

## 5. Motion in Inclined Planes

When studying a stochastisity of disk-crossing orbits Hunter (2002) has used various surfaces of section as well as delay plots of angular momenta. We try to introduce an additional way of analyzing orbits. Namely we use surface of section in Lindblad diagram, i.e., in energy-angular momentum $EL^2$ space. When crossing the disk the usual radius-vector **r** and the velocity **v** change continuously, but the angular momentum vector **L** as well as Laplace vector **A** [eqs. (10), (11)] suffer leaps because the radius-vector **s** does so [eq. (1)].

One stage of our calculations contains circular starts at the points on $y$ axis on different distances from the center. Since $E$ is constant the section points jump along segments parallel to the $L^2$ axis. One edge of each segment lies on the envelope [eq. (5)] whose points $L_c^2$, $E_c$ correspond to circular orbits. To describe a behaviour of a stochastic quantity $L^2$, we have built the series of histograms for highly inclined initial planes (Figure 1). $N \in [26,\ 200]$ is the

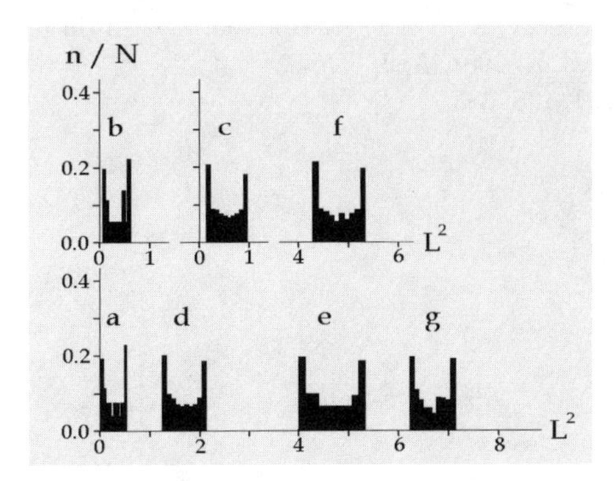


Figure 1. Angular momenta distribution.

number of crossing, $n$ is an absolute frequency, number of bins is 10 everywhere. Absolute values of negative energy are as follows—a: 0.735, b: 0.695, c: 0.588, d: 0.405, e: 0.298, f: 0.233, g: 0.191. The angular momenta distributions are U-shaped and change moderately with energy.

## 6. Conclusions

Nowadays the problem of motion in the bicentral gravitational field is of great importance. We have presented preliminary results. There are many unsolved questions in the problem. We intend to continue investigations in analytical aspect as well as in numerical one.

**Acknowledgments.** It is a great pleasure for the author to thank Professor C. Hunter for the courtesy to private communication and Dr L. P. Ossipkov for stimulating interest to the work and the reviewer for useful remarks.

## References

Hunter, C. 2002, Disk-crossing orbits (private communication, a talk in Athens, to be published by Springer in their Lecture Notes in Physics series), 1

Kutuzov, S. A. 2001, in All-Russia astron. conf. abstracts, ed. V. V. Ivanov (St. Petersburg: St. Petersburg Univ. Press), 107 (in Russian)

Kutuzov, S. A., & Ossipkov, L. P. 1980, AZh, 57, 28 (in Russian)

Kutuzov, S. A., & Ossipkov, L. P. 1988, AZh, 65, 468 (in Russian)

Kuzmin, G. G. 1953, Izvestiya AN ESSR, 2, 368 [Tartu astron.observ.Contr. N 2] (in Russian)

Kuzmin, G. G. 1956, AZh, 33, 27 [Tartu astron.observ.Contr. N 2] (in Russian)

Kuzmin, G. G., & Malasidze, G. A. 1969, Publ.Tartu astrophys.observ., 38, 181 (in Russian)

Kuzmin, G. G., & Malasidze, G. A. 1990, in Voprosy Neb. Mekh. i zvezd. dinamiki (Alma-Ata), 107 (in Russian)

Lynden-Bell, D. 2002, Astro-ph/0210417

Muñoz, G. 2003, Physics/0303106

Ossipkov, L. P., & Kutuzov, S. A. 1987, Astrofizika, 27, 523 (in Russian)

Tohline, A., & Voyages, K. 2001, ApJ, 555, 524

Toomre, A. 1963, ApJ, 138, 385

# Part II

# The Few-Body Problem

*Order and Chaos in Stellar and Planetary Systems*
*ASP Conference Series, Vol. 316, 2004*
*G. Byrd, K. Kholshevnikov, A. Mylläri, I. Nikiforov, and V. Orlov, eds.*

# Statistical Approach to the Three-Body Problem

M. Valtonen

*Dept. of Physics and Tuorla Observatory, University of Turku, Finland, and Dept. of Physics, The University of The West Indies, Trinidad*

A. Mylläri

*Dept. of Information Technologies, University of Turku, Finland*

V. Orlov and A. Rubinov

*Sobolev Astronomical Institute, St. Petersburg State University, Universitetskij pr. 28, Staryj Peterhof, St. Petersburg 198504, Russia*

**Abstract.** Strongly bound triple systems often evolve in a way that can be described as a deterministic chaos. Therefore the most useful description of the final state of the three-body evolution in many situations is purely statistical. One may assume that the phase space of initial configurations is uniformly covered in some sense, and that in the end of the dynamical evolution, when the triple has broken up into a binary and a single escaping body, the systems are again uniformly distributed in the allowable phase space. In this paper we study the validity of this assumption. Dynamical evolution of 100,000 equal-mass triple systems is investigated. The systems posess net angular momentum which is quantified by the parameter $w = -L_0^2 E_0/G^2 m_0^5$, where G is the gravitational constant, $m_0$ is the mass of a body, and $L_0$ and $E_0$ are the angular momentum and the total energy of the triple system. We consider the values of $w = 0.1,\ 1,\ 2,\ 4,\ 6$ which covers the range of angular momenta for strongly interacting, initially bound systems. For each $w$, 20,000 triple systems are studied. The initial coordinates and velocities of the components are chosen in two different ways: the first one assumes a hierarchical structure initially, the second one does not. The evolution of each triple system is calculated until either the escape of one of the bodies occurs or the time exceeds 1000 mean crossing times of the system. The statistical escape theory is based on the assumption of ergodicity, i.e., that the only information on the initial conditions remaining at the time of the escape of the third body is contained in the conserved total energy, total angular momentum and the mass values. The distributions of various quantities are derived from the allowable phase space volumes. We consider as an example the distributions of escape angle predicted by theory and found from numerical simulations. Those are in agreement. The escape directions are preferentially perpendicular to the total angular momentum vector, the more so the greater is the angular momentum.

## 1. Introduction

The dynamical evolution of three-body gravitational systems has been studied by many authors since the pioneering work by Agekian & Anosova (1967). For more recent reviews, see, e.g., Anosova & Orlov (1985), Valtonen (1988), and Valtonen & Mikkola (1991). In addition to numerical orbit integrations, the dynamical escape from a triple systems has been studied in the framework of statistical escape theories (Heggie 1975; Monaghan 1976a,b; Nash & Monaghan 1978). These theories are based on the supposition that the phase trajectory of a triple system is quasi-ergodic in the region of strong interaction of the bodies, called a close triple approach. Then the probability of escape with definite orbital elements of the final binary and the escaper is calculated. After the close triple approach one may assume an absence of interactions between the escaper and the final binary.

Numerical simulations of rotating triple systems have shown that the angular momentum is an important parameter in the description of the final state (see, e.g., Anosova 1969; Standish 1972; Saslaw, Valtonen, & Aarseth 1974; Valtonen 1974; Anosova, Bertov, & Orlov 1984; Mikkola & Valtonen 1986; Anosova & Orlov 1986). Using a statistical theory Monaghan (1976b) and Nash & Monaghan (1978) showed that the limitations on the phase space put by the constant angular momentum explain many of the findings of the numerical simulations but contradict others. For example, a trend towards orthogonality between the escape velocity vector and the angular momentum vector (i.e., the angle $\theta$ between these vectors being close to $90°$) is expected to become stronger when the angular momentum grows (Valtonen 1974).

In Monaghan's (1976b) theory, the ensemble of three-body systems is assumed to possess a constant total energy $E_0$ and a constant angular momentum $L_0$. The volume of the available phase space is calculated under these restrictions, and it is used to calculate the statistical properties of the ensemble after the break-up of the three-body systems. The theory contains a free parameter, the radius of the sphere of interaction $R$ which enters into the final distributions. Even when the value of the parameter is optimised, the distributions of escape energy do not have quite the right shape, and the correct dependence on angular momentum is not reproduced. Moreover, it is not possible to carry out the calculations through analytically which makes it difficult to see how the theory could be adjusted for a better agreement with numerical experiments.

An alternative approach, introduced by Mikkola & Valtonen (1986), is to combine analytical theory with numerical orbit calculations in a more straightforward theory. In this theory the ensemble of triple systems possesses a common total energy $E_0$ but the total angular momentum is allowed to vary within reasonable limits. Alternatively, it may be required that the ensemble has a common value of total angular momentum $L_0$ but the total energy is allowed to vary within certain limits. The phase space limitation arising from the second parameter is introduced afterwards using numerical experiments and with the guidance of some basic principles. This leads to a simple description of the results of the experiments.

Many of the details are described in our other papers (Valtonen et al. 2003, 2004, Papers I and II correspondingly). Here we give a short summary of the new computer simulations and the theoretical framework. In particular, we consider the distributions of of the escape angle as an example.

## 2. Parameters of the Problem

It is useful to describe the state of rotation of the triple system by the dimensionless parameter

$$w = -L_0^2 E_0 / G^2 m_0^5, \tag{1}$$

where $G$ is the gravitational constant, $m_0$ is the mean mass of the three bodies (Marchal et al. 1984), $L_0$ and $E_0$ are the angular momentum and the total energy of the triple system (Mikkola & Valtonen 1986). In the simulations we have put the masses of all three bodies equal: $m_0 = m_a = m_b = m_s = 1$, but this is not a limitation of the general theory.

Two different methods of generating initial conditions are used (see Paper I for details). In the first one the dynamical evolution is started from an elliptic encounter of a single body and a close binary. The binding energy of the outer binary, composed of the single body and the close binary, is small in comparison with the close binary energy. Thus the triple system has a hierarchical structure initially.

The second method does not necessarily assume a hierarchy. The initial configuration is randomly chosen within so called region $D$ of the Agekian & Anosova map (see, e.g., Figure 1 in Paper I). In the Agekian–Anosova map, the positions of the two bodies with the greatest mutual separation are mapped at points $(-0.5, 0)$ and $(+0.5, 0)$ , respectively. The third body is placed somewhere within unit radius of the first body, and closer to the second body than to the first body. The region $D$ so defined contains all possible configurations of triple systems (Agekian & Anosova 1967). Initial velocities are distributed isotropically when the virial ratio is randomly distributed within the interval $(0, 1)$.

The quantity $w$ (Eq. 1) is given one of several fixed values, within the tolerance of 0.01 units. The values used are $w = 0.1,\ 1,\ 2,\ 4,\ 6$. For each $w$ and each method of generating initial data, we study the evolution of 10,000 triple systems. Thus the total number of runs is 100,000. Calculations are performed until the escape of one of the bodies occurs or until the time equals $1000T_{\rm cr}$, where

$$T_{\rm cr} = \frac{GM^{5/2}}{|2E_0|^{3/2}} \tag{2}$$

is mean crossing time. Here $M = m_a + m_b + m_s$ is total mass of triple system. The length unit is the mean size

$$d = \frac{G(m_a m_b + m_a m_s + m_b m_s)}{|E_0|}. \tag{3}$$

The velocities are given in units of $d/T_{\rm cr}$.

Here we will consider the following parameters of the systems after the escape:

- number $N$ of escapes;
- life-time $T$ [in units of $T_{\rm cr}$ (2)]; as a quantitative measure we use the half-life time $T_{1/2}$ of the systems; at this time 50% of the systems have broken up;
- semi-major axis $a$ of the binary;

- eccentricity $e$ of the binary;
- asymptotic relative velocity $V_s$ of the escaper with respect to the binary;
- angle $\theta$ between the angular momentum vector of the triple system and the velocity vector of the escaper;
- angle $\lambda$ between the angular momentum vectors of the (inner) binary and the outer binary. The latter is formed by the centre of mass of the (inner) binary and the escaper;
- ratio $\beta$ of the angular momenta of the (inner) binary and the outer binary.

The detailed analysis of these parameter distributions has been described in Papers I and II. Here we give a brief description of the main results. They are summarised in Table 1. It contains the medians of escape parameters mentioned above for different values of $w$. Last column corresponds to the systems with life-times $T > 10T_{cr}$ at $w = 6$.

Table 1. Medians of escape parameters. The symbols H and $D$ in the first column correspond to two choices of initial conditions: hierarchical and from region $D$ (see Section 2 for details).

| | $w$ | | | | | |
|---|---|---|---|---|---|---|
| | 0.1 | 1 | 2 | 4 | 6 | 6 $T > 10$ |
| (1) | (2) | (3) | (4) | (5) | (6) | (7) |
| $N$ H | 9814 | 9543 | 8700 | 4591 | 883 | 511 |
| $D$ | 9839 | 9705 | 9447 | 7269 | 4692 | 618 |
| $T$ H | 16.5 | 26.7 | 36.9 | 60.3 | 160 | 430 |
| $D$ | 14.6 | 36.4 | 52.5 | 56.3 | 4.7 | 409 |
| $a$ H | 0.124 | 0.142 | 0.149 | 0.159 | 0.165 | 0.165 |
| $D$ | 0.106 | 0.136 | 0.143 | 0.152 | 0.130 | 0.164 |
| $e$ H | 0.758 | 0.736 | 0.754 | 0.756 | 0.651 | 0.636 |
| $D$ | 0.744 | 0.742 | 0.750 | 0.750 | 0.747 | 0.681 |
| $V_s$ H | 1.25 | 0.90 | 0.73 | 0.45 | 0.23 | 0.23 |
| $D$ | 0.87 | 0.55 | 0.47 | 0.35 | 0.61 | 0.15 |
| $\theta$ H | 71.7 | 82.8 | 84.6 | 86.3 | 87.3 | 87.3 |
| $D$ | 73.2 | 83.2 | 85.1 | 86.9 | 87.7 | 88.0 |
| $\lambda$ H | 121.9 | 90.1 | 81.9 | 59.5 | 36.0 | 35.6 |
| $D$ | 116.0 | 83.0 | 71.8 | 53.9 | 73.1 | 25.1 |
| $\beta$ H | 0.688 | 0.300 | 0.218 | 0.171 | 0.175 | 0.179 |
| $D$ | 0.728 | 0.281 | 0.205 | 0.150 | 0.111 | 0.148 |

## 3. Analytical Theory

### 3.1. Phase Space Volume

In Monaghan's (1976a,b) theory the three-body systems are studied after one of the bodies has escaped (Figure 1). A body of mass $m_s$ escapes from the binary of component masses $m_a$ and $m_b$. The position vector of the third body relative to the barycentre of the binary is $\mathbf{r}_s$ while the binary components are separated by the vector $\mathbf{r}$. We call the binary mass $m_B = m_a + m_b$ and the total mass $M = m_B + m_s$, and the reduced masses $\mathcal{M} = m_a m_b / m_B$ and $m = m_B m_s / M$. If we neglect the potential energy of the escaper relative to the binary, the total energy is

$$E_0 = \frac{1}{2} m \dot{r}_s^2 + \frac{1}{2} \mathcal{M} \dot{r}^2 - G \frac{m_a m_b}{r} \tag{4}$$

or

$$E_0 = E_s + E_B, \tag{5}$$

where the energy is divided into the binary energy $E_B$ and the energy of the escaper relative to the barycentre of the binary $E_s$.

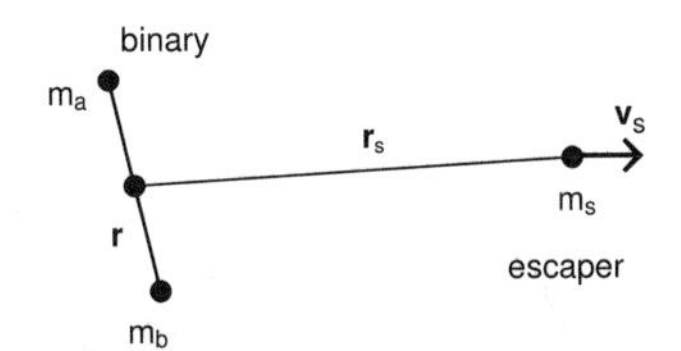


Figure 1. The basic configuration of three bodies after the body with mass $m_s$ has escaped from the binary with component masses $m_a$ and $m_b$. The current speed of the escaper relative to the binary centre of mass is $v_s$ and the current distance $r_s$. The separation of the binary components is $r$ at this time.

The density $\sigma$ of escape configurations in the phase space per unit energy is obtained by integrating a $\delta$-function over the phase space coordinates $\mathbf{r}$, $\mathbf{r}_s$, $\mathbf{p}$ and $\mathbf{p}_s$ where $\mathbf{p}$ and $\mathbf{p}_s$ are the momenta of the two relative motions. Thus

$$\sigma = \int\!\!\int\!\!\int\!\!\int \delta\left(\frac{p_s^2}{2m} + E_B - E_0\right) d\mathbf{r}_s\, d\mathbf{p}_s\, d\mathbf{r}\, d\mathbf{p}, \tag{6}$$

where we have put $E_s = p_s^2/2m$.

Monaghan (1976a) integrates over $\mathbf{r}_s$ and $\mathbf{p}_s$ assuming that the escaper orbit is radially outward from the barycentre of the binary. We assume instead that the orbits of the escapers come from inside a "loss cone" of base radius $7a$ where $a$ is the semi-major axis of the binary (Figure 2). The number seven is a practical choice based on numerical orbit calculations. The semi-major axis is related to the binary energy by

$$a = -\frac{G m_a m_b}{2 E_B}. \tag{7}$$

The loss cone directions contain the fraction $\pi(7a)^2/(4\pi r_s^2) = 12.25(a/r_s)^2$ of the whole sphere of radius $r_s$ surrounding the escaper.

Since the $\delta$-function does not depend on $\mathbf{r}_s$ we can immediatelly perform the first integration taking account of the loss cone

$$49\pi \int_0^R \left(\frac{a}{r_s}\right)^2 r_s^2\, dr_s = 49\pi a^2 R = \frac{12.25\pi (G m_a m_b)^2}{E_B^2} R. \tag{8}$$

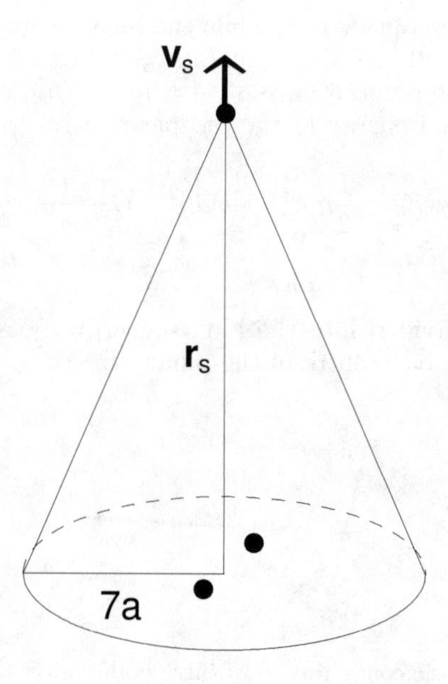


Figure 2. The loss cone. The escaping body, at the apex of the cone, must have come from inside the cone in order for there to have occurred a strong interaction with the binary in the past. In reverse, bodies falling into the cone will interact efficiently with the binary and will be perturbed to new orbits. Thus these orbits are "lost".

where the upper limit of the range of $r_s$ is an unspecified parameter $R$.

The integration over $\mathbf{p}_s$ is carried out as in Monaghan (1976a) with the result

$$\sigma = 49\sqrt{2}\pi^2 G^2 (m_a m_b)^2 R m^{3/2} \int\int \frac{\sqrt{E_0 - E_B}}{E_B^2}\, d\mathbf{r}\, d\mathbf{p}. \tag{9}$$

We would like to simplify the integrand in order to have a pure power-law dependence on $E_B$. We note that the quantity inside the square root is the escaper energy $E_s$. In the limit of parabolic motion it is equal to the potential energy (which we have so far neglected) $Gm_B m_s/R = E_s$. Substituting this inside the square root we get

$$\sigma = 49\sqrt{2}\pi^2 (GMR)^{1/2} m^2 (Gm_a m_b)^2 \int\int \frac{d\mathbf{r}\, d\mathbf{p}}{|E_B|^2}. \tag{10}$$

Even though the last step is only approximate, it turns out that the power-law dependence on $E_B$ obtained in this way gives a good account of three-body experiments.

In order to make direct comparisons with experiments, it is convenient to transform the remaining variables of integration and to carry out some of the integrations. Following the steps of Monaghan (1976a) we get

$$\sigma = 49\pi^5 (Gm_a m_b)^{11/2} R^{1/2} m_B^{3/2} M^{-3/2} m_s^2 \mathcal{M} \int\int \frac{dE_B}{|E_B|^{9/2}} e\, de. \tag{11}$$

The functions inside the integral give the distributions of binary energy

$$f(|E_{\rm B}|)\,d|E_{\rm B}| = 3.5|E_0|^{7/2}|E_{\rm B}|^{-9/2}\,d|E_{\rm B}| \tag{12}$$

and binary eccentricity

$$f(e)\,de = 2e\,de. \tag{13}$$

in a three-body break-up. The same result has been obtain by Heggie (1975) using the principle of detailed balance in a stellar system.

The calculation may be repeated for planar systems with the result

$$f(|E_{\rm B}|)\,d|E_{\rm B}| = 2|E_0|^2|E_{\rm B}|^{-3}\,d|E_{\rm B}|, \tag{14}$$

and

$$f(e)\,de = e(1-e^2)^{-1/2}\,de. \tag{15}$$

As far as eccentricities go, these results are identical to those of Monaghan (1976a), but the energy distributions have steeper power-laws. Numerical experiments have confirmed that indeed the present modification of the work of Monaghan (1976a) does represent an improvement. Monaghan's negative power-law index $n = 2$ is too shallow, while $n = 4.5$ of Eq. 12 agrees well with data.

### 3.2. Angular Momentum Dependence

Mikkola & Valtonen (1986) have shown that the power-law index is actually a function of the total angular momentum $L_0$. In general we may write

$$f(|E_{\rm B}|)\,d|E_{\rm B}| = (n-1)|E_0|^{n-1}|E_{\rm B}|^{-n}\,d|E_{\rm B}|, \tag{16}$$

where

$$n - 3 = 18L^2. \tag{17}$$

The quantity $L$ is the normalised total angular momentum, i.e., $L_0$ divided by

$$L_{\rm max} = 2.5G(m_0^5/|E_0|)^{1/2}. \tag{18}$$

Here $m_0$ is the average mass of the three bodies defined by

$$m_0 = \sqrt{\frac{m_{\rm a}m_{\rm b} + m_{\rm a}m_{\rm s} + m_{\rm b}m_{\rm s}}{3}}. \tag{19}$$

We see that $L$ goes from 0 to 1 while $w$ goes from 0 to 6.25. To give an idea of the magnitude of $L$ in a well known three-body system, we note that for the Lagrangian equilateral equal mass triangle with circular orbits we have approximately $L = 0.85$. This is definitely a system of relatively high angular momentum.

Even though the values of $n$ in Eq. 17 are based on numerical data, we may justify them by the following arguments. When $L = 0$, the motion is restricted to a plane, and therefore we would expect the result derived for the planar systems, $n = 3$. When the angular momentum $L$ is very high, the vector sum $\mathbf{L}_0 = \mathbf{L}_{\rm B} + \mathbf{L}_{\rm s}$ of the angular momenta of the binary and the escaper should be large, as a rule. This is possible only if $\mathbf{L}_{\rm s}$ makes most of the contribution to $\mathbf{L}_0$

and $\mathbf{L}_B$ is relatively small. The small values of $\beta = L_B/L_s$ at large $w$ in Table 1 confirm this. The angular momentum of the binary $L_B$ and its energy $E_B$ are connected by

$$L_B^2 = \mathcal{M}\frac{(Gm_a m_b)^2}{2|E_B|}(1 - e^2). \tag{20}$$

Therefore a small angular momentum generally (for a constant eccentricity) means that $|E_B|$ has to be large. The probability that the binding energy is greater than $|E_B|$ is proportional to $|E_B|^{-7/2}$, according to Eq. 12. When Eq. 11 is multiplied by this factor we obtain

$$\sigma \propto \int\int \frac{d|E_B|}{|E_B|^8} e\, de. \tag{21}$$

Therefore we would expect the power-law index $n = 8$ at high angular momentum. But as Eq. 17 shows, this is not the maximal value of $n$, but it becomes as great as $n = 16$ for the Lagrangian case, and possibly even greater when $L \to 1$.

### 3.3. Escape Angle

The escape angle $\theta$ (see Figure 3) is primarily controlled by the available phase space for different combinations of angular momenta. Therefore, we calculate the phase space volume for an ensemble of escape orbits which have a common magnitude of the total angular momentum $L_0$.

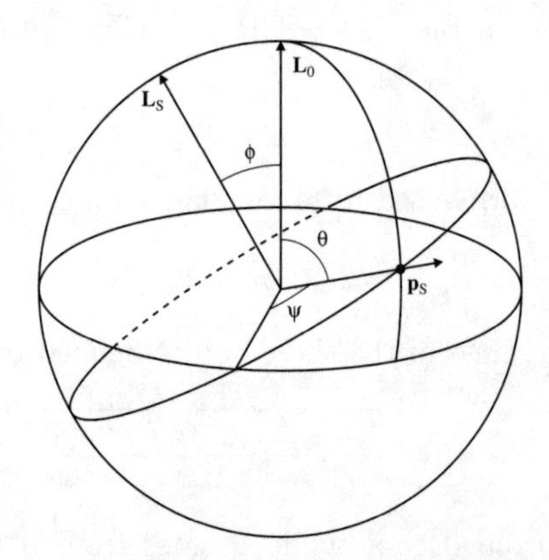


Figure 3. The geometry of the vectors $\mathbf{L}_0$, $\mathbf{p}_s$ and the angles between them. The momentum vector $\mathbf{p}_s$ lies in a plane perpendicular to $\mathbf{L}_s$ while this plane is at an angle $\phi$ relative to the fundamental plane perpendicular to $\mathbf{L}_0$. The direction of $\mathbf{p}_s$ is specified by the two angles $\theta$ and $\psi$.

The total angular momentum $\mathbf{L}_0$ is the vector sum of the binary angular momentum $\mathbf{L}_B$ and the angular momentum of the escaping body $\mathbf{L}_s$. Let the angle between $\mathbf{L}_s$ and $\mathbf{L}_0$ be $\phi$ (Figure 4). Then

$$L_B^2 = L_0^2 + L_s^2 - 2L_0 L_s \cos\phi \tag{22}$$

or

$$L_0 = L_s \cos\phi \left[1 + \sqrt{1 + (L_B^2 - L_s^2)/L_s^2 \cos^2\phi}\right]. \tag{23}$$

Here we consider only the larger one of two roots of $L_0$. The inclusion of the other root is not important to the final result (Paper II). The squares of the angular momenta are natural phase space coordinates. Thus we label $x = L_s^2$ and $y = L_B^2$.

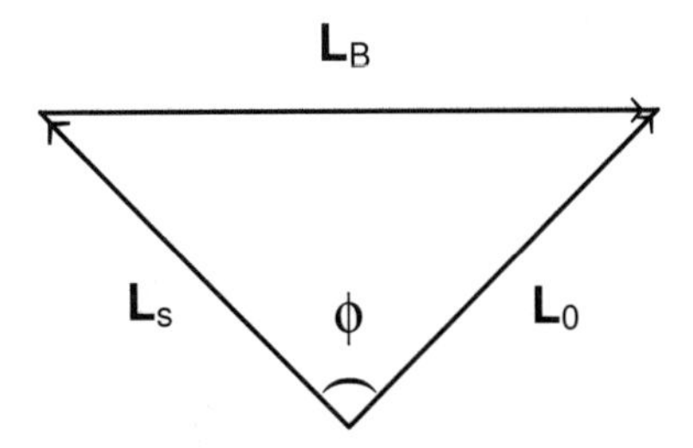


Figure 4. The relation between vectors $\mathbf{L}_0$, $\mathbf{L}_s$ and $\mathbf{L}_B$. $\mathbf{L}_0$ is the vector sum of $\mathbf{L}_s$ and $\mathbf{L}_B$, and the angle between $\mathbf{L}_s$ and $\mathbf{L}_0$ is $\phi$.

The escape direction may be specified by angles $\phi$ and $\psi$ (Figure 3) whereby

$$\cos\theta = \sin\phi \sin\psi. \tag{24}$$

Let us first consider the special case of $\psi = \pi/2$. Then the momentum vector of the escaper $\mathbf{p}_s$ lies in the plane defined by vectors $\mathbf{L}_s$ and $\mathbf{L_0}$ and $\theta = \pi/2 - \phi$. The natural phase space coordinate associated with $\phi$ is $\sin\phi = \cos\theta$ since $\cos\theta$ is uniformly distributed between $-1$ and $+1$ if the escape direction is random. We call this phase space coordinate $\zeta$: $1 - \zeta^2 = \cos^2\phi$. Using these natural coordinates

$$L_0 = \sqrt{x}\sqrt{1-\zeta^2}\left\{1 + \sqrt{1 + (y-x)/[x(1-\zeta^2)]}\right\}. \tag{25}$$

The available phase space volume may now be written as

$$\sigma = \int\int\int \delta\left[\sqrt{x}\sqrt{1-\zeta^2}\left\{1 + \sqrt{1 + (y-x)/[x(1-\zeta^2)]}\right\} - L_0\right] dx\, dy\, d\zeta. \tag{26}$$

The limits of integration are: $x$ from 0 to $\infty$, $\zeta$ from 0 to 1, and $y$ from 0 to a finite maximum value

$$y_{\max} = \mathcal{M}(Gm_a m_b)^2/2|E_B|. \tag{27}$$

The quantity inside the square root of Eq. 25 should not be negative. Thus

$$\zeta^2 \leq y/x. \tag{28}$$

This is a serious limitation to the available phase space. When $x \gg y$, $|\zeta|$ is confined to the neighbourhood of $|\zeta| \approx 0$. Therefore escape directions tend to concentrate in the plane perpendicular to the angular momentum $\mathbf{L_0}$ (Saari 1974).

We may avoid the negative square root by introducing a new variable $k$ in place of $y$ by

$$y/x = (k^2 + 1)\zeta^2, \tag{29}$$

where $k$ is a real number. Transforming to this variable we have

$$\sigma = \int\int\int \delta\left[\sqrt{x}\left(\sqrt{1-\zeta^2} + k\zeta\right) - L_0\right] 2xk\zeta^2\, dx\, dk\, d\zeta. \tag{30}$$

The integrations over $x$ and $k$ may be carried out in a straightforward manner, with the result

$$\sigma = \int F(\zeta)\, d\zeta, \tag{31}$$

where

$$F(\zeta) = \frac{2}{3} L_0^3 \frac{\sqrt{1-\zeta^2} + 3k_0\zeta}{(\sqrt{1-\zeta^2} + k_0\zeta)^3}. \tag{32}$$

Here $k_0$ is the lower limit of integration over $k$, the upper limit being $\infty$. The parameter $k_0$ is viewed here as a free parameter, to be determined on the basis of numerical experiments.

So far we have dealt with the special case of $\theta = \pi/2 - \phi$. But in fact for every value of $\phi$ there is a whole range of escape directions from $\theta = \pi/2 - \phi$ to $\theta = \pi/2$ (here we neglect $\theta > \pi/2$ because of symmetry). Thus the distribution of Eq. 32 is akin to the accumulated distribution of $\theta$. In the first approximation, $F(\zeta_0)$ gives the relative weight of all escape orbits up to $\zeta \leq \zeta_0$. That is,

$$\frac{F(\zeta_0) - F(0)}{F(1) - F(0)} \approx \int_0^{\zeta_0} f(\zeta)\, d\zeta. \tag{33}$$

Here $f(\zeta)$ is the distribution of the escape directions $\zeta = \cos\theta$.

In order to make $F(\zeta)$ in Eq. 32 a proper accumulated distribution, it has to be normalised. It can be done simply if we approximate $\zeta^2 \approx 0$ and consider $k_0 \gg 1$. Then the form of the accumulated distribution becomes

$$F(\zeta) = \frac{(1+k_0)^2}{k_0(2+k_0)}\left[1 - \frac{1}{(1+k_0\zeta)^2}\right], \tag{34}$$

and the corresponding distribution function of $\theta$ is

$$f(\theta) = \frac{(1+k_0)^2}{2+k_0} \frac{2\sin\theta}{(1+k_0\cos\theta)^3}. \tag{35}$$

From numerical experiments we find that the parameter $k_0$ depends on the normalized angular momentum (17) as

$$k_0 = 9L^{1.25}. \tag{36}$$

When we consider the case $L \to 0$, the distribution $f(\zeta)$ is flat, as it should be since there cannot be any preferred escape direction relative to $\mathbf{L_0} = \mathbf{0}$. The greater the value of $L$, the greater is the concentration towards $\theta = \pi/2$. In Figure 5 we compare numerical experiments with Eq. 35. The full phase space calculation by Nash & Monaghan (1978) tends to give distributions of $\theta$ which are too wide.

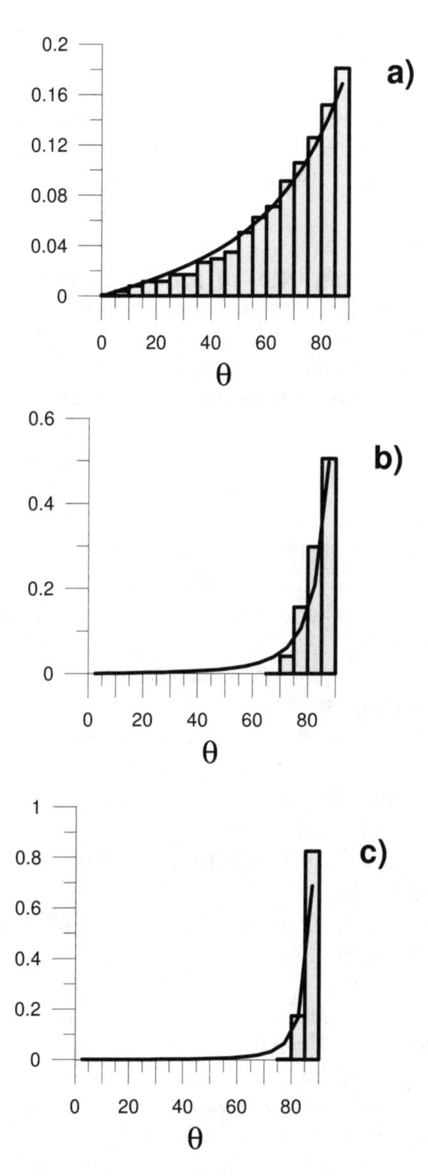


Figure 5. Comparison of distributions $f(\theta)$ for $w = 0.1$ (a), $w = 2$ (b), $w = 6$ (c).

## 4. Conclusions

We have demonstrated that the ergodic principle is a useful tool in the description of the statistical results of the three-body break-up. This work is an extension of the theory of Monaghan and associates, and its general principles are confirmed. Therefore it appears that the final state of the three-body system can be described in a (semi-)analytical way even for strongly interacting systems. These solutions are necessarily of statistical nature. In many astrophysical applications the exact initial conditions for the three bodies are not known, or even required, and statistical distributions are exactly what is needed (Heggie 1975).

**Acknowledgments.** Two authors (V.O. and A.R.) thank for the support Leading Scientific School (grant NSh-1078.2003.02), Russian Foundation for Basic Research (grant 02-02-17516) and program "Universities of Russia" of the Ministry of Education of Russian Federation (grant UR.02.01.027). A part of work was carried out during visit of one author (V.O.) in the University of Turku in frameworks of exchange program between St. Petersburg State University and the University of Turku.

## References

Agekian, T. A., & Anosova, J. P. 1967, AZh, 44, 1261

Anosova, J. P. 1969, Trudy Ast.Observ.Leningr.Univ., 26, 88

Anosova, J. P., Bertov, D. I., & Orlov V. V. 1984, Astrofizika, 20, 327

Anosova, J. P., & Orlov V. V. 1985, Trudy Ast.Observ.Leningr.Univ., 40, 66

Anosova, J. P., & Orlov, V. V. 1986, AZh, 63, 643

Heggie D. C., 1975, MNRAS, 173, 729

Maschal, C., Yoshida, J., & Sun, Y. S. 1984, CeM, 34, 65

Mikkola, S., & Valtonen, M. 1986, MNRAS, 223, 269

Monaghan, J. J. 1976a, MNRAS, 176, 63

Monaghan, J. J. 1976b, MNRAS, 177, 583

Nash, P. E., & Monaghan, J. J. 1978, MNRAS, 184, 119

Saari, D. G. 1974, CeM, 9, 175

Saslaw, W. C., Valtonen, M. J., & Aarseth, S. J. 1974, ApJ, 190, 253

Standish, E. M. 1972, A&A, 21, 185

Szebehely, V. 1972, CeM, 6, 84

Valtonen, M. 1974, in The stability of the solar system and of small stellar systems, ed. Y. Kozai (Dordrecht: D. Reidel), 211

Valtonen, M. 1988, Vistas Ast., 32, 23

Valtonen, M., & Aarseth, S. J. 1977, Revista Mex.Ast.Ap., 3, 163

Valtonen, M., & Mikkola, S., 1991, ARA&A, 29, 9

Valtonen, M., Mylläri, A. A., Orlov, V. V.., & Rubinov, A.V. 2003, Ast.Lett., 29, 50 (Paper I)

Valtonen, M., Mylläri, A. A., Orlov, V. V.., & Rubinov, A.V. 2004, in preparation (Paper II)

*Order and Chaos in Stellar and Planetary Systems*
*ASP Conference Series, Vol. 316, 2004*
*G. Byrd, K. Kholshevnikov, A. Mylläri, I. Nikiforov, and V. Orlov, eds.*

# Symbolic Dynamics and Chaos in the Three-Body Problem

A. Mylläri, H. Lehto, M. Valtonen, and P. Heinämäki

*University of Turku, Finland*

A. Rubinov, A. Petrova, and V. Orlov

*St. Petersburg State University, Russia*

A. Martynova

*Forest Technical Academy, St. Petersburg, Russia*

A. Chernin

*Moscow State University, Russia and University of Turku, Finland*

**Abstract.** Recent results concerning some chaotic and regular features of triple system dynamics are given. We consider the free-fall equal-mass three-body problem. We study the stochastic and regular regions constructing symbolic sequences during the evolution of a triple system. These symbolic sequences are used to calculate entropy-like characteristics to describe the behavior of triple systems before escape of one body. We study the effect of initial conditions on the entropies and reveal some isoentropic regions. The regularity zones correspond to fast escapes, whereas the stochastic sets have a structure like a Cantor set.

## 1. Introduction

Symbolic dynamics is widely used in studies of dynamical systems (see, e.g., Alekseev 1969; Katok & Hasselblatt 1995). The transition from studies of trajectories of dynamical systems in the (often high-dimensional) phase space to studies of shifts in the space of symbols allows to simplify considerations. One of long-standing problems in dynamics is the three-body problem. It has a history of more than three centuries long. Symbolic dynamics was applied in studies of some special cases of this problem: Sitnikov problem (Alekseev 1969), the rectilinear problem (Tanikawa & Mikkola 2000), the isosceles problem (Zare & Chesley 1998; Chesley 1999). We attempt to use methods of symbolic dynamics in studies of the more general case.

We consider the equal-mass free-fall three-body problem, i.e., masses of all bodies are the same (equal to 1), and initial velocities are 0. In this case, motion takes place in the plane. All possible configurations could be represented by the homology region $D$ (Agekian & Anosova 1967; see also Chernin & Valtonen 1998) which is shown in Figure 1.

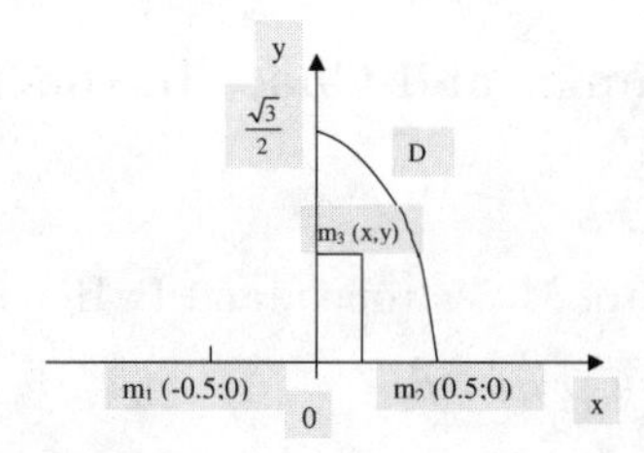


Figure 1. Homology region $D$.

We give a brief introduction to symbolic dynamics in Section 2. Applications to the above mentioned cases of the three-body problem are described in Section 3. Our results are given in Section 3 too.

## 2. Symbolic Dynamics

A symbolic dynamical system consists of three parts—$(X, \Omega, \sigma)$, where $\Omega$ is a finite alphabet, $X$ is the space of infinite sequences $\{\omega_i\}_{i \in Z}$, $\omega_i \in \Omega$, $\sigma$ is the shift transformation $\omega \mapsto \omega'$, $\omega'_i = \omega_{i+1}$.

For example, consider $\Omega = \{0, 1\}$. A fixed point:

$$...0, 0, 0, 0, 0, 0, 0, ...$$

Trajectory approaching this fixed point:

$$...1, 0, 1, 1, 0, 0, 0, 0, 0, ...$$

Periodic trajectories:

$$...0, 1, 0, 1, 0, 1, 0, 1, 0, 1, ...;$$

$$...0, 1, 1, 0, 1, 1, 0, 1, 1, 0, 1, 1, ...$$

"Dense" trajectory:

$$...0, 1, 0, 0, 0, 1, 1, 0, 1, 1, 0, 0, 0, 0, 0, 1, 0, 1, 0, 0, 1, 1, 1, 0, 0, 1, 0, 1, 1, 1, 0, 1, 1, 1, ...$$

(all possible combinations of length 1, all possible combinations of length 2, 3, etc.).

Sometimes we can establish a correspondence (homeomorphism): Classical Dynamical System $\leftrightarrow$ Symbolic Dynamical System. It could be done, e.g., in the following way: consider a classical dynamical system in discrete moments of time $t = t_j$ ($j$ is integer); let $x(t)$ be a trajectory of the Classical Dynamical System in the phase space $M$. We can construct a corresponding symbolic sequence $\omega$ making a partitioning of the phase space: if $M = \cup_{i=1}^{n} A_i$ then $\omega_j = k$ for $x(t_j) \in A_k$. For the part of the trajectory shown on Figure 2 we will get following sequence: $...1, 5, 2, 5, 3, 2, 3, 4, 5, ...$

The complexity of these symbolic sequences could be characterized using entropy-like parameters: $h_1 = -\sum_{i,j} q_i p_{ij} \log p_{ij}$ and $h_2 = -\sum_i q_i \log q_i$ where $q_i$ is a frequency of the symbol "$i$" in the sequence, and $p_{ij}$ is a frequency of transitions from $i$ to $j$ (see, e.g., Alexeev 1976).

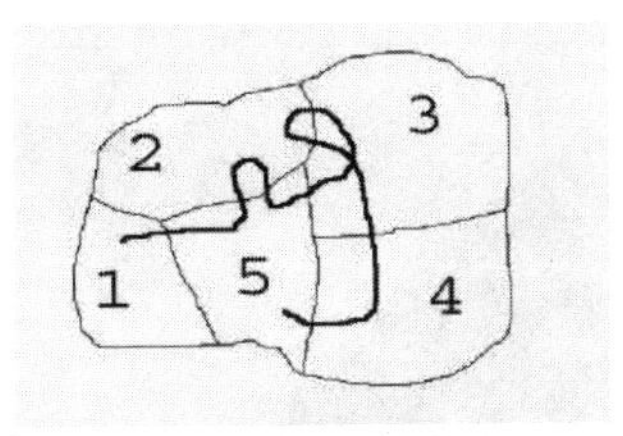


Figure 2. Example of phase trajectory for symbolic sequence.

## 3. Symbolic Dynamics and Three-Body Problem

Alexeev (1969) considered a particular case of the three-body problem—Sitnikov's problem. He showed for this case an existence of the homeomorphism three-body system $\leftrightarrow$ Symbolic Dynamical System. He also showed that $h_1$ and $h_2$ correspond to the topological entropies in the cases of Markov's and Bernoulli's shifts (for details, see Alexeev 1969, 1976). Since we consider the more general case it is very hard to prove analytically that such a homeomorphism exists. Some numerical arguments in favor of the existence of this homeomorphism will be given below. Moreover, since the life-time of the typical three-body system is limited (see Figure 3 below) we will have sequences approaching a fixed point in both directions (at both ends of infinite sequences we will have repetitions of the same symbols).

We scan a region $D$ (see Figure 1) of all possible initial conditions for the free-fall equal-mass three-body problem. The life-time of the system in dependence on initial conditions is shown in Figure 3. Darker points correspond to the longer life-times. We use the homology region $D$ to introduce the partitioning of the phase space: at any moment of the evolution we could project the geometric configuration of three points to the region $D$. This projection corresponds to the ordering of three points in accordance with mutual distances. In this way we get six different partitions; when the trajectory enters to the partition of number $k$, we put partition number to the symbolic sequence and calculate the time that the system spends in this state.

We scan the area $[0, 0.5] \times [0, 1]$ with step of 0.001 along both coordinates and estimate values $h_1$ and $h_2$ along the trajectory. We trace the partition number along the trajectory. We fix partition number and the moment of time when partition changes. Typical sequences in symbolic dynamics are of infinite length. In our case, the triple systems are unstable and disrupt at the end of the evolution. So the rest of the symbolic sequence will be infinite repetition of two symbols (due to the rotation of final binary). Therefore, should we calculate entropies $h_1$ and $h_2$ for the very long interval of time, we would obtain same values for (almost) all initial conditions. For each initial condition we fix maximum values of $h_1^{\max}$ and $h_2^{\max}$ and values of time $t_1^{\max}$ and $t_2^{\max}$ when they are achieved. Actually the parameters $h_1$ and $h_2$ are characteristics of the mixing in the system, and the values of time we fix correspond to the moments of active interplay. The results are given in Figures 4 and 5. Since we consider the area larger than the region $D$ we observe symmetric reflection with respect

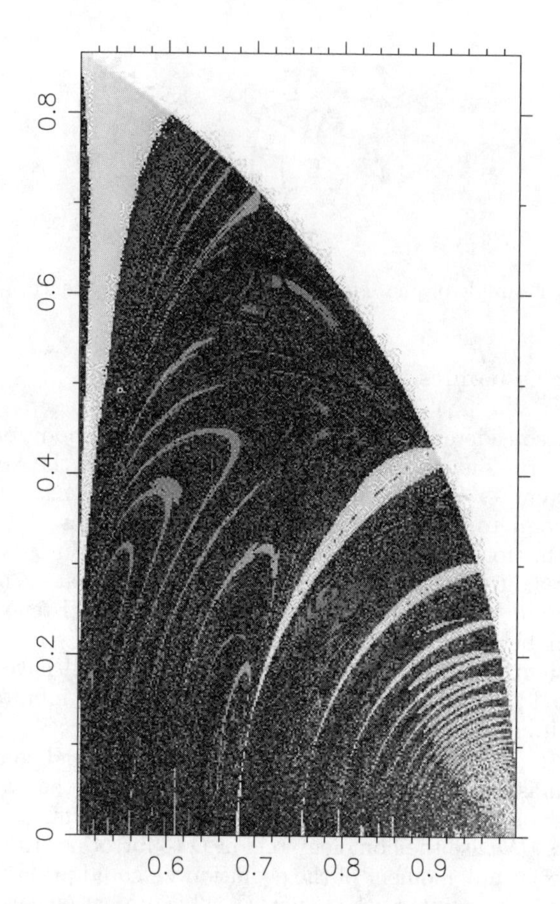


Figure 3. Life-time as a function of initial conditions in the homology region.

to the circle of radius 1 centered at the point $(-0.5, 0)$. In these figures, the brighter points correspond to larger values of $h_1^{\max}$ ($h_2^{\max}$) and to larger values of $t_1^{\max}$ ($t_2^{\max}$). One can see that general structure of regions corresponding to the different values of "entropies" and corresponding moments of time is similar to the structures seen in Figure 3. At the same time new features are revealed.

To put some basis for using methods of symbolic dynamics we plot Figure 6 which shows the initial configuration and the consecutive changes of the configurations using the next five symbols in the constructed sequence. Different configuration numbers are represented by different shades of gray. One can see that the mixing is quite efficient and the area of regions corresponding to the same sequences decreases very fast.

The value of the entropy should not depend on the method of partitioning. We used a few other ways of partitioning with 3 and 2 partitions and obtained

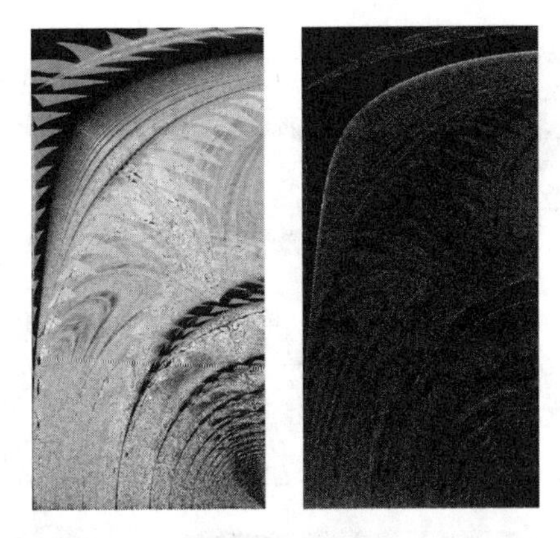

Figure 4. The maximum entropy $h_1$ (left) and time of its achievement (right) as a function of initial conditions in the region $D$.

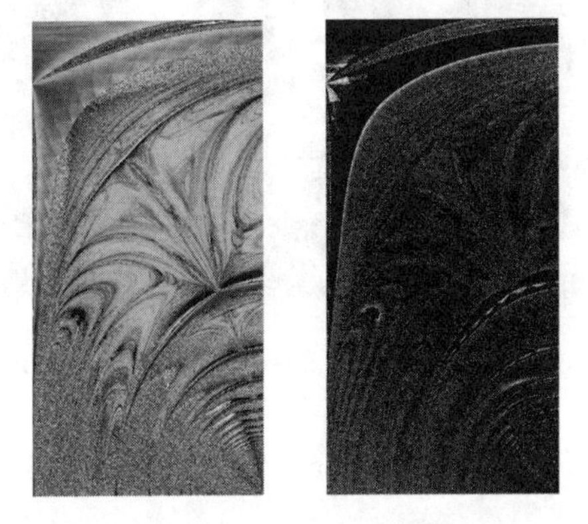

Figure 5. The same as in Figure 4, but for the entropy $h_2$.

pictures revealing similar structures. This is an additional argument for the validity of using these methods.

Thus, the symbolic dynamics approach can be applied to plane three-body systems and provides one more tool to study the structure of chaos. In this case we consider configuration space partitioning instead of fixing some definite states (e.g., double collisions). At the same time, in the general case we could also construct a symbolic sequence using, e.g, the number of the distant body at the moment of double encounter. We will realize such an approach in further studies.

**Acknowledgments.** This work was partly supported by the RFBR (Grant 02-02-17516), Program *Universities of Russia* (Grant UR.02.01.027), the Leading Scientific School (Grant NSh-1078.2003.02), and the programme of scientific

Figure 6. The mixing of symbolic sequences in the region $D$.

cooperation between St. Petersburg State University and the University of Turku for 2002–2006.

## References

Agekian, T. A., & Anosova, J. P. 1967, AZh, 44, 1261

Alexeev, V. M. 1969, Mathematics of the USSR, Sbornik, 7, 1

Alexeev, V. M. 1976, Symbolic Dynamics, 11th Summer mathematical school (Kiev: Mathematics Institute of the Ukrainian Academy of Sciences)

Chernin, A. D., & Valtonen, M. J. 1998, New Ast.Rev., 42, 41

Chesley, S. 1999, CeMDA, 73, 291

Katok, A., & Hasselblatt, B. 1995, Introduction to the Modern Theory of Dynamical Systems (Cambridge Univ. Press)

Tanikawa, K., & Mikkola, S. 2000, Chaos, 10, 649

Zare, K., & Chesley, S. 1998, Chaos, 8, 475

*Order and Chaos in Stellar and Planetary Systems*
*ASP Conference Series, Vol. 316, 2004*
*G. Byrd, K. Kholshevnikov, A. Mylläri, I. Nikiforov, and V. Orlov, eds.*

# Collinear Three-Body Problem with Non-Equal Masses by Symbolic Dynamics

Masaya Masayoshi Saito and Kiyotaka Tanikawa

*National Astronomical Observatory/Department of Astronomical Science, SOKENDAI, Mitaka, Tokyo, JAPAN*

**Abstract.** We study the collinear three-body problem with the non-equal masses using symbol dynamics. In asymmetric cases, folds of triple collision curves, not observed in the equal mass case, appear.

## 1. Introduction

The collinear three-body problem is one of the general three-body problems. Hietarinta & Mikkola (1993) studied this problem for various masses with the help of a Poincaré map. They found that the initial value plane (equivalent to the Poincaré section) is divided into three regions: fast escape regions, the quasiperiodic region (the Schubart region), and the chaotic scattering region. The Poincaré map scarcely resolves the structure of the chaotic scattering region. Tanikawa & Mikkola (2000) found that this region is divided by an infinite number of triple collision curves by introducing symbolic dynamics into this problem. They also found a rule of transitions among the regions separated by these curves.

The authors apply symbolic dynamics to the case of non-equal masses, and study the structure of the Poincaré section for various masses. We pickup one sample each from symmetric and asymmetric cases in this report.

## 2. The Method of Calculation

### 2.1. Poincaré Sections

We call three particles on a line $m_1$, $m_0$, and $m_2$ from the left, and take $q_1$ (the distance between $m_1$ and $m_0$), $q_2$ (the distance between $m_0$ and $m_2$) as coordinate variables. In this selection, the Hamiltonian of the system is

$$H = K - U; \qquad K = \frac{1}{2}\left(\frac{1}{m_1} + \frac{1}{m_0}\right) p_1^2 + \frac{1}{2}\left(\frac{1}{m_0} + \frac{1}{m_2}\right) p_2^2 - \frac{p_1 p_2}{m_0},$$
$$U = \frac{m_1 m_0}{q_1} + \frac{m_0 m_2}{q_2} + \frac{m_1 m_2}{q_1 + q_2}. \qquad (1)$$

A Poincaré section $(\theta, R)$ for the isoenergetic surface, $H = E_0$, is introduced as follows. In the collinear three-body problem, there is a special solution, the so-called homothetic solution, that repeats triple collision retaining the relation,

$$q_1/q_2 = \dot{q}_1/\dot{q}_2 = \tau(m_i). \qquad (2)$$

The constant $\tau$ depends on masses, and is unity in the symmetric case. For general initial conditions, we take the set of $(q, \dot{q})$ that satisfies relation (2) as a Poincaré section, and define $R$ by

$$R = (q_1 + q_2)/2 \quad \text{when} \quad q_1/q_2 = \tau. \tag{3}$$

A freedom of distributing the kinetic energy among the particles still remains. The expression of kinetic energy in $(\dot{q}_1, \dot{q}_2)$ is

$$K = A\dot{q}_1^2 + B\dot{q}_2^2 + C\dot{q}_1\dot{q}_2,$$

$$A = \frac{m_1(m_0 + m_2)}{2M}, \quad B = \frac{m_2(m_0 + m_1)}{2M}, \quad C = \frac{m_2 m_0}{2M}, \quad M = \sum_i m_i. \tag{4}$$

The function $K$ can be diagonalized by a linear transformation $\bar{v}_i = a_{ij}\dot{q}_j$:

$$K = \bar{v}_1^2 + \bar{v}_2^2. \tag{5}$$

A variable $\theta$ is introduced as

$$\sqrt{K} \sin\theta = \bar{v}_1, \quad \sqrt{K} \cos\theta = \bar{v}_2. \tag{6}$$

### 2.2. Symbolic Dynamics

A *symbol sequence* is usually defined as the sequence of the names of regions that an orbit has visited. However Tanikawa & Mikkola (2000) introduced symbol sequences using binary collisions. The symbol "1" is assigned to a collision between $m_1$ and $m_0$, "2" is to $m_0$ and $m_2$, and "0" is to a triple collision. In general, the symbol sequence is written in the form,

$$(\ldots n_{-2}n_{-1}.n_1n_2\ldots); \quad n_i \in \{0, 1, 2\}.$$

This means that binary collisions occur in the order of $n_1$, $n_2$, ... to the future, and $n_{-1}$, $n_{-2}$, ... to the past.

Suppose that orbits starting from $(\theta_1, R_1)$ and $(\theta_2, R_2)$, have the same symbols for the first $n - 1$ digits, but the former has "1" and the latter has "2" at the $n$-th digit. Then there must exist $(\theta, R)$ between $(\theta_1, R_1)$ and $(\theta_2, R_2)$, for which the central particle collides with the other particles simultaneously at the $n$-th digit due to the continuity of orbits with respect to the initial condition. So, comparing two symbol sequences of neighboring points on the $(\theta, R)$-plane for all $n$, we obtain step by step initial points whose orbits end in triple collision.

### 2.3. Specification of Calculations

We integrate orbits which start from the lattice points of the $(\theta, R)$-plane for each mass ratio, and obtain corresponding symbol sequences. The specifications of masses, grid size and symbol sequence are shown in Table 1. Total energy $E_0$ is fixed to $-1$.

Table 1. The parameters determining the specification of calculation

| | |
|---|---|
| Masses | (a) Equal Masses<br>(b) Symmetric Case: $m_1 = 1.4$, $m_0 = 0.2$, $m_2 = 1.4$<br>(c) Asymmetric Case: $m_1 = 0.9$, $m_0 = 1.0$, $m_2 = 1.1$ |
| Poincaré Section | $\{(\theta, R) \mid 0 \leq \theta \leq 180°, \Delta\theta = 180°/1800, \Delta R = R_{\max}/1000\}$ |
| Symbol Sequence | To the future, until 64 digits |

## 3. The Results of Calculations

### 3.1. Triple Collision Curves on the $(\theta, R)$-Plane

According to the specification given in Section 2.3, we calculate symbol sequences on the $(\theta, R)$-plane. Figure 1 shows triple collision curves [the set of $(\theta, R)$ whose orbits end in a triple collision] on the $(\theta, R)$-plane. The horizontal and vertical axes represent $\theta$ and $R$, respectively. Three cases we consider are labeled as (a), (b), and (c) in Table 1. We use these labels throughout the paper. In any case, the Poincaré section is divided into fast escape regions [black; symbol sequence is $\ldots(a)^\infty$ or $\ldots(b)^\infty$], the Schubart region [gray; $(21)^\infty$] and the chaotic scattering region (between two above)(Hietarinta & Mikkola 1993).

At first, let us look at the equal mass case. The chaotic scattering region is separated by an infinite number of triple collision curves. These curves and fast escape regions construct four *scallops* (Tanikawa & Mikkola 2000). Triple collision curves converge at four points on the $\theta$-axis. Here we call each convergence point as a *root*, and the curves converging to the *root* as a *foot*. Note that all triple collision curves begin at one *root* and end at another *root*.

When the central mass becomes light in the symmetric case, the number of *scallops* increases, though overall features of triple collision curves seem unchanged. The lighter the central body, the more numerous *scallops* appear: this has been confirmed by additional integrations.

In a slightly asymmetric case, there exist in the middle *scallop* triple collision curves that begin and end at the same *root*. We describe this phenomenon as *a fold*. In addition, for the cases (a) and (c), we divide the Poincaré section into sub-regions with different symbol sequences (Figure 2). In the equal mass case, there exist regions $I_{12}$ : $1.2121(a)^\infty$ and $I_8$ : $1.2121(a)^n$. *Roots* of their border are the leftmost ($\theta = 0°$) and the righthmost ones, and its symbol sequence is 1.21210. However if masses become slightly asymmetric, $I_{12}$ vanishes and the right *foot* of 1.21210 rises up and folds back to the left root. The region $I_8$ becomes smaller and is wrapped in the border. (However, this border is no longer the border between $I_8$ and $I_{12}$, but $I_3$ and $I_8$.)

## 4. Interpretation of Results

The phenomena observed in the preceding Section are related to *roots*. In order to understand these phenomena, we introduce McGehee's variables, and we integrate limiting orbits of triple close encounter. Let us enumerate minimal features of the variables necessary for later discussion.

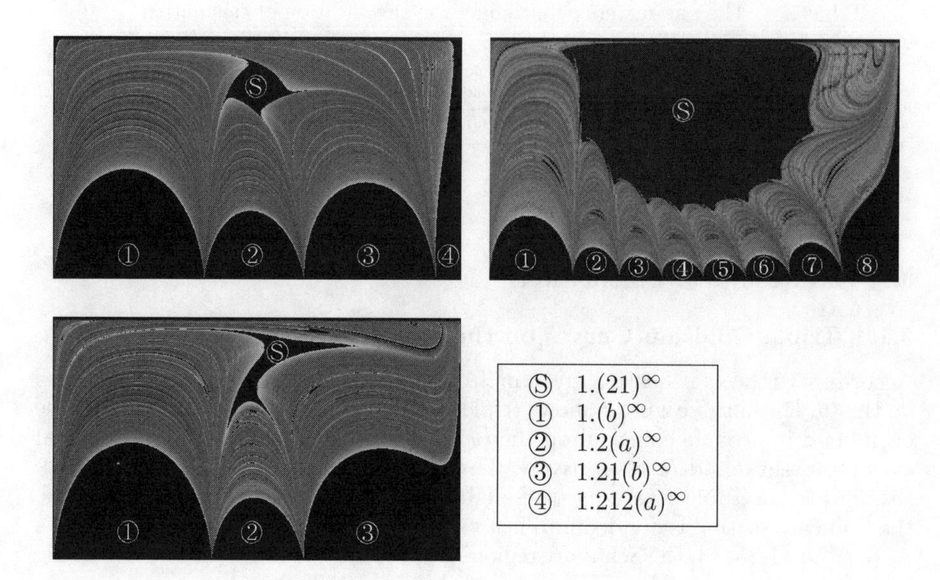


Figure 1. Triple collision curves on $(\theta, R)$-plane. Top left (a): equal masses; top right (b): $m_1 = 1.4$, $m_0 = 0.2$, $m_2 = 1.4$; bottom left (c): $m_1 = 0.9$, $m_0 = 1.0$, $m_2 = 1.1$.

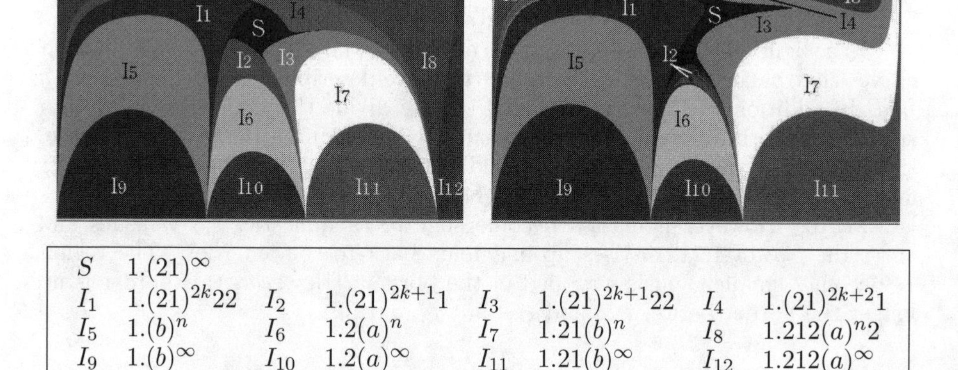


| | | | | | | | |
|---|---|---|---|---|---|---|---|
| $S$ | $1.(21)^\infty$ | | | | | | |
| $I_1$ | $1.(21)^{2k}22$ | $I_2$ | $1.(21)^{2k+1}1$ | $I_3$ | $1.(21)^{2k+1}22$ | $I_4$ | $1.(21)^{2k+2}1$ |
| $I_5$ | $1.(b)^n$ | $I_6$ | $1.2(a)^n$ | $I_7$ | $1.21(b)^n$ | $I_8$ | $1.212(a)^n2$ |
| $I_9$ | $1.(b)^\infty$ | $I_{10}$ | $1.2(a)^\infty$ | $I_{11}$ | $1.21(b)^\infty$ | $I_{12}$ | $1.212(a)^\infty$ |

Figure 2. The partition of the Poincaré section by symbol sequence. Left (a): equal masses; right (c): $m_1 = 0.9$, $m_0 = 1.0$, $m_2 = 1.1$.

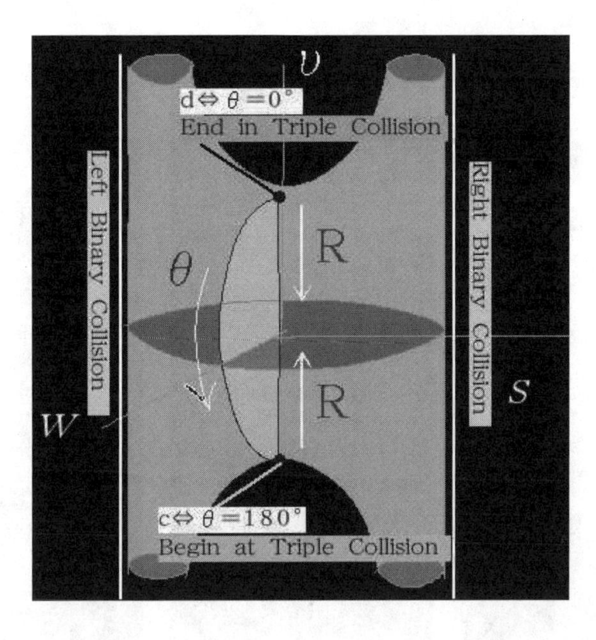


Figure 3. The triple collision manifold (TCM).

- McGehee's variables are $r$, $s$, $w$, and $v$, where $r$ represents the scale of a system, and $s$ the relative configuration. The collision between left and central particles occurs at $s = -1$, and the collision between central and right ones at $s = +1$. $w$ and $v$ are variables for the velocity.
- For $r = 0$, the time evolution of $(w, s, v)$ is well defined. The $r = 0$ defines a manifold, which is called the *triple collision manifold* (hereafter TCM), and its shape is shown in Figure 3. An orbit on TCM is called a *virtual orbit.*
- There exist two fixed points $c$ and $d$ in TCM.
- The flow on TCM is gradient-like with respect to $g(0, w, s, v) = -v$.

Figure 3 shows also the Poincaré section embedded as the plane $s = 0$. The $\theta$-axis is the intersection of TCM and $s = 0$. Two virtual orbits (thick and thin lines) and the Poincaré section for each mass are shown in Figure 4. We follow Simó's method for integration of virtual orbits (Simó 1980). In the symmetric cases, the virtual orbits for the lighter central mass wind around TCM more tightly, so intersect more frequently with the Poincaré section. Because these intersections correspond to *roots* in Figures 1 or 2, the fact that the number of *scallops* increases with decrease of the central mass can be understood. We assign symbol sequences to virtual orbits also. This assignment is necessary because some symbol sequences disappear in the asymmetric case, and disappearance is related to roots. In (a), points $e$, $f$, and $g$ are intersections of the orbit and the $\theta$-axis, and corresponding symbols are shown when orbits intersect with $s = -1$("1") or $s = +1$("2"). Symbol sequences of these orbits are shown in Table 2. If orbits start in other points on the $\theta$-axis, they shall wind up

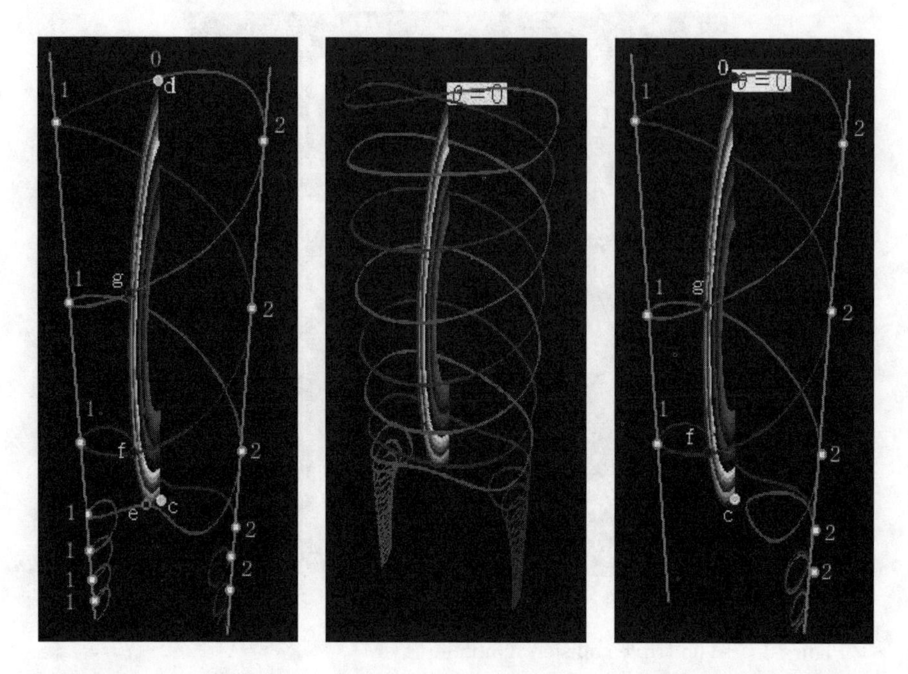


Figure 4. Virtual orbits ending in a triple collision. Left (a): equal masses; middle (b): $m_1 = 1.4$, $m_0 = 0.2$, $m_2 = 1.4$; right (c): $m_1 = 0.9$, $m_0 = 1.0$, $m_2 = 1.1$.

around one of the arms of TCM (they correspond to fast escape orbits). Symbol sequences of these escape orbits are also shown in Table 2. Comparing Table 2 and $I_9$–$I_{12}$ of Figure 2, one can find that symbol sequences of the $\theta$-axis are the limits of the symbol sequences of non-collision orbits.

Now we show how the form of virtual orbits changes continuously by increasing the asymmetry of masses from (a) to (c). The point $e$ approaches and reaches fixed point $c$ at $(m_2 - m_1)/2 \risingdotseq 0.065$. As the asymmetry becomes larger, the point $e$ vanishes, and the winding number of the thick-lined orbit decreases by 1. Then, we arrive at (c). Consequently symbol sequences of $e$ and $e$–$c$ vanish, and all symbol sequences of orbits beginning between $f$ and $c$ become $1.21(2)^\infty$. So the regions which touch the $\theta$-axis at the rightmost root and which include $I_8$, etc., must part from it.

Finally region $I_2$, which has remarkable *folds* in Figure 2c, can be obtained by mapping of $I_8$ to the past (Figure 5).

## 5. Conclusions

We have calculate the collision sequences of orbits begin at the Poincaré section. The chaotic scattering region is separated by the layer of triple collision curves also in non-equal mass cases as well as in equal one. In this report, we have

Table 2. Symbol sequences of virtual orbits

| | | |
|---|---|---|
| d | | |
| | $1.(b)^\infty$ | $\to I_9$ |
| $g$ | 1.20 | |
| | $1.2(a)^\infty$ | $\to I_{10}$ |
| $f$ | 1.210 | |
| | $1.21(b)^\infty$ | $\to I_{11}$ |
| $e$ | 1.2120 | |
| | $1.212(a)^\infty$ | $\to I_{12}$ |
| c | | |

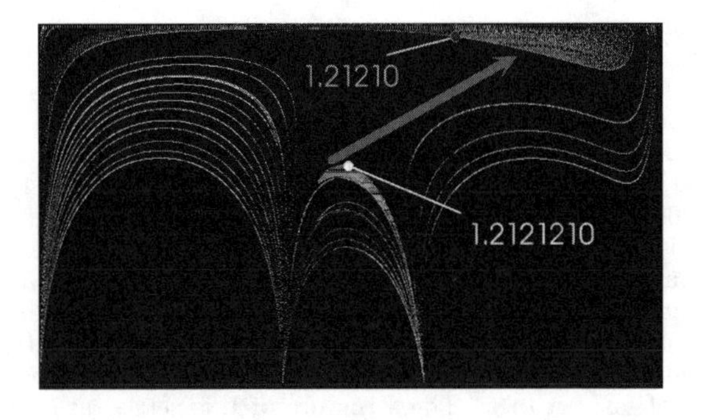


Figure 5. The Poincaré map of $I_2$ for the past.

picked up one sample for each symmetric or asymmetric non-equal mass case, and interpreted their structure associating with virtual orbits.

We summarize the results of the present research:

1. **Symmetric case.** Systems with a lighter central mass have more *scallops*. This can be understood from the fact that the virtual orbit winds around the triple collision manifold more tightly in this mass case.

2. **Asymmetric case.** The rightmost *foot* rises up from the $\theta$-axis. Introducing the symbol sequence to the virtual orbit, one can understand what these regions are detached from the $\theta$-axis, due to the change of the winding number around TCM. As these regions are mapped to the past, the most significant *fold* appears.

**Acknowledgments.** We have used the code implemented by Dr S. Mikkola for integration of orbits of collinear three-body problem. We would like to thank him.

## References

Hietarinta, J., & Mikkola, S. 1993, Chaos, 3, 183

Tanikawa, K., & Mikkola, S. 2000, Chaos, 10, 649

McGehee, R. 1974, Inventiones Mathematicae, 27, 191

Simó, C. 1980, CeM, 21, 25

*Order and Chaos in Stellar and Planetary Systems*
*ASP Conference Series, Vol. 316, 2004*
*G. Byrd, K. Kholshevnikov, A. Mylläri, I. Nikiforov, and V. Orlov, eds.*

# Metastable Trajectories in Free-Fall Three-Body Problem

V. Orlov, A. Petrova, and A. Rubinov

*Sobolev Astronomical Institute, St. Petersburg State University, Universitetskij pr. 28, Staryj Peterhof, St. Petersburg 198504, Russia*

A. Martynova

*Forest Technical Academy, Institutskij per. 5, St. Petersburg 194018, Russia*

**Abstract.** We consider the free-fall equal-mass three-body problem. The dynamical evolution of about 300,000 triple systems was considered. We found about 9,000 triple systems where the motions are limited within a finite region during a long time. These regions are concentrated to the zones of regular motions in the vicinities of stable periodic orbits. Such triple systems were named as metastable. Triple system leaves the metastable regime after some time and its evolution is ended by escape of one body. The set of initial configurations corresponding to metastable systems occupies a whole homology region, besides the zones of fast escapes. The structure of this set is rather complicated. In particular, there are some fractal clumps of initial configurations corresponding to metastable systems.

## 1. Introduction

Two basic types of trajectories in dynamical systems can be identified: regular and stochastic. The motions are regular in integrable systems. The various types of orbits are possible in non-integrable systems. The regular trajectories with finite motions are placed in vicinities of stable periodic orbits. An alternation of regularity and stochasticity regions takes place. Some unstable periodic trajectories appear near the boundaries of regularity zones surrounding the stable periodic orbits.

The complicated motions in vicinities of invariant tori were described by Chirikov (1979, 1991). He noted a "stickiness" phenomenon in some dynamical systems when a phase trajectory spent a long time near the chaos border. Morbidelli & Giorgilli (1995) have theoretically studied the dynamics in the neighbourhood of an invariant torus in a nearly integrable system. They have provided an upper bound to the diffusion speed $\exp[-\exp(1/\rho)]$ where $\rho$ is the distance from the invariant torus, i.e., this speed is superexponentially slow. We explore some similar phenomena in dynamical systems far from integrable cases, e.g., in the three-body problem.

So Dvorak et al. (1998) have revealed the property of phase trajectories ("stickiness" to regular motion regions around the stable periodic orbits) in the Sitnikov's restricted three-body problem (1960), when the zero-mass body moves in the gravitational potential of binary system. This body moves along the straight line perpendicular to the binary orbital plane and crossing its center of mass.

It is of interest to clarify, if such a phenomenon takes place in the general three-body problem where all three masses are non-zero? In order to study this question, we consider dynamical evolution of equal-mass free-fall three-body systems. In this case, the angular momentum of triple system is equal to zero.

## 2. Periodic and Metastable Orbits

Three simple stable periodic orbits are well-known for triple systems with zero angular momentum (von Schubart 1956; Broucke 1979; Moore 1993). These orbits are shown in Figure 1 in barycentric coordinate frame. The "Eight" orbit was also found analytically by Chenciner & Montgomery (2000). One can expect that these orbits are surrounded by regularity zones, and phase trajectories of unstable triple systems can "stick" to these domains. We will name such "sticking" trajectories as metastable (let us note that similar effect in theory of dynamical systems was called as "hamiltonian intermittency"). The first such trajectory was found by chance (Orlov, Petrova, & Martynova 2001), when we investigated a correlation between triple approach and following ejection parameters. The distant ejections and close triple encounters were absent in this system during a long time (about 300 crossing times). The motions of bodies were within a finite compact manifold. The metastable state had a temporary character and took only part of a whole triple system evolution time.

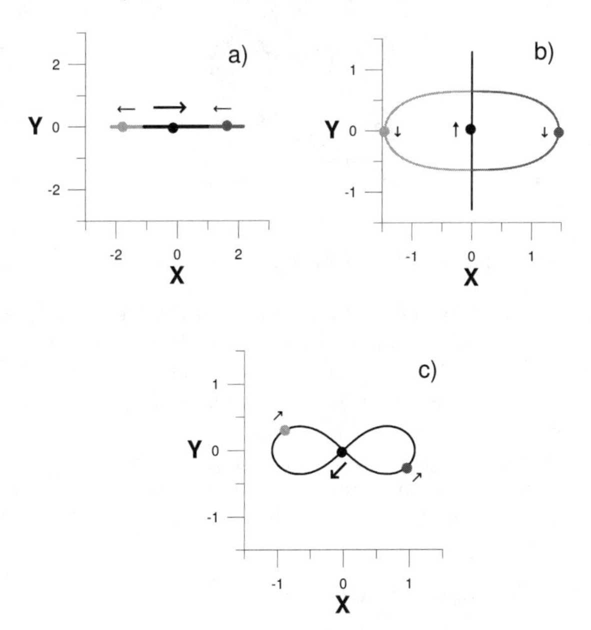


Figure 1. Simple periodic orbits in free-fall three-body problem: (a) von Schubart (1956) orbit, (b) Broucke (1979) orbit, (c) Moore (1993) orbit.

Here we search for suchlike metastable systems and study their dynamics. We scan the region $D$ (Figure 2) which was firstly introduced by Agekian and Anosova (1967). Here the maximum separation between bodies is assumed to be unity. These bodies are placed on $\xi$-axis and have the coordinates $(-0.5, 0)$ and $(+0.5, 0)$. The third body is located inside the region $D$ limited by the coordinate axes and the circle arc of unit radius with the centre in the point $(-0.5, 0)$. This region contains all possible configurations of triple systems. It was named as a homology region (see, e.g., Chernin & Valtonen 1998), because every triangle has the similar one within the region $D$. Scanning along both coordinates $(\xi, \eta)$ was carried out with the same steps $\Delta\xi = \Delta\eta = 0.001$. More than 300,000 systems were considered. We choose the systems where the ejections longer than twice the ejection radius (see Agekian & Martynova 1973) did not observe during at least 100 crossing times. We have found more than 9,000 such systems (about 3% of the sample). Those are suspicious for metastability. The initial coordinates $(\xi, \eta)$ for these systems in region $D$ are shown in Figure 2.

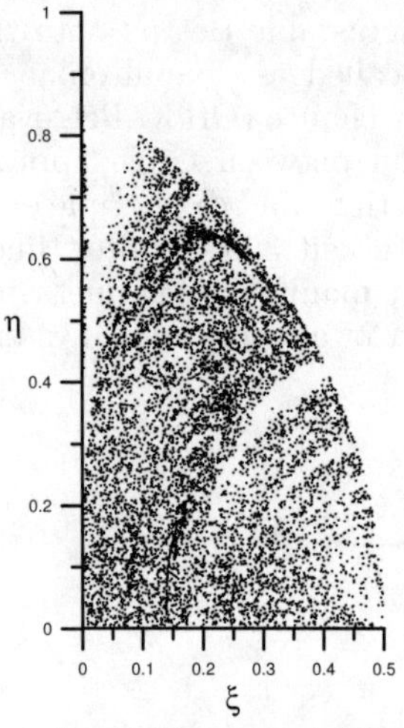


Figure 2. The initial coordinates for metastable systems.

The points are distributed around a whole region $D$, excluding a few families of "voids". One such family is a set of concentric arcs with center in the point $(\xi, \eta) = (0.5, 0)$. Moreover, these arcs become thicker to the right boundary of region $D$. These families correspond to short-living triple systems. Their evolution is finished by fast escape of one body.

The distribution of the points corresponding to metastable systems is inhomogeneous. Here we observe the clumps of points. As a rule, these clumps are stretched along the arcs corresponding to fast escapes.

## 3. Character of Metastable Motions

Let us consider the character of motions in metastable regimes in detail. In order to reveal the features of motion, we made an animation of many metastable trajectories. Two examples of metastable orbits are shown in Figure 3.

We have found that the orbit "sticks" to one of three stable periodic orbits mentioned above (see Figure 4). Thus, we can propose a classification of motions in metastable regimes: Schbart-like, Eight-like, and Broucke-like orbits.

Often, a change of motion types happens during the evolution—the trajectory can "bounce" from one periodic orbit to another. Especially, such alternation is typical for Eight-like and Broucke-like motions (see Figure 5). Sometimes, the system may "visit" the metastable regimes a few times. In the intervals between these "visits", the triple encounters and ejections occur in the triple system.

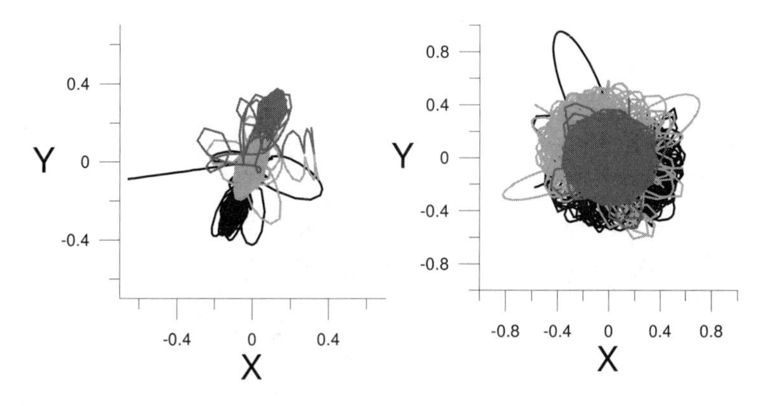


Figure 3. Two examples of metastable trajectories.

## 4. Methods for Study of Metastable Regimes

In order to investigate the trajectory behaviour in metastable state, one may apply different approaches. A visual information about the motion may be obtained from pictures and animations. An additional qualitative and quantitative knowledge can be found from statistics of recurrences, symbolic dynamics, wavelet analysis, entropy calculation, etc.

As an example, let us consider the symbolic dynamics methods which are well developed in the theory of dynamical systems. For example, one can fix a sequence of double encounters and define a number of third distant component at the moment of each encounter. The symbolic sequence of these numbers reflects a trajectory affinity with the periodic orbit. For instance, the symbolic sequence for Schubart-like orbit looks like

$$\ldots 121212 \ldots \text{ or } \ldots 131313 \ldots \text{ or } \ldots 232323 \ldots$$

The Eight-like orbit may have the following sequence

$$\ldots 123123123 \ldots \text{ or } \ldots 213213213 \ldots$$

One of three realizations

$$\ldots 111111 \ldots \text{ or } \ldots 222222 \ldots \text{ or } \ldots 333333 \ldots$$

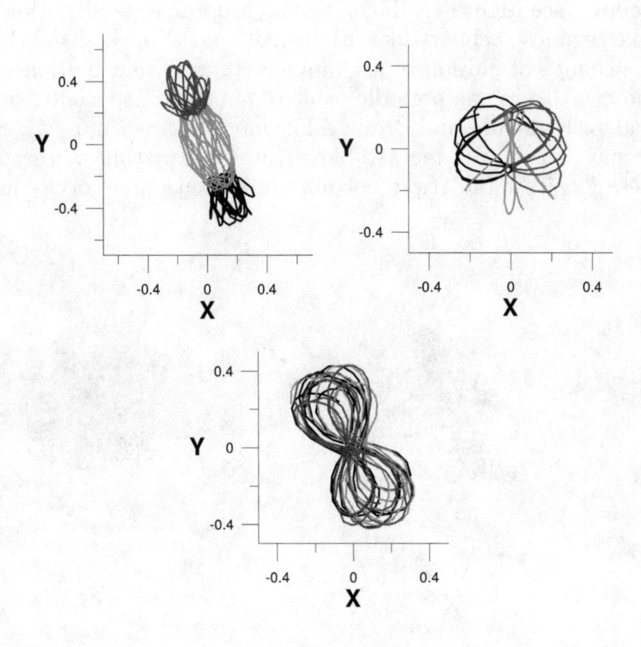


Figure 4. Motions of metastable trajectories in vicinities of periodic orbits.

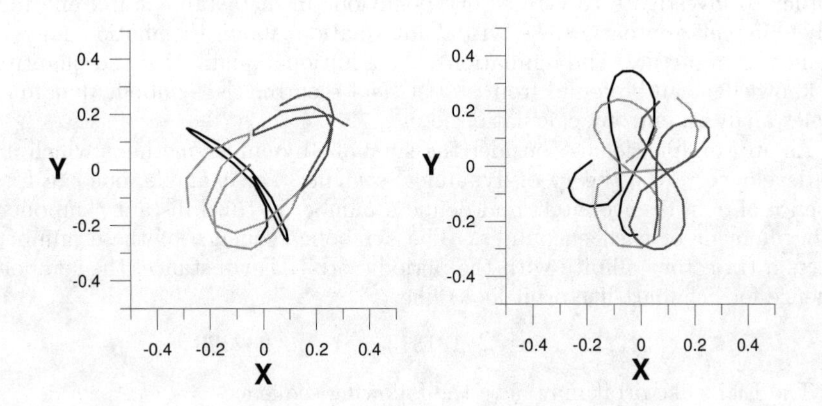


Figure 5. Change of motion types in one trajectory.

is valid for Broucke orbit, depending on the number of the body that oscillates between two other components. Let us note that the Broucke orbit is practically not realized. Usually it is combined with Eight-like or Schubart-like motions.

## 5. Conclusions

Let us formulate the main conclusions from our study:

1. In equal-mass free-fall three-body problem, a "stickiness" of trajectory to regularity zones around stable periodic orbits occurs.
2. The captures in these resonance regions have a temporary character. The dynamical evolution of metastable systems is finished by escape of one body. The metastable triple systems are still unstable.

**Acknowledgments.** This work was partly supported by the RFBR (Grant 02-02-17516), Program *Universities of Russia* (Grant UR.02.01.027), and the Leading Scientific School (Grant NSh-1078.2003.02).

## References

Agekian, T. A., & Anosova, J. P. 1967, AZh, 44, 1261

Agekian, T. A., & Martynova, A. I. 1973, Vestnik Leningr.Univ., 1, 122

Broucke, R. 1979, A&A, 73, 303

Chenciner, A., & Montgomery, R. 2000, Ann.Math., 52, 881

Chernin, A. D., & Valtonen, M. J. 1998, New Astron.Rev., 42, 41

Chirikov, B. V. 1979, Phys.Rep., 52, 263

Chirikov, B. V. 1991, Chaos, Solitons and Fractals, 1, 79

Dvorak, R., Contopoulos, G., Efthymiopoulos, Ch., & Voglis, N. 1998, Planetary and Space Sci., 46, 1567

Moore, C. 1993, Phys.Rev.Lett, 70, 3675

Morbidelli, A., & Giorgilli, A. 1995, J.Stat.Phys., 78, 1607

Orlov, V. V., Petrova, A. V., & Martynova A. I. 2001, Ast.Lett., 27, 549

Sitnikov, K. A. 1960, Doklady Sov.Acad.Sci., 133, 303

von Schubart, J. 1956, Ast.Nachr., 283, 17

*Order and Chaos in Stellar and Planetary Systems*
*ASP Conference Series, Vol. 316, 2004*
*G. Byrd, K. Kholshevnikov, A. Mylläri, I. Nikiforov, and V. Orlov, eds.*

# Boundaries of the Chaotic Stable Motion in the Restricted 3-Body Problem

N. P. Pitjev and L. L. Sokolov

*Sobolev Astronomical Institute, St. Petersburg State University, Universitetskij pr. 28, Staryj Peterhof, St. Petersburg 198504, Russia*

**Abstract.** Motions in the 1:1 orbital resonance the restricted circular 3-body problem have been considered. Regular and chaotic trajectories were investigated. Domains of initial conditions $\Omega$, $i$ with regular and chaotic behavior of small body eccentricity are constucted.

We investigate the resonance motion in the restricted spatial three-body problem for the case of 1:1 resonance. The properties of motion near triangular libration points and their boundaries (stability and so on) are sufficiently well known (Szebehely 1967).

Recently Brasser, Heggie, & Mikkola (2004) have discovered interesting phenomenon of "stable chaos" for the spatial restricted circular three-body problem with commensurability 1:1. The motion starts in the triangular libration point, the initial velocity corresponds to a circular inclined orbit. If the inclination is not large, there are regular small oscillations of elements. But if the initial inclination ranges up to the critical value $i_{\rm cr} \simeq 61^{\circ}\!.5$, there is rapid increase of maximum eccentricity value, and oscillations of elements become chaotic. But, the commensurability 1:1 maintains (semi-major axis is near to its initial value), there are no approaches of planets to each others, and eccentricity oscillates in the limited region. It is a stable chaos. Brasser et al. (2004) have explained this phenomenon as jumping to chaos using analytical solution of approximate system of the equations of motion.

Now we discuss the same spatial restricted three-body problem with commensurability 1:1 for different initial positions of the zero-mass point on the "Jupiter's" orbit: we change longitude of ascending node and some other parameters. If 1:1 resonance maintains then the motions of bodies have co-orbital features (Namouni 1999). In addition we discuss an elliptical restricted spatial three-body problem taking into account the influence of "Jupiter's" eccentricity.

Our results have been obtained numerically. To solve the equations of motion, we used the Everhart integrator of the 23th order with a double precision. We have traced the trajectories over 50000–500000 "Jupiter's" revolution times. This time roughly corresponds to the time of trajectory divergence due to instability for the cases of chaotic motion.

In Figure 1a we present three types of regions in the plane $(i, \Omega)$, $i$ is the initial inclination, $\Omega$ is the initial longitude of ascending node of small body. These types correspond to a) small regular oscillations, b) stable chaos, and c) escape from resonance. The last case corresponds to escape of small body. The trajectories in the central regular "island" (Figures 1a and 1b) around point $(i = 120^{\circ}, \Omega = 180^{\circ})$ have integrated during 800000 "Jupiter's" revolution times.

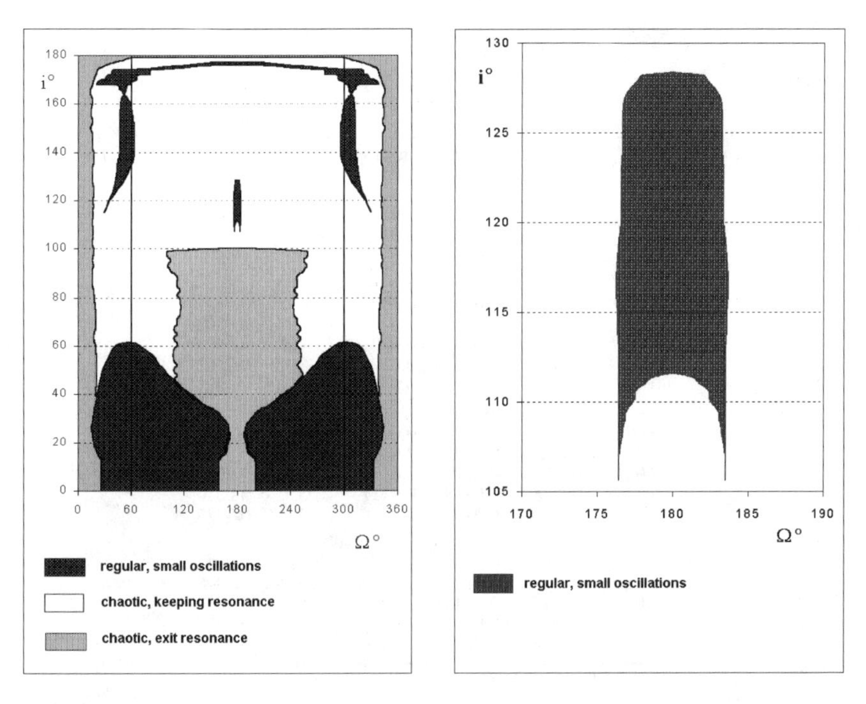


Figure 1. Left (a): Regions of regular and chaotic traectories. Right (b): Central "island".

Our calculations confirm the results of Brasser et al. (2004). Their results correspond to the vertical lines $\Omega = 60°$ and $\Omega = 300°$ in Figure 1a. The segments, corresponding to regular oscillations and stable chaos on these lines are given by Brasser et al. (2004).

We have revealed two types of stable chaotic eccentricity oscillations: the first one was described by Brasser et al. (2004). This is a region of moderate eccentricities $e \sim 0.5$. The second one is region of large eccentricities $e \sim 0.6 \div 0.8$.

For spatial elliptical restricted three-body problem when eccentricity of "Jupiter's" orbit is small, but non-zero, we have similar properties of motion. The critical value of inclination ($i_{\rm cr} \simeq 61^\circ\!.5$ for circular problem) separating regular and chaotic motions decreases with increasing "Jupiter's" orbital eccentricity. This dependence is shown in Figure 2. The regular trajectories exist when $e_{\rm Jup} < 0.33$.

Maximum amplitude of regular eccentricity oscillations, depending on the eccentricity of "Jupiter's" orbit $e_{\rm Jup}$ and on initial value of the small body eccentricity $e_0$ for different values of initial inclination $i_0$ is shown in Table 1.

The same dependence for other "Jupiter's" mass $\mu$ is given in Table 2.

When the inclination is not large, the approximate relationship between eccentricities $e_{\max} \approx 2e_{\rm Jup} - e_0$ holds quite well. For larger inclinations this relation becomes less suitable.

Table 1. Maximum value of small body eccentricity $e_{\max}$ ($\mu = 0.001$)

| $e_{\text{Jup}} \backslash e_0$ | 0.00 | 0.01 | 0.02 | 0.03 | 0.04 |
|---|---|---|---|---|---|
| | | Initial inclination $i_0 = 0°$: | | | |
| 0.00 | 0.00 | 0.0201 | 0.0401 | 0.0602 | 0.0802 |
| 0.01 | 0.0101 | 0.01 | 0.0301 | 0.0501 | 0.0701 |
| 0.02 | 0.0201 | 0.0201 | 0.02 | 0.0401 | 0.0601 |
| 0.03 | 0.0302 | 0.0301 | 0.0301 | 0.03 | 0.0501 |
| 0.04 | 0.0403 | 0.0402 | 0.0401 | 0.0401 | 0.04 |
| | | Initial inclination $i_0 = 10°$: | | | |
| 0.00 | 0.0001 | 0.0201 | 0.0402 | 0.0602 | 0.0801 |
| 0.01 | 0.0101 | 0.0101 | 0.0301 | 0.0502 | 0.0702 |
| 0.02 | 0.0203 | 0.0201 | 0.0201 | 0.0401 | 0.0602 |
| 0.03 | 0.0304 | 0.0303 | 0.0301 | 0.0302 | 0.0501 |
| 0.04 | 0.0405 | 0.0403 | 0.0402 | 0.0402 | 0.0402 |
| | | Initial inclination $i_0 = 30°$: | | | |
| 0.00 | 0.0006 | 0.0211 | 0.0417 | 0.0625 | 0.0831 |
| 0.01 | 0.0109 | 0.0109 | 0.0310 | 0.0517 | 0.0724 |
| 0.02 | 0.0217 | 0.0210 | 0.0215 | 0.0410 | 0.0616 |
| 0.03 | 0.0326 | 0.0318 | 0.0311 | 0.0320 | 0.0510 |
| 0.04 | 0.0434 | 0.0426 | 0.0419 | 0.0414 | 0.0425 |
| | | Initial inclination $i_0 = 50°$: | | | |
| 0.00 | 0.0018 | 0.0248 | 0.0481 | 0.0712 | 0.0943 |
| 0.01 | 0.0141 | 0.0132 | 0.0343 | 0.0574 | 0.0805 |
| 0.02 | 0.0282 | 0.0243 | 0.0251 | 0.0443 | 0.0669 |
| 0.03 | 0.0424 | 0.0381 | 0.0351 | 0.0371 | 0.0550 |
| 0.04 | 0.0566 | 0.0522 | 0.0486 | 0.0463 | 0.0491 |

$\Omega_0 = 300°$, $N = 10000$, $a_0 = a_{\text{Jup}} = 5.2$.

Table 2. Maximum value of small body eccentricity $e_{\max}$ ($\mu = 0.0195$)

| $e_{\text{Jup}} \backslash e_0$ | 0.00 | 0.01 | 0.02 | 0.03 | 0.04 | 0.05 |
|---|---|---|---|---|---|---|
| | | Initial inclination $i_0 = 0°$: | | | | |
| 0.00 | 0.00 | 0.0132 | 0.0269 | 0.0413 | 0.0568 | 0.0743 |
| 0.01 | 0.0231 | 0.01 | 0.0232 | 0.0370 | 0.0515 | 0.0674 |
| 0.02 | 0.0466 | 0.0332 | 0.02 | 0.0333 | 0.0472 | 0.0619 |
| 0.03 | 0.0706 | 0.0567 | 0.0432 | 0.03 | 0.0434 | 0.0574 |
| 0.04 | 0.0954 | 0.0808 | 0.0668 | 0.0533 | 0.04 | 0.0535 |
| 0.05 | 0.1227 | 0.1057 | 0.0910 | 0.0769 | 0.0634 | 0.05 |

$\Omega_0 = 300°$, $N = 10000$, $a_0 = a_{\text{Jup}} = 5.2$.

Table 3. Maximum amplitude of oscillation $\Delta i = |i - i_0|$ of small body

| $e_{\text{Jup}} \backslash i_0$ | 4° | 8° | 12° | 16° |
|---|---|---|---|---|
| | "Jupiter" mass $\mu = 0.001$: | | | |
| 0.02 | 0°.0036 | 0°.0070 | 0°.0112 | 0°.0168 |
| 0.03 | 0°.0078 | 0°.0150 | 0°.0220 | 0°.0308 |
| | "Jupiter" mass $\mu = 0.002$: | | | |
| 0.02 | 0°.0044 | 0°.0085 | 0°.0136 | 0°.0210 |
| 0.03 | 0°.0095 | 0°.0178 | 0°.0255 | 0°.0356 |

$\Omega_0 = 300°$, $N = 10000$, $e_0 = 0$, $a_0 = a_{\text{Jup}} = 5.2$.

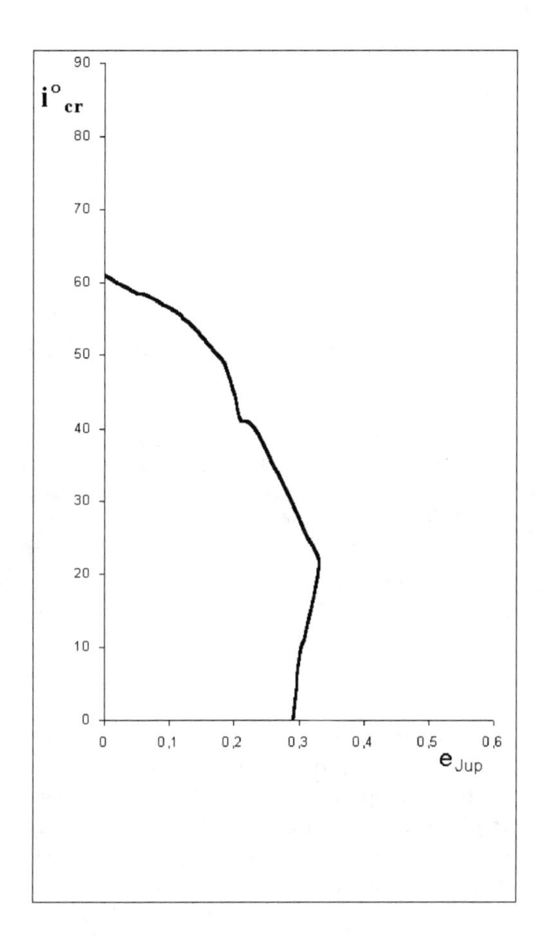


Figure 2. Dependence $e_{\rm Jup}$–$i_{\rm cr}$.

In Table 3 we show the dependence between inclination amplitude and initial inclination $i_0$ of small body circular orbit (for regular oscillations). For different values of eccentricity $e_{\rm Jup}$ and "Jupiter" mass $\mu$ approximate dependence is close to linear one with coefficient to be proportional to $\mu$ and $e_{\rm Jup}$.

**Acknowledgments.** This work was partly supported by RFBR (Grant 02-02-17516), Program "Universities of Russia" (Grant UR.02.01.027) and the Leading Scientific School (Grant NSh-1078.2003.02).

## References

Brasser, R., Heggie, D. C., & Mikkola, S. 2004, CeMDA, 88, 2, 123

Namouni, F. 1999, Icarus, 137, 293

Szebehely, V. 1967, Theory of orbits: The restricted problem of three bodies (New York and London: Academic Press)

*Order and Chaos in Stellar and Planetary Systems*
*ASP Conference Series, Vol. 316, 2004*
*G. Byrd, K. Kholshevnikov, A. Mylläri, I. Nikiforov, and V. Orlov, eds.*

# Dynamics of Capture in the Restricted Three-Body Problem

Sergey Astakhov,[1] Andrew Burbanks,[2] Stephen Wiggins,[2] and David Farrelly[1]

[1]*Department of Chemistry & Biochemistry, Utah State University, Logan, Utah 84322-0300, USA*

[2]*School of Mathematics, University of Bristol BS8 1TW, UK*

**Abstract.** We propose a new dynamical model for capture of irregular moons which identifies chaos as the essential feature responsible for initial temporary gravitational trapping within a planet's Hill sphere. The key point is that incoming potential satellites get trapped in chaotic orbits close to "sticky" KAM tori in the neighbourhood of the planet, possibly for very long times, so that the chaotic layer largely dictates the final orbital properties of captured moons.

## 1. Introduction

The often puzzling properties of the irregular satellites of the giant planets—most of which have been discovered during the last six years (see Gladman et al. 2001; Hamilton 2003; Sheppard & Jewitt 2003; references therein and IAU Circular 8193)—provide a window into conditions in the early Solar System.

The general mechanism by which the irregular satellites were captured is thought to involve the following steps (Heppenheimer & Porco 1977; Pollack et al. 1979; Murison 1989; Peale 1999; Gladman et al. 2001); (i) temporary trapping close to the planet in a region roughly demarked by the Lagrange points $L_1$ and $L_2$; (ii) gradual energy loss through dissipation (e.g., gas drag or planetary growth) which translates temporary trapping into permanent capture; and (iii) possible fragmentation due to collisions at much later times. The hypothesis that the observed clustering among populations of irregular moons may be a result of fragmentation (Pollack et al. 1979; Gladman et al. 2001) contradicts, however, the fact that the orbits of known irregulars are clustered in inclination but not necessarily in eccentricity or other orbital elements (Nesvorný et al. 2003). Although there have been extensive studies of how the systems of irregular satellites have formed (see also Hénon 1970; Colombo & Franklin 1971; Huang & Innanen 1983; Saha & Tremaine 1993; Gor'kavyi & Taidakova 1995; Marzari & Scholl 1998; Namouni 1999; Viera Neto & Winter 2001; Winter & Viera Neto 2001; Carruba et al. 2002; Carruba et al. 2003; Nesvorný et al. 2002; Nesvorný et al. 2003; Winter et al. 2003) a coherent *dynamical* picture of capture has not emerged; e.g., it has been widely held that the propensity for retrograde motion among Jupiter's irregulars is simply due to the well known enhanced stability of retrograde orbits with large semimajor axes $a$ (Nesvorný et al. 2003). Gladman et al. (2001) have called this, and the alternative "pull-down" (Heppenheimer

& Porco 1977) capture mechanism, into question based on the following observation: while the bulk of Jupiter's irregular moons are retrograde and lie distant from the planet, Saturn's cortege contains a more even mix of prograde and retrograde moons even though they have similarly large semimajor axes $a$ when expressed in planetary radii. Here, we study capture in the circular restricted three-body problem (CRTBP) in two and three-dimensions (3D) taking the Sun–Jupiter–moon system as the specific example: the dynamical picture that emerges in the Hill limit (Murray & Dermot 1999; Simó & Stuchi 2000) is, however, rather similar for the other giant planets.

## 2. The Hamiltonian

In a coordinate system rotating with the mean motion of the primaries, but with origin transformed to the planet, the CRTBP Hamiltonian is given by

$$\begin{aligned} H = E \;&=\; \frac{1}{2}\mathbf{p}^2 - (x\,p_y - y\,p_x) - \frac{\mu}{\sqrt{x^2+y^2+z^2}} - \\ &\quad \frac{1-\mu}{\sqrt{(1+x)^2+y^2+z^2}} - (1-\mu)x + \alpha, \end{aligned} \tag{1}$$

where $a$ is scaled to 1, $\mu = m_1/(m_1+m_2)$; $m_1$ and $m_2$ are the masses of the primaries and $\alpha$ is a collection of inessential constants retained for consistency in relating the energy $E$ to the Jacobi constant $C_J = -2E$; $\mathbf{r} = (x, y, z)$ and $\mathbf{p} = (p_x, p_y, p_z)$ are the coordinates and momenta of the potential satellite. This Hamiltonian is obtained from the standard CRTBP Hamiltonian (Murray & Dermot 1999) by the canonical transformation $x \to x' + (1-\mu)$, $p_y \to p'_y + (1-\mu)$ (and dropping the primes). Angular momentum, $\mathbf{h} = (h_x, h_y, h_z)$, where $h_z = x\,p_y - y\,p_x$, is now defined with respect to the planet as is natural for a study of capture.

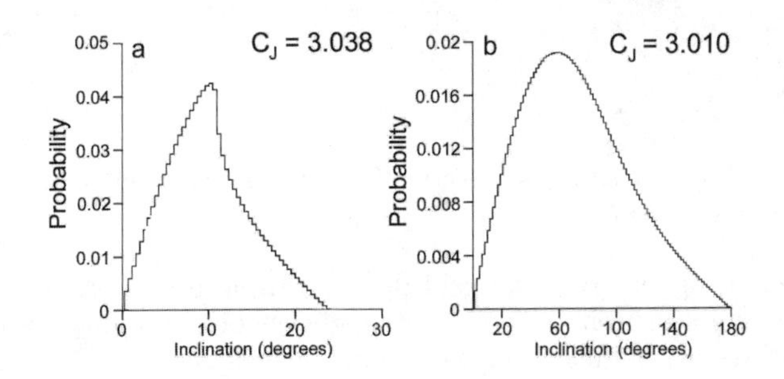


Figure 1. Histograms of inclination distribution for $10^8$ test particles originating at the Hill sphere for indicated values of energy (Jacobi constant).

## 3. Simulations

Figure 1 shows computed orbital inclination distributions [$I = \arccos(h_z/h)$] for a flux of $10^8$ test particles as they pass through the Hill sphere [radius $R_H = a(\mu/3)^{1/3}$, Murray & Dermot 1999] at two energies. The key observation is that at low energy (Figure 1a) only *prograde* orbits can enter (or exit) the capture zone between Lagrange points $L_1$ and $L_2$ whereas at higher energies (Figure 1b) the distribution shifts to include both senses of $h_z$. This is because not all parts of the Hill sphere are energetically accessible at low energies. Figure 1 thus suggests that the statistics of capture might be expected to depend on initial $C_J$ and $\mathbf{h}$.

To investigate this we consider first the structure of phase space in the planar limit ($z = p_z = 0$). Figure 2 displays a series of Poincaré surfaces of section (SOS) for randomly chosen initial conditions inside the Hill radius $R_H$. The hypersurface is the $x$–$y$ plane with units rescaled to $R_H = 1$ and points colored according to the sign of angular momentum (grey, retrograde $h_z < 0$; black, prograde $h_z > 0$) as they intersect the surface with $p_x = 0$ and $dy/dt > 0$.

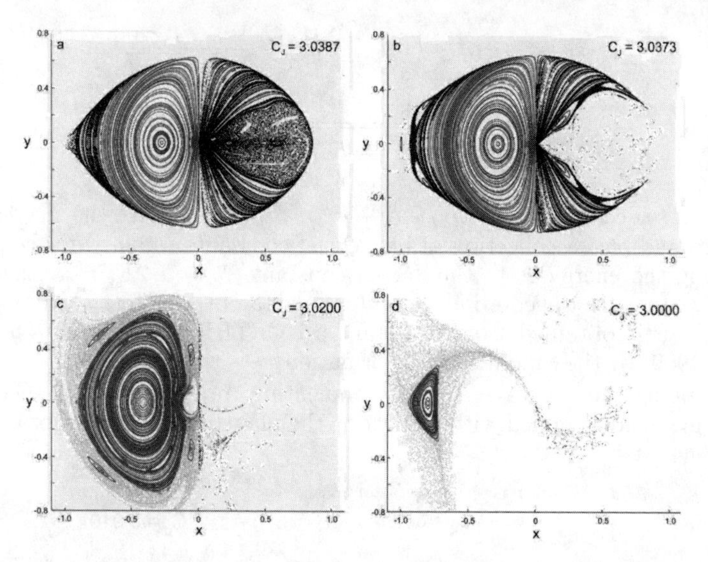


Figure 2. Poincaré surfaces of section for Sun–Jupiter–satellite 2D CRTBP with increasing energy.

With increasing energy a coherent dynamical picture emerges; in Figure 2a the prograde orbits exhibit regions of strongly chaotic motion whereas *all* the retrograde orbits are regular (quasiperiodic). This forces incoming prograde orbits to remain prograde because they cannot penetrate the KAM regions in Figure 2a. After the gateway at $L_2$ has opened in Figure 2b the chaotic "sea" of prograde orbits visible in Figure 2a rapidly disappears except for a thin residual front of chaos which sticks to the KAM tori, separating them from the growing basin of direct scattering. As energy increases further this front moves from prograde to retrograde motion while the tori steadily erode. KAM tori are "sticky" and chaotic orbits near them can appear locally near-integrable, i.e., they are trapped in almost regular orbits for very long times. Note especially that the KAM tori in Figure 2c and 2d exist at energies well *above* $L_1$ and $L_2$. Permanent capture happens if dissipation is sufficient to switch long lived chaotic orbits into KAM regions which means that chaotic orbits can be permanently captured, even above the saddle points by relatively weak dissipation.

The SOS also reveal that large distance from the planet need not imply retrograde motion: e.g., the orbits visible inside two KAM islands centered at $x > 0$, $y = 0$ in Figure 2a are large, almost circular, periodic *prograde* orbits but whose centers are displaced from the origin. At higher energies capture (through dissipation) is into retrograde KAM surfaces nested around the circular, retrograde orbit ($x < 0, y = 0$) (Hénon 1970; Winter & Viera Neto 2001), and which remains almost perfectly centered on the planet.

In 3D, initial conditions in $\mathbf{r}$ were chosen uniformly and randomly on the Hill sphere with random velocities. The Jacobi constant was also chosen randomly and uniformly $C_J \in (2.995, C_J^{L1})$. Trajectories were integrated until they exited the Hill sphere, came within 2 planetary radii of the origin (Carruba et al. 2002) or survived for a predetermined cutoff time $t_{\text{cut}}$. Computational limitations restricted us to 80 million test particles, of which 23000 survived for $t_{\text{cut}} = 20000$ years. Figure 3a shows a clear trend from prograde to retrograde capture with increasing energy. Three main islands stand out in the archipelago visible in Figure 3a; the prograde (low inclination) island shrinks noticeably with increasing $t_{\text{cut}}$ reflecting the lower probability of prograde capture.

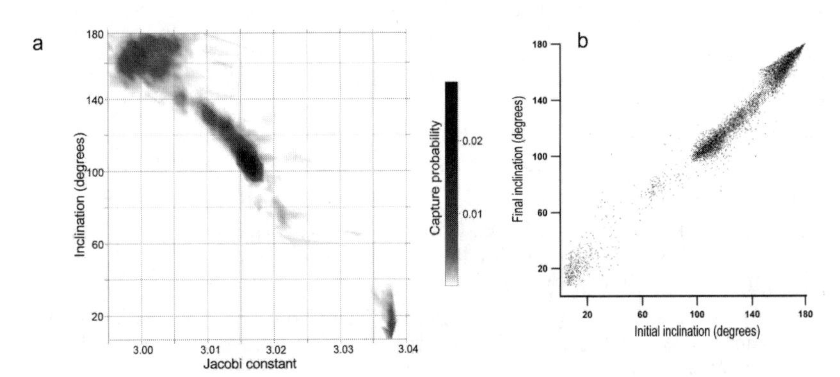


Figure 3. (a) The normalized capture probability distribution from Monte Carlo simulations (Jupiter), and (b) correlation between initial and final inclination of the test particles trapped in the capture zone.

The most noticeable feature is the large island at $I \approx 100°$ whose stability is related to the Kozai resonance centered at $I = 90°$ (Kozai 1962; Innanen et al. 1997; Carruba et al. 2002; Carruba et al. 2003; Nesvorný 2003). Unlike in 2D direct injection into KAM regions is possible but in near-integrable regions occurs exponentially slowly which seems to be why incoming particles are excluded from the center of the resonance itself. Likely these particles are trapped in a chaotic separatrix layer of the Kozai resonance. Both the "Kozai island" and the smaller, very high energy island of retrograde motion in Figure 3a are stable for extremely long times. Figure 3b demonstrates a very strong correlation between final and initial inclination (i.e., "inclination memory", Astakhov et al. 2003) for long lived orbits which is consistent with the picture in 2D.

We have confirmed numerically that these distributions are similar for the other giant planets; are robust in the presence of gas drag of different forms. The specific details of capture depend sensitively on how this dynamical mechanism intersects local environment around the planet. One of the factors is the part of the Hill sphere occupied by the massive regular moons. This observation is important, because prograde orbits penetrate much deeper towards the planet than do most of the retrograde orbits (see Figures 2b, 2d and also Simó & Stuchi 2000). Therefore, to be permanently captured, prograde satellites must survive close encounters or collisions with "influential" regular moons. In our Monte Carlo simulations with dissipation (Astakhov et al. 2003) we eliminated test particles that crossed the orbits of Titan at Saturn and Callisto at Jupiter. These simulations indicate that Saturn/Titan in tandem have a clear tendency to capture a higher ratio of prograde to retrograde moons as compared to Jupiter/Callisto. This is because Callisto's orbit represents a larger fraction of the Hill sphere than does Titan's orbit. Thus the relative scarcity of jovian prograde irregulars may be due to potential prograde satellites having been swept away more efficiently by Jupiter's Galilean moons (Astakhov et al. 2003).

**Acknowledgments.** This work was funded by grants from the US National Science Foundation and Petroleum Research Fund to D.F.

## References

Astakhov, S. A., Burbanks, A. D., Wiggins, S., & Farrelly, D. 2003, Nature, 423, 264

Carruba, V., Burns, J. A., Nicholson P. D., & Gladman, B. J. 2002, Icarus, 158, 434

Carruba, V., Nesvorný, D., Ćuk, M., Burns, J. A., & Rand, R. 2003, BAAS, 35(4), DPS 35th Meeting, abstr. No. 15.01; Carruba, V., Nesvorný, D., Burns, J. A., & Ćuk, M. 2003, BAAS, 35(4), DDA 34th Meeting, abstr. No. 10.01

Colombo, G., & Franklin, F. A. 1971, Icarus, 15, 186

Gladman, B. J., Kavelaars, J. J., Holman, M., Nicholson P. D., Burns, J. A., Hergentorher, C. W., Petit, J. M., Marsden, B. J., Jacobson, R., Gray, W., & Grav, T. 2001, Nature, 412, 163

Gor'kavyi, N. N., & Taidakova T. A. 1995, Ast.Lett., 21, 846

Hamilton, D. P. 2003, Nature, 423, 235

Huang, T.-Y., & Innanen, K. A. 1975, AJ, 80, 290

Hénon, M. 1970, A&A, 9, 24

Heppenheimer, T. A., & Porco, C. 1977, Icarus, 30, 385

Innanen, K. A., Zheng, J. Q., Mikkola, S., & Valtonen, M. J. 1997, AJ, 113, 1915

Kozai, Y. 1962, AJ, 67, 591

Marzari, F., & Scholl, H. 1998, Icarus, 131, 41

Murray C. D., & Dermot S. F. 1999, in Solar System Dynamics (Cambridge: Cambridge Univ. Press)

Murison, M. A. 1989, AJ, 98, 2346

Namouni, F. 1999, Icarus, 137, 293
Nesvorný, D., Thomas, F., Ferraz-Mello, S., & Morbidelli, A. A. 2002, CeMDA, 82, 323
Nesvorný, D., Alvarellos, J. L. A., Dones, L., & Levison, H. F. 2003, AJ, 126, 398
Peale, S. J. 1999, ARA&A, 37, 533
Pollack, J. R., Burns, J. A., & Tauber, M.E. 1979, Icarus, 37, 587
Saha, P., & Tremaine, S. 1993, Icarus, 106, 549
Sheppard, S.S., Jewitt, D. C. 2003, Nature, 423, 261
Simó, C., & Stuchi, T. J. 2000, Physica D, 140, 1
Viera Neto, E., & Winter, O. C. 2001, AJ, 122, 440
Winter, O. C., & Viera Neto, E. 2001, A&A, 377, 1119
Winter, O. C., Viera Neto, E., & Prado, A. F. B. A. 2003, Adv.Space Res., 31, 2005

*Order and Chaos in Stellar and Planetary Systems*
*ASP Conference Series, Vol. 316, 2004*
*G. Byrd, K. Kholshevnikov, A. Mylläri, I. Nikiforov, and V. Orlov, eds.*

# Statistical Laws in Chaotic Dynamics of Multiple Stars

A. Rubinov, A. Petrova, and V. Orlov

*Sobolev Astronomical Institute, St. Petersburg State University, Universitetskij pr. 28, Staryj Peterhof, St. Petersburg 198504, Russia*

**Abstract.** Statistical analysis of the modeled stellar systems dynamical evolution is performed. The initial global parameters (amount of stars, system size, virial ratio, mass spectrum) are varied. Final state distribution, final binaries and stable triples orbital elements are analysed. It is shown that the probability of the stable triple formation is rather high (about 10–15%). The eccentricity distribution of the final binaries satisfies the $f(e) = 2e$ law. The hierarchy in the stable triple systems is rather strong (the mean ratio of the outer and inner binary semimajor axes is about 20 : 1). In stable triples the eccentricities of internal binaries are in average greater than the ones of external binaries ($\overline{e_{\rm in}} \approx 0.7$, $\overline{e_{\rm ex}} \approx 0.5$). Stable triples with prograde motions are preferable.

## 1. Introduction

It is well known that the result of dynamical system evolution in gravitational $N$-body problem is rather unpredictable if close encounters or cumulative effect of rather wide encounters take place. Small variations of initial conditions can lead to the significant modifications in the result of dynamical evolution. Due to the series of close binary and multiple encounters of bodies the system forgets initial conditions after several crossing times.

If one considers the dynamical evolution of small non-hierarhical stellar groups one can see all features mentioned above. But when the statistical analysis of dynamical evolution products is made then one can reveal some statistical laws insensitive to initial conditions of these groups.

By the end of 1960s van Albada (1968) put forward a hypothesis that wide binary and multiple stellar systems were formed via disintegration of small stellar groups consisting of a few or a few tens of stars. He and also Harrington (1974, 1975) carried out series of numerical experiments to model the dynamical evolution of small stellar groups. They integrated a set of multiple systems and carried out the statistical analysis of dynamical evolution products. The basic results of their investigations were as follows:

- Soft binaries and triples indeed may be a result of small stellar group dynamical decay.
- Binary eccentricity distribution may be well described by $f(e) = 2e$ law.

It should be mentioned that the $f(e) = 2e$ law was found by Ambartsumian (1937) for the stellar field where the amount of forming binaries per time unit is equal to the amount of disrupting binaries per time unit. The same law was derived by Monaghan (1976) in the statistical theory of triple system disintegration.

Sterzik & Durisen (1995, 1998) carried out numerical simulations of small stellar groups decay consisting of 3, 4, 5 stars. They revealed some results insensitive to the choice of system scale and initial mass spectrum:

- The dominant decay mode is binary + $N-2$ singles but other configurations (e.g., triple + $N-3$ singles) are not negligible.
- The effect of dynamical biasing (i.e., the final configuration consists of most massive stars in the system) is observed.
- The distribution of binary binding energy follows $f(E_b) \sim |E_b|^{-\frac{7}{2}}$ law.
- Typical single star velocities are rather high (about 5 km/s).

In present paper the dynamical evolution of systems consisting of 3–18 stars is considered. Initial global parameters (system size, virial ratio, mass spectrum) are varied and the statistical analysis of the dynamical evolution end-products is performed. Final state distribution, final binary and stable triple orbital elements are analysed.

## 2. Model

Our approach to the investigation of small stellar groups dynamical evolution is the numerical integration of $N$-body point-mass gravitational problem. CHAIN regularization algorithm (Mikkola & Aarseth 1993) was used to treat close binary and multiple encounters correctly.

The physical collisions of stars were taken into account using the results of stellar collision simulations by Benz & Hills (1987, 1992). Also the escapes of single and binary stars were fixed.

Initial conditions were chosen as follows. The positions of stars were randomly uniformly distributed inside a sphere with the radius $R$. Stellar velocities were chosen to satisfy the virial ratio $k$ assumed; the velocity distribution was isotropic. The following values of $R$ and $k$ parameters were considered: $R = 3$, 10, 100, 1000 A.U. and $k = 0.001$, 0.1, 0.5, 0.9.

The following variants of initial mass spectrum were considered:

1. Equal masses: $m_i = 1 M_\odot$.
2. Salpeter (1955) mass spectrum: $f(m) \sim m^{-2.35}$, $m \in [0.4, 10] M_\odot$.
3. Clump mass spectrum: $f(m) \sim m^{-1.5}$, $m \in [0.9, 10] M_\odot$.

The choice of clump initial mass spectrum can be motivated by the fact that clumpy and possibly fractal structure of molecular cloud cores implies fragment mass function power index to be in the interval $[-2.0, -1.5]$ (see, e.g., Elmegreen & Falgarone 1996).

The systems containing different number of stars $N = 3$, 6, 9, 12, 15, 18 at the beginning of the integration were considered. For each set of the initial conditions, 500 runs were integrated. The dynamical evolution was traced during 300 initial crossing times of the system $T_{\rm cr}$. Also the integration process was stopped when all bodies were escaped except final binary.

## 3. Results

### 3.1. Final State Distribution

The manifold of the final configurations was divided into several final states. There are final binary with positive total energy (2 single stars), final binary with negative total energy, stable triple system, unstable triple system, system of multiplicity higher than three (high multiplicity). The stability of the final triples was studied using Golubev (1967) analytical stability criterion. If a triple system is stable according to Golubev's criterion then the hierarchy of components will be conserved.

Table 1. The final state distribution at the moment $t = 300T_{\rm cr}$ for different values of the virial ratio $k$ ($N = 6$, $R = 100$ A.U.)

| $k$ | mass spectrum | final binary | two singles | stable triple | unstable triple | high mult. |
|---|---|---|---|---|---|---|
| 0.001 | EM | 0.60 | 0.05 | 0.13 | 0.10 | 0.12 |
| | SM | 0.65 | 0.01 | 0.11 | 0.17 | 0.06 |
| 0.1 | EM | 0.60 | 0.04 | 0.14 | 0.08 | 0.14 |
| | SM | 0.60 | 0.01 | 0.11 | 0.17 | 0.10 |
| 0.5 | EM | 0.57 | 0.07 | 0.06 | 0.15 | 0.14 |
| | SM | 0.55 | 0.01 | 0.18 | 0.18 | 0.08 |
| 0.9 | EM | 0.68 | 0.08 | 0.15 | 0.07 | 0.02 |
| | SM | 0.73 | 0.04 | 0.10 | 0.11 | 0.02 |

Table 1 demonstrates the final state distribution for equal masses and Salpeter mass spectrum cases for different values of the virial ratio $k$. Let us outline the properties that are rather robust with respect to the choice of the initial conditions. The dynamical evolution of multiple systems is usually ended by the final binary formation. The probability of the stable triple formation is rather high (about 10–15%). Although the greater part of stellar systems decays during 300 initial crossing times, the significant part (about 10%) is undecayed (see last column).

Also one can see that the virial ratio equal to 0.5 corresponds to the slowest speed of the system disintegration [one can compare the amount of the end-products of dynamical evolution (binaries and stable triples) and the amount of unstable triples and systems of higher multiplicity]. The mean rate of system disintegration with unequal-mass components is higher than that one for the equal-mass systems.

Let us note that if the initial system size and the amount of components are varied the main features of dynamical evolution are the same. Growth of the multiplicity leads to the decrease of disintegration rate.

### 3.2. Final Binaries

The probability that dynamical evolution is ended by the final binary formation is about $\frac{1}{2}$. Thus it is of interest to study the properties of final binaries.

The average values of the final binary semimajor axes are about several tenths of the initial median system size. Growth of the virial ratio and/or initial system size leads to the increase of the average value of the binary semimajor axes.

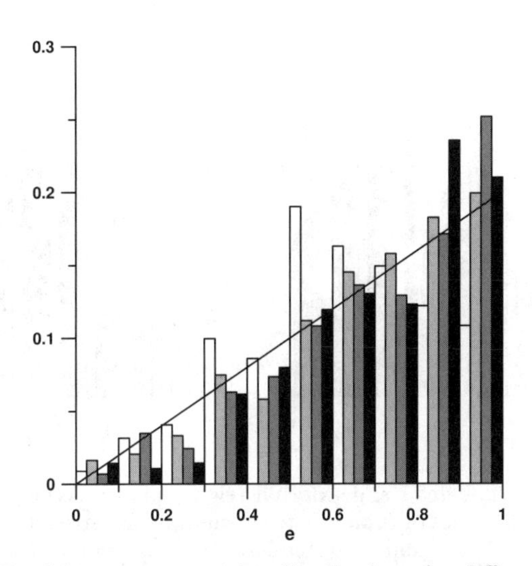


Figure 1. Final binary eccentricity distributions for different values of $R$ parameter in equal-mass case ($N = 6$, $k = 0.5$). White, light-grey, dark-grey, and black columns correspond to $R = 3$, 10, 100, and 1000 A.U., respectively. Solid line shows the $f(e) = 2e$ law.

Figure 1 demonstrates the final binary eccentricity distributions for different values of the initial system size $R$. The final binary eccentricity distributions can be well described by the theoretical law $f(e) = 2e$ if the initial system size is not very small. The eccentricity distribution is robust to the variations of initial virial ratio and choice of the initial mass spectrum.

For small initial sizes one can see a lack of elongated binary systems. This feature may be connected with the increase of merging probability for the elongated systems in the pericenters of their orbits.

### 3.3. Stable Triples

The fraction of the stable triples is rather high (about 10–15%). Thus it is interesting to analyse the basic characteristics of the final stable triple systems. One can imagine a stable triple system as a superposition of two binaries: external and internal. The inner binary consists of two nearest components of the triple. The external binary consists of the internal binary mass center (an object in the inner binary mass center, the mass of this object equals to the sum of inner binary component masses) and the distant component of the triple.

The average values of the inner binaries semimajor axes are about 10–100 times smaller than the initial system size. The average values of the outer binaries semimajor axes are comparable with the initial system size. Variations of the initial conditions lead to the same tendencies as for final binary semimajor axis average values (the increase of $R$ and $k$ parameters leads to the growth of the semimajor axes average values). Let us note that stable triples in the equal-mass case are in average closer than the ones in the case of unequal masses.

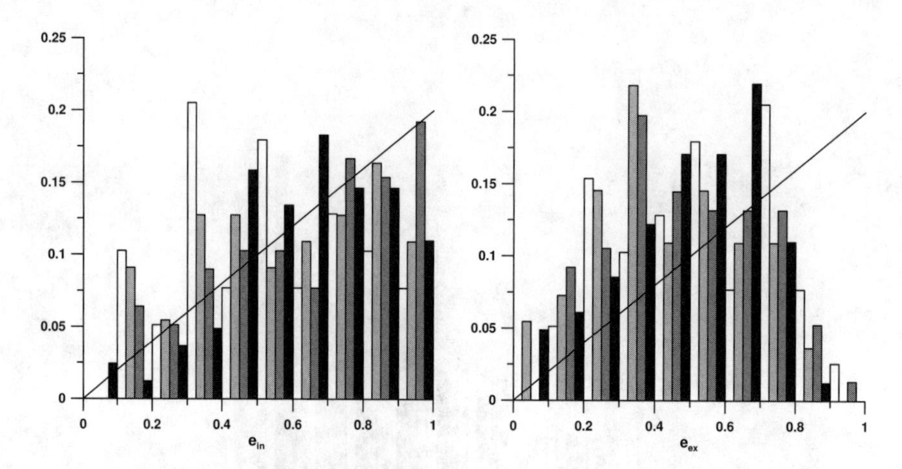


Figure 2. Internal (left) and external (right) binary eccentricity distributions for different values of $R$ parameter in clump mass spectrum case ($N = 6$, $k = 0.5$). Levels of grey and solid line have the same meaning as in Figure 1.

Figure 2 demonstrates the inner and outer binary eccentricity distributions for different values of initial system size. One can see a disagreement between the theoretical and the modeled eccentricity distributions. In case of small initial system sizes this disagreement is stronger for inner binaries; maybe it is due to the increase of the merging probability (as for the final binaries). The variations of other parameters do not lead to the remarkable tendencies. The inner binaries are in average more elongated than the outer binaries. The average values of external and internal binary eccentricities are as follows: $\overline{e_{ex}} \approx 0.5$, $\overline{e_{in}} \approx 0.7$.

The hierarchy of the stable triple systems is rather strong. Mean inner and outer binary semimajor axis ratio is about 1 : 20 and is rather robust with respect to the variation of initial conditions.

Figure 3 illustrates the dependence of angle $i$ between outer and inner binary orbital momentum vectors on the initial mass spectrum. One can mention that orbits of inner and outer binaries are usually non-coplanar. Although the shapes of the distributions are similar, one can see an asymmetry of the modeled distributions to the systems with the prograde motions. Lack of the systems with the retrograde motions may be connected with the fact that according to Golubev's stability criterion stable triples with the retrograde motions should have stronger hierarchy than the ones with the prograde motions. One can mention that in the case of equal masses, the mean value of the angle $i$ is slightly greater than in the case of unequal masses.

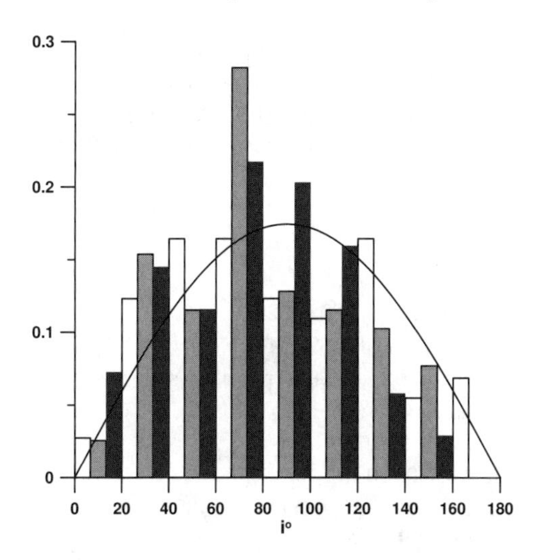


Figure 3. Distribution of the angle between internal and external binary orbital momentum vectors for different inital mass spectra ($N = 6$, $R = 100$ A.U., $k = 0.5$). White, grey, and black columns correspond to the equal-mass case, the clump mass spectrum, and the Salpeter mass spectrum, respectively. Solid line shows the distribution of $i$ angle in the case of random vector orientations [$f(i) = \frac{1}{2}\sin i$].

## 4. Conclusions

The character of dynamical evolution of small non-hierarchical stellar groups with negative total energy is chaotic and regular simultaneously. One can not predict the final state of the individual multiple system, but one can evaluate the statistical characteristics of the dynamical evolution for a sample of runs. This work is devoted to the investigation of the statistical laws in multiple non-hierarchical systems dynamical evolution which are robust to the variation of initial conditions. Let us outline these features:

- The dynamical evolution is usually ended by the formation of a final binary. The fraction of stable triples is rather high (about 10–15%).
- The eccentricity distribution of the final binaries satisfies the $f(e) = 2e$ law.
- In stable triples the eccentricities of the inner binaries are in average greater than the ones of the outer binaries ($\overline{e_{\rm in}} \approx 0.7$, $\overline{e_{\rm ex}} \approx 0.5$).
- The hierarchy in stable triple systems is rather strong (the mean ratio of the outer and inner binary semimajor axes is about $20 : 1$).
- The orbits of the inner and outer binaries in stable triples are non-coplanar. Stable triples with the prograde motions are preferable.

**Acknowledgments.** This work was partly supported by the RFBR (Grant 02-02-17516), Program *Universities of Russia* (Grant UR.02.01.027), and the Leading Scientific School (Grant NSh-1078.2003.02). A.R. is thankful to the Russian Ministry of Education for support (Grant M03-2.3K-220).

## References

Ambartsumian, V. A. 1937, AZh, 14, 207

Benz, W., & Hills, J. G. 1987, ApJ, 323, 628

Benz, W., & Hills, J. G. 1992, ApJ, 389, 546

Elmegreen, B. G., & Falgarone, E. 1996, ApJ, 471, 816

Golubev, V. G. 1967, Dokl.AN SSSR, 174, 767

Harrington, R. S. 1974, CeM, 9, 465

Harrington, R. S. 1975, AJ, 80, 1081

Mikkola, S., & Aarseth, S. J. 1993, CeMDA, 57, 439

Monaghan, J. J. 1976, MNRAS, 176, 63

Salpeter, E. E. 1955, ApJ, 121, 161

Sterzik, M. F., & Durisen, R. H. 1995, A&A Lett., 304, 9

Sterzik, M. F., & Durisen, R. H. 1998, A&A, 339, 95

van Albada, T. 1968, Bull.Ast.Neth., 19, 479

*Order and Chaos in Stellar and Planetary Systems*
*ASP Conference Series, Vol. 316, 2004*
*G. Byrd, K. Kholshevnikov, A. Mylläri, I. Nikiforov, and V. Orlov, eds.*

# On the Structure and Evolution of the "Basins of Attraction" in Ring-Type $N$-Body Formations

Maria Croustalloudi and Tilemahos Kalvouridis

*National Technical University of Athens, Faculty of Applied Sciences, Department of Mechanics, 5 Heroes of Polytechnion Ave, 157 73, Athens, Greece*

## 1. Introduction

The ring problem (Kalvouridis 1999) describes the motion of a small body $S$ in the combined gravitational field produced by $\nu = N - 1 > 2$ big bodies of equal masses, which are arranged on a circle in equal distances and which rotate with constant angular velocity around a $N$-th body of a different mass, located at the center of this circular arrangement (Figure 1a). The system is characterized by two parameters, the number of the peripheral primaries $\nu$ and the mass ratio $\beta = m/m_0$.

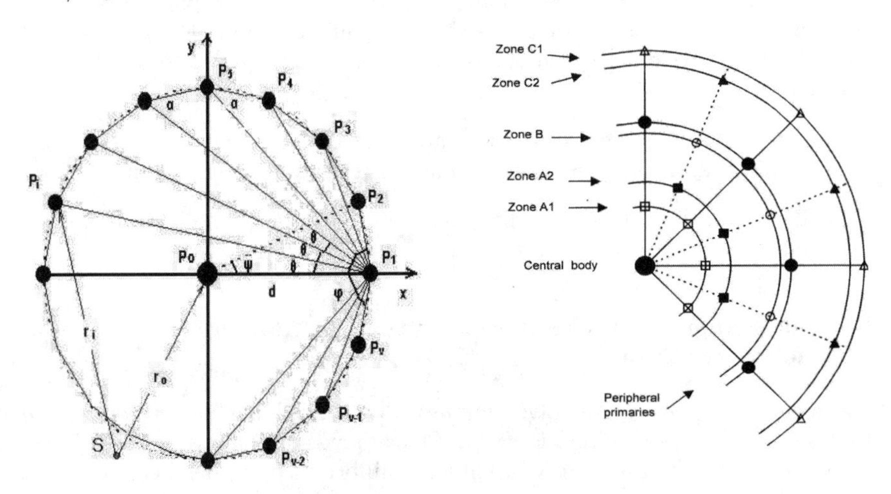


Figure 1. Left (a): The ring problem. Right (b): The equilibrium zones.

## 2. Particle Equilibrium Positions and Configurations in the Ring-Type Formation. The Effect of the Mass Parameter on the Equilibrium Zones

There are several axes of symmetry in the primaries' configuration. One of them is taken as the $x$-axis of a synodic coordinate system $Oxyz$. The plane of the primaries' revolution is $z = 0$. In this system, the equations of the planar motion of the particle are

$$\begin{array}{rcl} \ddot{x} - 2\dot{y} & = & U_x, \\ \ddot{y} + 2\dot{x} & = & U_y, \end{array} \quad \text{and for an equilibrium position} \quad \begin{array}{rcl} U_x & = & 0, \\ U_y & = & 0, \end{array} \tag{1}$$

where $U_\eta = \frac{\partial U}{\partial \eta}$, $\eta = x, y$, and

$$U(x,y) = \frac{1}{2}\left(x^2 + y^2\right) + \frac{1}{\Delta}\left[\frac{\beta}{r_0} + \sum_{i=1}^{\nu} \frac{1}{r_i}\right]$$

with

$$\Delta = M(\Lambda + \beta M^2), \quad \Lambda = \sum_{i=2}^{\nu} \frac{\sin^2\theta \cos(\frac{\nu}{2} + 1 - i)\theta}{\sin^2(\nu + 1 - i)\theta}, \quad \mathrm{M} = [2(1 - \cos\psi)]^{1/2},$$

$r_0$ and $r_i$ , $i = 1, 2, 3, \ldots, \nu$ are the distances of the particle from the primaries and $\psi$ is the angle formed between two successive peripheral primaries with $\theta = \psi/2$, $\psi = 2\pi/\nu$ (Figure 1a).

The planar equilibrium positions in the general gravitational case of the ring problem, are arranged in five circular zones, named $A_1$, $A_2$, $B$, $C_2$, and $C_1$ (Figure 1b) (Kalvouridis 1999). The points that belong to a particular zone are characterized by the same values of the Jacobian constant. The number of the equilibrium points of each zone (provided they exist) equals the number of the primaries:

- If $0 < \beta < l_\nu$, where $l_\nu$ is the marginal value different for each $\nu$, all the equilibrium zones exist. The number of the equilibria is $5\nu$ in this case.

- If $\beta \geq l_\nu$, then $A_2$ and $B$ disappear and only $A_1$, $C_2$ and $C_1$ remain. The number of the equilibria is $3\nu$ in this particular case.

## 3. Determination of the attracting regions. Remarks on their parametric evolution

The algebraic system (1) is solved numerically by applying an iterator, provided that an initial approximation is given. This approximation represents a point on plane $(x, y)$. The iterator stops when an equilibrium point has been found with the predetermined accuracy. From this point of view, an equilibrium position (whatever the real state of stability is) can be considered as an "attractor" (Peitgen et al. 1992). We call the set of the initial points that lead to the points of a particular equilibrium zone, an "attracting region".

The attracting regions have been determined by double scanning of the plane $(x, y)$ with a step equal to 0.005. We have recorded in separate groups the initial points that lead to the equilibrium points of each zone.

The attracting regions of each zone present, as it is expected, all the symmetry elements of the primaries' arrangement. They generally consist of some compact parts, namely areas every point of which leads to an equilibrium positions of this particular zone. Furthermore, a lot of dispersed points that lie on the boundaries of the "compact" areas of this or other zones appear. We must stress that these boundaries are not clearly defined. In general, the attracting regions are organized in formations that sometimes present the self-similarity of fractal structures. A major conclusion that we can elicit is that the attracting regions in zones $A_1$ and $C_2$ are more expanded than those in the rest of the zones, for all values of $\beta$. The corresponding area of $A_2$, provided it exists (for $0 < \beta < l_\nu$), is less expanded.

As regards the shape of the diagrams, we can note the following:

- The attracting region of zone $A_1$ comprises "compact", diamond-shaped parts with vague boundaries and many isolated points diffused all over the plane as well as on the boundaries of the "compact" region of this or of the other zones (Figure 2). The "compact" parts develop between the central primary and each of the peripheral ones. Inside these areas lie the equilibrium positions of that zone. The more $\beta$ augments, the more these areas expand in a way that their radial dimension remains steady, while their maximum transverse dimension increases.

- The attracting region of $B$ consists of "compact" areas that develop between the consecutive peripheral primaries and many dispersed points. The form of the "compact" areas is similar to that of the "compact" areas of $A_1$ (Figure 2). As $\beta$ augments (always for $0 < \beta < l_\nu$), their radial dimension is compressed, while their transverse dimension remains almost stable. When $\beta$ increases, the attracting region of $B$ (both "compact" areas and scattered points) is reduced.

- Between the "compact" areas of zones $A_1$ and $B$ develops the attracting region of $A_2$, thus creating three-legged formations, one part of which evolves toward the direction of the radius that connects the origin of the axes with an equilibrium position of that zone, while the other two develop toward the directions connecting the particular position with two neighboring peripheral primaries (Figure 2). The "compact" areas are characterized by self-similarity and dispersed points, which accumulate around them and evolve in a similar way, frame them. As $\beta$ increases (but remains less than $l_\nu$), the attracting region of $A_2$ reduces.

- The attracting region of $C_1$ is extremely complex. It generally consists of $\nu$ formations. Each of them comprises two basic "compact" parts, the biggest of which contains the equilibrium point (Figure 2). The dispersed points are densely distributed around the two basic "compact" parts, as well as on the boundaries of the "compact" parts of the other zones. As $\beta$ augments, we observe a small increase in the dispersed points and a decrease in the "compact" areas. The final result of these two conflicting processes is a total reduction of the attracting region of $C_1$.

- In the plane areas that lie between the "compact" parts of zones $C_1$ and $B$ stretch the "compact" areas of the attracting region of zone $C_2$, which have the form of a mushroom and contain the equilibrium points of this zone (Figure 2). The scattered points densely surround the "compact" parts, but also diffuse on the boundaries of the "compact" areas of the other zones. As $\beta$ increases (for $0 < \beta < l_\nu$), the "compact" areas are slightly

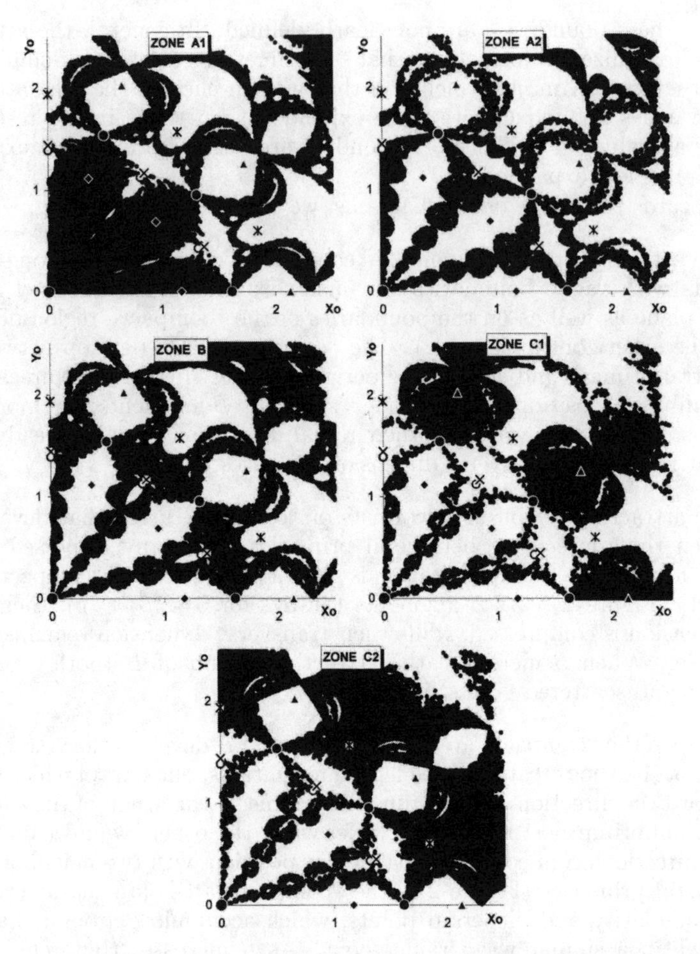


Figure 2. Part of the attracting regions of the equilibrium zones $A_1$ (♦), $A_2$ (○), $B$ (×), $C_2$ (✻), and $C_1$ (▲); • correspond to primaries.

reduced, while the dispersed points get denser. Near the marginal value of $\beta = l_\nu$ we observe an abrupt increase in these points, which are densely distributed in the former attracting regions of the no longer existing zones $A_2$ and $B$.

## References

Kalvouridis, T. J. 1999, Ap&SS, 260(3), 309

Croustalloudi, M., & Kalvouridis, T. 2003, in Recent Advances in Mechanics and the Related Fields (University of Patras, Greece), 89

Peitgen, H., Jurgens, H., & Saupe, D. 1992, Chaos and fractals (Springer Verlag)

# Part III

# The Solar System and Extrasolar Planetary Systems: Dynamics, Evolution, Formation

*Order and Chaos in Stellar and Planetary Systems*
*ASP Conference Series, Vol. 316, 2004*
*G. Byrd, K. Kholshevnikov, A. Mylläri, I. Nikiforov, and V. Orlov, eds.*

# Behaviour of a Weakly Perturbed Two-Planetary System on a Cosmogonic Time-Scale

K. V. Kholshevnikov

*Sobolev Astronomical Institute, St. Petersburg State University, Universitetskij pr. 28, Staryj Peterhof, St. Petersburg 198504, Russia*

E. D. Kuznetsov

*Astronomical Observatory, Urals State University, Lenin pr. 51, Ekaterinburg 620083, Russia*

**Abstract.** Orbital evolution of planetary systems similar to our Solar one represents one of the most important problems of Celestial Mechanics. In the present work we use Jacobian coordinates, introduce two systems of osculating elements, construct the Hamiltonian expansions in the Poisson series in all elements for the planetary three-body problem (including the problem of Sun–Jupiter–Saturn). Further we construct the averaged Hamiltonian by the Hori–Deprit method with accuracy upto second order with respect to the small parameter, the generating function, change of variables formulae, and right-hand sides of averaged equations. The averaged equations for the Sun–Jupiter–Saturn system are integrated numerically at the time-scale of 10 Gyr. The motion turns out to be almost periodical. The low and upper limits for eccentricities are 0.016, 0.051 (Jupiter), 0.020, 0.079 (Saturn), and for inclinations to the ecliptic plane (degrees) are 1.3, 2.0 (Jupiter), 0.73, 2.5 (Saturn). It is remarkable that the evolution of the ascending node longitudes may be secular (Laplace's plane) as well as librational one (ecliptic plane).

Estimates of the Liapunovian Exponents for the Sun–Jupiter–Saturn system are obtained. The corresponding Liapunovian Time turns out to be 14 Myr (Jupiter) and 10 Myr (Saturn).

## Introduction

Orbital evolution of a planetary system similar to our Solar one represents one of the most important problems of Celestial Mechanics. This work continues researches beginning in (Kholshevnikov, Greb, & Kuznetsov 2001, 2002).

We consider the planetary three-body problem for Sun–Jupiter–Saturn. Let us assume $m_0$, $\mu m_0 m_1$, $\mu m_0 m_2$ as masses of the Sun, Jupiter and Saturn respectively. Small parameter $\mu$ is equial to $10^{-3}$. In this case the dimensionless masses $m_1$ and $m_2$ are the unity order values ($m_1 \approx 1$, $m_2 \approx 1/3$).

We use Jacobian coordinates as best-fitting. Let us represent the Hamiltonian as a sum of the unperturbed part $h_0$ and the perturbed one $\mu h_1$:

$$h = h_0 + \mu h_1. \tag{1}$$

The first term depends on semi-major axes only

$$h_0 = -\frac{Gm_0m_1}{2a_1} - \frac{Gm_0m_2}{2a_2},$$

$G$ being the gravitational constant. Here and below the subscripts 1 and 2 for coordinates and elements correspond to Jupiter and Saturn respectively. The second term (1) may be thought of as a constant factor having dimension of velocity squared and a dimensionless part $h_2$

$$h_1 = \frac{Gm_0}{a_0}h_2, \qquad h_2 = h_3 + h_4, \tag{2}$$

where $a_0$ is an arbitrary parameter with the length dimensionality,

$$h_3 = \frac{m_2a_0}{\mu}\left(\frac{1}{r_2} - \frac{1}{\rho}\right) = \frac{m_2a_0\left[2\frac{m_1}{1+\mu m_1}\mathbf{r}_1\mathbf{r}_2 + \mu\left(\frac{m_1}{1+\mu m_1}\right)^2 r_1^2\right]}{r_2\rho(r_2+\rho)},$$

$$h_4 = -\frac{m_1m_2a_0}{\Delta}, \qquad \rho = \left|\mathbf{r}_2 + \frac{\mu m_1}{1+\mu m_1}\mathbf{r}_1\right|, \qquad \Delta = \left|\mathbf{r}_2 - \frac{1}{1+\mu m_1}\mathbf{r}_1\right|.$$

The disturbing function $h_2$ is presented as Poisson series

$$h_2 = \sum A_{kn}x^k \cos ny. \tag{3}$$

Here $x = \{x_1, \ldots, x_6\}$ are positional elements, $y = \{y_1, \ldots, y_6\}$ are angular ones, $A_{kn}$ are numerical coefficients, $k = \{k_1, \ldots, k_6\}$ and $n = \{n_1, \ldots, n_6\}$ are multi-indices. The summation is taken over non-negative $k_s$ and integer $n_s$.

Introduce two systems of osculating elements. The first system is close to the Keplerian one:

$$x_{3s-2}^{(1)} = \widetilde{a}_s, \quad x_{3s-1}^{(1)} = e_s, \quad x_{3s}^{(1)} = \widetilde{I}_s,$$

$$y_{3s-2}^{(1)} = \alpha_s, \quad y_{3s-1}^{(1)} = \beta_s, \quad y_{3s}^{(1)} = \gamma_s. \tag{4}$$

Here $\widetilde{a} = (a-a^0)/a^0$, $\widetilde{I} = \sin(I/2)$, $\alpha = l+g+\Omega$, $\beta = g+\Omega$, $\gamma = \Omega$ are expressed in terms of Keplerian elements $a$, $a^0$, $e$, $I$, $l$, $g$, $\Omega$: semi-major axis and its mean value, eccentricity, inclination, mean anomaly, argument of pericenter, longitude of ascending node. The index $s$ changes from 1 to the number of planets $N = 2$.

The second system realizes simplifications due to the homogeneity of the perturbation function with respect to semi-major axes. In this system denominators arising in a process of averaging transforms are extremly simple. On the other hand, it has a deficiency, mixing several elements of all planets

$$x_{3s-2}^{(2)} = z_s, \quad x_{3s-1}^{(2)} = e_s, \quad x_{3s}^{(2)} = \widetilde{I}_s,$$

$$y_{3s-2}^{(2)} = \alpha_s, \quad y_{3s-1}^{(2)} = \beta_s, \quad y_{3s}^{(2)} = \gamma_s. \tag{5}$$

Here $z_1 = \omega_1^0/\omega_1 - 1$, $z_2 = (\omega_1^0\omega_2)/(\omega_2^0\omega_1) - 1$, $\omega_s = \kappa_s a_s^{-3/2}$ are mean motions of the planets, $\omega_s^0$ are constants close to mean values $\omega_s$, $\kappa_s^2 = Gm_0m_s/M_s$ are gravitational parameters of the planets, reduced masses are $M_s = m_s(1+\mu m_1 + \ldots + \mu m_{s-1})/(1 + \mu m_1 + \ldots + \mu m_s)$, $s = 1, 2$.

It is well known (Charlier 1927; Subbotin 1968) that

$$n_1 + n_2 + n_3 + n_4 + n_5 + n_6 = 0, \qquad n_3 + n_6 = \text{even},$$

$$k_s = |n_s| + \text{non-negative even} \quad (s = 2,\ 3,\ 5,\ 6). \tag{6}$$

The simplest restriction on $k$ is

$$k_1 + k_2 + \ldots + k_6 \leq d. \tag{7}$$

Let us choose $d$ ensuring the Hamiltonian accurate up to $\mu^\sigma$. Taking into account that eccentricities and inclinations have the same order $\mu_1 \approx \sqrt{\mu}$ ($e_1 \approx 0.05$, $e_2 \approx 0.05$, $\widetilde{I}_1 \approx 0.01$, $\widetilde{I}_2 \approx 0.02$) we obtain the dependence $d(\sigma)$:

$$d(1) = 1, \quad d(2) = 6, \quad d(3) = 11, \quad d(4) = 16, \quad \ldots \tag{8}$$

As to $n$ it is sufficient to demand

$$0 \leq n_1 \leq c, \quad |n_4| \leq c \tag{9}$$

for ensuring the finite number $N(d, c)$ of the sum (3) terms. Choice of $c(\sigma)$ is determined by the convergence rate of (3) for the planar circular orbits ($\mu_1 = 0$):

$$c(2) = 13, \quad c(3) = 25, \quad c(4) = 37. \tag{10}$$

Taking into account (6), (8), (10) we may estimate the number $N(d, c)$ of terms in (3). In particular, $N(6, 13) \approx 4 \times 10^4$, $N(11, 25) \approx 4 \times 10^6$, $N(16, 37) \approx 8 \times 10^7$.

To calculate $A_{kn}$ we ought to fix the values of the following constants: $a_0$, $m_1$, $m_2$, $a_1^0$, $a_2^0$. We put the scale factor $a_0$ equal to the astronomical unit. The planetary masses one taken from (Moisson 1999): $m_1 = (1047.3486\mu)^{-1}$, $m_2 = (3497.90\mu)^{-1}$.

On the stage of the series (3) construction initial data are not required. It is enough to fix roughly the constants $a_1^0$ and $a_2^0$ near to mean values of semi-major axes on long-time interval. We take them from (Simon et al. 1994). Process of averaging Lie transforms leads to appearance of small divisors due to nearness of mean motions $\omega_1$, $\omega_2$ to commensurability 5:2. In our problem there exists only one small divisor $2\omega_1^0 - 5\omega_2^0$. To reduce the influence of the small divisor the frequencies $\omega_i^0$ corresponding to average values of semi-major axes $a_i^0$ are chosen as far as possible (conserving 5/2 as one of the convergents of the continued fraction for the ratio of the frequencies) from the resonance values that lead $F = \left(2\omega_1^0 - 5\omega_2^0\right)/\omega_1^0$ to zero. Let us pose $a_1^0 = 5.215$ a.u., $a_2^0 = 9.530$ a.u. In this case $F = -0.023331$.

## Expansion of Disturbing Hamiltonian into Poisson Series

The Poisson series processor PSP (Brumberg 1995; Ivanova 1995) is used to construct the expansion of disturbing Hamiltonian $h_2$ into Poisson series (3). Classical expansions of celestial mechanics (Subbotin 1968) for functions $r/a$, $(r/a)^2$, $\cos\theta$ and $\sin\theta$ are used, $\theta$ being the true anomaly.

The rational version of the PSP is used to decrease round-off errors during calculations of the coefficients $A_{kn}$. The expansion of disturbing Hamiltonian is processed up to $\mu^2$. The summation is taken over $k_1 + \ldots + k_6 \le 6$, $|n_s| \le 15$ $(s = 1, \ldots, 6)$. For each of the osculating elements system two variants of the expansion are constructed. The first variant deals with numerical values of parameters (masses, mean values of semi-major axes, etc.) corresponding to the system Sun–Jupiter–Saturn. The second one deals with their symbolic expressions depending on parameters of the system. The expansions with numerical data contain 61086 terms. The expansions with symbolic parameters contain 182744 terms for the first system and 183227 terms for the second one.

## Averaged Planetary Three-Body Problem

The Hori–Deprit method (Lie transforms method) is used to construct the averaged Hamiltonian $H(X, Y)$. This method is based on Poisson brackets that allows us to use non-canonical elements writing down the Poisson brackets in the corresponding system of phase variables (Kholshevnikov & Greb 2001).

For a stationary change of variables without confusion of impulses $p = (p_1, \ldots, p_6)$ and coordinates $q = (q_1, \ldots, q_6)$ the Poisson bracket is written down through partial brackets (Kholshevnikov & Greb 2001) as

$$\{f, g\} = V_{jk}(f, g)_{jk}, \qquad V_{jk} = \frac{\partial x_j}{\partial p_i}\frac{\partial y_k}{\partial q_i}, \qquad (f, g)_{jk} = \frac{\partial f}{\partial x_j}\frac{\partial g}{\partial y_k} - \frac{\partial f}{\partial y_k}\frac{\partial g}{\partial x_j}.$$

Summation is made on repeating indices from 1 to 6.

The matrix $\mathcal{V}$ having elements $V_{jk}$ for the first system of elements (4) was obtained in Kholshevnikov & Greb (2001). It is a block diagonal one

$$\mathcal{V}^{(1)} = \begin{pmatrix} \mathcal{V}_1^{(1)} & \mathcal{O} \\ \mathcal{O} & \mathcal{V}_2^{(1)} \end{pmatrix}, \qquad \mathcal{V}_s^{(1)} = \frac{1}{c_s} \begin{pmatrix} \frac{2a_s}{a_s^0} & 0 & 0 \\ \frac{\eta_s^2 - \eta_s}{e_s} & -\frac{\eta_s}{e_s} & 0 \\ -\frac{\sin\frac{I_s}{2}}{2\eta_s} & -\frac{\sin\frac{I_s}{2}}{2\eta_s} & -\frac{1}{4\eta_s \sin\frac{I_s}{2}} \end{pmatrix}.$$

Here $\mathcal{O}$ is null $3 \times 3$ matrix, $\eta_s = \sqrt{1 - e_s^2}$, $c_s = M_s \kappa_s \sqrt{a_s}$.

For the second system (5) Poisson matrix is more complex

$$\mathcal{V}^{(2)} = \begin{pmatrix} \frac{3(1+z_1)^{2/3}}{c_1} & 0 & 0 & 0 & 0 & 0 \\ \frac{\eta_1^2-\eta_1}{c_1 e_1} & -\frac{\eta_1}{c_1 e_1} & 0 & 0 & 0 & 0 \\ -\frac{\sin\frac{I_1}{2}}{2c_1\eta_1} & -\frac{\sin\frac{I_1}{2}}{2c_1\eta_1} & -\frac{1}{4c_1\eta_1\sin\frac{I_1}{2}} & 0 & 0 & 0 \\ \frac{3(1+z_2)}{c_1(1+z_1)^{1/3}} & 0 & 0 & -\frac{3(1+z_2)^{4/3}}{c_2(1+z_1)^{1/3}} & 0 & 0 \\ 0 & 0 & 0 & \frac{\eta_2^2-\eta_2}{c_2 e_2}\zeta & -\frac{\eta_2\zeta}{c_2 e_2} & 0 \\ 0 & 0 & 0 & -\frac{\sin\frac{I_2}{2}}{2c_2\eta_2}\zeta & -\frac{\sin\frac{I_2}{2}}{2c_2\eta_2}\zeta & -\frac{\zeta}{4c_2\eta_2\sin\frac{I_2}{2}} \end{pmatrix}$$

Here $\zeta = (1+z_2)^{1/3}(1+z_1)^{-1/3}$. The matrix $\mathcal{V}^{(2)}$ is triangular.

The Hamiltonian $h(x, y)$ (1), as well as the averaged Hamiltonian $H$, the generating function, change of variables formulae, and right-hand sides of averaged equations of motion are presented by power series in the small parameter $\mu$ (upto $\mu^2$ for $H$ and right-hand sides), and each term is the echeloned Poisson series. For calculations we use the rational version of the echeloned Poisson series processor EPSP (Ivanova 2001) to reduce the round-off errors. Transformations are made for both systems of elements with numerical parameters corresponding to the Sun–Jupiter–Saturn system.

The average procedure allows to increase the step of integration of motion equations more than thousand times. For the Sun–Jupiter–Saturn system one increases upto several hundreds or thousands years.

## Behaviour of the Sun–Jupiter–Saturn System on a Cosmogonic Time-Scale

The averaged equations are integrated numerically at the time-scale of 10 Gyr. The equations for slow variables are integrated by Everhart and Runge–Kutta high order methods. The equations for fast variables are integrated by spline interpolation method. Accuracy of integration is detected by computation of integrals of energy and area. To calculate area integrals the properties of form conservation in Jacobian coordinates (Charlier 1927) and under averaging transform (Kholshevnikov 1992) are used.

The motion turns out to be almost periodical (Figure 1). The low and upper limits for averaged eccentricities are 0.016, 0.051 (Jupiter), 0.020, 0.079 (Saturn), and for averaged inclinations are $1^\circ\!.3$, $2^\circ\!.0$ (Jupiter), $0^\circ\!.73$, $2^\circ\!.5$ (Saturn).

Evolution of the arguments of pericentre turns out to be secular. Evolution of the ascending node longitudes (Figure 2) depends on the base plane and may be secular (Laplace plane) as well as librational one (ecliptic plane). Difference between the ascending nodes longitudes of Jupiter and Saturn with respect to Laplace plane is equal to 180° exactly in accordance with Jacobi theorem (Charlier 1927).

Estimates of the Liapunovian Exponents for the system Sun–Jupiter–Saturn are obtained. The corresponding Liapunovian Time turns out to be 14 Myr (Jupiter) and 10 Myr (Saturn).

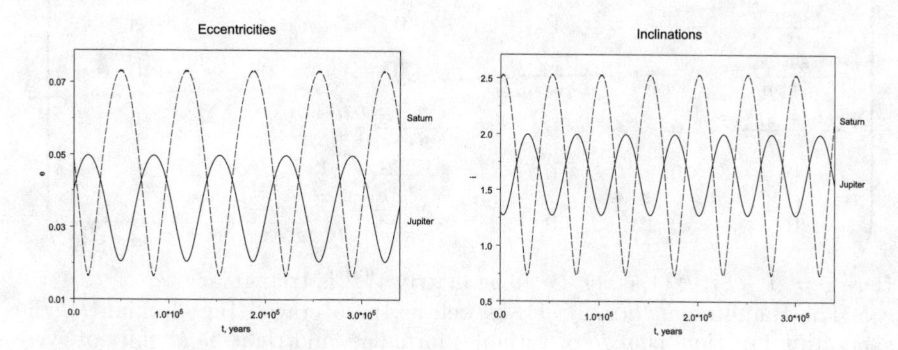


Figure 1. Evolution of the eccentricities and inclinations.

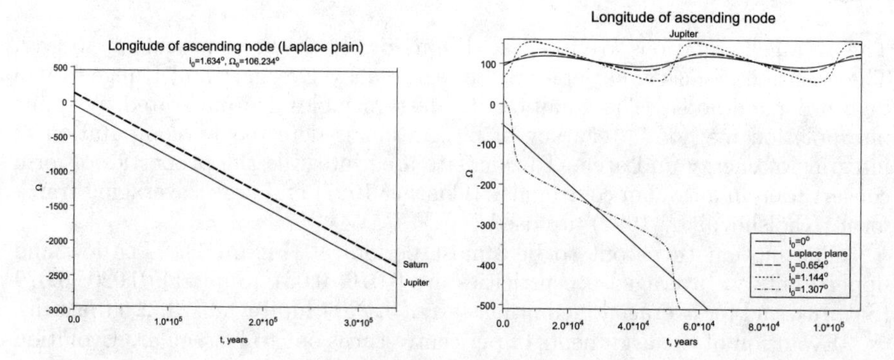


Figure 2. Evolution of the ascending nodes longitude with respect to several planes.

Our results on the stability of the system are qualitatively in agreement with ones obtained by Laskar (1994), Murray & Dermott (1999), Ito & Tanikawa (2002). The estimates of maximum eccentricity and inclination values for Jupiter and Saturn are adjusted with Laskar (1988) results where all Solar System planets was take into account: maximum eccentricity of Jupiter orbit is 0.061, Saturn orbit is 0.085, maximum inclinations are $2.^{\circ}1$ and $2.^{\circ}6$ correspondingly. Our estimates of the Liapunovian Time are corresponded with results of Laskar (1994) (5 Myr for inner planets) and Sussman & Wisdom (1988) (20 Myr for Pluto).

**Acknowledgments.** This work was partly supported by the RFBR (Grant 02-02-17516), Program *Universities of Russia* (Grant UR.02.01.027), and the Leading Scientific School (Grant NSh-1078.2003.02).

## References

Brumberg, V. A. 1995, Analytical Techniques of Celestial Mechanics (Heidelberg: Springer)

Charlier, C. L. 1927, Die Mechanik des Himmels (Berlin, Leipzig: Walter de Gruyter & Co.)

Ito, T., & Tanikawa, K. 2002, in Proceedings IAU 8 Asian-Pacific Regional Meeting, Vol. II, 45

Ivanova, T. V. 1995, in IAU Symp. 172, Dynamics, Ephemerides and Astrometry of the Solar System, ed. S. Ferraz-Mello, B. Morando & J.-E. Arlot (Kluwer)

Ivanova, T. 2001, CeMDA, 80, 167

Kholshevnikov, K. V. 1992, AZh, 68, 660

Kholshevnikov, K. V., Greb, A. V., & Kuznetsov, E. D. 2001, Solar System Research, 35, 243

Kholshevnikov, K. V., & Greb, A. V. 2001, Solar System Research, 35, 457

Kholshevnikov, K. V., Greb, A. V., & Kuznetsov, E. D. 2002, Solar System Research, 36, 68

Laskar, J. 1988, A&A, 198, 341

Laskar, J. 1994, A&A, 287, L9

Moisson, X. 1999, A&A, 341, 318

Murray, C. D., & Dermott, S. F. 1999, Solar System Dynamics (Cambridge: Univ. Press)

Simon, J. L. et al. 1994, A&A, 282, 663

Subbotin, M. F. 1968, Introduction to the Theoretical Astronomy (Moscow: Nauka)

Sussman, G. J., & Wisdom, J. 1988, Science, 241, 433

*Order and Chaos in Stellar and Planetary Systems*
*ASP Conference Series, Vol. 316, 2004*
*G. Byrd, K. Kholshevnikov, A. Mylläri, I. Nikiforov, and V. Orlov, eds.*

# Regular Coordinates Correction on Computing the Ephemeris of Small Celestial Bodies of Solar System

S. M. Poleshchikov

*Syktyvkar forest institute, Lenin st. 35, Syktyvkar 167000, Russia*

A. A. Kholopov

*Syktyvkar state university, Oktjabrsky pr. 55, Syktyvkar 167001, Russia*

**Abstract.** We consider the two-body motion equations in regular coordinates for small celestial bodies. The problem of enlarging an integration step in Runge–Kutta–Fehlberg method under the preservation of a given accuracy is solved in plane motion case.

## 1. Problem Statement

Consider the equation of perturbed motion in two-body problem

$$\ddot{\mathbf{x}} + \frac{\mu}{r^3}\mathbf{x} = -\frac{\partial V}{\partial \mathbf{x}} + \mathbf{F}, \quad r = |\mathbf{x}|, \quad \mathbf{x}(t_0) = \mathbf{x}_0, \quad \dot{\mathbf{x}}(t_0) = \dot{\mathbf{x}}_0, \tag{1}$$

where $\mathbf{x} = (x_1, x_2, x_3)^\top$ is the radius-vector of a particle with mass $m$ with respect to the point with mass $M$, $\gamma$ is the gravity constant, $\mu = \gamma(M + m)$.

Three-dimensional physical vectors are expanded to four-dimensional ones by adding zero as a fourth coordinate. Consider an $L$-matrix $L(\mathbf{u})$ described in (Poleshchikov & Kholopov 1999, 2000a) and put

$$\mathbf{x} = L(\mathbf{u})\mathbf{u}, \quad \mathbf{x},\, \mathbf{u} \in \mathbf{R}^4, \quad dt = r\, d\tau. \tag{2}$$

Denoting by stroke a differentiation along $\tau$ and choosing

$$\mathbf{x}_0 = L(\mathbf{u}_0)\mathbf{u}_0, \quad \mathbf{u}'_0 = \frac{1}{2}L^\top(\mathbf{u}_0)\dot{\mathbf{x}}_0, \tag{3}$$

we obtain the system of equations

$$\begin{cases} \mathbf{u}'' = -\frac{h_{\mathrm{v}}}{2}\mathbf{u} - \frac{1}{2}L^\top(\mathbf{u})\Big(\frac{\partial(rV)}{\partial \mathbf{x}} - r\mathbf{F}\Big), \\ h'_{\mathrm{v}} = -r\Big(\frac{\partial V}{\partial t} + \mathbf{F}^\top\dot{\mathbf{x}}\Big), \\ t' = r, \end{cases} \tag{4}$$

where $h_{\mathrm{v}} = \dfrac{\mu - 2|\mathbf{u}'|^2}{|\mathbf{u}|^2} - V$. In initial time, the variables $h_{\mathrm{v}}$ and $t$ have the values

$$h_{\mathrm{v}}(0) = \frac{\mu - 2|\mathbf{u}'_0|^2}{|\mathbf{u}_0|^2} - V(\mathbf{x}_0, t_0), \quad t_0 = 0.$$

The system (4) has following properties:

1. The equations have no singularities caused by attraction of central point.

2. The solutions of the system are stable in Liapunov's sense in absense of perturbations. The value $h_{\mathrm{v}}$ is a parameter.

3. The equations are invariant with respect to orthogonal transformations of regular coordinates by the following way. Under change of variables it is possible to use in (2) an another $L$-matrix $\widehat{L}(\mathbf{u}) = L(S\mathbf{u})S$, where $S$ is an orthogonal matrix (Poleshchikov & Kholopov 2000a). As a result, we come to the system of equations of kind (4) where physical quantities have the same numerical values at the same time $\tau$ as in system (4), only the variables $\mathbf{u}$, $\mathbf{u}'$ and matrix $L(\mathbf{u})$ are different.

The last property can be used for the modification of numerical integration of equations (4) by Runge–Kutta–Fehlberg method (RKF) with different order of accuracy. After each RKF-step we may perform some orthogonal transformation of regular coordinates and continue the integration. We are interested in enlarging the integration step in RKF-method under the preservation of a given integration accuracy.

According to Fehlberg (1970), the value of integration step is controlled by the increment of two formulae of Runge–Kutta method with different orders of accuracy. The controlling inequality for system (4) to be accurate within the first order of smallness of the integration step $\eta$ becomes (Poleshchikov & Kholopov 2000b):

$$\begin{aligned} &|\eta| c_\gamma \frac{|\mathbf{u}'|}{|\mathbf{u}|} \max\left\{ \frac{|b_1|}{|a_1|}, \dots, \frac{|b_4|}{|a_4|}, \Omega \frac{|a_1 + (L^\top(\mathbf{a})\mathbf{P})_1|}{|b_1|}, \dots, \right. \\ &\left. \Omega \frac{|a_4 + (L^\top(\mathbf{a})\mathbf{P})_4|}{|b_4|}, \frac{2}{|\dot{\mathbf{x}}|} \frac{r|\frac{\partial V}{\partial t} + \mathbf{F}^\top \dot{\mathbf{x}}|}{|h_{\mathrm{v}}|}, \frac{2}{|\dot{\mathbf{x}}|} \frac{r}{|t|} \right\} < \varepsilon, \end{aligned} \tag{5}$$

where

$$\Omega = \frac{2|h_{\mathrm{v}}|}{|\dot{\mathbf{x}}|^2}, \quad \mathbf{a} = \frac{\mathbf{u}}{|\mathbf{u}|}, \quad \mathbf{b} = \frac{\mathbf{u}'}{|\mathbf{u}'|}, \quad \mathbf{P} = \frac{1}{h_{\mathrm{v}}} \left( \frac{\partial (rV)}{\partial \mathbf{x}} - r\mathbf{F} \right),$$

$c_\gamma$ is a positive coefficient containing the parameters of RKF-method, $\varepsilon$ is the local relative error of the method. It follows from this inequality that the smaller maximum is in (5) the larger integration step $\eta$ may be chosen.

The two last fractions in (5) are expressed by the only physical variables. Therefore these two fractions do not depend on orthogonal transformations of the regular coordinate system and we may change the maximum in (5) by choosing the parameters of $L$-matrix so that the first eight functions are taken into account. In undisturbed case ($\mathbf{P} = 0$), this maximum will have a minimum (Poleshchikov & Kholopov 2000b) if

$$|a_i| = |b_i|, \quad i = 1, \dots, 4. \tag{6}$$

Corresponding orthogonal transformations were considered in (Poleshchikov & Kholopov 1999; Poleshchikov & Kholopov 2000b).

In the disturbed case, the functions at maximum in (5) have a more complicated character. If perturbations are small in comparison with the central acceleration $-\frac{\mu \mathbf{x}}{r^3}$ then the small summand $(L^\top(\mathbf{a})\mathbf{P})_i$ in (5) may be neglected for the simplification of the problem. We can suppose in this case that the rotation matrix that leads to optimal position of the couple of vectors $\mathbf{a} + L^\top(\mathbf{a})\mathbf{P}$ and $\mathbf{b}$ coincides with the matrix for undisturbed case. If $\mathbf{P}$ is big (as, for example, in approach of comet with the Jupiter or in close passage of an asteroid near the Earth) then (6) is wrong. In this case, the minimizing function will be

$$\Phi = \max\left\{\frac{|b_1|}{|a_1|}, \ldots, \frac{|b_4|}{|a_4|}, \Omega\frac{|a_1 + (L^\top(\mathbf{a})\mathbf{P})_1|}{|b_1|}, \ldots, \ \Omega\frac{|a_4 + (L^\top(\mathbf{a})\mathbf{P})_4|}{|b_4|}\right\}. \quad (7)$$

The relationship $S\mathbf{a} + L^\top(S\mathbf{a})\mathbf{P} = S(\mathbf{a} + \widehat{L}^\top(\mathbf{a})\mathbf{P})$ gives that the angles between vectors $\mathbf{a}$, $\mathbf{b}$, $\mathbf{a} + L^\top(\mathbf{a})\mathbf{P}$ do not change under orthogonal transformations of the regular coordinates $\mathbf{u}$. In other words, this triplet of vectors rotates as a single object under the transformation $S$. Thus we come to the problem of finding such positions of vector triplets $\mathbf{a}$, $\mathbf{b}$, $\mathbf{a} + L^\top(\mathbf{a})\mathbf{P}$ for which the function (7) has a minimum. These positions of triplets are defined as the optimum ones.

We found the optimal positions of tripltes in case of plane motion, where we have only two coordinates: $\mathbf{a}, \mathbf{b} \in \mathbf{R}^2$. There are six ordinary and seven exclusive types of optimal triplets. Corresponding formulae are rather cumbersome to be given here.

## 2. Numerical Experiments

Consider the results of numerical experiments with the system (4) describing the perturbed motion of a satellite. The mass of a satellite is assumed to be zero. Therefore the geocentric gravity constant is equal to $\mu = 398601.3$ km$^3$/s$^2$. For all satellites from Tables 1 and 2 the initial coordinate and velocity values are chosen so that the motion starts from pericentre. Those are taken to be $x_1 = 6600$ km, $x_2 = 0$ km, $\dot{x}_1 = 0$ km/s. The values of the second velocity component vary as in the Tables. The origin of coordinates is placed in the center of Earth mass. As a perturbation source we use the second zonal harmonic of the geopotential and the attraction of the third celestial body with the Moon parameters. This third body describes a path along the circle $a_1 = \rho \sin \sigma t$, $a_2 = \rho \cos \sigma t$, $\rho = 384400$ km with angular velocity $\sigma = 2.6653167 \times 10^{-6}$ s$^{-1}$.

The calculations were fulfilled twice: with fixed $L$-matrix and with $L$-matrix whose parameters $l_1$, $l_2$ were recalculated in computation process by found formulae (the $L$-matrix correction). We used two pairs of RKF-method with orders $5, 4$ and $8, 7$ denoted as RKF5(4) and RKF8(7). The method RKF5(4) has 6 stages, the method RKF8(7) has 13 ones. The parameters of these methods may be found in (Forsythe et al. 1977). All calculations were made in system *Maple* with 21 significant digits. The inequality controlling an integration step has a relative local error $\varepsilon = 10^{-13}$.

The integration was done in fictitious time interval $[0, \tau_{100}]$. The value of $\tau_{100}$ for every satellite was calculated by formula $\tau_{100} = 100\pi\sqrt{\frac{2}{h_v(0)}}$. It corresponds to the time when the satellite passages 100 winds around the Earth without perturbations. A comparison of integration results was made using three quantities: $\delta|\mathbf{x}| = \frac{|\mathbf{x}-\mathbf{x}_s|}{|\mathbf{x}_s|}$, $\delta|\dot{\mathbf{x}}| = \frac{|\dot{\mathbf{x}}-\dot{\mathbf{x}}_s|}{|\dot{\mathbf{x}}_s|}$, $N_f$, where $\mathbf{x}$, $\dot{\mathbf{x}}$ are the calculation

Table 1. Integration by pair RKF5(4)

| | | | Without correction | | | With correction | | |
|---|---|---|---|---|---|---|---|---|
| n | $\dot{x}_2$ | $T$ | $\delta\lvert\mathbf{x}\rvert$ $\times 10^{-9}$ | $\delta\lvert\dot{\mathbf{x}}\rvert$ $\times 10^{-9}$ | $N_f$ | $\delta\lvert\mathbf{x}\rvert$ $\times 10^{-9}$ | $\delta\lvert\dot{\mathbf{x}}\rvert$ $\times 10^{-9}$ | $N_f$ |
| 1 | 8.0 | 6.76 | 13 | 12 | 294925 | 13 | 12 | 222455 |
| 2 | 8.5 | 8.56 | 15 | 13 | 301458 | 15 | 13 | 226643 |
| 3 | 9.1 | 12.36 | 18 | 13 | 305684 | 18 | 13 | 232361 |
| 4 | 9.9 | 26.57 | 20 | 14 | 310236 | 20 | 14 | 242177 |
| 5 | 10.3 | 51.16 | 16 | 13 | 313860 | 16 | 13 | 249275 |

data; $\mathbf{x}_s$, $\dot{\mathbf{x}}_s$ are standard data of vectors $\mathbf{x}$, $\dot{\mathbf{x}}$ in the integration interval end; $N_f$ is a number of subroutine calls for the calculation of right-hand members in differential equations. The standard orbit was calculated by RKF8(7) method with relative local error $\varepsilon = 10^{-18}$ and without $L$-matrix correction.

The third column of Tables includes the approximate values of physical time $T$ (in days) of satellite total motion corresponding to the length of integration interval $\tau_{100}$ in fictitious time.

Table 2. Integration by pair RKF8(7)

| | | | Without correction | | | With correction | | |
|---|---|---|---|---|---|---|---|---|
| n | $\dot{x}_2$ | $T$ | $\delta\lvert\mathbf{x}\rvert$ $\times 10^{-13}$ | $\delta\lvert\dot{\mathbf{x}}\rvert$ $\times 10^{-13}$ | $N_f$ | $\delta\lvert\mathbf{x}\rvert$ $\times 10^{-13}$ | $\delta\lvert\dot{\mathbf{x}}\rvert$ $\times 10^{-13}$ | $N_f$ |
| 1 | 8.0 | 6.76 | 58 | 54 | 36777 | 60 | 56 | 36557 |
| 2 | 8.5 | 8.56 | 67 | 55 | 36920 | 70 | 58 | 36609 |
| 3 | 9.1 | 12.36 | 74 | 54 | 38296 | 78 | 59 | 36882 |
| 4 | 9.9 | 26.57 | 92 | 43 | 44542 | 44 | 61 | 38325 |
| 5 | 10.3 | 51.16 | 96 | 70 | 48880 | 64 | 84 | 41306 |

The results show that the computation volumes decrease in all numerical experiments with correction. The more decreasing takes place in the methods of smaller order. Besides the integration accuracy is preserved. Note that the perturbation accelerations are not recalculated during the regular coordinates correction.

**Acknowledgments.** This research was supported by Russian Foundation for Basic Research, the grant No. 01-02-16918.

## References

Poleshchikov, S. M., & Kholopov, A. A. 1999, Theory of $L$-matrices and regularization of motion equations in Celestial Mechanics (Syktyvkar: SFI) (in Russian)

Poleshchikov, S. M., & Kholopov, A. A. 2000, Solar Syst.Res., 34, 342

Poleshchikov, S. M., & Kholopov, A. A. 2000, Solar Syst.Res., 34, 517

Fehlberg, E. 1970, Computing, 6, 61

Forsythe, G., Malcolm, M., & Moler, C. 1977, Computer methods for mathematical computations (Prentice-Hall, Inc.), 280

*Order and Chaos in Stellar and Planetary Systems*
*ASP Conference Series, Vol. 316, 2004*
*G. Byrd, K. Kholshevnikov, A. Mylläri, I. Nikiforov, and V. Orlov, eds.*

# Distance between Two Arbitrary Unperturbed Orbits

K. V. Kholshevnikov and R. V. Baluyev

*Sobolev Astronomical Institute, St. Petersburg State University, Universitetskij pr. 28, Stary Peterhof, St. Petersburg 198504, Russia*

**Abstract.** The problem of finding critical points of the distance function between two keplerian elliptic orbits (hence finding distance between them in a sense of set theory) is reduced to determination of all real roots of a trigonometric polynomial of degree eight (Kholshevhikov & Vassiliev 1999). A polynomial of smaller degree with such properties does not exist in non-degenerate cases. Here we extend the results to all 9 cases of conic section ordered pairs. Note, that ellipse–hyperbola and hyperbola–ellipse cases are not equivalent as we exclude the variable marking the position on the second curve.

## 1. Introduction

The problem of finding a distance between two arbitrary Keplerian ellipses $E$, $E'$ (in a sense of set theory—the minimal distance between two points lying on $E$, $E'$) emerged four centuries ago together with discovery of Keplerian laws. An optimal solution was found in the paper (Kholshevhikov & Vassiliev 1999), containing also the detailed discussion and all necessary bibliography. The problem is reduced to solving an equation $g(u) = 0$, $g$ being a trigonometric polynomial in one variable of degree 8 sharp, and it cannot be diminished in non-degenerate cases. Here we extend these results to all types of conic sections.

## 2. Elliptic Orbits

Let $E$, $E'$ be two confocal elliptic orbits with Keplerian elements $a$, $e$, $i$, $\Omega$, $\omega$ for $E$ and the same with a stroke for $E'$. In terms of eccentric anomaly $u$ the position vector $\mathbf{r}$ on $E$ is

$$\mathbf{r}/a = \mathbf{P}(\cos u - e) + \mathbf{S}\sin u, \tag{1}$$

where $\mathbf{S} = \sqrt{1-e^2}\mathbf{Q}$ and components of the orthogonal unit vectors $\mathbf{P}, \mathbf{Q}$ *for all types of conic sections* are

$$\begin{aligned}
P_x &= \cos\omega\cos\Omega - \cos i\sin\omega\sin\Omega,\\
P_y &= \cos\omega\sin\Omega + \cos i\sin\omega\cos\Omega,\\
P_z &= \sin i\sin\omega,\\
Q_x &= -\sin\omega\cos\Omega - \cos i\cos\omega\sin\Omega,\\
Q_y &= -\sin\omega\sin\Omega + \cos i\cos\omega\cos\Omega,\\
Q_z &= \sin i\cos\omega.
\end{aligned}$$

Using (1) one deduces for a normalized squared distance function

$$\rho(u, u') = \frac{|\mathbf{r} - \mathbf{r}'|^2}{2aa'}$$

an expression

$$\begin{aligned} \rho \; = \; & \rho_0 + (PP'e' - \alpha e)\cos u + P'Se'\sin u + \\ & (PP'e - \alpha' e')\cos u' + PS'e\sin u' - PP'\cos u \cos u' - \\ & PS'\cos u \sin u' - P'S\sin u \cos u' - SS'\sin u \sin u' + \\ & (\alpha/4)e^2\cos 2u + (\alpha'/4)e'^2\cos 2u'. \end{aligned} \qquad (2)$$

Here

$$\rho_0 = \frac{\alpha + \alpha'}{2} + \frac{\alpha e^2 + \alpha' e'^2}{4} - PP'ee', \qquad (3)$$

$$\alpha = \frac{a}{a'}, \qquad \alpha' = \frac{a'}{a}, \qquad (4)$$

$PP'$, $PS'$, $P'S$, and $SS'$ are scalar products of corresponding vectors. Function $\rho$ receives its minimal and maximal value at one of the critical points satisfying equations

$$A\,\sin u' + B\,\cos u' = C, \qquad M\,\sin u' + N\,\cos u' = K\,\sin u' \cos u'. \qquad (5)$$

Here

$$\begin{aligned} A &= PS'\,\sin u - SS'\,\cos u, \\ B &= PP'\,\sin u - P'S\,\cos u, \\ C &= e'\,B - \alpha e \sin u\,(1 - e\cos u), \\ M &= PP'\,\cos u + P'S\,\sin u + \alpha' e' - PP'\,e, \\ N &= PS'\,e - SS'\,\sin u - PS'\,\cos u, \\ K &= \alpha' e'^2 \end{aligned} \qquad (6)$$

are trigonometric polynomials in $u$ of degree 0, 1 or 2.

The system (5) can be reduced (Kholshevhikov & Vassiliev 1999) to an equation in one variable

$$g(u) = 0, \qquad (7)$$

$g$ being a trigonometric polynomial of degree 8

$$\begin{aligned} g(u) \; = \; & K^2(A^2 - C^2)(B^2 - C^2) + 2\,KC\left[NA(A^2 - C^2) + MB(B^2 - C^2)\right] - \\ & (A^2 + B^2)\left[N^2(A^2 - C^2) + M^2(B^2 - C^2) - 2\,NMAB\right]. \end{aligned} \qquad (8)$$

After solving (7) we obtain from the first of equations (5)

$$\cos u' = \frac{BC + mA\sqrt{D}}{A^2 + B^2}, \qquad \sin u' = \frac{AC - mB\sqrt{D}}{A^2 + B^2} \qquad (9)$$

with

$$D = A^2 + B^2 - C^2, \qquad m = \pm 1. \qquad (10)$$

We extend now the results above to all types of conic sections. Marking elliptic, hyperbolic and parabolic cases in alphabetic order by symbols 1, 2, 3 we obtain 9 cases $\mathcal{K}_{jk}$ and correspondingly 9 functions $g_{jk}(u)$. For example, $\mathcal{K}_{23}$ means that $E$ is a hyperbola and $E'$ is a parabola. Evidently $g_{23} \neq g_{32}$ and the function (8) coincides with $g_{11}$.

## 3. Hyperbolic Cases

1. Let begin with hyperbolic-elliptic case $\mathcal{K}_{21}$. Eccentricity of $E$ is greater than 1, $a$ is negative, the eccentric anomaly and $\sqrt{1-e^2}$ have imaginary values. Instead of (1) we have

$$\mathbf{r}/a = \mathbf{P}(\cosh u - e) - \mathbf{S}\sinh u, \tag{11}$$

$u \in (-\infty, \infty)$ being a hyperbolic analogue of eccentric anomaly, $\mathbf{S} = \sqrt{e^2-1}\mathbf{Q}$.

Any expression in old (real for an ellipse) quantities $\mathbf{S}, u$ can be easily rewritten in new ones (real for a hyperbola) using replacement

$$\mathbf{S} \longmapsto i\,\mathbf{S}, \qquad u \longmapsto i\,u, \tag{12}$$

$i$ being the imaginary unit and never being mixed with the inclination. In particular, this replacement converts (1) to (11) and converts (2) to

$$\begin{aligned}\rho \;=\;& \rho_0 + (PP'e' - \alpha e)\cosh u - P'Se'\sinh u + \\ & (PP'e - \alpha' e')\cos u' + PS'e\sin u' - PP'\cosh u\cos u' - \\ & PS'\cosh u\sin u' + P'S\sinh u\cos u' + SS'\sinh u\sin u' + \\ & (\alpha/4)e^2\cosh 2u + (\alpha'/4)e'^2\cos 2u'.\end{aligned} \tag{13}$$

Relations (5) and function (8) are homogeneous with respect to $A$, $B$, $C$ and $M$, $N$, $K$. Hence, after the substitution (12) we may multiply $A$, $B$, $C$ or $M$, $N$, $K$ by any number, not equal to zero. In particular, we may multiply $A$, $B$, $C$ by $i$ and change (6) by

$$\begin{aligned}A &= PS'\sinh u - SS'\cosh u, \\ B &= PP'\sinh u - P'S\cosh u, \\ C &= e'B - \alpha e\sinh u\,(1 - e\sinh u), \\ M &= PP'\cosh u - P'S\sinh u + \alpha' e' - PP'e, \\ N &= PS'e + SS'\sinh u - PS'\cosh u, \\ K &= \alpha' e'^2.\end{aligned} \tag{14}$$

Function $g_{21}$ expression in variables (14) coincides with the expression (8) for $g_{11}$, as well as (9,10) for $\cos u'$, $\sin u'$, $D$ hold true.

2. In the case $\mathcal{K}_{12}$ we need the replacement

$$A \longmapsto i\,A, \qquad N \longmapsto i\,N.$$

New functions $A, \dots, K$ retain their meaning (6), meanwhile the expressions (2), (5), (8) change

$$\begin{aligned}\rho &= \rho_0 + (PP'e' - \alpha e)\cos u + P'Se'\sin u + \\ &(PP'e - \alpha'e')\cosh u' - PS'e\sinh u' - PP'\cos u\cosh u' + \\ &PS'\cos u\sinh u' - P'S\sin u\cosh u' + SS'\sin u\sinh u' + \\ &(\alpha/4)e^2\cos 2u + (\alpha'/4)e'^2\cosh 2u',\end{aligned} \tag{15}$$

$$-A\sinh u' + B\cosh u' = C, \quad M\sinh u' + N\cosh u' = K\sinh u'\cosh u', \tag{16}$$

$$\begin{aligned}g_{12}(u) &= -K^2(A^2+C^2)(B^2-C^2) + \\ &2KC\left[NA(A^2+C^2) + MB(B^2-C^2)\right] + \\ &(A^2-B^2)\left[N^2(A^2+C^2) + M^2(B^2-C^2) + 2NMAB\right].\end{aligned} \tag{17}$$

Instead of (9), (10) we have

$$\cosh u' = \frac{BC + mA\sqrt{D}}{B^2 - A^2}, \qquad \sinh u' = \frac{AC + mB\sqrt{D}}{B^2 - A^2} \tag{18}$$

with

$$D = A^2 + C^2 - B^2, \qquad m = \pm 1. \tag{19}$$

3. In the case $\mathcal{K}_{22}$ we ought to replace

$$A \longmapsto -A, \qquad B \longmapsto i\,B, \qquad C \longmapsto i\,C, \qquad N \longmapsto i\,N,$$

and for new quantities $A, \dots, K$ the expressions (14) hold true as in the case $\mathcal{K}_{21}$. On the contrary, the expressions (16)–(19) hold true as in the case $\mathcal{K}_{12}$. Finally,

$$\begin{aligned}\rho &= \rho_0 + (PP'e' - \alpha e)\cosh u - P'Se'\sinh u + \\ &(PP'e - \alpha'e')\cosh u' - PS'e\sinh u' - PP'\cosh u\cosh u' + \\ &PS'\cosh u\sinh u' + P'S\sinh u\cosh u' - SS'\sinh u\sinh u' + \\ &(\alpha/4)e^2\cosh 2u + (\alpha'/4)e'^2\cosh 2u'.\end{aligned} \tag{20}$$

## 4. Parabolic Cases

The best way to treat a parabolic orbit is to present it as a limiting case ($\varepsilon \to 0$) of an ellipse

$$a = \frac{q}{2\varepsilon^2}, \qquad e = 1 - 2\varepsilon^2, \qquad u = 2\varepsilon\sigma, \tag{21}$$

$q$, $\sigma$ being the pericentric distance and the tangent of the half of the true anomaly. The substitution (21) converts (1) to

$$\mathbf{r}/q = \mathbf{P}\left[(1-\sigma^2) + \mathcal{O}(\varepsilon^2)\right] + \mathbf{Q}\left[2\sigma + \mathcal{O}(\varepsilon^2)\right]. \tag{22}$$

1. In the cases $\mathcal{K}_{31}$ and $\mathcal{K}_{32}$ we may deal with

$$\rho_{31} = \rho_{32} = \frac{|\mathbf{r} - \mathbf{r}'|^2}{qa'} = \frac{\rho}{\varepsilon^2},$$

$$\beta = \frac{q}{a'} = 2\varepsilon^2\alpha, \qquad \beta' = \frac{a'}{q} = \frac{\alpha'}{2\varepsilon^2}, \qquad \mathbf{S} = 2\varepsilon\mathbf{Q}\left[1 + \mathcal{O}(\varepsilon^2)\right].$$

Taking (21) into account we have after passing to the limit ($\varepsilon \to 0$)

$$\begin{aligned}\rho_{31} \;=\;& \beta\left(1+\sigma^2\right)^2 + \beta'\left(1 + \frac{1}{2}e'^2\right) + 2PP'e'\left(1-\sigma^2\right) + \\ & 4\sigma e'P'Q - 2\left[(1-\sigma^2)PP' + 2\sigma P'Q + \beta'e'\right]\cos u' - \\ & 2\left[(1-\sigma^2)PS' + 2\sigma QS'\right]\sin u' + \frac{1}{2}\beta' e'^2\cos 2u'. \end{aligned} \tag{23}$$

The system (5) holds true if we devide $A$, $B$, $C$ by $2\varepsilon$ and $M$, $N$, $K$ by $2\varepsilon^2$ and then pass to the limit. So we have for $\mathcal{K}_{31}$

$$\begin{aligned} A &= PS'\sigma - QS', \\ B &= PP'\sigma - P'Q, \\ C &= e'B - \beta\sigma(1+\sigma^2), \\ M &= PP'(1-\sigma^2) + 2\sigma P'Q + \beta'e', \\ N &= -PS'(1-\sigma^2) - 2\sigma QS', \\ K &= \beta'e'^2. \end{aligned} \tag{24}$$

Due to the homogeneity of $g$ we may use the expression (8) for $g_{31}$, as well as (9), (10) for $u'$.

2. The case $\mathcal{K}_{32}$ may be deduced from $\mathcal{K}_{12}$ by the similar procedure. So

$$\begin{aligned}\rho_{32} \;=\;& \beta\left(1+\sigma^2\right)^2 + \beta'\left(1 + \frac{1}{2}e'^2\right) + 2PP'e'\left(1-\sigma^2\right) + \\ & 4\sigma e'P'Q - 2\left[(1-\sigma^2)PP' + 2\sigma P'Q + \beta'e'\right]\cosh u' + \\ & 2\left[(1-\sigma^2)PS' + 2\sigma QS'\right]\sinh u' + \frac{1}{2}\beta' e'^2\cosh 2u'. \end{aligned} \tag{25}$$

Relations (24) are valid and we may use the expression (17) for $g_{32}$, as well as (18), (19) for $u'$.

3. Cases $\mathcal{K}_{13}$ and $\mathcal{K}_{23}$ are more complicated and we omit them.

4. In the case $\mathcal{K}_{33}$ we introduce

$$\rho_{33} = \frac{|\mathbf{r} - \mathbf{r}'|^2}{qq'} = \frac{\rho_{31}}{2\varepsilon^2},$$

$$\gamma = \frac{q}{q'} = 2\frac{\beta}{2\varepsilon^2}, \qquad \gamma' = \frac{q'}{q} = 2\varepsilon^2\beta', \qquad \mathbf{S}' = 2\varepsilon\mathbf{Q}\left[1 + \mathcal{O}(\varepsilon^2)\right].$$

After passing to the limit

$$\begin{aligned} \rho_{33} &= \gamma\left(1+\sigma^2\right)^2+\gamma'\left(1+\sigma'^2\right)^2-2PP'\left(1-\sigma^2\right)\left(1-\sigma'^2\right)- \\ & 4\sigma\left(1-\sigma'^2\right)P'Q-4\left(1-\sigma^2\right)^2\sigma'PQ'-8\sigma\sigma'QQ'. \end{aligned} \tag{26}$$

One uses (24) for calculations of $A,\ldots,K$. Relations

$$B\sigma'^2-2A\sigma'+C=0, \qquad K\sigma'^3+M\sigma'+N=0,$$

and

$$\begin{aligned} g_{33} &= K^2C^3+2K\left[NA(4A^2-3BC)+MC(2A^2-BC)\right]+ \\ & B^2\left(N^2B+M^2C+2MNA\right) \end{aligned}$$

are taken to find $\sigma$, $\sigma'$.

## 5. Conclusions

We have proposed 9 functions $g_{jk}$ solving the problem of finding distance between $E$ and $E'$ in all possible combinations of conic sections. In the non-parabolic cases ($j \le 2$, $k \le 2$) functions $g_{jk}$ are trigonometric or hyperbolic polynomials of degree 8. Remember that the degree of a corresponding algebraic polynomial must be multiplied by a factor 2. In the cases when only one of the orbits is parabolic functions $g_{31}$, $g_{32}$, $g_{13}$, $g_{23}$ are trigonometric or hyperbolic polynomials of degree 6 or algebraic polynomial of degree 12. Hence, all of them can be reduced to an algebraic polynomial of degree 12. In the case when the both orbits are parabolic, function $g_{33}$ represents an algebraic polynomial of degree 9 only.

For non-diagonal elements of $\mathcal{K}_{jk}$ it is useful to choose the simplest function between $g_{jk}$ and $g_{kj}$. Supposing a trigonometric equation being simpler than a hyperbolic one, we recommend the function $g_{12}$ for the case ellipse–hyperbola. Supposing an algebraic equation being simpler than a trigonometric or hyperbolic one, we recommend the functions $g_{31}$ and $g_{32}$ for the cases parabola–ellipse and parabola–hyperbola.

**Acknowledgments.** This work is supported by Russian Foundation for Basic Research (Grant 02-02-17516), Program *Universities of Russia* (Grant UR.02.01.027), and Leading Scientific School (Grant NS-1078.2003.2).

## References

Kholshevhikov, K. V., & Vassiliev, N. N. 1999, CeMDA, 75, 75

*Order and Chaos in Stellar and Planetary Systems*
*ASP Conference Series, Vol. 316, 2004*
*G. Byrd, K. Kholshevnikov, A. Mylläri, I. Nikiforov, and V. Orlov, eds.*

# Catalogue of Almost Coinciding Orbits in the Solar System

K. V. Kholshevnikov and I. S. Bessmertny

*Sobolev Astronomical Institute, St. Petersburg State University, Universitetskij pr. 28, Staryj Peterhof, St. Petersburg 198504, Russia*

**Abstract.** In a set of the Solar System body orbits there are many pairs possessing close ones. Here we put a problem of finding all pairs of asteroids possessing similar orbits.

## Introduction

In a set of the Solar System body orbits there are many pairs possessing close ones. Such are the orbits of meteoroids—members of a meteor stream, orbits of meteoroids and a parent comet, orbits of several pairs of asteroids. Sometimes this proximity is accidental, but sometimes it marks a physical connection between bodies and even the common origin.

Here we pose a problem of finding all pairs of asteroids possessing similar orbits. As a criterion of the proximity we choose the distance $\rho(E_1, E_2)$ between the Keplerian orbits $E_1$, $E_2$ (Kholshevnikov 2001). It represents a mean-squared distance in the 5-dimensional space $\mathcal{E}$ of elliptic orbits, each of them is regarded as a point in $\mathcal{E}$. In particular, $\rho(E_1, E_2) = 0$ if and only if $E_1 = E_2$. For almost coinciding orbits the distance is close to zero.

Let us give a definition:

$$\rho^2(E_1, E_2) = \min_{v \in [0, 2\pi]} \frac{1}{2\pi} \int_0^{2\pi} d^2[Q_1(u), Q_2(u+v)]\, du\,.$$

Here $Q_{k+1}(u + kv)$ is a point on $E_{k+1}$ with eccentric anomaly $u + kv$, $k = 0, 1$, and $d$ is the Euclidian distance between $Q_1$, $Q_2$.

All axioms of a metric space are fulfilled (Kholshevnikov 2001).

## Results

We calculate the distance above between all pairs of 58000 asteroidal orbits presented in the catalogue by Montenbruck & Pfleger (2000), see also updatable electronic version by Bowell (2003), and select the pairs having $\rho < \varepsilon$. Denote $N(\varepsilon)$ the number of selected pairs. We find $N(0.05) = 23863$, $N(0.025) = 1089$, $N(0.01) = 24$, $N(0.004) = 1$, distances being measured in AU. The electronic version of the catalogue containing $N(0.05)$ pairs is under preparation. We intend to place it on the Astronomical Institute site (Bessmertny 2004). In the Table 1 below we bring the catalogue of $N(0.008) = 8$ pairs: first two columns contain the number or the preliminary designation of asteroids, the last gives the distance between their orbits.

Table 1.

| | | |
|---|---|---|
| 8898 | 1999$RH$118 | 0.00382 |
| 10437 | 1998$UW$12 | 0.00509 |
| 1270 | 1999$UZ$6 | 0.00582 |
| 4127 | 1997$GL$16 | 0.00596 |
| 1995$TG$2 | 1998$OW$3 | 0.00606 |
| 1988$XL$2 | 1999$TX$27 | 0.00672 |
| 1997$WC$10 | 1999$XL$34 | 0.00757 |
| 5428 | 7544 | 0.00783 |

**Acknowledgments.** This work is supported by Russian Foundation for Basic Research (Grant 02-02-17516), Program *Universities of Russia* (Grant UR.02.01.027), and Leading Scientific School (Grant NSh-1078.2003.2).

## References

Kholshevnikov, K. V. 2001, in Proc. 30th Int. Winter Astr. School (Ekaterinburg: Ural Univ. Press), 145 (in Russian)

Montenbruck, O., & Pfleger, T. 2000, CD-application to "Astronomy on the PC (+CD)" (Berlin–Heidelberg: Springer)

Bowell, T. 2003, ftp://ftp.lowell.edu/pub/elgb/astorb.dat

Bessmertny, I. S. 2004, http://www.astro.spbu.ru

*Order and Chaos in Stellar and Planetary Systems*
*ASP Conference Series, Vol. 316, 2004*
*G. Byrd, K. Kholshevnikov, A. Mylläri, I. Nikiforov, and V. Orlov, eds.*

# The Orbital Evolution of Near-Earth Asteroids in the 3:1 Mean Motion Resonance

E. I. Timoshkova

*Central Astronomical Observatory at Pulkovo, Pulkovskoe sh. 65/1, St. Petersburg 196140, Russia; e-mail: elenatim@gao.spb.ru*

**Abstract.** The paper examines the orbital evolution of a group of 12 near-Earth asteroids with the mean motions being in resonance 3:1 with Jupiter, by the numerical integration over 100 000 years. The computation of six osculating elements has been done by forward integration taking into account the perturbations from all 9 major planets. The behaviour of the osculating semi-major axis, eccentricity and inclination has been analysed for all time of integration.

## 1. Introduction

The problem of long timescale motion of the near-Earth asteroids (NEAs) remains one of the central themes of the recent research. As numerical experiments have demonstrated, the orbital evolution of the NEAs is very complex and varied. At the present moment we have a general understanding of the major dynamic mechanisms that result in the emergence of chaotic movements or, alternatively, contribute to relative stability of asteroid orbits. Among these mechanisms an important role belongs to close encounters with major planets and also to various resonance phenomena. These results have been examined in a number of general studies (see Michel et al. 1997; Michel, Froeschlé & Farinella 1996; Froeschlé, Michel, & Froeschlé 1999). At the same time it has become clear that these asteroids have no typical dynamic evolution or typical lifespan. Many problems related to the issues of origin and evolution of the NEAs require further clarification.

The aim of this paper is to study the orbital evolution of a selected group of NEAs on the basis of their numerical models of movement. The choice of this group of 12 asteroids has been determined mainly by the fact that at the present moment their orbits are localised in a narrow region of resonance for mean motions 3:1 with Jupiter. Remarkably, among a numerous population of NEAs at the beginning of 2000 year there were only 12 numbered asteroids with mean motions comparable to the mean motion of Jupiter as 3:1. It should be noted that 3:1 resonance is considered to be responsible for an increase of an orbit's eccentricity almost up to 1 that is often result an asteroids falls on the Sun. Thus we could ask whether all these asteroids or at least some of them could have common origins and for how long they could remain in the indicated region of osculating orbital elements? Therefore comparative analysis of orbital evolution of the selected group of NEAs could be of a great interest even for a relatively small time interval.

## 2. Constructing Numerical Models of Motion

The numerical model of motion of each asteroid has been constructed by integrating the system of differential equations of motion in rectangular coordinates when perturbations from nine major planets at the time span of 100 thousand years are taken into account. The integration has been done by Bulirsch–Stoer method. For initial integration data we have taken the values of osculating elements for 12 asteroids and 9 major planets for the standard epoch JD 2451800.5 = 2000 September 13.0 ET. The elements are given with respect to the ecliptic and equinox J2000.0 and have been taken from The Ephemerides of Minor Planets for 2000 (Batrakov et al. 1999). The initial values of osculating elements of 12 asteroids under investigation are given in Table 1.

Table 1.

| No. | $a$ | $e$ | $i$ | $\omega$ | $\Omega$ | $M$ | $n$ |
|---|---|---|---|---|---|---|---|
| 887 | 2.484078 | 0.563455 | 9.3097 | 350.0502 | 110.7085 | 278.7645 | 0.251742 |
| 1915 | 2.541913 | 0.571781 | 20.4155 | 347.8372 | 163.0378 | 305.2910 | 0.241913 |
| 2608 | 2.505498 | 0.576801 | 15.0063 | 35.3934 | 168.6797 | 256.8152 | 0.248521 |
| 3360 | 2.466146 | 0.742692 | 21.7196 | 60.6695 | 245.4406 | 332.7035 | 0.254493 |
| 4179 | 2.510299 | 0.634348 | 0.4696 | 274.6895 | 128.3761 | 347.1308 | 0.251029 |
| 6318 | 2.510593 | 0.465051 | 25.9572 | 12.0964 | 71.2777 | 164.5389 | 0.247764 |
| 6322 | 2.516934 | 0.473160 | 28.1947 | 296.7812 | 174.3483 | 150.2945 | 0.246829 |
| 6489 | 2.514885 | 0.598853 | 2.2914 | 65.1190 | 212.3114 | 110.6051 | 0.251418 |
| 6491 | 2.511053 | 0.587434 | 5.5349 | 317.9937 | 306.0179 | 116.8048 | 0.247696 |
| 7092 | 2.525323 | 0.701755 | 17.8260 | 91.9344 | 59.6090 | 39.0620 | 0.245600 |
| 8201 | 2.531837 | 0.708992 | 9.5894 | 24.9864 | 164.2967 | 211.3533 | 0.244652 |
| 8709 | 2.536902 | 0.483276 | 3.4974 | 153.3069 | 119.5074 | 189.8653 | 0.243920 |

The asteroid number is given in the first column of the Table. For the osculating elements we use their standard designations. The values of angular variables are always given in degrees and the values of mean motions in degrees/days. According to the Table of the initial data, the selected group of asteroids gets into a narrow region of the change in the mean motions: $0.24 < n < 0.26$, and the value of $n = 0.249$ per day corresponds to the exact commensurability 3:1 with Jupiter. We should also note that among 12 asteroids one half have the orbits with an angle to the plane of ecliptic more than 15 and three of them have very elongated orbits (eccentricity $e > 0.7$). Applying numerical integration we have reconstructed the correlations of the major evolutionary parameters of an orbit—the semi-major axis $a$, the eccentricity $e$, and the inclination $i$—from the period of time at the interval of 100 000 years for all 12 asteroids. As it could be expected, the behaviour of all three parameters is quite characteristic for the 3:1 type resonance orbits. Here we observe chaotic changes of the semi-major axis $a$ for almost all asteroids under consideration, large amplitudes in the variations of eccentricities $e$ and inclination $i$. The Table 2 gives maximum and minimum values of osculating elements $a$, $e$ and $i$ for the whole interval of integration.

As you could see from this Table, the semi-major axis $a$ of all asteroids experience big variations and in most cases it passes many times through the resonance value $a$. These cases are marked with the note "Remains in resonance" in the "Comments" column of the Table 2. It does not mean, however, that an asteroid remains in the resonance zone for the whole interval of integration: it is possible that the asteroid leaves this zone at some relatively small time intervals (about several thousands years). An example of this type of behaviour of the semi-major axis can be seen for asteroid 6489 Golevka. According to the general theory of NEA motion (Milani et al. 1989) big variations of $a$ are determined mainly by the resonance perturbation from Jupiter. Asteroid 6322 1991 CQ is a slight exception. The semi-major axis of its orbit has considerably smaller amplitude of changes and at all interval of integration it does not exceed the value corresponding to the exact commensurability of the mean motion $n = 0.249$ degrees/days.

Table 2.

| Asteroid | | $a$ | | $e$ | | $i^o$ | | Comments |
|---|---|---|---|---|---|---|---|---|
| | | max | min | max | min | max | min | |
| 887 | Alinda | 2.60 | 2.44 | 0.92 | 0.54 | 19.8 | 3.7 | Remains in 3:1 |
| 1915 | Quetzalcoatl | 2.54 | 2.46 | 0.95 | 0.52 | 50.3 | 8.2 | Remains in 3:1 |
| 2608 | Seneca | 2.68 | 2.44 | 0.79 | 0.54 | 21.1 | 5.1 | Exits |
| 3360 | 1981 VA | 2.55 | 2.40 | 0.89 | 0.57 | 54.2 | 18.4 | e, i synchronized |
| 4179 | Toutatis | 2.58 | 2.42 | 0.95 | 0.51 | 18.5 | 0.5 | Remains in 3:1 |
| 6318 | Cronkite | 2.55 | 2.45 | 0.93 | 0.38 | 29.1 | 4.2 | Remains in 3:1 |
| 6322 | 1991 CQ | 2.54 | 2.51 | 0.55 | 0.18 | 38.4 | 24.1 | Remains in 3:1 |
| 6489 | Golevka | 2.57 | 2.43 | 0.97 | 0.48 | 15.5 | 0.5 | Remains in 3:1 |
| 6491 | 1991 OA | 2.56 | 2.44 | 0.74 | 0.33 | 12.5 | 3.8 | Remains in 3:1 |
| 7092 | Cadmus | 2.62 | 2.45 | 0.74 | 0.46 | 39.0 | 15.1 | Remains in 3:1 |
| 8201 | 1994 AH | 2.58 | 2.42 | 0.99 | 0.71 | 55.1 | 0.2 | e=1 for t=28000 ys |
| 8709 | Kadlu | 2.58 | 2.42 | 0.82 | 0.41 | 5.5 | 0.3 | Remains in 3:1 |

Virtually all asteroids under consideration are also characterised by a broad scope of changes of their eccentricity and inclination, while one half of our cases has the maximum value $e > 0.9$. For asteroid 8201 1994 AH the increase of eccentricity was especially fast so that for $t = 28000$ years $e$ becomes equal to 1. For three asteroids (3360, 6322 and 7092) we observe a quite regular change of eccentricity and inclination for all period of integration. Moreover, these changes are synchronised in such a way that when $e$ reaches its maximum, $i$ takes its minimum value, and vice versa. As you can see from Table 1, the orbits of these asteroids at the present moment have relatively big inclinations respect to the ecliptic and their values remain bigger than 15 degrees for all the interval of integration. It is an impact of the so-called Kozai resonance, which explains the synchronised character of changes in eccentricities and inclination of these asteroids. We should also note similarities in the evolutions of orbit eccentricities of the asteroids 887 Alinda and 1915 Quetzalcoatl for the interval of time from the initial moment of integration up to approximately 60 000 years. The situation is quite similar for variations of $e$ and $i$ for the asteroids 4179 Toutatis and 6489 Golevka, which orbits at the present moment are situated almost in the plane of ecliptic.

## 3. Conclusions

Our comparative analysis has demonstrated that the behaviour of the main evolutionary parameters of the group of asteroids under consideration that has been described above is quite characteristic for the NEAs with commensurability of their mean motions to Jupiter as 3:1. Their orbital evolution during the interval of time equal to 100 000 years seems to belong to the dynamic type of the Alinda class—following the classification of the PROJECT SPACEGUARD (Milani et al. 1989).

This classification is based on a study of the dynamic behaviour of the Earth-crossing asteroids' orbital parameters. The study was conducted mainly by the methods of numerical integrations over a time scale ranging from 10 000 to some 100 000 years. Asteroids are defined as belonging to the Alinda class when resonance perturbations caused by Jupiter are dominant; when these resonance perturbations from Jupiter cause rapid and profound changes in the asteroid's semimajor axis $a$; when we observe a very rapid secular evolution of the eccentricity $e$.

A visual inspection of the plots for the orbital elements would suffice for placing asterods into the Alinda class.

It should be stressed, however, that for a detailed research of the orbital evolution we would also need to analyse the behaviour of the angular variables: the argument perihelion $\omega$ and the longitude of ascending node $\Omega$, and to compare the character of their evolution with the variations of eccentricities and the inclinations of the orbits. Moreover, considering particular sensitivity of the resonance orbits towards small changes of the initial data it is essential to study the numerical models for a certain set of the orbits, which initially have very small differences. We plan to conduct such research in future.

## References

Batrakov, Yu. V. et al. 1999, Ephemerides of Minor Planets for 2000 (St. Petersburg: Institute of Applied Astronomy RAS)

Michel, P. et al. 1997, CeMDA, 69, 133

Michel, P., Froeschlé, Ch., & Farinella, P. 1996, Earth, Moon and Planets, 72, 151

Froeschlé, Ch., Michel, P., & Froeschlé, C. 1999, in Evolution and Source Regions of Asteroids and Comets, Proc. IAU Coll. 173 , ed. J. Svoren, E. M. Pittich & H. Rickman, 87

Milani, A. et al. 1989, Icarus, 78, 212

*Order and Chaos in Stellar and Planetary Systems*
*ASP Conference Series, Vol. 316, 2004*
*G. Byrd, K. Kholshevnikov, A. Mylläri, I. Nikiforov, and V. Orlov, eds.*

# The Orbital Distributions of Particles Resulting from Multiple Close Encounters with Planet in a Steady-State Approximation

Valeri Dikarev[1,2] and Eberhard Grün[1,3]

[1] *MPI für Kernphysik, Postfach 103980, 69117 Heidelberg, Germany*

[2] *Sobolev Astronomical Institute, St. Petersburg State University, Russia*

[3] *HIGP, University of Hawaii, Honolulu, USA*

**Abstract.** Dynamics of meteoroids and dust grains of cometary origin are investigated both analytically and numerically. The gravity of the Sun and one planet is taken into account. A simple analytical formula for the orbital distribution of particles in the regime of close encounters with Jupiter is suggested supposing ergodicity on a limited region of phase space.

## 1. Introduction

Due to their dusty tails Jupiter family comets (JFCs) are an obvious source of meteoroids in the interplanetary space. Studies of the origin of comets in the Solar system have shown that most of these bodies come from the Edgeworth–Kuiper belt via cascade encounters with the giant planets (Levison & Duncan 1997). Some of the short-period comets could have come from the Oort cloud as well (see, e.g., Valtonen et al. 1998). Because of a number of loss mechanisms, such as ejection from the Solar system by planets and fading out, very few comets are displayed at a time. Moreover, catalogues of comets are prone to observational biases since the cometary nuclei are revealed by gas and dust shed at higher intensity preferentially at the low perihelion distances. Imperfect removal of these biases and low-number statistics have impact on quality of dust source distribution based on the catalogues, and thereby on quality of the meteoroid models derived from the source.

In this paper, we begin the exploration of the orbital distributions of dust from Jupiter-family and other comets on a large scale. We simulate the motion of test particles emitted by every possible source in the region of encounters with Jupiter, rather than by a limited number of pre-selected comets—the choice made in many previous works (Liou et al. 1995; Cremonese et al. 1997; Landgraf et al. 2002). Assuming the ergodic hypothesis holds true in the region of intensive encounters with planet, we obtain a simple analytical expression for the orbital distributions of test particles that resembles the results of our numerical simulations closely.

## 2. The Orbital Distributions of Scattered Particles

Consider the system of the Sun, planet and dust particle which constitute the three-dimensional restricted circular three-body problem, assuming that the planet moves along a circular orbit about the Sun and the particle does not affect the motion of the two major bodies. The equations of the particle motion can be found in many textbooks on Celestial Mechanics, e.g., in Danby (1992). The problem admits one constant of the motion—the Jacobi integral $C$ which at a separation from the planet long enough for its gravity to vanish is closely represented by the Tisserand quantity

$$T = 1/a + 2\sqrt{a(1-e^2)}\cos i, \tag{1}$$

where $a$ is the semimajor axis measured in the orbital radius of planet, $e$ and $i$ are the eccentricity and inclination of the osculating heliocentric orbit of the particle.

The equations of motion in the three-body problem obey the Liouville theorem. Under the assumption of the ergodic hypothesis, the phase density should therefore be a function of the integral of motion, the Jacobi constant $C$

$$n(x, y, z, \dot{x}, \dot{y}, \dot{z}) = n(C) \approx n(T). \tag{2}$$

It is quite clear, however, that the ergodic hypothesis cannot be true everywhere in the phase space. Trajectories confined to the regions very close to the Sun or very far from the Sun-planet pair, where the planetary influence is negligibly weak, are well-described by the bound Keplerian orbits, the solutions of the two-body problem for the Sun and particle, stable forever or on the time scales well beyond the dust particle lifetimes (e.g., against collisions with other particles). The confinement of trajectories to limited regions of phase space, or their boundness like in the case of the pure two-body problem, contradicts to the ergodic hypothesis. There is a large subspace of the phase space, however, where the intensity of dynamics can be sufficient to enable the ergodic hypothesis on this subspace. This is the region of close encounters with planet.

Indeed, the equations of the motion determine continuous time derivatives and therefore unique solution everywhere excluding the locations of the Sun and planet. At the latter locations, in contrast, the gravitational potential approaches infinity, and multiple solutions are admitted in the singularities of Sun and planet. Even a close encounter with planet can modify the heliocentric orbit of a particle remarkably, and the high gradient of the force near the planet makes the outcome of such a flyby highly sensitive to the initial trajectory. Thus a small and inexpensive adjustment of interplanetary spacecraft trajectory before a flyby allows to switch between distant post-encounter orbits—the idea of "gravity-assist" technique. Uncontrolled encounters with planet just scatter the orbits of natural bodies, helping visit vastly distant points of the subspace in a short time.

Let $I_S$ be the indicator function of the subspace where the close encounters are dominant. The particles involved in this regime of dynamics will then be distributed in accord with (2) multiplied by $I_S$. Transformation to the elements of the heliocentric orbit and integration over the argument and longitude of perihelion and the mean anomaly gives the distribution of scattered particles in semimajor axis, eccentricity and inclination

$$g(a,e,i) = n_6(T)\, I(a,e,i)\, \frac{\sqrt{a}e\sin i}{2}, \tag{3}$$

where $I(a,e,i)$ is the integral of the indicator function $I_S$ over the argument and longitude of perihelion and the mean anomaly.

The indicator function $I_S$ is not easy to construct. Only numerical solution of the equations of motion can help constrain the region of motion of the scattered particles in the absence of independent integrals other than the Jacobi constant since the latter admits infinite region of motion when $C < 3$.

We have performed the simulations and found that, in fact, a very simple $I_S$ approximates the numerical results very well, in a certain sense. The indicator function equal one on the heliocentric orbits that cross the torus along the planet's orbit of the section radius 0.1 planetary orbit radii, and zero elsewhere, turns out to be extraordinarily good core of the distribution function (3) as comparison with the numerical data shows. The idea behind this definition is simple, it is in this torus where the heliocentric orbits of the dust particles can encounter Jupiter and jump into another orbit, also passing through the torus and preserving the Jacobi constant $C$.

The numerical simulations were organized as follows. For the sake of convenience, the Jupiter encounter speed $U = \sqrt{3-T}$ will be used to describe the experiments hereafter. In each experiment, a thousand of massless particles was injected into the system of Sun and Jupiter-on-a-circular-orbit for every value of the encounter speed from 0.2 to 2.0 by step 0.2. The encounter velocities (see, e.g., Carusi et al. 1990) were chosen randomly distributed over the sphere of $U =$ const, the orbital angles (the argument and longitude of perihelion, and the mean anomaly) were also uniformly distributed random numbers.

The equations of the motion were solved by the MERCURY software package (Chambers 1999), using the energy-conserving Bulirsch–Stör integration method. The time of integration was 200,000 Jovian years, the output step 20 Jovian years. All the intermediate positions were stored to accumulate statistics of the orbital elements of the simulated particles. The results of numerical simulations are quite "mature" since most of the particles that had not been sorted out finished their orbital evolution—they were ejected from the Solar system. The comparison of predictions of the approximate analytical formula and numerically obtained statistics are shown in Figures 1 and 2.

The orbits with semimajor axis spending too much time (set to 20,000 Jovian years, or one tenth of the integration period) with non-jumping (staying within any 0.5 AU-wide range) semimajor axis below 10 AU were sorted out, however. The constancy of semimajor axis during that long period is very improbable for a particle with the orbital angles uncorrelated with Jovian anomaly. The maximum Öpik time of collision with a scattering sphere around Jupiter of the radius 0.52 AU (our torus section radius) is below 10,000 Jovian years even for the longest semimajor axis in the range of interest, 10 AU. A correlation of the angles, on the other hand, implies a resonance.

Each resonance is yet another breaker of the ergodicity assumption, since even bound periodic orbits exist there. Thus a separate theory of the particle

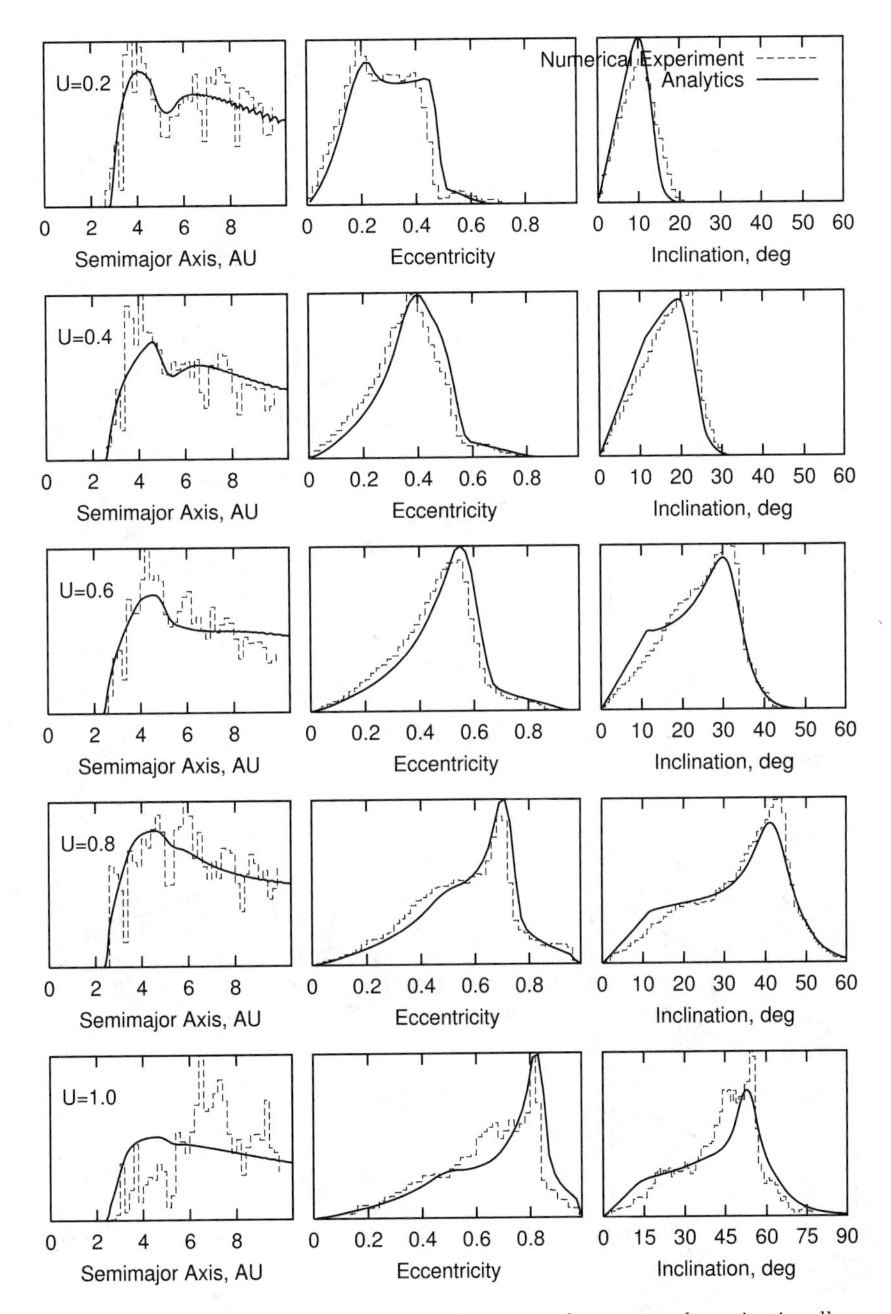


Figure 1. The orbital distributions of the particles scattered gravitationally by a Jupiter-sized planet on a circular orbit 5.2 AU from the Sun.

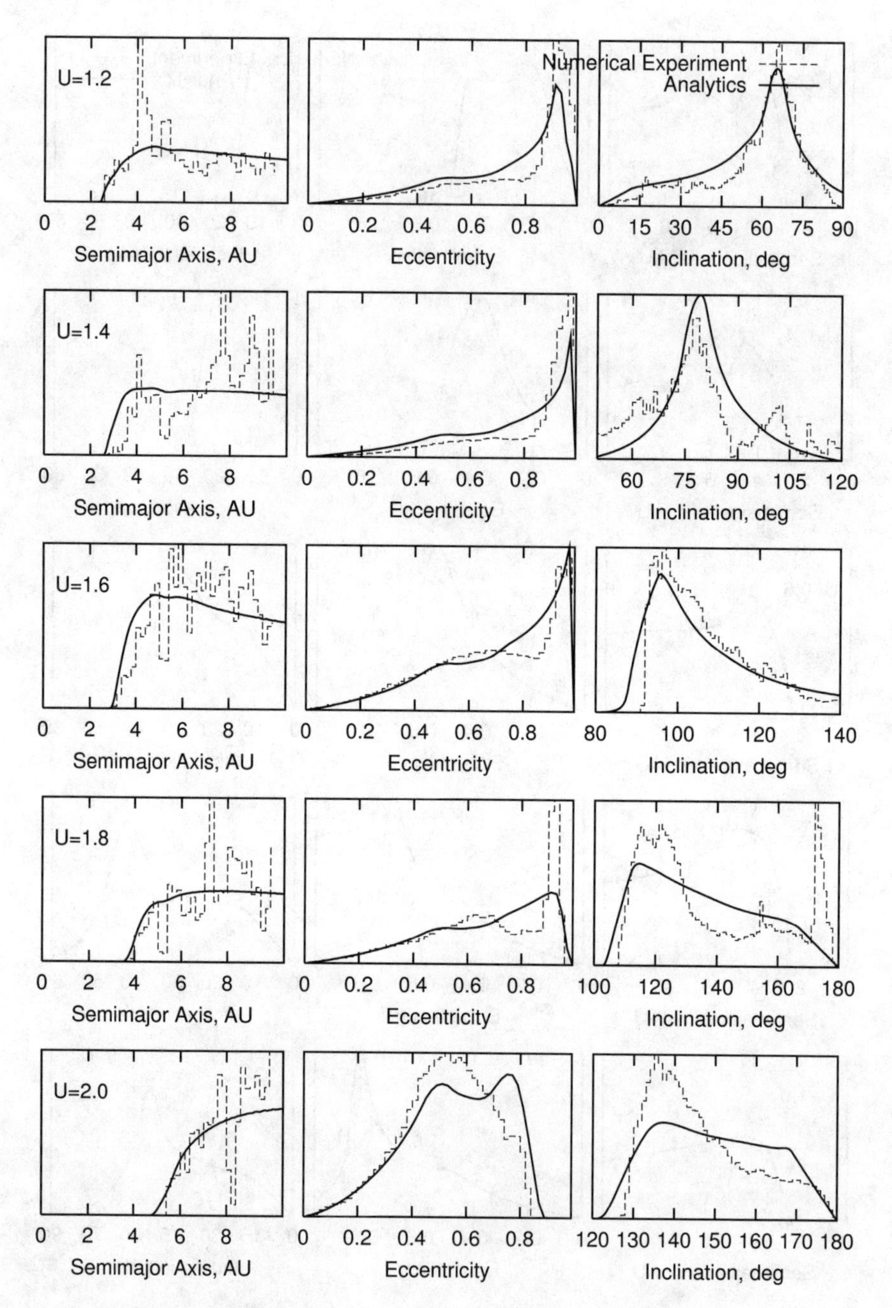


Figure 2. The orbital distributions of the particles scattered gravitationally by a Jupiter-sized planet on a circular orbit 5.2 AU from the Sun.

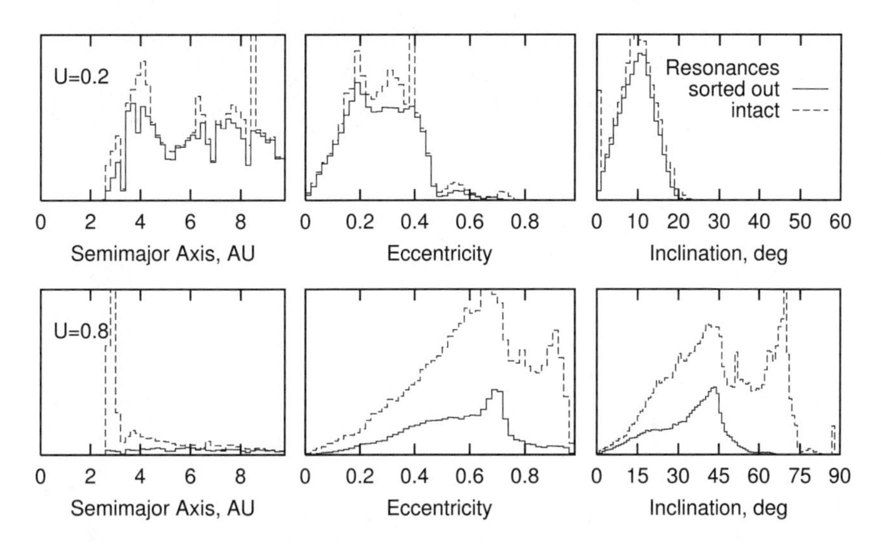


Figure 3. The effect of removal of the trajectories locked in resonances with planet on the orbital distributions of Jupiter-encountering particles.

motion near resonances should be applied to describe the resonant clouds in the three-body problem and their contribution to the overall distribution function. The numerically obtained orbital distributions before and after sorting out the particles locked in resonances for a long time are displayed in Figure 3.

At the smallest encounter speed $U = 0.2$, the effect of removal of the resonant trajectories is minimal. In contrast, at the encounter speed $U = 0.8$, about one half of trajectories was laid off because of resonances. Note, however, that the relative distribution in semimajor axis is most affected, while the inclination and eccentricity profiles are altered softer.

## 3. Conclusions

We have demonstrated that the orbital distributions of the particles scattered gravitationally by a Jupiter-sized planet can be described by a very simple analytical formula stated out under the ergodicity hypothesis on a subspace of the phase space, provided that the test particles falling in the resonant interaction with planet for a long time are laid off. The resonant clouds of particles in the three-body problem are to be modeled separately since the ergodicity is broken there. The formula gives the distributions in semimajor axis, eccentricity and inclination as a function of one parameter instead of three, of the Jacobi integral. Within the hyperplane of $C$ = const the close encounters with Jupiter determine the steady-state distribution. The reduction of dimensionality is a great simplification of the inverse problem, when constructing a meteoroid model.

**References**

Carusi, A., Valsechi, G. B., & Greenberg, R. 1990, CeMDA, 49, 111
Chambers, J. E. 1999, MNRAS, 304, 793
Cremonese, G., Fulle, M., Marzari, F., & Vanzani, V. 1997, A&A, 324, 770
Danby, J. M. A. 1992, Fundamentals of celestial mechanics, 2nd edn. (Richmond: Willman-Bell)
Landgraf, M., Liou, J. C., Zook, H. A., & Grün, E. 2002, AJ, 123, 2857
Levison, H. F., & Duncan, M. J. 1997, Icarus, 127, 13
Liou, J. C., Dermott, S. F., & Xu, Y. L. 1995, Planet.Space Sci., 43, 717
Valtonen, M. J., Zheng, J. Q., Mikkola, S., & Nurmi, P. 1998, CeMDA, 69, 89

*Order and Chaos in Stellar and Planetary Systems*
*ASP Conference Series, Vol. 316, 2004*
*G. Byrd, K. Kholshevnikov, A. Mylläri, I. Nikiforov, and V. Orlov, eds.*

# Fine-Scale Irregular Structure in Saturn's Rings

Evgeny Griv, Michael Gedalin, and Edward Livertz

*Department of Physics, Ben-Gurion University, Beer-Sheva, Israel*

Chi Yuan

*Institute of Astronomy, Academia Sinica, Taipei, Taiwan, ROC*

**Abstract.** In view of the possibility of employing CASSINI's experiments for the diagnosis of Saturn's rings, a semi-review is given of recent studies of morphology and dynamics of low and moderately high optical depth regions of Saturn's rings with special emphasis on its fine-scale radial structure (irregular cylindric structures of the order of 100 m or so). We predict that forthcoming (2004) CASSINI spacecraft high-resolution images of Saturn's rings may reveal this kind of radial structure.

## 1. Introduction

VOYAGER images of Saturn's rings have shown evidence of a great deal of structure: ranging from a few kilometres down to the several hundrends metres resolution of the spacecraft's camera (see Goldreich & Tremaine 1982 and Esposito 1993, 2002 as reviews). The best resolution demonstrated structures at all scales in the rings, down to the limit of resolution. Most of the structures are irregularly spaced and do not correspond to resonances with known satellities. It is important that the VOYAGER's stellar occultation data revealed some indirect evidence for structuring in the densest central parts of Saturn's B ring down to 100 m length scale. One cannot exclude the existence of a fine-scale irregular structure of this kind in other, low and moderately high optical depth regions of this ring system of mutually gravitating and rarely colliding particles.

A quasi-linear kinetic theory of the system under study has been developed (Griv 1996; Griv & Chiueh 1996; Griv & Yuan 1996; Griv, Yuan, & Chiueh 1997; Griv et al. 2000; Griv, Gedalin, & Yuan 2003a,b). The theory predicts that as a direct result of the gravitational instability of small gravity disturbances the A, B, and C rings of the Saturnian ring system with optical depth $\tau \lesssim 1$ may be divided into numerous irregular spiral ringlets of the order of $2\pi$ times the local thickness $h = 5 \div 30$ m. Let us describe $N$-body computer simulations in order to verify the validities of the theory.

## 2. Local $N$-Body Simulations

Modeling an entire ring by particles is still beyond the capability of any modern computer. (See, however, Griv et al. 2003c.) Therefore, the dynamical behavior

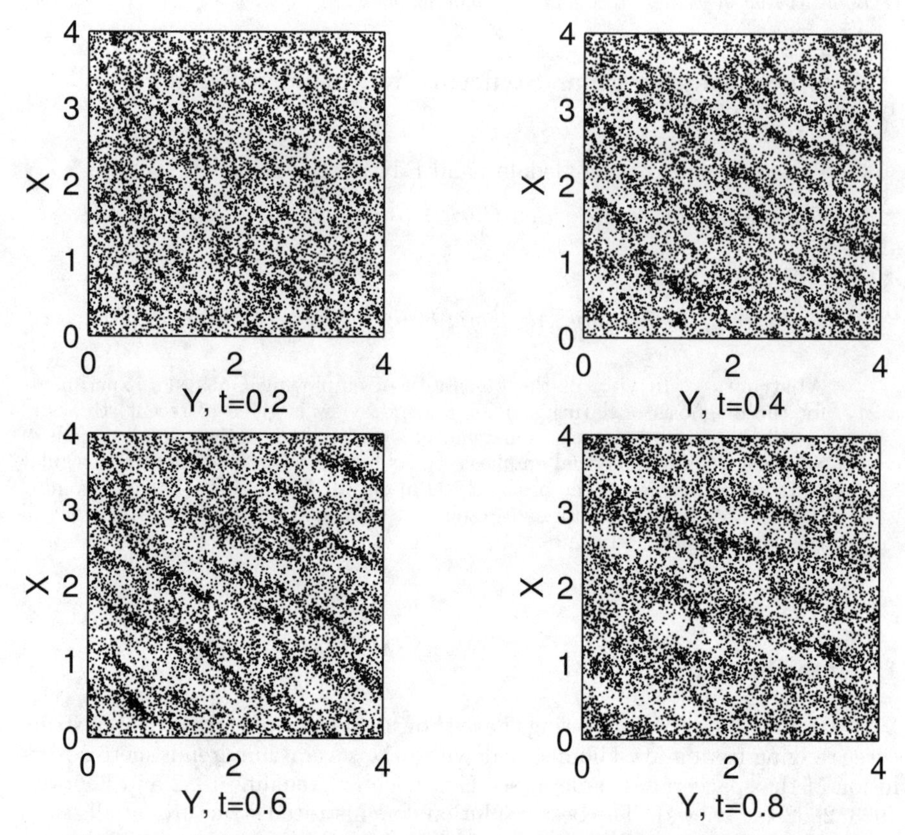


Figure 1. $N$-body gravitational simulation snapshots at normalized times $t$ for the Saturnian ring disk model (face-on view). The direction of disk rotation is taken to be clockwise and the direction of the planet is downward. The radius of a particle is 30 cm, the thickness of the model is 10 m, and each side of the $XY$-calculation cell is 400 m.

of planetary rings has been studied via simplified $N$-body simulations of an orbiting patch of the ring (Salo 1992; Richardson 1994; Osterbart & Willerding 1995; Griv 1998). In these $N$-body experiments in a *local* or in a Hill approximation dynamics of particles in small regions of the particulate disk are assumed to be statistically independent of dynamics of the particles in other regions. The local nimerical model thus simulates only a small part of the system and more distant parts are represented as copies of the simulated region.

The linearized Newton's equations for the three-dimensional motion of $N = 18,000$ particles were solved by the Runge–Kutta integration. A rotating Cartesian coordinate system was chosen, the $X$-axis points radially outward, the $Y$-axis points in the direction of the rotation, while the $Z$-axis points in the direction normal to the plane. The particles were initially placed on nearly circular orbits with an anisotropic Schwarzschild distribution of small chaotic velocities. The initial distribution was generated by means of random number generator placing particles in the box uniformly in real space. In turn, the box should be thought of as being embedded in the ring disk which has a constant angular velocity gradient in the $X$ direction. To maintain the system under the shearing stress in a steady state, the cyclic boundary conditions were used (Wisdom & Tremaine 1988; Salo 1995; Daisaka & Ida 1999).

The results of the simulation are shown in Figure 1. At the beginning particle distribution in the $XY$ plane is random, which can be seen in Figure 1 at times $t < 0.4$. At times $t \geq 0.4$ most of the particles are accumulated inside the *spiral*, almost irregularly spaced *trailing* filaments with a typical pitch angle $\psi = 25°$. As the simulations show, we see fine structure at all scales around 100 m.

We conclude that computer experiments confirm in general the predictions of the theory developed by Griv and co-workers. That is, the Saturnian ring system is gravitationally unstable against the spiral trailing modes of collective vibrations. In such a system the fine-scale spiral structure develops in the equatorial plane with size and spacing of the order of 100 m and the pitch angle $\psi \approx 25°$. CASSINI spacecraft observations should reveal this kind of structure.

**Acknowledgments.** Our thanks are to Tzi-Hong Chiueh, Baruch Rosenstein, Irina Shuster, and Rafael Steinitz for clarifying remarks on the subject. The work reported was supported in part by the Israel Science Foundation, the Binational Israel–USA Science Foundation, and the Academia Sinica in Taiwan.

## References

Daisaka, H., & Ida, S. 1999, EP&S, 51, 1195

Esposito, L. W. 1993, ARE&PS, 21, 487

Esposito, L. W. 2002, Rep.Prog.Phys., 65, 1741

Goldreich, P., & Tremaine, S. 1982, ARA&A, 20, 249

Griv, E. 1996, P&SS, 44, 579

Griv, E. 1998, P&SS, 46, 615

Griv, E., & Chiueh, T. 1996, A&A, 311, 1033

Griv, E., Gedalin, M., Eichler, D., & Yuan, C. 2000, P&SS, 48, 679

Griv, E., Gedalin, M., & Yuan, C. 2003a, A&A, 400, 375

Griv, E., Gedalin, M., & Yuan, C. 2003b, MNRAS, 342, 1102

Griv, E., Liverts, E., Gedalin, M., & Yuan, C. 2003c, in IAU Symp. 208, Astrophysical Supercomputing using Particle Simulations, ed. J. Makino & P. Hut (San Francisco: PASP), in press

Griv, E., & Yuan, C. 1996, P&SS, 44, 1185

Griv, E., Yuan, C., & Chiueh, T. 1997, P&SS, 45, 627

Osterbart, R., & Willerding, E. 1995, P&SS, 43, 289

Richardson, D. C. 1994, MNRAS, 269, 493

Salo, H. 1992, Nature, 359, 619

Salo, H. 1995, Icarus, 117, 287

Wisdom, J., & Tremaine, S. 1988, AJ, 95, 925

*Order and Chaos in Stellar and Planetary Systems*
*ASP Conference Series, Vol. 316, 2004*
*G. Byrd, K. Kholshevnikov, A. Mylläri, I. Nikiforov, and V. Orlov, eds.*

# Effects of Planetesimal Dynamics on the Formation of Terrestrial Planets

Roman R. Rafikov
*Institute for Advanced Study, Einstein Drive, Princeton, NJ, 08540, USA*

**Abstract.** Formation of terrestrial planets by agglomeration of planetesimals in protoplanetary disks sensitively depends on the velocity evolution of planetesimals. We describe a novel semi-analytical approach to the treatment of planetesimal dynamics incorporating the gravitational scattering by massive protoplanetary bodies. Using this method we confirm that planets grow very slowly in the outer Solar System if gravitational scattering is the only process determining planetesimal velocities, making it hard for giant planets to acquire their massive gaseous envelopes within $\lesssim 10^7$ yr. We put forward several possibilities for alleviating this problem.

## 1. Introduction

Current paradigm of planetary origin (Ruden 1999) assumes that terrestrial planets have formed in protoplanetary nebulae out of swarms of planetesimals—rocky or icy bodies with initial sizes of several kilometers. The same process is thought to account for the growth of solid cores of giant planets in the core instability scenario which postulates that huge gaseous envelopes of gas giants were acquired as a result of instability-driven gas accretion on preexisting cores made of solids (Mizuno 1980).

Our understanding of planetesimal accretion dates back to pioneering works by Safronov (1969) who (1) proposed to use the methods of kinetic theory for investigating the behavior of large number of planetesimals, and (2) included planetesimal dynamics in the picture of their gravitational agglomeration. Gravitational scattering between planetesimals tends to excite their random motions increasing the velocities with which they approach each other. This can have an important effect on their merging because of the phenomenon of gravitational focusing—an enhancement of collision cross-section of two bodies through the deflection of their orbits caused by their mutual gravitational interaction. Gravitational focusing increases collision cross-section by a factor $1 + v_{\rm esc}^2/v_{\rm rel}^2$ over its geometrical value $\pi(R_1 + R_2)^2$, where $R_{1,2}$ are the physical radii of colliding planetesimals, $v_{\rm esc}$ is their mutual escape velocity, and $v_{\rm rel}$ is the relative velocity of planetesimals at infinity.

From this formula it can be seen that gravitational focusing is important only provided that $v_{\rm rel}$ is significantly below $v_{\rm esc}$. Safronov's original assumption (1969) was that the biggest bodies in the system would be able to quickly increase the velocities of surrounding small-mass planetesimals to $v_{\rm esc}$ thus rendering further accretion of planetesimals by these massive protoplanets inefficient. As a result, typical timescale for forming the Earth at 1 AU from the Sun is very long—about $10^8$–$10^9$ yr. This timescale rapidly increases as one goes further out in the Solar System and reaches $\sim 10^{11}$ yr at 10 AU from the Sun (roughly present location of Saturn). This timescale is in stark contrast with the age of the Solar System (about 4.5 Gyr) implying that the Safronov's assumption of $v_{\rm rel} \sim v_{\rm esc}$ is faulty.

Wetherill & Stewart (1989) pointed out that at least initially planetesimal velocities in protoplanetary disks are not that large ($v_{\rm rel} \ll v_{\rm esc}$) and are moderated by mutual planetesimal scattering rather than by a small number of very massive bodies (which contain too little mass). They showed that in this case planetesimal accretion by massive bodies proceeds in a self-accelerating manner when most massive objects exhibit fastest growth; as a result, a *single* massive object detaches itself from the continuous mass spectrum of planetesimals. This so called "runaway" accretion allows Moon or Mars sized objects to appear on rather short timescale (typically $10^4$–$10^5$ yr) in the terrestrial zone. It also seemed to have rescinded the timescale problem for the giant planets by enabling their solid cores to grow within the gaseous nebula lifetime of several Myr at 5–10 AU from the Sun thus allowing them to accrete gas.

The runaway growth scenario was challenged by Ida & Makino (1993) who demonstrated using $N$-body simulations that massive protoplanetary "embryos" are in fact able to *locally* couple dynamically to the planetesimal disk after reaching some threshold mass. The major results of their study were that (1) massive embryo can strongly "heat up" planetesimal velocities within several Hill radii of its orbit (dynamically "heated zone"), and (2) embryo tends to repel planetesimal orbits away from its own orbit thus decreasing the surface density of small bodies at its location. The former effect decreases the role of gravitational focusing while the latter lowers the amount of mass which can be accreted by the massive body. Both of them act to reduce the accretion rate of the embryo and this stops its rapid runaway growth. This accretion regime was termed "oligarchic growth" since in this picture one embryo would reign inside its own heated zone, while there can be *many* such embryos (and their corresponding heated zones) growing within the local patch of the disk.

Straightforward $N$-body simulations are neither very well suited for determining the threshold mass at which transition from runaway to oligarchic growth occurs as a function of planetesimal disk properties, nor they can follow the evolution of the system for long enough. Although they can treat gravitational interactions between planetesimals and the embryo directly, without simplifications, they are too time consuming and not very flexible. Thus it is important to come up with alternative approaches which would be better suited for treating this important problem.

## 2. Planetesimal Dynamics in the Vicinity of Protoplanetary Embryo

Given the large number of planetesimals present in protoplanetary disks it is natural to employ the statistical approach in studying their dynamics. At the same time, presence of inhomogeneities in the planetesimal disk induced by embryo's gravity calls for inclusion of spatial dimension into consideration, something which conventional one-zone coagulation simulation are lacking.

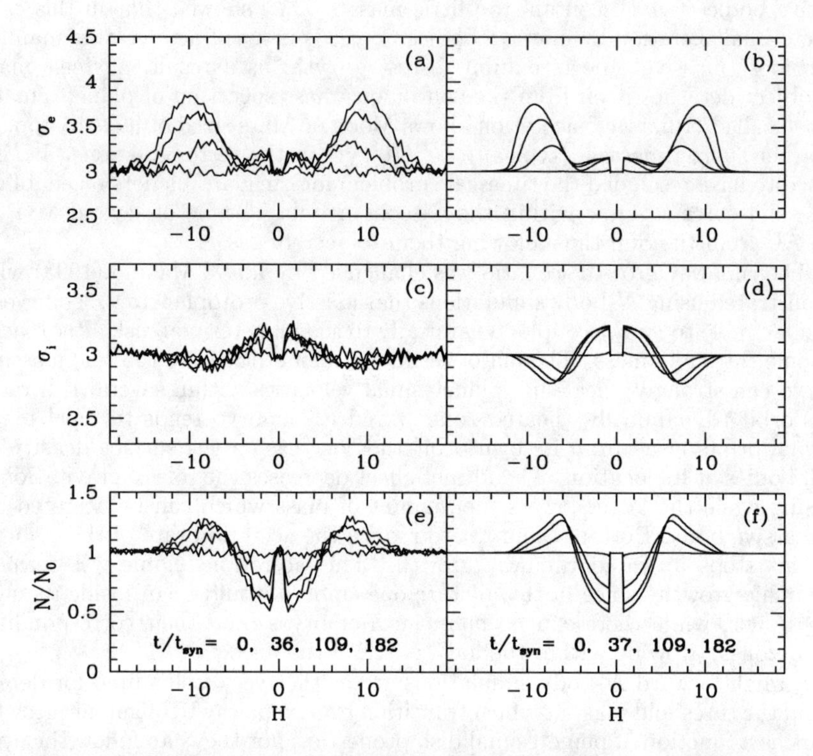


Figure 1. Planetesimal disk evolution driven by the presence of a protoplanetary embryo. The plots contain numerical (*left row*) and analytical (*right row*) time sequences of profiles of $\sigma_e$ (a, b), $\sigma_i$ (c, d), and dimensionless surface density normalized by its value at infinity (e, f). See text for details (from Rafikov 2003b).

Rafikov (2003a,b) came up with an analytical statistical prescription for treating planetesimal—planetesimal and embryo–planetesimal gravitational interactions. He assumed that the distribution function of planetesimal velocities in the disk has a Schwarzschild form, which allows one to considerably simplify the collision operator in the Boltzmann equation (describing the evolution caused by gravitational perturbations). In this approximation, planetesimal disk is unambiguously described by just three quantities for each mass population—surface number density $N$, dispersion of eccentricities $\sigma_e$, and dispersion of inclinations $\sigma_i$—which are the functions of radial coordinate in the disk (all these quantities are assumed to be azimuthally averaged). A set of three integro-differential equations self-consistently describes the evolution of these quantities in time and space as mutual planetesimal perturbations and gravitational scattering by massive protoplanetary embryos cause them to evolve.

Equations significantly simplify in two regimes: (1) *shear-dominated*, when the planetesimal random velocities are small compared to the differential shear in the disk across the corresponding Hill radius, and (2) *dispersion-dominated*, when these velocities are larger than the shear across Hill radius. First case can only be appropriate for the embryo–planetesimal interactions. When it is realized, scattering of planetesimals has a deterministic character which substantially simplifies its treatment. In the dispersion-dominated regime, appropriate for planetesimal–planetesimal scattering and in most cases for embryo–planetesimal scattering, Fokker–Planck expansion can be performed on the collision operator and evolution equations reduce to a set of partial differential equations. Intermediate regime, when planetesimal random velocities are comparable to the shear across the Hill radius can be treated by interpolation between the two limiting cases.

To check the performance of this approach we have compared the results of its application to studying the embryo–planetesimal scattering in different velocity regimes with the outcome of numerical orbit integrations (local) of the same problem (Rafikov 2003b). Representative results for the evolution of $\sigma_e$, $\sigma_i$ and $N$ in the dispersion-dominated case (initially $\sigma_e = \sigma_i = 3$ in Hill units) are shown in Figure 1. Different curves represent different moments of time [displayed in panels (e) and (f) by corresponding numbers]. One can see that analytical theory (right panels) is in excellent agreement with numerical simulations (left panels)—they follow each other with considerable quantitative accuracy even in minor details. This makes one confident that semi-analytical approach of Rafikov (2003a,b) is quite robust. It is numerically inexpensive: analytical calculations displayed in Figure 1 took about 1 minute to run on a conventional desktop, while the orbit integrations (following several $10^5$ particles for several hundred approaches to the embryo to achieve enough accuracy) required about 1 month on the same hardware. Planetesimal–planetesimal scattering was neglected in this calculation (reasonable assumption for large embryo masses), and only a single mass planetesimal population was considered. One can clearly see both the local excitation of planetesimal velocities and the development of a gap in the surface density of planetesimal orbits near the embryo, in agreement with previous $N$-body simulations.

This semi-analytical approach thus represents powerful and efficient tool for studying the dynamics of planetesimal disks. It can be easily extended to

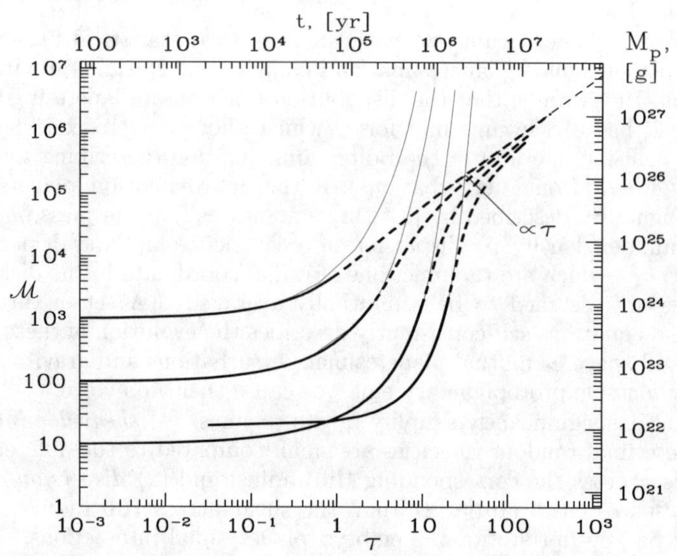


Figure 2. Growth of the embryo's mass in the Jupiter region of proto-Solar nebula for different initial conditions. See text for details (from Rafikov 2003c).

include the planetesimal–planetesimal scattering (see below), planetesimal mass spectrum and its evolution, dynamics of *several* embryos, dissipative effects, etc.

## 3. Growth of an Isolated Embryo

As a particular application of this approach we have studied the growth of an isolated embryo in a single mass planetesimal disk evolving dynamically and spatially under the action of *both* embryo–planetesimal and planetesimal–planetesimal gravitational interactions (Rafikov 2003c). Planetesimal–planetesimal scattering acts as an effective viscosity in the planetesimal disk opposing the embryo's tendency of clearing a gap. Masses of planetesimals in a disk are assumed not to vary, a simplification justified when embryo's mass increases faster than planetesimals grow. This would be the case in the runaway growth regime, as well as in the oligarchic case (Ida & Makino 1993). At the same time, embryo's mass was allowed to increase by accreting small planetesimals. Accretion rate was calculated analytically taking into account inhomogeneity of planetesimal disk near the embryo and has been checked against the numerical orbit integrations (Rafikov 2003b).

Several calculations for different initial starting embryo masses and planetesimal velocities were carried out and embryo's mass as a function of time is displayed in Figure 2. Properties of the planetesimal disk are those to be expected at 3.6 AU with planetesimal mass set to $6 \times 10^{20}$ g ($\mathcal{M}$ is embryo's mass scaled by the planetesimal mass, $\tau$ is time scaled by the synodic period of planetesimals separated by a Hill radius; corresponding physical values are displayed on the right and upper axes). Thin solid lines display the mass evolution tracks which embryo follows if its dynamical effect on its surroundings is neglected. As expected, these tracks exhibit unimpeded runaway growth and embryo reaches very large mass on a rather short timescale of $\sim 10^6$ yr. Thick lines represent tracks with the same initial conditions but with embryo's local perturbations taken into account. One can see that they initially follow the runaway tracks (solid portions). This is the result of small embryo's mass, which makes it incapable of perturbing planetesimals around it: their velocity dispersions increase independently of embryo's growth and planetesimal–planetesimal scattering is strong enough to smooth any inhomogeneities around the embryo. However, when embryo grows beyond some threshold mass ($\sim 10^{24}$ g in this case) it takes over the control of planetesimal dynamics around it (dashed portions of thick curves): planetesimals are being pushed away from the embryo's orbit, clearing a gap (similar to the simple calculation described in §2), and their velocities increase in accord with the embryo's growth. As a result, rapid runaway growth changes to a slower power-law increase of embryo's mass with time, roughly linearly with $t$. Thus, it would require a considerably longer timescale (by a factor of $\sim 10$) to reach $1 M_{\oplus}$ than simple runaway picture would predict. At 5 AU from the Sun this would stretch the formation timescale of giant planet cores to $\sim 10^8$ yr which is unacceptable from the point of view of core instability scenario of giant planet formation given short ($\lesssim 10^7$ yr) lifetimes of gaseous nebulae.

## 4. Discussion

Simple problem described above clearly demonstrates the difficulty (encountered by conventional scenarios of planet formation) of producing solid cores of giant planets in the outer Solar System on reasonable timescales. The primary reason for this is the strong dynamical coupling between massive protoplanetary bodies and surrounding planetesimals, which causes their gravitational focusing to decrease with time making accretion less and less efficient. In conclusion we want to suggest several possibilities for curing this problem (Rafikov 2003d).

Embryos likely not have evolved in *complete isolation*—as they grow in mass their heated zones overlap and they start affecting each other's environment. This would likely reduce the tendency for gap formation around embryo orbits, keeping planetesimal disks homogeneous enough to provide the steady supply of planetesimals.

*Dissipative processes* such as *gas drag* and *inelastic collisions* between planetesimals counteract the tendency of planetesimal velocities to increase under the action of embryo's perturbations. And one does expect gas to be naturally present during the formation of solid cores of gas giants (and initial stages of core formation of ice giants). This damping would not allow embryos to go back to runaway growth, but it would still let them grow faster than if gravity were the only force affecting planetesimal velocities.

*Fragmentation* of planetesimals in energetic collisions can grind them down to small sizes in the vicinity of massive bodies. Planetesimals would then be strongly affected by dissipative processes and their velocities could be considerably reduced allowing embryos to grow faster.

Closer look at these processes would hopefully help us in resolving the issue of planet formation timescale in the outer Solar System.

**Acknowledgments.** The financial support of this work by Charlotte Elizabeth Procter Fellowship and W. M. Keck Foundation is gratefully acknowledged.

## References

Ida, S., & Makino, J. 1993, Icarus, 106, 210

Mizuno, H. 1980, Prog.Theor.Phys., 64, 544

Rafikov, R. R. 2003a, AJ, 125, 906

Rafikov, R. R. 2003b, AJ, 125, 922

Rafikov, R. R. 2003c, AJ, 125, 942

Rafikov, R. R. 2003d, submittied to AJ, astro-ph/0311440

Ruden, S. P. 1999, in The Origins of Stars and Planetary Systems, ed. C. J. Lada & N. D. Kylafis (Dordrecht: Kluwer), 643

Safronov, V. S. 1969, Evolution of the Protoplanetary Cloud and Formation of the Earth and Planets (Moscow: Nauka)

Wetherill, G. W., & Stewart, G. R. 1989, Icarus, 77, 330

*Order and Chaos in Stellar and Planetary Systems*
*ASP Conference Series, Vol. 316, 2004*
*G. Byrd, K. Kholshevnikov, A. Mylläri, I. Nikiforov, and V. Orlov, eds.*

# Towards a Formation of the Solar System by Gravitational Instability in a Disk of Rarely Colliding Planetesimals

Evgeny Griv

*Department of Physics, Ben-Gurion University of the Negev,*
*P.O. Box 653, Beer-Sheva 84105, Israel*

**Abstract.** During the early evolution of a protoplanetary rotationally flattened gas–dust disk it is believed that the dust particles coagulate into numerous kilometer-sized rocky planetesimals. The dynamics of a three-dimensional disk of mutually gravitating and rarely colliding planetesimals is investigated. $N$-body simulations of a gravitational Jeans-type instability of small gravity disturbances for such a system are presented. The possibility of obtaining the Titius–Bode "law" of planetary distances on the basis of the concept of Jeans instability in sufficiently flat, rapidly rotating, and gravitationally unstable systems is examined.

## 1. Introduction

Attempts to find a plausible explanation of the origin of the solar system in the framework of idea of core accretion have not yet been quantitatively successful (Safronov 1972; Cameron 1988; Weidenschilling 2000). Accordingly, planetary formation is thought to start with dust particle settling to the central plane of a rotating nebula to form a thin dust layer. During the early evolution of such a flat disk it is believed that the dust particles coagulate into numerous kilometer-sized rocky planetesimals—bodies large enough to decouple from the turbulence and the gas drag. The planets formed by accretion of planetesimals, with or without the presence of gas during the late planetary formation.

The core accretion mechanism remained the most popular until recently, when it was criticized (Boss 2002). The main problem is the timescale, which is longer than estimates of the lifetime of planet-forming disks. Contrary, one may suggest that the gas *giant* planets were created by disk instability: a gravitationally unstable disk fragments directly into self-gravitating clumps of gas and dust that can contract and become giant gaseous protoplanets. The disk instability process itself is quite fast, and could form planets in $\sim 10^4$ yr.

Here the evolution of the self-gravitating disk of rarely colliding planetesimals is investigated. Similar to Boss (2002), the Jeans instability of gravity perturbations in sufficiently flat rapidly rotating systems is studied. Unlike Boss, however, I advocate below the idea that *all* planets formed directly from a hierarchy of *planetesimals*. The giant planets accreted a significant amount of gas subsequently from the nebula after accumulating solid cores of about 5–10 Earth masses. The fact that Mercury, like the Moon, had suffered a "late heavy bombardment" after its formation (Wetherill 1989) probably supports this idea.

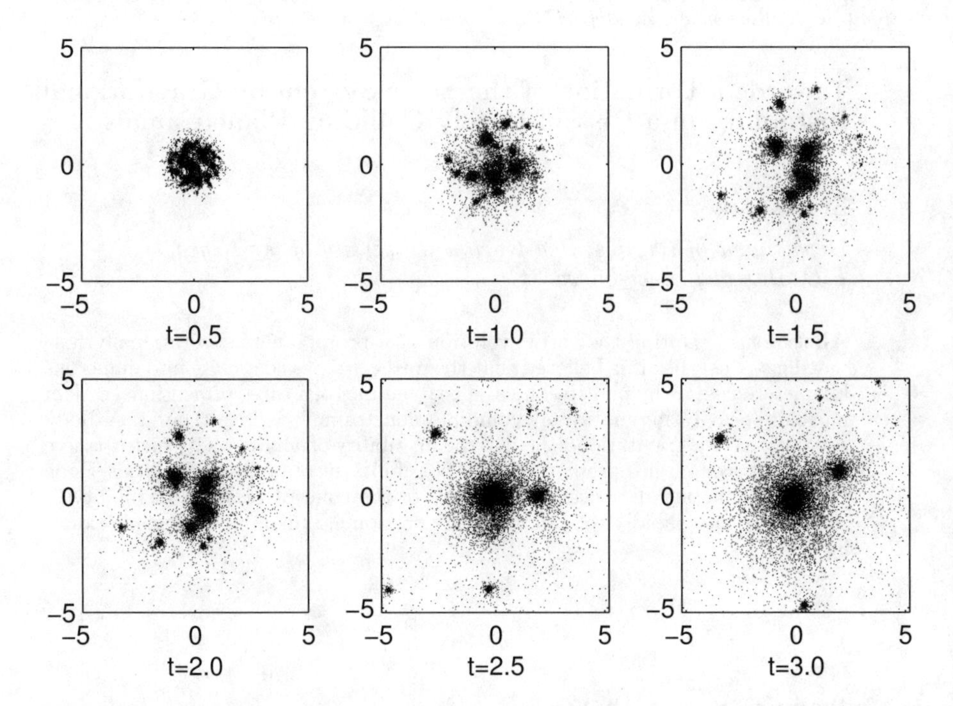


Figure 1. The time evolution (face-on view) of a Jeans-unstable disk of planetesimals $N = 250\,000$. The time is normalized so that the time $t = 1$ corresponds to a single revolution of the initial disk, and the rotation is taken to be counterclockwise.

## 2. *N*-Body Simulations

The required number of initial planetesimals is very large ($\sim 10^{10}$ bodies). To study the orbital evolution of so many bodies, it is necessary to use the methods of the molecular theory of gases, in a similar way to that used in stellar and Saturn's rings dynamics (e.g., Shu 1970; Griv, Gedalin, & Yuan 2002, 2003; Griv & Gedalin 2003). Alternatively, one can use $N$-body simulations.

I investigate a disk of planetesimals by direct integration of Newton's equations of motion (see Griv & Zhytnikov 1995 and Griv & Chiueh 1998 for a discussion). In general, the numerical procedure is first to seek stationary solutions of the Boltzmann kinetic equation in the self-consistent field approximation for the equilibrium parameters of the system, and then to determine the stability of those solutions to gravity perturbations. Rapidly rotating model disks with different surface mass density variations are considered. As a solution of a time-independent collisionless Boltzmann equation, at the start of the $N$-body integration, our similation initilizes the particles on a set of concentric circular rings with a circular velocity $\mathbf{V}_{\rm rot}$ of disk rotation in the equatorial plane; the system is isolated. Then the position of each particle is slightly perturbed by applying a random number generator. The Maxwellian-distributed chaotic velocities $\mathbf{v}$ are added to the initial circular velocities $\mathbf{V}_{\rm rot}$, and $|\mathbf{v}| \ll |\mathbf{V}_{\rm rot}|$. Using the modern 128-processor SGI Origin 2000 supercomputer in Israel, I make long simulation runs with a large number of particles, $N = 250\,000$.

Figure 1 displays a series of snapshots from a simulation run. The initial surface density of the disk $\sigma = \sigma(0)\left(1 - r^2/R^2\right)^{3/2}$, the squared angular velocity

$$\Omega^2(r) = (3\pi^2/4R)G\sigma(0)\left[1 - (3/4)(r^2/R^2)\right],$$

and the widely used Toomre's stability parameter $Q = 0.5$, where $\sigma(0)$ is the surfase density at $r = 0$ and $R$ is the outer radius of an initial model. As calculations show, interparticle collisions are very rare and, therefore, the treatment of collisions can be grossly simplified. The main physical effect there is a self-gravity collective effect. We adopted the standard "post-interaction" hard-sphere collision model (e.g., Griv & Gedalin 2003). The figures include only a face-on view of the simulation region. In accordance with the theoretical explanation (Griv et al. 2002, 2003; Griv & Gedalin 2003), the effects of the Jeans instability of spontaneous gravity perturbations appear quickly in the simulation. During the first rotation, Jeans-unstable perturbations break the system into several macroscopic fragments. At the end of the third rotation, one can see a quasi-stationary system of the very massive "sun" and 4 minor "planets." Interestingly, the distances of the planets from the sun are described rather well by the Titius–Bode rule. A possible reason for this is advanced. Briefly, planetary positions correspond to the maxima of the surface density perturbations in the particulate disk, caused by the gravitational instability. Also, the distance between planets is the wavelength of the most unstable perturbations at the given point of the disk. This may be suggested as a theoretical explanation of the Titius–Bode rule. Clearly that a deeper quantitative study would be desired.

**Acknowledgments.** The author thanks Michael Gedalin and Yury Lubarsky for helpful discussions and the encouragement of David Eichler, Shlomy Pistiner, and Raphael Steinitz. This work was performed in part under the auspices of the Israel Science Foundation and the German–Israeli Science Foundation.

## References

Boss, A. P. 2002, ApJ, 576, 462

Cameron, A. G. W. 1988, ARA&A, 26, 441

Griv, E., & Chiueh, T. 1998, ApJ, 503, 186

Griv, E., & Gedalin, M. 2003, P&SS, 51, 899

Griv, E., Gedalin, M., & Yuan, C. 2002, A&A, 383, 338

Griv, E., Gedalin, M., & Yuan, C. 2003, MNRAS, 342, 1102

Griv, E., & Zhytnikov, V. V. 1995, Ap&SS, 226, 51

Safronov, V. S. 1972, Evolution of the Protoplanetary Cloud and Formation of the Planets (Jerusalem: Israel Program for Scientific Translations)

Shu, F. H. 1970, ApJ, 160, 99

Weidenschilling, S. J. 2000, Space Sci.Rev., 92, 295

Wetherill, G. W. 1989, in The Formation and Evolution of Planetary Systems, ed. H. A. Weaver & L. Danly (Cambridge: Cambridge Univ. Press), 1

*Order and Chaos in Stellar and Planetary Systems*
*ASP Conference Series, Vol. 316, 2004*
*G. Byrd, K. Kholshevnikov, A. Mylläri, I. Nikiforov, and V. Orlov, eds.*

# Disastrous Events in Phanerozoic History of the Earth and the Sun Motion in the Galaxy

G. Goncharov

*Geochemistry Department, St. Petersburg State University, St. Petersburg, Russia*

V. Orlov

*Sobolev Astronomical Institute, St. Petersburg State University, Universitetskij pr. 28, 198504 St. Petersburg, Russia*

**Abstract.** Analysis of a correlation in chronology of some global intermittent disastrous events (mass extinctions of marine fauna, impacts) and geomagnetic reversals frequency in Phanerozoic history of the Earth and the solar system motion in the Galaxy is carried out. We found the recurrence step of 13 from 16 Phanerozoic extinctions equal to 183±3 Myr that corresponds to the anomalistic period (the time interval between two successive passages of the solar galactocentric orbit apocentre). The intensity of mass extinctions and energy output at impacts are systematically greater when the Sun moves from apocenter to pericenter than vice versa. A systematic decreasing of organic carbon content in total mass of carbon hidden in deposits is observed between the epochs of most extensive mass extinctions. Thus, one can say something about various connections between repeated global events in Phanerozoic history and the solar motion in the Galaxy.

## 1. Introduction

The periods of mass extinctions in the Earth's geologic history are characterized by global perturbations of an environment what evidence a powerful stimulating source. The literature data do not allow one to make a definite opinion on nature of this source. Nevertheless, a correlation between the periods of mass extinctions and the largest impacts from one side, as well as between these periods and geomagnetic reversals frequency splashes from other side, may indicate that these phenomena were connected with falling of large celestial bodies to the Earth's surface. Besides, the orbits of these bodies might be periodically perturbed by tidal galacitic field.

A chronologic correlation analysis between some global repeated events (marina mass extinctions, impacts, and geomagnetic reversals frequency splashes) in Phanerozoic history of the Earth and the solar system motion in the Galaxy for five axisymmetric models of regular galactic gravitational field is carried out.

## 2. Chronology of Mass Extinctions, Impacts, and Geomagnetic Reversals and the Solar Motion in the Galaxy

Repeated events are separated with the following time intervals (see Table 1 in Goncharov & Orlov 2003, thereafter GO03): $N_2^2 - J_{1p}$ (185); $P_2^3 - T_{3n}$ (175); $K_{2m} - P_{2t}$ (185); $K_{2c} - P_{1s}$ (179); $K_{1a} - C_{3g}$ (178); $J_{3tt} - C_{1s}$ (177); $J_{1p} - D_{3f}$ (180); $T_{3n} - S_{2ld}$ (201); $P_{2t} - O_{3as}$ (189); $C_{3g} - O_{2l}$ (179).
We found the recurrency step of 13 from 16 Phanerozoic extinctions (Figure 1). The mean interval from these ten differences is equal to $183 \pm 3$ Myr that corresponds to the anomalistic period (the time interval between two successive passages of the solar galactocentric orbit apocentre) in the Allen & Martos' (1986) model . The minima and maxima positions in Gaussians fitting the distributions of geomagnetic reversals in time are also in agreement with the solar passages through apocenter and pericenter of its orbit in this model (Figure 2 in GO03).

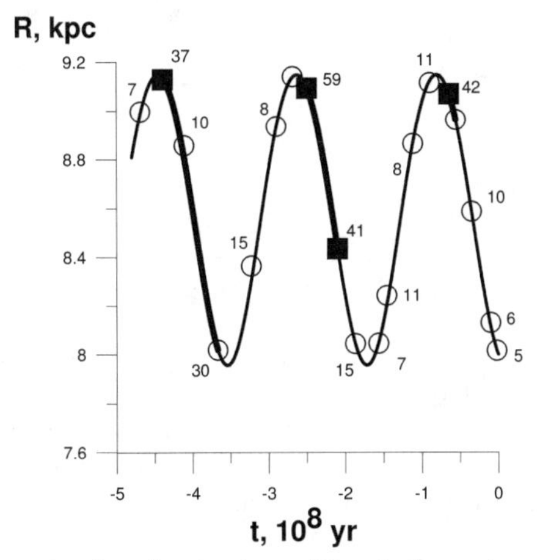


Figure 1. The solar motion in galactic plane. The circles and squares correspond to mass extinction epochs. The numbers are the percentages of marina species extinctions. The thick trajectory parts may correspond to the Solar system passages through star formation regions.

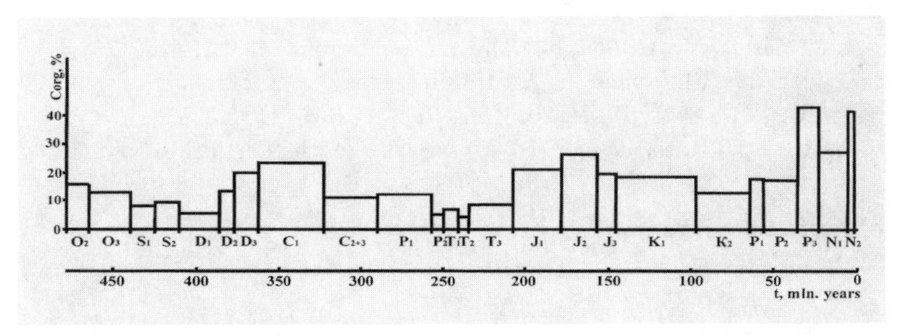


Figure 2. Changes of organic carbon contents in total mass of carbon hidden in Phanerozoic deposits calculated using data of Ronov (1993).

The maximum position in distribution of differences between mass extinctions and impact dates is near zero what follows from agreement of many these events (see Figure 4 in GO03). Commonly, impact events are followed by extinctions. More often the impacts of large cosmic bodies have taken place when the solar system intersects galactic plane. Whereas mass extinctions occur more often at some separation from this plane (about 30–40 pc).

As a rule, the times of geomagnetic reversals frequency splashes are in agreement with impacts of large cosmic bodies. However the splashes dates do not indicate any agreement with mass extinctions. The intensity of mass extinctions and energy output at impacts are systematically greater when the Sun moves from apocenter to pericenter than vice versa (see Figure 1 in this paper and Table 4 in GO03).

Thus, one can say about various connections between repeated global events in Phanerozoic history and the solar motion in the Galaxy. The long-term variations of geomagnetic reversals agree with the orbital solar motion, whereas their splashes agree with impacts. The mass extinctions may be connected with impacts of large cosmic bodies perturbed by interstellar clouds which are concentrated to the galactic plane as well as by the shock wave front of Perseus spiral arm (the solar system may cross this arm), giving the most significant perturbations.

On short parts of the the solar galactic orbit between the epochs of most extensive mass extinctions ($O_{3as} - D_{3f}$, $P_{2t} - T_{3n}$, $K_{2m} - P_1^2$), a systematic reduction of organic carbon content in total mass of carbon hidden in deposits (Figure 2) is established. The bulk of organic carbon in deposits is a product of a metabolism as well as phyto- and zooplankton decomposing which are rather sensitive to the cosmic radiation UV-component (Belousov 1991).

The reaction of geologic events to the solar system crossing through the shock front in the Perseus spiral arm may be a sequence of its construction. Entrances into the gas-dust matter band are fixed by amplification of energy output intensities in impact events and mass extinctions. The passage through star formation regions behind a shock front where young hot OB stars and supernovae events may take place results in amplification of hard cosmic radiation and phyto- and zooplankton depression in the Earth hydrosphere without any essential impacts increasing.

## References

Allen, C., & Martos, M. A. 1986, Rev.Mex.Astron.Astrofis., 13, 137

Belousov, V. V. 1991, Informations AN USSR, Biology, no.2, 242

Goncharov, G. N., & Orlov, V. V. 2003, Ast.Rep., 47, 925 (GO03)

Ronov, A. B. 1993, Stratisphere or sedimentary shell of the Earth (Moscow: Nauka)

*Order and Chaos in Stellar and Planetary Systems*
*ASP Conference Series, Vol. 316, 2004*
*G. Byrd, K. Kholshevnikov, A. Mylläri, I. Nikiforov, and V. Orlov, eds.*

# Motions in Extrasolar Planetary Systems: between Regularity and Chaos

L. Sokolov and N. Pitjev

*Sobolev Astronomical Institute, St. Petersburg State University, Universitetskij pr. 28, Staryj Peterhof, St. Petersburg 198504, Russia*

**Abstract.** Most of the discovered extrasolar planets are massive, Jupiter-sized bodies, moving about their central stars along considerably eccentric orbits. Low-mass planets like the Earth still cannot be observed in these systems. However, their existence is especially interesting in view of the search for extraterrestrial life as well as planetary formation theories. In this work, we examine the orbital stability of low-mass planets in the systems with "Jupiters" on eccentric orbits, in the frame of restricted elliptical three-body problem. The initial orbit of the massless particle representing an Earth-sized planet is circular. The regions of regular and chaotic motion are studied, a simple analytical model of the regular motion is presented. If the distance between the orbits of "Jupiter" and massless particle is sufficiently small, the motion becomes chaotic. In the transition region between the regularity and chaos the trajectory properties are highly sensitive to the initial positions of planets on their orbits. These trajectories are investigated in more detail. In contrast to the region of regular motion, we found that low-eccentricity orbits exist in the transition region, some of which are stable on the time scale of $10^6$ orbital revolutions of "Jupiter."

The discovery of extrasolar planetary systems is an outstanding achivement of science. Many new problems connected with extrasolar planets arise in astronomy. For example, new planetary systems differ sufficiently from the Solar system, and known theories of planet formation cannot explain formation of the extrasolar planets.

Now we will discuss one dynamical problem. Many massive extrasolar planets have large orbital eccentricities, in contrast to Jupiter (Schneider 2003). We cannot observe low-mass planets like the Earth or Mars. The problem considered in this paper is the influence of the massive planet with the large orbital eccentricity on possible motions of small planets. What can we say about planetary system design, if extrasolar "Jupiter" has a large orbital eccentricity? In particular, the existence of quasi-circular orbits of small planets like the Earth is very interesting. We constrain ourselves to model planetary systems similar to the solar system. The mass of the star is equal to the solar mass, semi-major axis of "Jovian" orbit is equal to 5.2 A.U., the low-mass planet has initially circular orbit. The only difference is a large eccentricity of Jovian orbit.

We investigate the low-mass planet motion within the framework of the restricted elliptical three-body problem. The consideration has been done for the planar problem, but if the inclination is small, our results hold true. We discuss mainly the "outer" problem, for which the distance between the star and the low-mass planet is longer than between a star and a massive planet. The results have been derived numerically; in order to verify them, we use different independent program tools, as well as analytical tools.

Let the extrasolar "Jupiter" parameters be fixed, its orbital eccentricity be not small, the initial eccentricity of the low-mass planet's orbit is equal to zero.

If the initial distance between orbits is sufficiently large (planet's semi-major axes differ sufficiently), then the motion is regular, not chaotic.

If the initial semi-major axes of planetary orbits become closer to each other, the regularity of motion reduces. Some trajectories stay regular, but some trajectories become stochastic.

If the difference between initial semi-major axes is sufficiently small, than all trajectories are chaotic, i.e., the motion is chaotic.

So, we have three regions for values of $a_0$ ($a_0$ is the initial value of semi-major axis of the small planet): the region corresponding to the regular motion, the region corresponding to the chaotic (stochastic) motion, and the region between regularity and chaos.

Let us discuss the properties of trajectories in each region.

The regular motion has the following properties:

1. In the planes $(a, e)$ and $(e \cos \pi, e \sin \pi)$ we have an invariant curve for each trajectory. Here $a$ is the semi-major axis, $e$ is the eccentricity, $\pi$ is the pericenter argument.

2. The orbital evolution does not depend on initial positions of planets on their orbits; we have the same invariant curve for different initial positions, excluding possible narrow resonance zones.

Invariant curves derived numerically are as follows: the segment $a = \text{const}$, $0 \leq e \leq e_{\text{max}}$ in the plane $(a, e)$ and the circle $(e \cos \pi - e_{\text{max}}/2)^2 + (e \sin \pi)^2 = e_{\text{max}}^2/4$ in Lagrangian variables $(e \cos \pi, e \sin \pi)$. The value of $e_{\text{max}}$ depends on $a_0$ and extrasolar "Jupiter" parameters.

If the initial value of $e_0 \neq 0$, regular trajectories are similar. We have segments $a = \text{const}$ in the plane $(a, e)$ and circles in Lagrangian variables with the center in the same point $(e_{\text{max}}/2, 0)$. Their radii depend on the initial values $(e_0, g_0)$. The point $(a_0, e_0 = e_{\text{max}}/2, \pi_0 = 0)$ is a stationary point.

To understand and verify these properties, we use the classical analytical solution—the theory of secular perturbations developed by Laplace and Lagrange (Charlier 1927). We have two integrals of the averaged equations of motion: $a = \text{const}$ and $(e \cos \pi - \tilde{a}/2)^2 + (e \sin \pi)^2 = \text{const}$. The approximate value of $\tilde{a}$ is $9a_J e_J/4a_0$. Here $a_J, e_J$ is the "Jupiter's" semi-major axis and eccentricity. So, the analytical and numerical results correspond to each other qualitatively. The difference between them for $e_{\text{max}}$ is less than 0.1 (Sokolov 2002).

For the "inner" problem, when $a_0 < a_J$, the regular trajectories are similar to the regular trajectories in the "outer" problem, but the dependence of $e_{\text{max}}$ on the parameters is different, more complicated.

If the eccentricity of massive planet orbit is large and the motion is regular, then usually the eccentricity of the low-mass planet orbit is large too. One can find similar results in (Malhotra 2002). So, the construction of extrasolar quasi-circular orbits is a serious problem.

Chaotic trajectories arise, when $a_0$ is sufficiently close to $a_J$. Usually such trajectories are finished by escape. It is interesting that the stochastic motion in Lagrangian variables infrequently seems more regular in the variables $(a, e)$. Roughly the equation $a(1 - e) = \text{const}$ hold true. In such a situation, initially circular orbit transforms to the one similar to a long-periodic comet orbit rather fastly (Sokolov 2002).

In the region between the regularity and chaos we have chaotic trajectories, as well as non-chaotic trajectories, depending on the initial positions of planets on their orbits. This dependence means that analytical methods based on averaging failed. The characteristic property of non-chaotic trajectories is $a \approx$ const. Some of them are similar to the trajectories in the region of the regular motion described above. Some others have crescent form in Lagrangian variables, and the rest of trajectories are quasi-circular (their eccentricity is less than 0.1). The next case of quasi-circular trajectories is especially interesting for the extrasolar planetary systems.

If the orbital eccentricity of "Jupiter" is equal to 0.5, then the region of the regular motion corresponds to $a_0 > 11.3$ A.U. and $a_0 < 1.8$ A.U., the region of chaotic motion corresponds to $a_0 < 8.9$ A.U. and $a_0 > 2.2$ A.U. If the "Jupiter" orbital eccentricity is equal to 0, then the region of the regular motion corresponds to $a_0 > 6.7$ A.U. and $a_0 < 4.0$ A.U., the region of chaotic motion corresponds to $a_0 < 6.2$ A.U. and $a_0 > 4.4$ A.U. Of course, this numerical result is valid only on restricted time interval for several thousand years. Below we discuss only the "outer" problem, $a_0 > a_J$.

In the region of regular motion the value of $y = e \sin \pi$ initially decreases (initial value $e = 0$). In the region between chaos and regularity for some trajectories $y$ initially decreases, and for some trajectories $y$ initially increases, depending on the initial position of zero-mass planet on its orbit. The quasi-circular trajectories are located between them.

Let us discuss trajectories corresponding to the initial position of the zero-mass planet equal to $v_0 = 0°, 3°, 6°, \ldots, 180°$ for $a_0 = 10.8$ A.U. and $a_0 = 10.9$ A.U. The initial value of eccentricity is equal to zero. The massive planet has $m_J/m_* = 10^{-3}$, $e_J = 0.5$, $a_J = 5.2$, its initial position corresponds to the pericenter, $v_{J0} = 0$. The time of evolution is equal to $10^4$ revolutions of "Jupiter." Due to the time reversibility we can restrict ourself by the segment $0° \leq v_0 \leq 180°$ for the initial position of the zero-mass planet. The trajectories, corresponding to $v_0 = 360° - \phi$ and $v_0 = \phi$, have the same properties.

If $a_0 = 10.8$ A.U., we have seven segments:

1) $0° \leq v_0 \leq 15°$: the non-chaotic trajectories, $\dot{y}_0 < 0$;

2) $18° \leq v_0 \leq 24°$: the chaotic trajectories, $\dot{y}_0 > 0$;

3) $27° \leq v_0 \leq 99°$: the non-chaotic trajectories, $\dot{y}_0 > 0$;

4) $102° \leq v_0 \leq 105°$: the chaotic trajectories, $\dot{y}_0 > 0$;

5) $108° \leq v_0 \leq 132°$: the non-chaotic trajectories, $\dot{y}_0 < 0$;

6) $135° \leq v_0 \leq 138°$: the chaotic trajectories, $\dot{y}_0 > 0$;

7) $141° \leq v_0 \leq 180°$: the non-chaotic trajectories, $\dot{y}_0 > 0$.

Between the segments number 1 and 2, 4 and 5, 5 and 6 the unstable quasi-circular trajectories are located.

If $a_0 = 10.9$ A.U., we have five segments:

1) $0° \leq v_0 \leq 12°$: the non-chaotic trajectories, $\dot{y}_0 < 0$;

2) $15° \leq v_0 \leq 18°$: the chaotic trajectories, $\dot{y}_0 > 0$;

3) $21° \leq v_0 \leq 99°$: the non-chaotic trajectories, $\dot{y}_0 > 0$;

4) $102° \leq v_0 \leq 138°$: the non-chaotic trajectories, $\dot{y}_0 < 0$;

5) $141° \leq v_0 \leq 180°$: the non-chaotic trajectories, $\dot{y}_0 > 0$.

Between the segments 1 and 2 the unstable quasi-circular trajectory is located; between segments 3 and 4, 4 and 5 the stable quasi-circular trajectories are located. The crescent–form trajectories are located in vicinities of the stable quasi-circular ones.

In Tables 1 and 2 the maximum values of the eccentricities for the stable quasi-circular orbits depending on the initial position of the zero–mass planet are presented. The semi-major axis is equal to 10.9 A.U. The time of evolution is equal to $10^6$ revolutions of "Jupiter."

Table 1. Maximum values of eccentricities for quasi-circular orbits at $v_0 \in [99\overset{\circ}{.}4, 100\overset{\circ}{.}2]$

| $v_0$ [deg.] | 99.4 | 99.5 | 99.6 | 99.7 | 99.8 | 99.9 | 100.0 | 100.1 | 100.2 |
|---|---|---|---|---|---|---|---|---|---|
| $e_{max}$ | 0.103 | 0.092 | 0.080 | 0.066 | 0.055 | 0.055 | 0.072 | 0.094 | 0.126 |

Table 2. Maximum values of eccentricities for quasi-circular orbits at $v_0 \in [138\overset{\circ}{.}3, 139\overset{\circ}{.}1]$

| $v_0$ [deg.] | 138.3 | 138.4 | 138.5 | 138.6 | 138.7 | 138.8 | 138.9 | 139.0 | 139.1 |
|---|---|---|---|---|---|---|---|---|---|
| $e_{max}$ | 0.095 | 0.079 | 0.065 | 0.056 | 0.055 | 0.062 | 0.072 | 0.082 | 0.091 |

It should be noted that we have observed the quasi-circular orbits only in the "outer" region between chaos and regularity.

So, the stable (as well as unstable) quasi-circular trajectories exist in the region between regularity and chaos for the case of the "outer" restricted elliptic three-body problem.

**Acknowledgments.** This study was supported by the Russian Foundation for Basic Research (grant 02-02-17516), by the program "Russian Universities" (grant UR.02.01.027), and by the Leading Scientific School (grant NSh-1078.2003.02). We are thankful to the referee of this paper.

## References

Schneider, J. 2003, Extra-Solar Planets Catalog, http://www.obspm.fr/encycl/cat1.html
Sokolov, L. L. 2002, Solar System Research, 36, 5, 403
Charlier, C. L. 1927, Die Mechanik des Himmels (Berlin, Leipzig)
Malhotra, R. 2002, ApJ, 575, L33.

*Order and Chaos in Stellar and Planetary Systems*
*ASP Conference Series, Vol. 316, 2004*
*G. Byrd, K. Kholshevnikov, A. Mylläri, I. Nikiforov, and V. Orlov, eds.*

# Habitability and Stability of Orbits for Earth-Like Planets in the Extrasolar System 47 UMa

Siegfried Franck, Werner von Bloh, and Christine Bounama

*Potsdam Institute for Climate Impact Research, P.O. Box 601203, 14412 Potsdam, Germany*

Manfred Cuntz

*Department of Physics, University of Texas at Arlington, Box 19059, Arlington, TX 76019, USA*

**Abstract.** We investigate wether Earth-type habitable planets can in principle exist in the planetary system of 47 UMa. The system 47 UMa consists of two Jupiter-size planets beyond the outer edge of the stellar habitable zone, and thus resembles our own Solar system most closely compared to all exosolar planetary systems discovered so far. We estimate if Earth-type habitable planets planets around 47 UMa are in principle possible by investigating if a distinct set of conditions is warrented. In the event of successful formation and orbital stability, two subjects of intense research, we find that Earth-type habitable planets around 47 UMa are in principle possible! The likelihood of those planets is increased if assumed that 47 UMa is relatively young (younger than approximately 6 Gyr) and has a relatively small stellar luminosity as permitted by the observational range of those parameters. We show that the likelihood to find a habitable Earth-like planet on a stable orbit around 47 UMa critically depends on the percentage of the planetary land/ocean coverage. The likelihood is significantly increased for planets with a very high percentage of ocean surface ("water worlds").

## 1. Introduction

Up to now, more than 100 planets have been detected around stars other than the Sun. Most of these planets are Jupiter-like planets or brown dwarfs, but no Earth-type planets have been found yet. The extrasolar system of 47 UMa (see Figure 1) consists of two Jupiter-size planets at respectable distances from the host star and thus resembles our own solar system most closely compared to all extrasolar systems discovered so far (Fischer et al. 2002).

Furthermore, it is known that there are no Jupiter-mass planets in the inner region around the star, which would otherwise thwart the formation of terrestrial planets at Earth-like distances around star (Laughlin et al. 2002) or would trigger orbital instabilities for those planets during inward migration. The central star has properties very similar to those of the Sun, including effective temperature, spectral type and metallicity (Henry et al. 1997). Metallicities not too dissimilar to our Sun are probably required for building up Earth-type habitable planets. The main question is, whether Earth-like planets harbouring life can exist around 47 UMa, i.e., planets within the habitable zone (HZ). Typically, stellar HZs

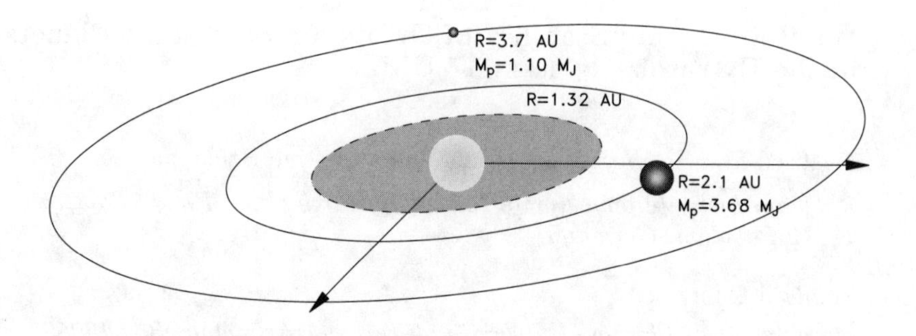


Figure 1. The extrasolar planetary system 47UMa. The dashed line denotes the outermost stable orbit of an Earth-like planet found by Jones & Sleep (2002).

are defined as regions near the central star, where the physical conditions are favourable for liquid water to be available at the planet's surface for a period of time long enough for biological evolution to occur. In the following, we adopt a definition of HZ previously used by (Franck et al. 2000). Here habitability (i.e., presence of liquid water at all times) does not just depend on the parameters of the central star, but also on the properties of the planetary climate model. In particular, habitability is linked to the photosynthetic activity of the planet, which in turn depends on the planetary atmospheric $CO_2$ concentration, and is thus strongly influenced by the planetary geodynamics. In principle, this leads to additional spatial and temporal limitations of habitability, as the stellar HZ (defined for a specific type of planet) becomes narrower with time due to the persistent decrease of the planetary $CO_2$ concentration.

## 2. Model Description

To estimate the habitability of an Earth-like planet, a so-called integrated Earth system approach is applied. Our model couples the stellar luminosity, the silicate rock weathering rate and the global energy balance to calculate the partial pressure of atmospheric and soil carbon dioxide, the mean global surface temperature, and the biological productivity as a function of time (Figure 2).

Planetary habitability requires orbital stability of the Earth-type planet over a biologically significant length of time in the HZ. There exist a variety of papers (Fischer et al. 2002; Noble et al. 2002; Jones & Sleep 2002; Gehman et al. 1996; Jones et al. 2001) discussing the orbital stability of (hypothetical) terrestrial planets in the 47 UMa system, which is strongly influenced by the masses, orbital positions and eccentricities of the two Jupiter-size planets in that system. Jones et al. (2001) explored the dynamical stability for terrestrial planets within HZs of four stars with detected gas giant planets. Without knowing about the second giant planet later discovered by Fischer et al. (2002), they concluded that 47 UMa is the best candidate to harbour terrestrial planets in orbits that could

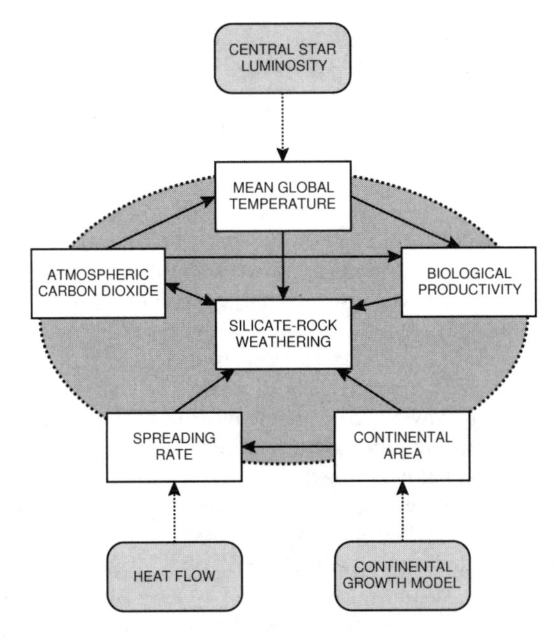


Figure 2. Box model of the integrated system approach. The arrows indicate the different forcings and feedback mechanisms (Cuntz et al. 2003).

remain confined to the HZ for biologically significant time spans. Jones et al. assumed a terrestrial planet of one Earth mass that was put at different initial positions of the stellar HZ. The orbital motion of that planet was followed to up to $1 \times 10^9$ yrs using a mixed variable symplectic integration method. Orbital instability was assumed if the terrestrial planet entered the so-called Hill radius of the innermost giant planet. They found that the outermost stable orbit is close to 1.32 AU. Jones et al. also considered non-zero inclinations between terrestrial and gas giant orbits up to $i = 10°$, which did however not seriously impact their results.

Previously, a less rigorous orbital stability limit was given by Gehman et al. (1996) using an analytical approach, and also assuming somewhat preliminary orbital data for the 47 UMa system. They argued that the outer boundary of the zone of orbital stability for the terrestrial planet is at 1.6 AU, which is not inconsistent with the results obtained by Jones et al. (2001). A study, which also considered the existence of the second Jupiter-type, has been given by Noble et al. (2002). The authors explored orbital stability for a one Earth-mass planet in the 47 UMa system initially placed at 1.13, 1.44, and 1.75 AU inside the stellar HZ. They identified orbital stability for the initial position at 1.13 AU, but no orbital stability for 1.44 and 1.75 AU was obtained, consistent with the findings by Jones et al. (2001). In case of the initial position at 1.44 AU, the terrestrial planet was found to wander between 1.185 and 1.617 AU in the first 700 years of orbital integration thwarting any possibility of habitability. In the paper by Fischer et al. (2002), in which the discovery of the second giant planet is announced, the authors also briefly discussed the effects of the secondary giant planet on the orbital stability of terrestrial planets. They argued that orbital stability of terrestrial planets is warranted as most test Earth-mass planets are found to survive within the HZ over $10^6$-year timescales. On the other hand, this study is only of limited use, as the authors did not communicate the extent of the stellar HZ used in those computations.

Subsequent work by Jones & Sleep (2002) also considered the presence of the two giant planets. They argued that the second giant planet noticeably reduces the range of orbital stability of Earth-mass planets in the HZ of 47 UMa. In some of their simulations, the outer radius of orbital stability was found to be as low as 1.2 AU. This value is however also affected by the possible mass and eccentricity ranges of the Jupiter-type planets taken into consideration. Nonetheless, the authors again concluded that Earth-type planets are still possible in the inner part of the present-day stellar HZ (note again the definition of HZ used here!), assumed that they stay away from mean-motion resonances invoked by the two giant planets and that certain extreme values for the masses and eccentricities of the giant planets are not realized.

A further paper, which analyses the orbital stability of Earth-mass planets, has been given by Goździewski (2002), based on the so-called MEGNO integration technique. He found that the HZ of 47 UMa is characterized by an alternation of narrow stable and unstable zones with the latter related to the mean motion and secular resonances with the giant planets. Beyond 1.3 AU, no stable zones were found. The positions and widths of the various unstable zones are sensitively depending on the masses and orbital parameters of the two giant planets, which are both uncertain. The author noted that his investigations did not include all possibilities of bounded orbital dynamics of hypothetical terrestrial planets, but rather provide a characteristic landscape filled with stable and unstable orbital evolutions. Therefore, we assume a representative value for the outer boundary of the orbital stability which is $R_{\rm max} = 1.25$ AU (Cuntz et al. 2003). In order to calculate the HZ within the framework of our model it is necessary to estimate the age and the luminosity of the central star 47 UMa. Following the discussion of Cuntz et al. (2003) the mean luminosity $L$ of 47 UMa is given as 1.54 $L_\odot$ based on Hipparcos data. As stellar age, we assume $6.32^{+1.2}_{-1.0}$ Gyr based on stellar evolution computations (Ng & Bertelli 1998) and the Ca II age-activity relation (Henry et al. 2000).

## 3. Results and Discussion

In Figure 3 we show the results of our calculations of the HZ for the likely value $L = 1.54\, L_\odot$ of the central star luminosity (the dotted rectangle area) and the grey shaded range of orbital stability, $R < R_{\rm max}$. The intersection of the two areas describes the interesting parameter range where an Earth-like planet on a stable orbit can exist within the HZ. It is evident that an almost completely ocean-covered planet ("water world") has the highest likelihood of being both habitable and orbitally stable. If the planet is covered with more than 50% continental area, then habitability and orbital stability cannot be found for the entire assumed range of stellar age. For a continental area of more than 90% of the total surface, no habitable solutions also meeting the requirement of orbital stability exist.

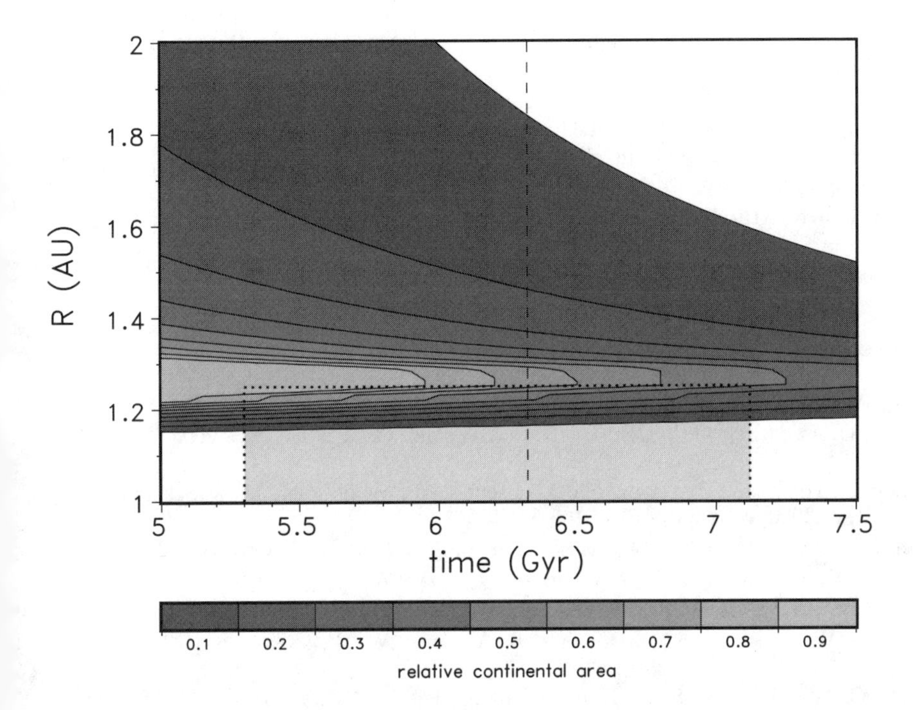


Figure 3. The habitable zone around 47 UMa for the likely value of luminosity $L = 1.54\,L_{\odot}$. The gray shaded areas indicate the extent of the HZ for different relative continental areas. The rectanlge area indicates the permissible parameter space as constraint by the stellar age and the orbital stability limit at 1.25 AU (Franck et al. 2003).

In general, we can state that finding an Earth-like habitable extrasolar planet is the more promising the younger the system and the lower its land coverage on its surface. Younger systems tend to be more geodynamically active and therefore contain more carbon dioxide in the planetary atmosphere. This leads to a stronger greenhouse effect and a broader HZ. As a consequence, habitability is maintained at larger distances from the stars, i.e., regions of lower stellar flux densities. In case of Earth-like planets around 47 UMa, the relevance of this effect is however seriously reduced due to the lack of orbital stability of planets beyond about 1.25 AU. Planets with a relative high percentage of land coverage show a stronger weathering and therefore an enhanced removal of carbon dioxide from the atmosphere. This leads to a weaker greenhouse effect and habitability ceases at smaller ages.

Finally, we want to emphasize that our planet Earth can be classified as a "water world" with a low relative continental area of about $^1/_3$. Therefore, our Earth would have a slight chance of being habitable on a stable orbit in the 47 UMa system.

**Acknowledgements.** This work was supported by the German Federal and Länder agreement (HSPN, grant number 24-04/235, 2000) (S.F., W.v.B., C.B.) and by the University of Texas at Arlington (M.C.).

## References

Cuntz, M., von Bloh, W., Bounama, C., & Franck, S. 2003, Icarus, 162, 214

Fischer, D. A., Marcy, G. W., Butler, R. P., Laughlin, G., & Voigt, S.S. 2002, ApJ, 564, 1028

Franck, S., von Bloh, W., Bounama, C., Steffen, M., Schöberner, D., & Schellnhuber, H.-J. 2000, J.Geophys.Res., 105 E1, 1651

Franck S., Cuntz, M., von Bloh, W., & Bounama, C. 2003, Int.J.Astrobiol., 2 (1), 35

Gehman, C. S., Adams, F. C., & Laughlin, G. 1996, PASP, 108, 1018

Goździewsky, K. 2002, A&A, 393, 997

Henry, G. W., Baliunas, S. L., Donahue, R. A., Soon, W. H., & Saar, S. H. 1997, ApJ, 474, 503

Henry, G. W., Baliunas, S. L., Donahue, R. A., Fekel, F. C., & Soon, W. 2000, ApJ, 531, 415

Jones, B. W., & Sleep, P. N. 2002, A&A, 393, 1015

Jones, B. W., Sleep, P. N., & Chambers, J. E. 2001, A&A, 366, 254

Laughlin, G., Chambers, J., & Fischer, D. 2002, ApJ, 579, 455

Ng, Y. K., & Bertelli, G. 1998, A&A, 329, 943

Noble, M., Musielak, Z. E., & Cuntz, M. 2002, ApJ, 572, 1024

*Order and Chaos in Stellar and Planetary Systems*
*ASP Conference Series, Vol. 316, 2004*
*G. Byrd, K. Kholshevnikov, A. Mylläri, I. Nikiforov, and V. Orlov, eds.*

# The 3:1 Resonance in the 55 Cancri

L. Y. Zhou,[1,2] H. J. Lehto,[1] and J. Q. Zheng[1]

[1] *Tuorla Observatory, Väisäläntie 20, Piikkiö 21500, Finland*

[2] *Department of Astronomy, Nanjing University, China*

**Abstract.** Twenty eight extra-solar planets are located in 13 "multiple planet systems", and some of them are in mean motion resonances (MMR). Recently, the inner two planets around the 55 Cnc, with masses $m_1 = 0.83\,M_{\rm Jupiter}$, $m_2 = 0.20 M_{\rm Jupiter}$ and semimajor axes $a_1 = 0.115$, $a_2 = 0.241$ AU (Fischer et al. 2003), were found to be in a possible 3:1 MMR (Ji et al. 2003). The three resonant angles $\theta_1 = \lambda_1 - 3\lambda_2 + 2\varpi_1$, $\theta_2 = \lambda_1 - 3\lambda_2 + 2\varpi_2$, and $\theta_3 = \lambda_1 - 3\lambda_2 + \varpi_1 + \varpi_2$ may librate around certain values, and the difference between the two periastrons $\Delta\varpi = \varpi_1 - \varpi_2$ may also be locked to a definite value. Our numerical integrations imply a complex orbital motion. Via a new expansion of the Hamiltonian of the planar three-body problem, we analyze the dynamics of this system.

## 1. Numerical Integrations

With a symplectic integrator (Wisdom & Holman1991; Mikkola & Palmer 2000), we integrate 400 sample systems to $10^6$ yr. One third (133) of them collapse during the integrations [the distance between any two planets approach the half of the "Hill stability" criterion (Gladman 1993)]. Since the Lyapunov time $T_L = 10^3$ yr separates these survival systems into two families with distinguishable final inclinations, those 38 survivors with $T_L \geq 10^3$ yr are regarded stable. They are all associated with the 3:1 MMR, and can be divided into three groups, with representative cases shown in Figure 1. Their final inclinations $\leq 0\overset{\circ}{.}1$, and $a_1$, $a_2$ vibrate with very small variations.

In case **a**, $\theta_{1,2,3}$ librate around 215°, 75° and 325°. Simultaneously $\Delta\varpi$ librates around 250°, which is called the *apsidal corotation.* The asymmetric configuration with $\Delta\varpi$ librating around angles other than 0° or 180° has been predicted by Beaugé, Ferraz-Mello, & Michtchenko (2003). In cases **b** and **c**, only $\theta_2$ librates with different centers (80° vs 180°) and amplitudes (40° vs 160°).

The majority of stable systems have configurations as case **a** (28) or **b** (7), and judging from the $T_L$, they are more stable than the only three examples as case **c**. In the rest of this paper, we only discuss the cases **a** and **b**.

## 2. Analytical Model

Ignoring the third planet, we describe this system as a planar three-body problem. To avoid the non-convergent property of the usual analytical models (Ferraz-Mello 1994), we apply a Hamiltonian developed by Beaugé & Michtchenko (2003). Near the 3:1 MMR, two resonant angles $(\sigma_1, \sigma_2)$ are de-

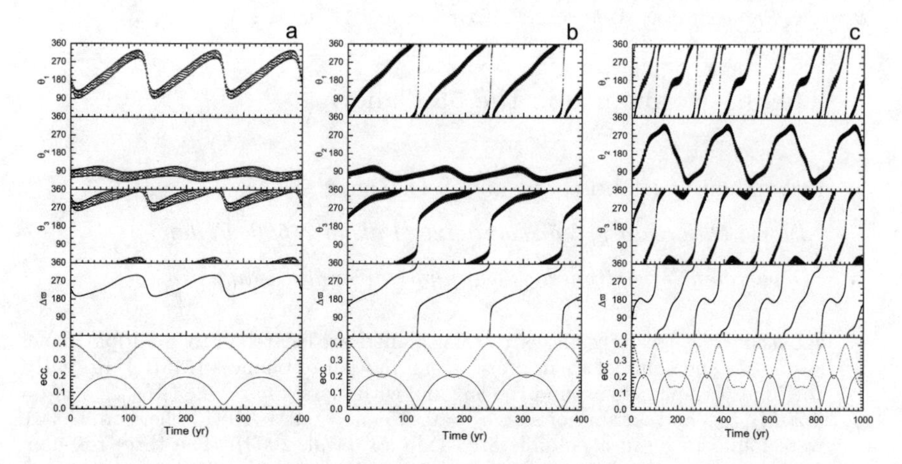


Figure 1. The temporal evolution of $\theta_1$, $\theta_2$, $\theta_3$, $\Delta\varpi$, $e_1$ and $e_2$ (dashed curve indicates $e_2$). Cases **a**, **b**, and **c** are from different initial conditions.

fined as $2\sigma_i = 3\lambda_2 - \lambda_1 - 2\varpi_i$. Their canonical conjugates $(I_1, I_2)$ are $I_i = L_i(1 - \sqrt{1 - e_i^2})$, where $L_i = m_i'\sqrt{\mu_i a_i}$ is the Delaunay variable. The Hamiltonian reads

$$H = -\sum_{i=1}^{2} \frac{\mu_i^2 m_i'^3}{2L_i^2} + \frac{Gm_1m_2}{a_2} \sum_{n=0}^{3} \sum_{j=0}^{j_{\max}} \sum_{k=0}^{k_{\max}} \sum_{u=0}^{u_{\max}} \sum_{l=-l_{\max}}^{l_{\max}} R_{n,j,k,u,l} \times (\alpha - \alpha_0)^n e_1^j e_2^k \cos[2u\sigma_1 + l(\sigma_2 - \sigma_1)], \tag{1}$$

where $\alpha = a_1/a_2$ and $\alpha_0$ is its value in the exact MMR. The coefficients $R_{n,j,k,u,l}$ can be determined beforehand. Besides the energy $H$, there are two other integrals of motion $J_1 = L_1 + \frac{1}{2}(I_1 + I_2)$ and $J_2 = L_2 - \frac{3}{2}(I_1 + I_2)$, generated from the angular momentum conservation and the averaging over the synodic angle.

The variations of $a_{1,2}$ are confined by $3J_1 + J_2 = \text{const}$. But here we simply assume $a_{1,2}$ constants, otherwise the resonance would be destroyed. Similarly,

$$J_{\text{sum}} = J_1 + J_2 = L_1\sqrt{1 - e_1^2} + L_2\sqrt{1 - e_2^2} \tag{2}$$

constrains the variations of $e_{1,2}$. Now we calculate the energy level curves of the Hamiltonian on the $(\sigma_1, \sigma_2)$-plane (Figure 2a). When $e_1$ evolves following Eq. (2), the thick curve in Figure 2a, case **a**, extends to an energy surface. This curve keeps the oblate shape on other sections of this surface (Figures 2a, cases **b**, **c**, **d**). The oblate profile implies $\sigma_1$ has a larger reachable range than $\sigma_2$. During this temporal evolution, $\sigma_1$ varies in a range of $(160°, 400°)$, while $\sigma_2 \in (20°, 60°)$. Correspondingly, $\theta_1$ and $\theta_2$ vary in ranges with widths of 480° (>360°, circulation) and 80° (<360°, libration). Another example indicated by a dashed curve exhibits the possibility of circulation of both $\theta_1$ and $\theta_2$.

Since $\sigma_2$ generally has a smaller variation amplitude, we can further simplify the Hamiltonian by setting $\sigma_2 \equiv \sigma_2^0$ when discussing the 3:1 MMR. Thus we can calculate the contour of Hamiltonian on the $(e_2, \Delta\varpi)$-plane (Figure 2b).

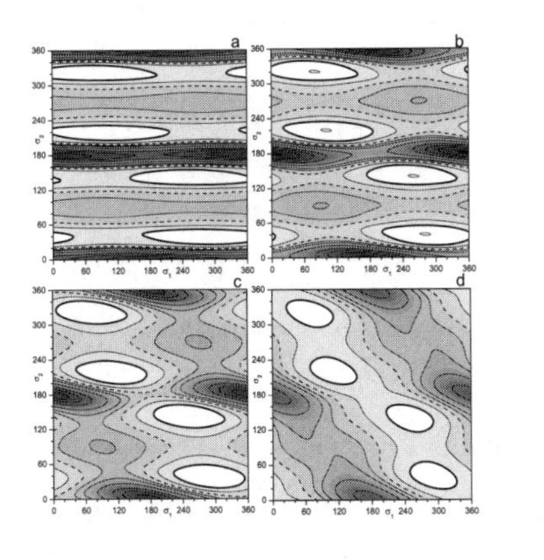


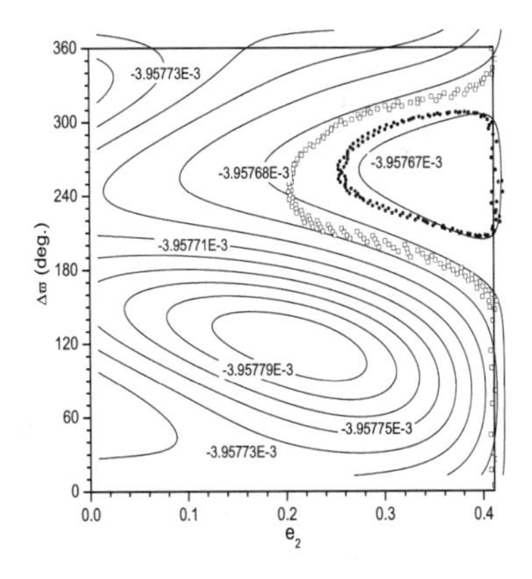


Figure 2. Left (a): The energy level curves of the Hamiltonian on the $(\sigma_1, \sigma_2)$-plane. **a**, **b**, **c**, and **d** are for $e_1 = 0.03$, 0.10, 0.15 and 0.20 respectively, with $e_2$ obtained from Eq. (2). Right (b): The contour of Hamiltonian on the $(e_2, \Delta\varpi)$-plane. Dots and squares represent the numerical results of cases **a** and **b**.

For the inner two planets in 55 Cnc, the initial eccentricities correspond to the right boundary of the box in Figure 2b. Two different fates of a system starting from these initial conditions are distinguished: one is characterized by a libration of $\Delta\varpi$ around $\sim 250°$ and the other is by a circulation of $\Delta\varpi$. In the latter case, the eccentricity $e_2$ can reach a smaller value (correspondingly $e_1$ reaches a higher value). The comparing plots in Figure 2b show the agreements between analytical and numerical results. Different behaviors of $\Delta\varpi$ (circulation or libration) have different varying directions of $\Delta\varpi$ along the right edge of the box in Figure 2b. By analyzing $d\Delta\varpi/dt$ there, we find that the apsidal corotation happens if $\Delta\varpi_0 \in (160°, 330°)$, while if $\Delta\varpi_0 \in (-30°, 160°)$, $\Delta\varpi$ circulates. This is consistent with the results from numerical integrations.

For more details of this work, see Zhou et al. (2003).

**Acknowledgments.** We thank Dr Beaugé for helpful discussions. This work was supported by Academy of Finland (Grant No. 44011). L.Y.Z. was also supported by the Natural Science Foundation of China (No. 10233020).

## References

Beaugé, C., Ferraz-Mello, S., & Michtchenko, T. 2003, ApJ, 593, 1124

Beaugé, C., & Michtchenko, T. 2003, MNRAS, 341, 760

Ferraz-Mello, S. 1994, CeMDA, 58, 37

Fischer, D. et al. 2003, ApJ, 586, 1394

Gladman, B. 1993, Icarus, 106, 247

Ji, J., Kinoshita, H., Liu, L., & Li, G. 2003, ApJ, 585, L139

Mikkola, S., & Palmer, P. 2000, CeMDA, 77, 305

Wisdom, J., & Holman, M. 1991, AJ, 102, 1528

Zhou, L., Lehto, H., Sun, Y., & Zheng, J. 2003, arXiv:astro-ph/0310121

*Order and Chaos in Stellar and Planetary Systems*
*ASP Conference Series, Vol. 316, 2004*
*G. Byrd, K. Kholshevnikov, A. Mylläri, I. Nikiforov, and V. Orlov, eds.*

# The Phenomenon of Double Star 61 Cygni: Some Hypothesis on its Satellites

D. L. Gorshanov, N. A. Shakht, A. A. Kisselev, and E. V. Polyakov

*Central Astronomical Observatory at Pulkovo, Pulkovskoe sh. 65/1, St. Peterbourg 196140, Russia*

**Abstract.** The double star 61 Cygni is known as one with a probable dark satellite. The results of photographic observations with 26″ refractor at Pulkovo observatory (1958–1997) consisted 240 plates with about 3600 individual positions and 39 normal places with mean error of one place equaled $0\farcs008$ are given. A small periodic wave in relative distances projected in RA and noticed by Deutsch & Orlova (1977) on the basis of Pulkovo observations remains in these observations which now have been completed by new data. This wave has the period and amplitude equaled 6.5 yr and $0\farcs010 \div 0\farcs020$ correspondly. A hypothesis on a satellite with the mass near $0.01\ M_\odot$ is discussed. Also some alternative explanations of these fluctuations with the attraction of control double stars series and with the suppositions about changes of the instrumental parameters are tested.

## 1. Introduction

This nearby double star 61 Cygni [$\pi = 0\farcs296, 5\fm4, 6\fm1$, K5V, K7V, $\rho = 30\farcs5, \theta = 150°$ (2000.0)] has attracted the attention of astronomers during many years in connection with its possible planetary satellite and some suppositions concerning to such satellite have been discussed. This problem has been examined in the principal works by Strand (1943), Deutsch & Orlova (1977), Deutsch (1978), Jostis (1983), Marcy & Chen (1992) and in others. It is known at present that the modern data and the space observations did not show any satellites of this double star. But no all observations were enough long-term and regular in order to reveal the influence of satellite with very low mass or with a long period. In any cases the testing of new observations is of use.

## 2. Results of Observations

The photographic observations of the star 61 Cygni have been made with Pulkovo 65-cm refractor since 1958. About 240 plates with 15 exposures on one plate and 39 mean yearly normal places have been obtained. The measurements by means of automatic machine "Fantasy" give the mean error of one yearly positions equaled to $0\farcs008$. The relative distances and positional angles for 39 normal yearly places are published in our article (Gorshanov et al. 2002).

The preliminary orbit of this visual double star with the elements $P =$ 730 yr, $a = 88.5$ a.e., $e = 0.76$, $\omega = 291°$, $i = 142°$, $\Omega = 61°$, $T_p = 1803.5$ has been determined on the basis of present observations, and the method of the Apparent Motion Parameters (Kisselev & Romanenko 1996) has been used. In this method we used the relative radial velocity of the components equaled 1.1 km/s, and the sum of masses $M_{ab}$ has been adopted as 1.3 $M_{\odot}$ according to $M$–$L$ relation.

## 3. Discussion

The possible perturbations in the relative orbital motion have been investigated. The residuals (O–C) in $X$ and $Y$ coordinates where $X = \rho\sin(\theta)$, $Y = \rho\cos(\theta)$ have been analysed. Two following hypotheses are considered:

1) the first one, which has been founded in the works mentioned earlier (Deutsch & Orlova 1977; Deutsch 1978) and which proposed the existence of dark satellite with mass near to $4 \div 10$ mass of Jupiter rotated about one of the components of the star with the period of 6.0 years, and

2) the second one, which has explained the origin of periodic oscillations by changes of astroclimatic conditions, in particular, by the anomal refraction exerted influence to the scale or by seasonal errors evoked periodic fluctuations too.

In the result of our analysis, it was found that the residuals $(O\text{–}C)_Y$ did not show any periods in the range 5–20 years, but $(O\text{–}C)_X$ admitted to yield a periodic wave with the period of about 6.4 years and with the amplitude from $0''.020$ to $0''.010$ which decreased to the end of this set. This period is repeated during all interval of observations but changes from 6.2 (in 1958–1977) to 6.7 years (in 1978–1997).

The distribution of the averaged residuals $\bar{R}_X$ and $\bar{R}_Y$ on the base of the admission of a circular orbit of a satellite with the preliminary period adopted as 6.5 yr is given in Figures 1a, 1b.

In spite of the revealed period we call in question this result and subject to the careful analysis this periodic wave taking into consideration possible instrumental and seasonal errors. The comparison of our series with two control series of ADS 14710 (1976–1997) and ADS 7251 (1960–1999) (21 and 35 annual positions with $0''.008$ and $0''.006$ m.e. correspondingly) received in the same conditions and in the same seasons was carried out. But the relative orbital motions of the control stars did not show any periodic perturbations in the limits of accessible precision.

Also we investigated the change of the geometrical scale over all the space of observations during about 40 years. No any periodic trends suitable to have any influence on the relative distances B–A and to be a cause of the periodic change of residuals $R_X$ mentioned above have been found. Then we tested some models of photocentric orbits, available for the explanation of perturbations in the orbital motion 61 Cygni.

One of the most satisfactory models yields the following elements: $\boldsymbol{P =}$ **6.4 yr**, $\boldsymbol{a = 0''.019}$, $\boldsymbol{e = 0.55}$, $\boldsymbol{i = \pm 54^\circ}$, $\boldsymbol{\omega = 310^\circ}$, $\boldsymbol{\Omega = 9^\circ}$, $\boldsymbol{T_0 = 1962.6}$ and the low limit the mass of possible satellite is 0.014 $M_{\odot}$. The corresponding ephemeris for $X$ coordinate is given in Figure 2a in the comparison with the annual residuals $R_X$.

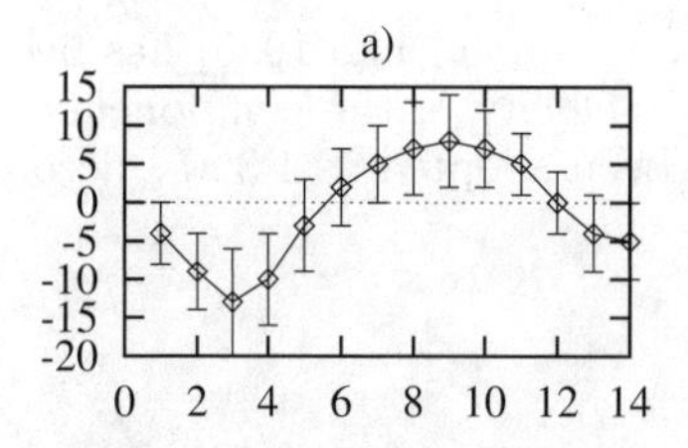


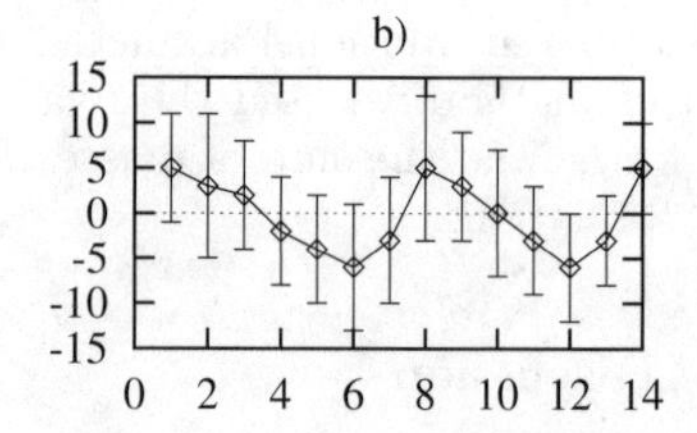


Figure 1. The averaged residuals a) $\bar{R}_X$ and b) $\bar{R}_Y$ reduced to one period 6.5 yr and given in 0."001 with mean errors. Abscissae are time in units of 0.5 yr.

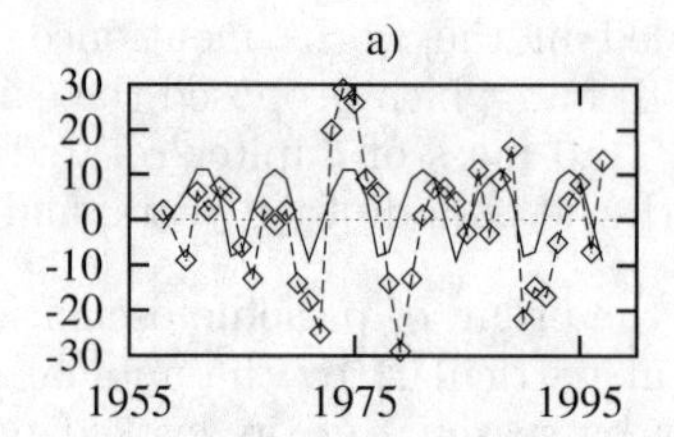


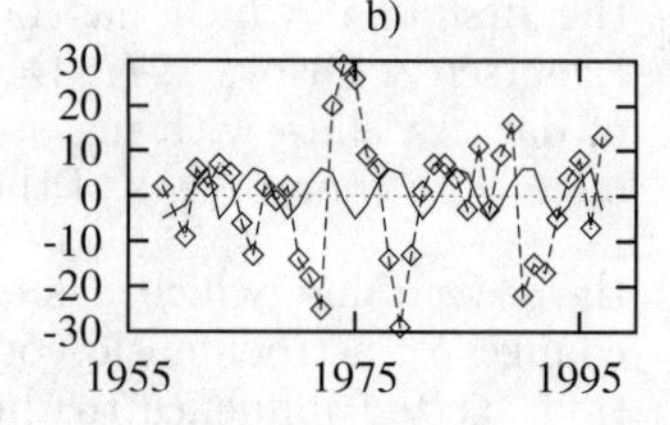


Figure 2. Residuals $R_x$ (dashed line) and ephemeris (solid line) calculated in admission of a satellite's presence: a) in accordance with our data and b) in accordance with Deutsch's data, given in 0."001.

The principal conclusion of our work consists in that fact as our new observations with 26″ refractor and reexamination of all data set obtained in 1958–1997 repeated a small periodic wave, now with the period of 6.5 years, in relative distances in $X$-coordinate, that does not contradict the work by Deutsch (1978), although our orbital elements differ from his model (see comparisons of our observations with the ephemeris corresponding to our and his results in Figures 2a, 2b). Our study will be continued and we hope to compare these results with Pulkovo Carte du Ciel astrograph's data which we are analysing now.

## References

Deutsch, A. N, & Orlova, O. N. 1977, AZh, 54, 2, 327 (in Russian)

Deutsch, A. N. 1978, Pis'ma AZh, 4, 95 (in Russian)

Gorshanov, D. L., Shakht, N. A., Kisselev, A. A., & Polyakov, E. V. 2002, Izv.GAO, 216, 100 (in Russian)

Jostis, P. J. 1983, Low Obs.Bull. 167, 16

Kisselev, A. A., & Romanenko, L. G. 1996, AZh, 32, 6, 875 (in Russian)

Marcy, G. W., & Chen, G. H. 1992, ApJ, 390, 550

Strand, K. A. 1943, PASP, 55, 322

# Part IV

# Star Clusters

*Order and Chaos in Stellar and Planetary Systems*
*ASP Conference Series, Vol. 316, 2004*
*G. Byrd, K. Kholshevnikov, A. Mylläri, I. Nikiforov, and V. Orlov, eds.*

# Kinematics of Stars in Old Open Cluster M67

A. V. Loktin

*Astronomical Observatory, Ural State University, Ekaterinburg, Russia*

**Abstract.** Preliminary results of the investigation of the kinematics of stars in open cluster M67 are discussed. This study is performed on the basis of star proper motions from a joint catalogue prepared for M67 cluster field. Cluster probable members are extracted by the use of modified Sanders method and photometric criteria. This provided us the sample of 484 probable cluster members. Radial dependence of mean radial component of stars proper motions on radius shows that most of the dependence can be attributed to the geometrical effect of convergemce to radiant. Neither halo nor core of the cluster experience statistically significant expansion or contraction. Mean tangential components of proper motions reveal some hints on rotation of cluster core, but the rotation of halo is undetectable. Some properties of velocity dispersions of cluster stars are considered.

## 1. Introduction

Proper motions of stars in open clusters carry important information about the inner kinematics of this kind of objects. But low accuracy of proper motions prevented the usage of this kind of information for the investigations of cluster stars kinematics. In present times a lot of determinations of proper motions are available for some clusters, but these proper motions are predominantly relative—presented not in one system. But there exist the possibility to combine such catalogues in one catalogue providing high precision set of homogeneous data. Such combined catalogue was recently made for M67 cluster, and it was shown that proper motions in this catalogue have mean errors less than 1 milliarcsecond/year. The method of combining of the catalogues will be published in one of issues of Astronomical and Astrophysical Transactions.

The open cluster M67 was chosen for the first investigation of inner kinematics for a good many of reasons. First, it is a well investigated cluster, for which a lot of catalogues of relative proper motions are published. Second, it is rich in stars which lead to reliable statistical estimates of kinematical parameters. Third, it is quite a distant cluster to estimate the limit of capabilities of proper motions for the investigation of the inner kinematics of cluster stars. Next, it is one of the oldest open clusters which lived during a dozens of relaxation times and the kinematics of its stars may be not complicated by group motions.

## 2. The Data

For the investigation of inner kinematics of M67 cluster the data from combined catalogue of proper motions in the field of this cluster was used. This catalogue was compiled from 6 catalogues including Tycho-2 catalogue as reference one. Modified Sanders method gives 484 probable cluster members in this list. Five stars were rejected from this list by photometric criterion, and seven stars because of the greatest values of errors of proper motions, which may be caused by misidentifications. Stellar density radial dependence for probable members is shown in Figure 1.

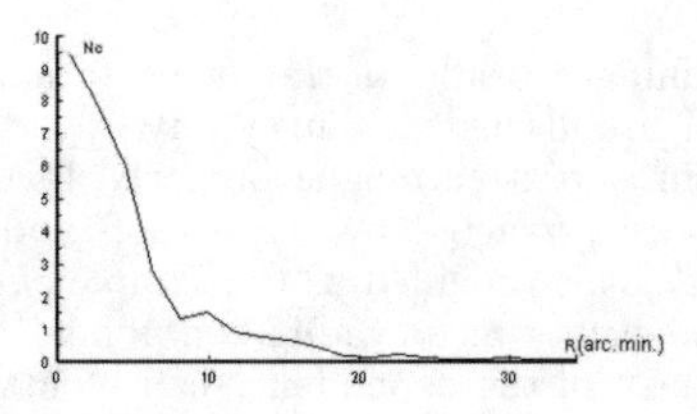


Figure 1. Projected star density for probable cluster members.

The value of $0\overset{\circ}{.}9$ can be taken as the radius for cluster halo which is equal to 14 pc for linear radius with cluster distance 908 pc, this value is equal to the estimate of tidal radius of the cluster for feasible value of total cluster mass of 1000 solar masses. Two probable members have greater distance from cluster center than accepted value of radius, these stars were discarded from the sample. For cluster core radius we can use the value of $9'$ (see Figure 1) which gives the value of linear radius 2.4 pc. Finally we have a sample of 452 stars, from this number 80% of stars have the errors of proper motions less than 1 mas/year.

## 3. Mean Velocities

First of all we subtracted the proper motion of cluster as a whole from all proper motions to derive residual proper motions. Components of cluster proper motion was derived as a modes of distributions of individual star proper motion components $\langle\mu_x\rangle = -7.62 \pm 0.04$ mas/year, $\langle\mu_y\rangle = -3.96 \pm 0.03$ mas/year. Then we rotate proper motions axes to get components in radial and tangential directions $\mu_R$, $\mu_T$ in reference to cluster center. Let us consider mean values of radial components of residual proper motions of cluster members. Figure 2 shows the radial dependence of mean values of these components of cluster stars proper motions where stars were divided in 9 radial groups.

Vertical bars show standard errors, bars for the rightmost point corresponds to 0.6 km/s and the distance of this point from cluster center is equal to 9.5 pc. Here horizontal line marks zero residual velocity, and solid curve shows the geometrical effect of convergence to radiant calculated for mean radial velocity of cluster $v_r = +33.6$ km/s from the expression $\mu_\theta = -0.4602 v_R \tan(\theta/2)/r$, where $\theta$ denotes angular distance from cluster center and $r$ is cluster distance from the Sun in pc. One can see from the figure that only three points out of nine lie above the solid curve, and one of these points is the most uncertain outer point. This can indicate overall contraction of the cluster. Three inner points

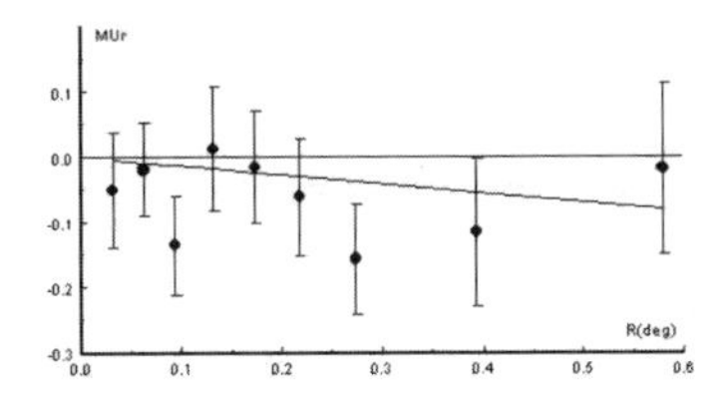


Figure 2. Mean radial components of proper motions of cluster members.

corresponding to cluster core give mean value of contraction $-0.26 \pm 0.15$ km/s, though the remaining 6 give the mean $-0.09 \pm 0.13$ km/s—the value obviously statistically insignificant. We can say that inner parts of cluster may slowly contract, but it is necessary to increase quality of the data to get unambiguous decision of expansion or contraction of the cluster or its parts.

Let consider tangential components of proper motions of probable cluster members. If rotation axis of cluster lies near the line of sight, then proper motions will reveal the effect of cluster rotation. Mean values of tangential components of proper motions for stars grouped in 9 radial groups are shown in Figure 3.

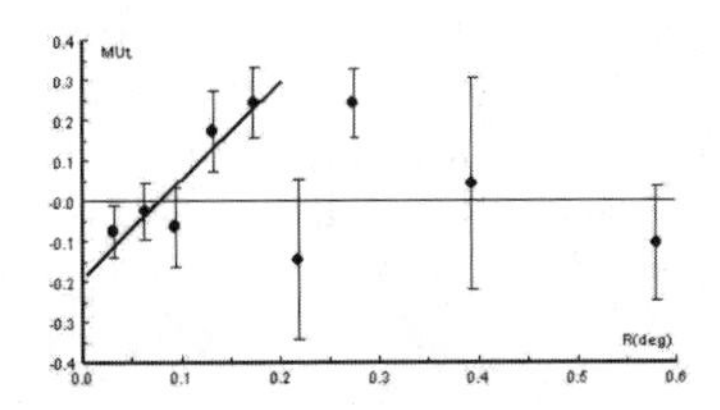


Figure 3. Mean tangential components of proper motions of cluster members.

Here we cannot see some rotation for the whole cluster, but as for core region (5 innermost points) there is some radial dependence. Least squares solution for straight line for these points gives $V_T = -0.81 + 0.66R$, with standard error for inclination coefficient 0.17. Statistical significance of this coefficient may indicate that there is some rotation of cluster core, and on the core boundary we get the value of linear rotational velocity of 1.6 km/s. This may mean that the core is kinematically detached from outer region of the cluster. If the rotational axis does not coincide with the line of sight then cluster rotation must appear in radial velocities. The azimuth dependence of radial velocities of cluster stars is shown in Figure 4, where one can not see any manifestation of rotation.

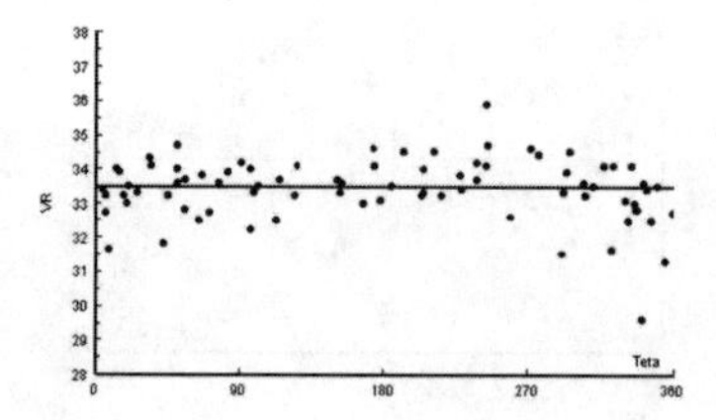

Figure 4. Azimuth run of radial velocities of cluster members.

## 4. Dispersions of Residual Velocities

Frequency distributions of residual velocities calculated from two components of proper motions are shown in Figure 5. We can notice the marked departures of these distributions from normal distribution. This prevents the approximation of distribution of cluster stars velocities by three-dimensional normal distribution. The dispersions of two distributions (standard deviations) shown in Figure 5 are equal to 3.26 km/s and 2.94 km/s, and accounting for the mean error of proper motions in our catalogue, corresponding to 2.23 km/s we have 2.37 km/s and 1.92 km/s as the estimates of true values of dispersions. F-criterion shows that that the difference between two estimates is statistically significant even for 99% level of significance. This means that velocity distribution of cluster stars has no explicit spherical symmetry.

For the discussion of velocity dispersion dependence on stellar masses the stars of the catalogue were divided in 5 groups according to their places on color-magnitude diagram. These groups are shown in Figure 6 and correspond to red giants (1st group), subgiants (2nd), double stars (4th), main sequence single stars (5th), and the area of turn-off point of main sequence where both double and single stars exist (3rd group).

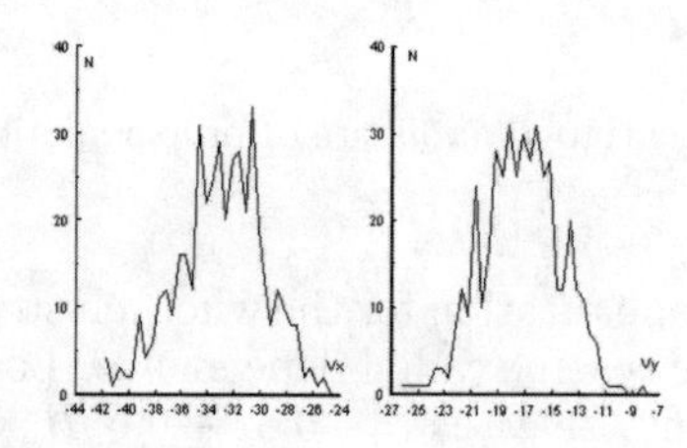

Figure 5. Frequency distributions of velocity components of star motions in two directions.

The estimates of velocity dispersions for these groups of stars are shown in the table.

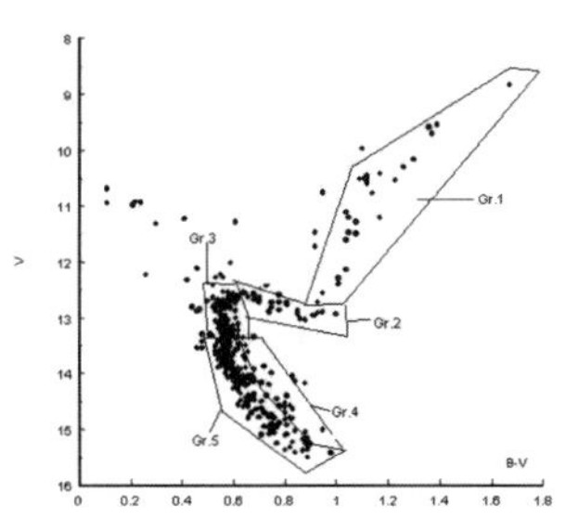


Figure 6. HR diagram for probable members of open cluster M67.

Table 1. Estimates of velocity dispersions for probable members of M67 cluster

| Group | N | Mass | $\sigma_R$ | $\sigma_t$ | S | $mV^2$ |
|---|---|---|---|---|---|---|
| 1 | 27 | 1.56 | 1.29 | 1.42 | 0.73 | 5.74 |
| 2 | 29 | 1.55 | 1.33 | 0.65 | 0.56 | 3.40 |
| 3 | 133 | - | 1.68 | 1.94 | 0.52 | - |
| 4 | 60 | - | 1.85 | 2.24 | 0.55 | - |
| 5 | 162 | 1.00 | 1.89 | 2.24 | 0.62 | 8.59 |

Columns in Table 1 contain group number (1), amount of stars in group (2), mean stellar mass $m$ in solar masses (3), values of radial and tangential dispersions in km/s with errors subtracted (4), mean errors for stars in groups (5) and mean values of $mV^2$ as a measure of mean kinetic energy, where $V^2 = \sigma_R{}^2 + \sigma_t{}^2$. Groups 3 and 4 have no estimates of mean mass because they are mixtures of single and double stars and to calculate mean masses we have to know the percentage of double stars and the distribution of mass ratios of double star components which we do not know.

The values in the Table 1 lead to some conclusions. We may expect that single and double stars interact differently with other stars. But both dispersions of single main sequence stars (group 5) are nearly the same as for double stars (group 4). The mean kinetic energy may be increased for stars of smaller masses. This may be associated with mass segregation in the cluster. The ratio of two dispersions nearly equals unity which means that orbits of stars in cluster are far from linear central orbits.

## 5. Conclusions

The results presented here have to be regarded as preliminary. Those are based on catalogues published after 1965. There are some older catalogues, and now our experience says that older catalogues have good enough quality to be included in such investigations. In near future precision of proper motions and number of members will be increased. It is very important for evaluation of velocity dispersions because the intrinsic nonequality of precision of proper motions leads to hard problems.

Cluster M67 is very old inplying a rather short interval of stellar masses we can investigate. Now our group works to prepare the catalogue of proper motions of Pleiades as much younger cluster for analogous investigations.

In any case we demonstrate that now for some not very distant from the Sun clusters we can use proper motions for the investigation of cluster stars kinematics.

*Order and Chaos in Stellar and Planetary Systems*
*ASP Conference Series, Vol. 316, 2004*
*G. Byrd, K. Kholshevnikov, A. Mylläri, I. Nikiforov, and V. Orlov, eds.*

# New Mechanism of Energy Transformation from Supernovae to Star Cluster

Vladimir Surdin

*Sternberg Astronomical Institute, Universitetskij pr. 13, Moscow 119992, Russia*

**Abstract.** A gravitational mechanism for the conversion of the energy of the expanding shell of a supernova into the energy of a star cluster is suggested. It is assumed that as long as the supernova shell continues to expand, it does not go beyond the cluster. After its expansion has completely stopped, the shell material gets mixed with the interstellar material of the cluster. Thus, the gas does not leave the limits of the cluster, but the "bubbles" pulsating in it convert some of the supernova energy into kinetic energy of the stars. It is demonstrated that frequent supernova bursts in a dense cluster rich in gas provide for a substantial inflow of mechanical energy and can perceptibly affect the evolution of the cluster as a whole.

## 1. Introduction

Supernova bursts in star clusters rich in gas are interesting first of all for the complex effects of interaction between the shells shed by the supernovae and the surrounding gas "at rest". In the traditional treatment of the dynamical aspect of this interaction, the kinetic energy stored in the shell of a supernova frequently proves to be sufficient for it to overcome the gravitational field of the cluster. In that case, the shell in its way outside "rakes up" the gas of the cluster and sweeps it out. The gravitational field being thus weakened, the cluster starts expanding; it loses its fastest stars and, given appropriate conditions, can completely disintegrate (Zwicky 1953; McCrea 1955; Surdin 1975).

But unexplored possibility exists that the supernova shell energy can be converted into the mechanical energy of the cluster with the gas being fully retained there.

Let us consider a simple problem of celestial mechanics on the motion of a probe particle around a massive body. If this body suddenly vanishes at the moment the particle was at a distance of $r_1$ from it, the particle would continue its free motion with an orbital velocity $v$. If, at some subsequent moment, the body appears in its initial place again, the particle, while moving at the same velocity, would find itself at a distance $r_2$ from it. At $r_2 > r_1$, the system acquires some energy; at $r_2 < r_1$ it loses it.

If the particle originally moved on an elliptical orbit and the time interval between the disappearance and the reappearance of the central body exceeded $2a/v_a$, where $a$ is the semimajor axis of the orbit and $v_a$ is the particle velocity at the apocenter, then the inflow of energy into the system would be positive. This inflow is provided by the mechanism responsible for the movement of the central body.

The example described above is but a crude analog of the expansion and subsequent contraction of the supernova shell in the gravitational field of a star cluster. This example demonstrates the principal possibility of conversion of the energy of a massive pulsating supernova shell into the mechanical energy of a star cluster. This mechanism was first suggested by the author when studying the formation of star clusters (Surdin 1975). For that specific problem estimates showed a relatively low efficiency of such an energy conversion. But new observational data obtained on the gas densities and supernova burst frequencies in galactic nuclei indicate that this mechanism should be taken into consideration when investigating the dynamical evolution of these nuclei.

## 2. Gravitational Interaction between the Massive Shell and Stars

Consider first a system free from dissipation of mechanical energy. We define the coefficient $\eta$ of conversion of the supernova shell energy into the cluster energy as the ratio between the energy imparted to the stars of the cluster and the initial shell energy $E$. By virtue of the conservative character of the system, we have

$$\eta = \frac{A_+ - A_-}{E}, \tag{1}$$

where $A_+$ is the work performed by the shell against the gravitational force up to the moment its expansion reaches its maximum and $A_-$ is the work done by the gravitational field to contract the shell.

To estimate $\eta$, we use a simple model: represent the star cluster as a homogeneous sphere of radius $R$ with a constant ratio of the total mass $M_{\rm s}$ of the stars to the total mass $M_{\rm g}$ of the gas. The total mass of the cluster being constant as well, $M = M_{\rm s} + M_{\rm g}$. In the initial state, the star and gas components have homogeneous densities, $\rho_{\rm s} = 3M_{\rm s}/(4\pi R^3)$ and $\rho_{\rm g} = 3M_{\rm g}/(4\pi R^3)$. The kinetic energy per unit mass of stars and gas is the same and equal to $V^2/2$, where $V$ is the root-mean-square velocity of the stars in the cluster. By the virial theorem, the internal kinetic energy of the cluster is equal to the half of its gravitational energy, $MV^2/2 = 3GM^2/(10R)$, whence we obtain $V = \sqrt{0.6GM/R}$.

Let the burst of a supernova take place at the center of the cluster, i.e., a spherically symmetric ejection of the supernova shell with the energy $E$ and initial mass small in comparison with the mass of the surrounding gas it has raked up in the course of expansion. Let the shell expand up to the maximum radius $R_{\rm m} < R$. Obviously, this assumption imposes a certain requirement on the cluster parameters and the shell energy, namely, $E < GMM_{\rm g}/R$ and we will take it that this requirement is fulfilled. The work done by the expanding shell against the gravitational force is made up of two components: the energy spent by the shell in overcoming the gravitational field of the cluster stars,

$$A_1 = \int_0^{R_{\rm m}} G\frac{M_{\rm sh}(r)M_{\rm s}(r)}{r^2}\,dr = \frac{16\pi^2}{45}G\rho_{\rm s}\rho_{\rm g}R_{\rm m}^5, \tag{2}$$

and the change of the gravitational energy of the shell itself in its own field,

$$A_2 = \int_0^{R_\mathrm{m}} \int_r^{R_\mathrm{m}} G\frac{M_\mathrm{sh}(r)}{r_1}\, dM_\mathrm{sh}(r)\, dr_1 = \int_0^{R_\mathrm{m}} G\frac{M_\mathrm{sh}^2(r)}{2r^2}\, dr = \frac{16\pi^2}{90} G\rho_\mathrm{g}^2 R_\mathrm{m}^5, \tag{3}$$

where $M_\mathrm{sh}(r) = 4\pi r^3 \rho_\mathrm{g}/3$ and $M_\mathrm{s}(r) = 4\pi r^3 \rho_\mathrm{s}/3$ are the current masses of the shell and the star cluster respectively. In that case, the total work done by the supernova shell by the moment its expansion reaches its maximum will be

$$A_+ = A_1 + A_2 = \frac{16}{90} G\pi^2 \rho_\mathrm{g} R_\mathrm{m}^5 \left(2\rho_\mathrm{s} + \rho_\mathrm{g}\right). \tag{4}$$

Considering that the shell has no other energy losses, we determine $R_\mathrm{m}$ such that the initial energy of the shell is equal to the work done by it,

$$R_\mathrm{m} = \left[\frac{90E}{16G\pi^2 \rho_\mathrm{g} \left(2\rho_\mathrm{s} + \rho_\mathrm{g}\right)}\right]^{1/5}. \tag{5}$$

So far we have assumed that the shell expands rapidly and the distribution of the stars in the cluster remains unchanged in this time. But, as the gas leaves the sphere of radius $R_\mathrm{m}$, the gravitational potential energy of the stars contained there increases, so that their virial distribution gets disturbed: following the gas, the star subsystem filling the shell also tends to expand. For this reason, the density of stars in the hollow of the shell drops to a certain value of $\rho'_\mathrm{s0}$ in the dynamic time $t_\mathrm{dyn} = R_\mathrm{m}/V$. From this point (which we take to be $t = 0$) onwards the shell starts falling back into its hollow under nonstationary gravitational field conditions. The change of the density of stars inside the shell is determined by their flow through the sphere of radius $R_\mathrm{m}$,

$$\frac{d}{dt} M_\mathrm{s}(R_\mathrm{m}) = \frac{dM_+}{dt} - \frac{dM_-}{dt}, \tag{6}$$

where

$$dM_+ = \rho_\mathrm{s} \frac{V}{2\sqrt{3}} S\, dt, \qquad dM_- = \nu(\rho'_\mathrm{s})\, dM_+,$$

$S$ is the surface area of the shell, and $\rho'_\mathrm{s}$ is the density of the star mass inside the shell. The density of the star mass outside the shell is assumed to be constant and equal to that of the star component of the unperturbed star cluster. The choice of the coefficient $\nu(\rho'_\mathrm{s})$ is governed by the normalization conditions for the outward star flow. We take it that $\nu(\rho'_\mathrm{s}) = \rho'_\mathrm{s}/\rho_\mathrm{s}$ and obtain from eq. (6) the following equation for the change of the density of stars in the hollow of the supernova shell:

$$\frac{d\rho'_\mathrm{s}}{dt} = \frac{V\sqrt{3}}{2R_\mathrm{m}} (\rho_\mathrm{s} - \rho'_\mathrm{s}). \tag{7}$$

The solution of this equation is

$$\rho'_\mathrm{s} = \rho_\mathrm{s} + (\rho'_\mathrm{s0} - \rho_\mathrm{s}) \exp\{-t/t_\mathrm{ef}\}, \tag{8}$$

where $t_\mathrm{ef} = 2R_\mathrm{m}/(\sqrt{3}V)$. Henceforward we will assume that $t_\mathrm{ef} = t_\mathrm{dyn}$.

The density $\rho'_{s0}$ of stars inside the shell at the beginning of the contraction stage can be determined considering that the dynamical behavior of the subcluster restricted to the size of the shell is independent, on time intervals of the order of $t_{dyn}$, of the dynamics of the remaining cluster. The radius of the expanded subcluster can then be easily found (Surdin 1975; Hills 1980):

$$R' = R_m \frac{M_s(R_m)}{M_s(R_m) - M_g(R_m)} = R_m \frac{\rho_s}{\rho_s - \rho_g}, \tag{9}$$

where $M_g(R_m)$ is the mass of the gas that filled the volume of the shell prior to its expansion. Consequently, for $\rho_g \le \rho_s$

$$\rho'_{s0} = \rho_s \left(1 - \frac{\rho_g}{\rho_s}\right)^3, \tag{10}$$

and for $\rho_g > \rho_s$ we will obviously obtain $\rho'_{s0} = 0$.

In the present work, we are considering for the first time the effect of pulsation of the interstellar medium within the limits of a star cluster in an attempt to get a simple analytical estimate of the efficiency of conversion of the energy of the gas component into the kinetic energy of stars. It is therefore quite justifiable to reduce the problem on the free fall of the gas in a region of variable mass density to that on the fall of the gas in a region of constant density equal to the average density within the limits of the shell in its free-fall time $t_{ff}$; we will designate this density as $\langle\rho'_s\rangle_{ff}$,

$$\langle\rho'_s\rangle_{ff} = \frac{1}{t_{ff}} \int_0^{t_{ff}} \rho'_s(t)\, dt = \rho_s + (\rho'_{s0} - \rho_s) \frac{t_{dyn}}{t_{ff}} \left[1 - \exp(-t_{ff}/t_{dyn})\right]. \tag{11}$$

In as much as the free-fall time inside the shell volume is close to the dynamical time, we put $t_{dyn} = t_{ff}$. In that case, the average density in the volume will be

$$\langle\rho'_s\rangle_{ff} = \rho_s + (\rho'_{s0} - \rho_s)\left(1 - \frac{1}{e}\right), \tag{12}$$

To estimate the coefficient of conversion of the kinetic energy of the shell into the energy of the cluster, we assume that the shell expansion and contraction processes are symmetrical,

$$\begin{cases} A_+ = \dfrac{16}{90} G\pi^2 \rho_s \rho_g R_m^5 \left(2 + \dfrac{\rho_g}{\rho_s}\right), \\ A_- = \dfrac{16}{90} G\pi^2 \langle\rho'_s\rangle_{ff} \rho_g R_m^5 \left(2 + \dfrac{\rho_g}{\langle\rho'_s\rangle_{ff}}\right). \end{cases} \tag{13}$$

In that case, assuming that the entire initial kinetic energy of the shell is spent to overcome gravitation, we get the energy conversion coefficient

$$\eta = \frac{A_+ - A_-}{E} = 1 - \frac{A_-}{A_+} = 1 - \frac{2\langle\rho'_s\rangle_{ff} + \rho_g}{2\rho_s + \rho_g}. \tag{14}$$

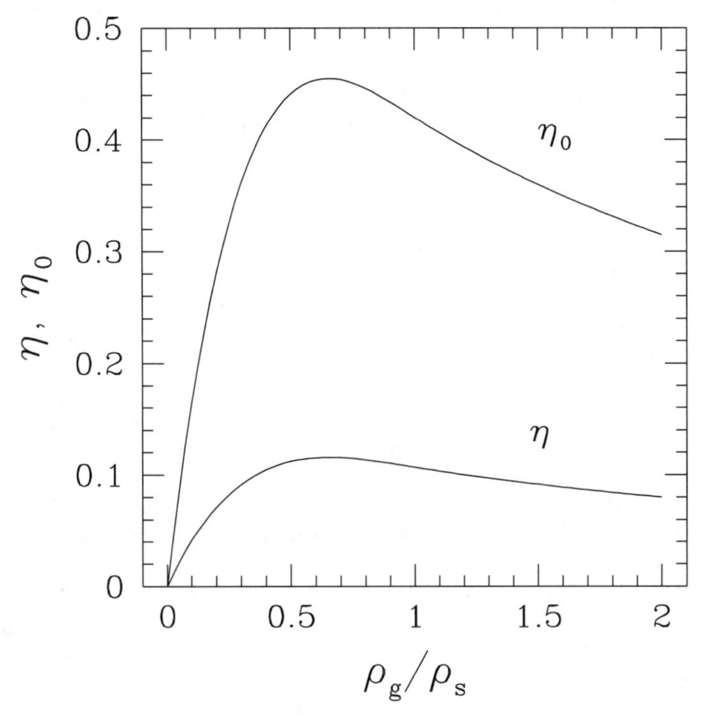


Figure 1. Coefficient $\eta$ of conversion of the supernova shell energy into the cluster energy versus the relative gas density ($\rho_g/\rho_s$): without allowance for radiative energy losses ($\eta_0$) and with due regard for the loss of the shell energy by radiation ($\eta = \varepsilon\eta_0$). It is assumed that the gas does not leave the cluster.

Substituting into eq. (14) expression (12) for the average density in the hollow formed, we obtain, subject to eq. (10), at $\rho_g \le \rho_s$,

$$\eta_0 = 0.63 \times \frac{3k - 3k^2 + k^3}{1 + 0.5k}, \tag{15}$$

where $k \equiv \rho_g/\rho_s$. At $\rho_g > \rho_s$, we will obviously get $\eta_0 = 0.63/(1 + 0.5k)$. The subscript 0 on the energy conversion coefficient indicates that formula (15) is obtained under the assumption that the kinetic energy of the shell is completely converted into its potential energy in the gravitational field of the stars and the gas. Obviously, this overestimates the efficiency of the process, because some of the shell energy is undoubtedly thermalized and emitted. We estimate the contribution from this process.

## 3. Allowance for the Radiation Emission by the Supernova Shell

The evolution of supernova shell in a homogeneous gas can be reduced to three consecutive stages:

1. The shell spreads freely; its initial mass exceeds the mass of the interstellar material raked by it; the entire energy of the shell is constituted by its initial kinetic energy $E_{\rm kin}$ and is conserved.

2. When the mass of the gas raked by the shell is several times greater than the initial mass of the shell, the kinetic energy of the shell starts to actively pass into the thermal energy of the gas being raked by it. The expansion remains adiabatic.

3. When the temperature of the gas reaches $\sim 5 \times 10^5$ K, it begins to intensively emit thermal energy. A cold shell front is thus formed, beyond which the rarefied hot gas continues to expand adiabatically. Once the shell has cooled down as a result of its radiative energy losses, the kinetic energy of the remainder of the supernova amounts to $\varepsilon = 0.2 \div 0.3$ of its initial energy.

At the moment of intensive cooling, the shell radius is

$$R_{\rm cool} = 20 \text{ pc} \times \left(\frac{E}{10^{51} \text{ erg}}\right)^{0.3} \left(\frac{n_{\rm g}}{1 \text{ cm}^{-3}}\right)^{-0.4}, \tag{16}$$

where $n_{\rm g}$ is the interstellar gas concentration. Obviously, at $R_{\rm m} < R_{\rm cool}$, we can disregard the dissipative processes and use formula (15) for the energy conversion coefficient. Otherwise, this coefficient can be estimated as $\eta \approx \varepsilon\eta_0 \approx 0.25\eta_0$. The condition for the passage from the first value of this coefficient to the second one is the equality $R_{\rm m} = R_{\rm cool}$. Let us find out when this equality is satisfied. We normalize formula (5)

$$R_{\rm m} = 400 \text{ pc} \times \left(\frac{E}{10^{51} \text{ erg}}\right)^{0.2} \left(\frac{n_{\rm g}}{1 \text{ cm}^{-3}}\right)^{-0.4} (1 + 2/r)^{-0.2}, \tag{17}$$

and obtain from (16) and (17) the ratio

$$\frac{R_{\rm m}}{R_{\rm cool}} = 20 \left(\frac{E}{10^{51} \text{ erg}}\right)^{-0.1} (1 + 2/r)^{-0.2}. \tag{18}$$

As can be seen, the condition $R_{\rm m} = R_{\rm cool}$ is fulfilled when $k \sim 10^{-6}$ and, accordingly, when the energy conversion coefficient $\eta_0 \sim 10^{-6}$, and so the process is of no practical interest. Consequently, the dissipative processes in our problem should be taken into consideration ($R_{\rm m} > R_{\rm cool}$) and the resultant energy conversion coefficient should be estimated as

$$\eta = \varepsilon\eta_0 \approx 0.16 \times \frac{3k - 3k^2 + k^3}{1 + 0.5k}. \tag{19}$$

This formula can be considered quite suitable for astrophysical estimation purposes.

## 4. Conclusion

The above-considered process of conversion of the supernova expanding shell energy into the star cluster energy can take place only in sufficiently dense clusters rich in gas, where supernova bursts occur intensively. Such clusters are galactic nuclei that meet all the conditions stipulated. Direct star-star collisions can be an important gas-supply process in a dense nucleus. In addition to the gas ejection, as a result of stellar collisions and mergence, supergiant stars can form that can then burst as supernovae.

In the course of stellar collisions, the galactic nucleus is filled with a vast amount of gas ($k \sim 1$), with supernova bursts frequently occurring there. As can be seen from formula (19), under such conditions $\eta \sim 10\%$. With the specific power of the bursts being $\sim 10^{50}$ erg/$M_{\odot}$, this leads to the transfer of such specific kinetic energies to the stars of the galactic nucleus as correspond to velocities of $\sim$1000 km/s. It is precisely such velocities that are typical of the nuclei of active galaxies, such as the Seyfert galaxies. Thus, at a certain stage in the evolution of a galactic nucleus, supernova bursts can provide, by the above-described mechanism, for a noticeable inflow of mechanical energy and thereby slow down the rate of its contraction and formation of a massive black hole in its central part.

**Acknowledgments.** This work was partly supported by RFBR (grant 03-02-16288).

## References

Zwicky, F. 1953, PASP, 65, 205

McCrea, W. H. 1955, Observatory, 75, 206

Surdin, V. G. 1975, On the evolution of globular clusters and the origin of halo stars, Thesis (Moscow State University)

Hills, J. G. 1980, ApJ, 235, 986

*Order and Chaos in Stellar and Planetary Systems*
*ASP Conference Series, Vol. 316, 2004*
*G. Byrd, K. Kholshevnikov, A. Mylläri, I. Nikiforov, and V. Orlov, eds.*

# Interaction of the Compact Stellar Cluster with an Accretion Disk in AGNs

Ch. T. Omarov,[1,2] R. Spurzem,[2] and E. Y. Vilkoviskij[1]

[1]*Fessenkov Astrophysical Institute, 480020 Almaty, Kazakhstan*

[2]*Astronomisches Rechen-Institut, Mönchhof-Strasse 12-14, 69120 Heidelberg, Germany*

**Abstract.** Dynamical evolution of stellar orbits interacting with an accretion disk in the active galactic nuclei is studied in this paper.

## 1. Introduction

The interaction of a single star with an accretion disk (AD) of AGN was considered in many works (e.g., Syer et al. 1991; Artymovicz et al. 1993; Vokrouhlicky & Karas 1998). Rauch (1995, 1999) had considered the evolution of the compact stellar cluster (CSC) with the AD, but he numerically calculated two limited cases, taking into account pure star-disk interaction or star-star interactions. Vilkoviskij & Czerny (2002) are the first who considered both interactions acting simultaneously and investigated a semi-analytical model of the system CSC+AD. The main result was that the stars of the CSC do not evolve strictly to the plane of the accretion disk, but preserve some velocity normal component in respect to the accretion disk. It is apparent that the dissipative interaction of the CSC with the AD plays an essential role in the stellar dynamics of the stellar cluster and, accordingly, in the AGN evolution as a whole. However, the problem of the evolution taking into account of both two-body gravitational interaction and the star-disk interaction is not yet solved, despite a big number of works were dedicated to the stellar dynamics in central parts of AGNs. And here we present the results of the $N$-body simulations of the evolution of the CSC interacting with a gas disk.

## 2. Numerical Results

The star-disk interaction is determined by the drag force

$$F_d = C_d \pi r_s^2 \rho_d V_{\rm rel}^2, \tag{1}$$

where $C_d \sim 1$, $\rho_d = \rho_0 \left[\frac{R(x,y)}{R_0}\right]^{-3/8} e^{\frac{-z^2}{2h_z^2}}$ is the gas density, $r_s$ is the stellar radius, $V_{\rm rel}$ is the local relative star-disk velocity,

$$\ddot{\vec{r}}_i = \sum_{i \neq j} G m_j \frac{\vec{R}_j}{R_j^3} + \frac{1}{m_i} C_d \rho_d \pi r_s^2 V_{\rm rel} \vec{V}_{\rm rel}, \tag{2}$$

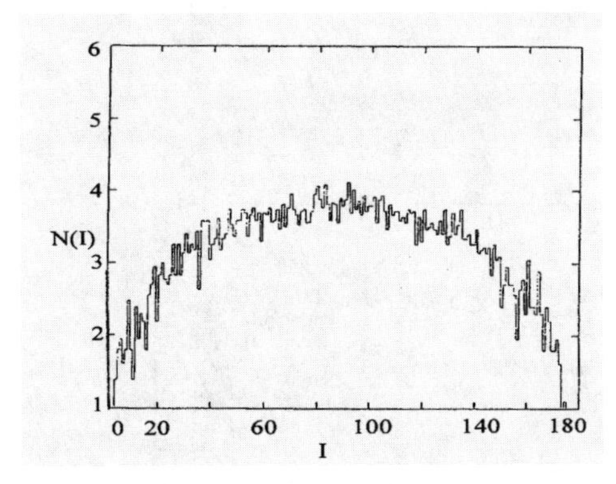


Figure 1. Initial distribution of the orbit inclinations.

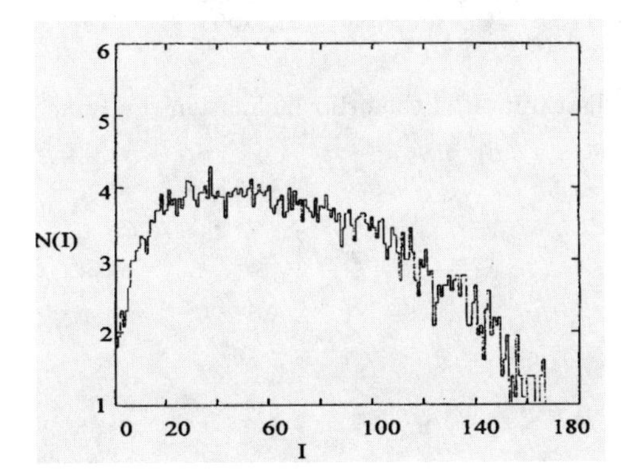


Figure 2. The evolved orbit inclinations in our calculations.

where $\vec{R}_j = \vec{r}_i - \vec{r}_j$; $i = 1, \ldots, N$. Giving some flat density distribution in the AD, we use an extension of the Nbody6++ code (Spurzem 1999) for the numerical simulations of the cluster evolution. We used a spherical CSC model with $N = 5000$ particles of equal masses. A result of our simulations is shown in Figures 1 and 2, and the results of Rauch (1995) is shown in Figures 3 and 4 for comparison.

One can see that the inclinations of the star orbits do not accumulate at zero value in our case, which coincides with the conclusion of the work by Vilkoviskij and Cherny (2002).

## 3. Discussion

In our model we took into consideration the finite thickness of the disk rather than infinitesimally thin disk approximation as it is usually considered in other works. We applied a fully self-consistent treatment of stellar interactions without consideration any additional assumptions. At the given stage, the efficiency of

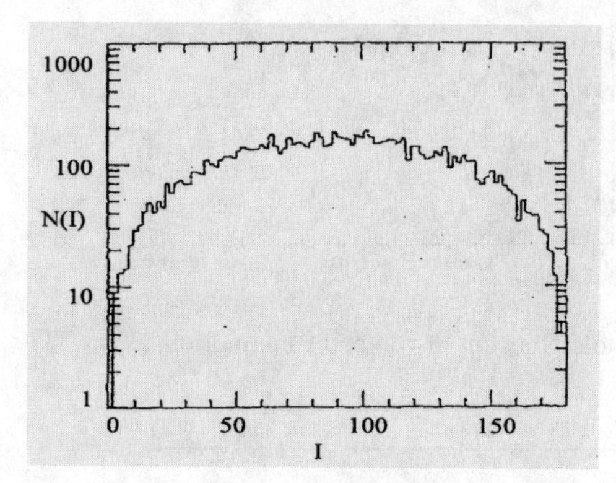


Figure 3. Initial distribution of the orbit inclinations by Rauch (1995).

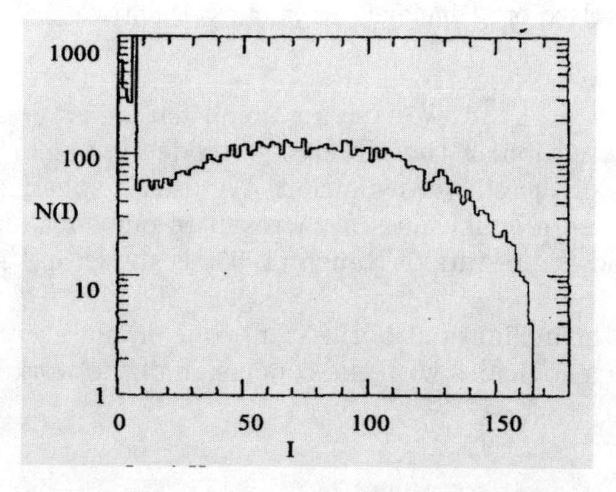


Figure 4. The evolved orbit inclinations in calculations by Rauch (1995).

the drag forces (which stars experience traveling through the disk matter) was artificially increased, such that a dissipative time of a typical star is comparable to the dynamical time of the stellar system. The analysis of a detailed orbit allowed us to study statistically in what way the interaction with a central accretion disk influences the kinematical and morphological parameters of the stellar system due to its differential effect on various orbits. In a next step we will increase a number of particles from 20 000 to 100 000 and investigate stationary behavior of the system on the long term by tuning the ratio of the drag force time scale and the two-body relaxation time scale in such a way that it will correspond to the realistic conditions of AGNs.

## References

Syer, D., Clarke, C. J, & Rees, M. J. 1991, MNRAS, 250, 505
Artymovicz, B., Lin, D. C, & Wampler, E. J. 1993, ApJ, 409, 592
Vokrouhlicky, D., & Karas, V. 1998, MNRAS, 298, 53
Rauch, K. P. 1995, MNRAS, 275, 628
Rauch, K. P 1999, ApJ, 514, 725
Vilkoviskij, E. Y., & Czerny, B. 2002, A&A, 387, 804
Spurzem, R. 1999, Journal Comp.Appl.Math. 109, 407

*Order and Chaos in Stellar and Planetary Systems*
*ASP Conference Series, Vol. 316, 2004*
*G. Byrd, K. Kholshevnikov, A. Mylläri, I. Nikiforov, and V. Orlov, eds.*

# Relaxation Time Estimates in Numerical Dynamical Models of Open Stellar Clusters

V. Danilov

*Astronomical Observatory, Ural State University, Ekaterinburg, Russia*

L. Dorogavtzeva

*Astronomical Observatory, Ural State University, Ekaterinburg, Russia*

**Abstract.** The relaxation time estimates in the spaces of the several star motion parameters were obtained for a number of the open cluster models. The relaxation time differences in these spaces are the greater for the more non-stationary cluster models. During the cluster models evolution their relaxation times increase in all considered spaces. During the violent relaxation the stars occupy all accessible for them regions at first in the space of the velocities moduli and then in the space of the distances from the cluster center. The tendency to the weakening of the dependence of the coarse grained phase density of cluster under the influence of small initial perturbations of the phase coordinates of stars in the cluster models for the time moments greater than the violent relaxation time and the two-body relaxation time was displayed.

During violent relaxation in models of open clusters (Danilov 2000, 2002a,b), the equilibrium distribution of stars is conserved in space of three parameters of stellar motion $\varepsilon$, $l$, $\varepsilon_\zeta$ (the energy, angular momentum, and energy of star's motion perpendicular to the Galactic plane per unit mass of the star). In these OC models, neither virial nor thermodynamical equilibrium are attained as $t$ increases, and the virial coefficient for the cluster models at $t > \tau_{vr}$ continues to oscillate with nearly constant amplitude and the period $P_r$. The distributions of $\varepsilon$, $l$, $\varepsilon_\zeta$ for the motions of individual stars remain well preserved over time intervals $\Delta t$ of the order of $P_r$ (where $\tau_{vr}$ is initial time scale for violent relaxation for the cluster model). Danilov (2002b) determined the equilibrium phase-space density function of the number of stars corresponding to the equilibrium of the OC models in $\varepsilon$. This equilibrium is incomplete, the coordinate and velocity distributions of the stars and the potential of the regular forces vary with the period $P_r$. Violent relaxation proceeds under the conditions of this equilibrium in phase space, and leads to the development of an "equilibrium" oscillatory process whose parameters vary little with time. In his analysis of stellar fluxes in phase space, space of $\varepsilon$, $l$, $\varepsilon_\zeta$, Danilov (2002a) found that there was an energy flux toward the cluster center, and transfer of energy from large-scale to small-scale motions within clusters. The distributions of the stellar fluxes in $\varepsilon$, $l$, $\varepsilon_\zeta$ allow to estimate the relaxation times of the OC models in these spaces (Danilov 2002a).

The aim of work was to analyze the characteristics of development of equilibrium in OC models that are nonstationary in the regular field, and to estimate the relaxation times in such systems.

Table 1. Parameters of open-cluster models

| $\mathcal{N}$ | $R_1/R_2$ | $N_1/N_2$ | $N_1$ | $R_2/R_t$ |
|---|---|---|---|---|
| 1 | 0.24 | 0.25 | 100 | 0.9 |
| 2 | 0.24 | 0.25 | 100 | 0.8 |
| 3 | 0.34 | 0.67 | 200 | 0.8 |
| 4 | 0.24 | 0.25 | 100 | 0.7 |
| 5 | 0.45 | 1.50 | 300 | 0.8 |
| 6 | 0.63 | 4.00 | 400 | 0.8 |

Table 2. Relaxation times for the open-cluster models

| $\mathcal{N}$ | $t_{r,c}^{(1)}$ | $t_{r,c}^{(2)}$ | $t_{r,h}$ | $\tau_\varepsilon$ | $\tau_l$ | $\tau_{\varepsilon_\zeta}$ | $\tau_r$ | $\tau_v$ |
|---|---|---|---|---|---|---|---|---|
| 1 | 0.5 | 1.0–1.1 | 1.2–1.3 | 5.2±0.8 | 2.2±0.3 | 1.2±0.1 | 1.9±0.5 | 0.8±0.1 |
| 2 | 0.5 | 1.2–1.4 | 1.5–1.6 | 4.7±0.5 | 3.1±0.3 | 1.1±0.1 | 1.8±0.2 | 0.7±0.1 |
| 3 | 0.5 | 1.1–1.2 | 1.1–1.4 | 4.2±0.5 | 2.8±0.4 | 0.8±0.1 | 1.3±0.2 | 0.60±0.04 |
| 4 | 0.5 | 1.0–1.2 | 1.2–1.3 | 4.4±0.4 | 3.2±0.5 | 0.9±0.1 | 1.6±0.2 | 0.7±0.1 |
| 5 | 0.5 | 1.2–1.4 | 1.4 | 3.5±0.5 | 2.9±0.4 | 0.80±0.05 | 1.0±0.1 | 0.60±0.05 |
| 6 | 0.65 | 1.4–1.5 | 1.9–2.0 | 2.7±0.5 | 3.2±0.3 | 1.1±0.1 | 0.9±0.1 | 0.7±0.1 |

Following Danilov (2000, 2002a,b), the cluster is modeled as a system of two gravitating spheres with coincident center of mass, imitating the halo and core. The initial parameters $R_1/R_2$ and $N_1/N_2$ for six models satisfy the observed relation $R_1/R_2 \simeq 0.39 \times (N_1/N_2)^{0.35}$ (Danilov & Seleznev 1994). Here $R_1$, $R_2$ and $N_1$, $N_2$ are the radii and the numbers of stars of the cluster core and halo respectively. The masses of the stars in the models are equal to $1M_\odot$. We analyze the motion of the cluster's stars in a rotating coordinate system $(\xi, \eta, \zeta)$ fixed to the cluster center of mass (Danilov 2000). We computed our OC models using the adopted stellar equations of the motion from Chandrasekhar (1942). We adopted the Galactic potential in the form suggested by Kutuzov & Osipkov (1980). The model of cluster does not rotate relative to external galaxies at $t = 0$, and initial stellar number densities at various points in the halo and core are approximately constant.

In Table 1 $R_t$ is the tidal radius for the stability of the cluster in the gravitational field of the Galaxy obtained using the criterion of King (1962).

Degree of nonstationarity of OC models in the regular field is determined by the amplitude of virial coefficient oscillations. Models 1–6 are listed in order of decreasing degree of nonstationarity.

Table 2 gives estimates of the local relaxation times $t_{r,c}^{(1)}$ and $t_{r,c}^{(2)}$ for the cluster core and $t_{r,h}$ for the cluster halo in units of $\tau_{vr} = 2.6\bar{t}_{cr}$. Here, $\bar{t}_{cr}$ is the mean initial cluster crossing time. These estimates are based on the instability of the phase-space density function relative to small initial perturbations of the stellar phase-space coordinates (Danilov & Dorogavtzeva 2003).

Table 2 gives also the estimates of relaxation times in $\varepsilon$, $l$, $\varepsilon_\zeta$, $r$, $v$ in units of $\tau_{vr}$. These estimates are based on the fluxes of the stars in $\varepsilon$, $l$, $\varepsilon_\zeta$, $r$, $v$, in accordance with the technique described in Danilov (2002a).

The following results were obtained:

(1) We have confirmed the conclusions of Danilov (2002a) that there is a balance of stellar fluxes in spaces of $\varepsilon$, $l$, $\varepsilon_\zeta$, $r$ and that there exist equilibrium distributions of the stars in $\varepsilon$, $l$, $\varepsilon_\zeta$ throughout the dynamical evolution of the clusters. The equilibrium phase-space density function corresponding to the balance of stellar fluxes in $\varepsilon$ determines the structure of the open cluster during both violent relaxation and the subsequent "equilibrium" oscillations of the density. The parameters of the equilibrium phase-space density function of a gas- and dust-free cluster can be determined at each stage of its dynamic evolution. These parameters can be used to analyze the stellar velocity distributions in such clusters and estimate the total masses and other dynamical parameters.

(2) The stellar fluxes were used to estimate the relaxation times of the OC models in $\varepsilon$, $l$, $\varepsilon_\zeta$, $v$, $r$. These times usually satisfy the inequalities $\tau_\varepsilon > \tau_l > \tau_{\varepsilon_\zeta} > \tau_v$ and $\tau_r > \tau_v$. The relaxation times increase during the evolution of the clusters in all parameter spaces considered. This behavior reflects a decrease in the rates of evolution in $\varepsilon$, $l$, $\varepsilon_\zeta$, $v$, $r$ spaces after violent relaxation.

(3) We have estimated the local violent relaxation times based on the instability of the phase-space density function relative to small initial perturbations of the stellar phase-space coordinates. Our estimates of the times $t_{r,c}^{(1)}$ for the cluster core agree well with estimates of $\tau_v$ derived from the stellar fluxes. During violent relaxation, stars with energies $\varepsilon$ occupy all domains of the cluster accessible to them, first in $v$ and then in $r$.

(4) The cores of OC models show dependence of the coarse-grained phase-space density function on small initial perturbations of the stellar phase-space coordinates to weaken at $t > t_{r,c}^{(2)}$. We also found a similar trend for the haloes of cluster models 1, 2, 5, 6 at $t > t_{r,h}$.

## References

Danilov, V. M. 2000, AZh, 77, 345
Danilov, V. M. 2002a, AZh, 79, 492
Danilov, V. M. 2002b, AZh, 79, 986
Danilov, V. M., & Seleznev A. F. 1994, A&A Trans., 6, 85
Chandrasekhar, S. 1942, Principles of Stellar Dynamics (Chicago)
Kutuzov, S. A., & Osipkov, L. P. 1980, AZh, 57, 28
King, I. R. 1962, ApJ, 67, 471
Danilov, V. M., & Dorogavtzeva L. V. 2003, AZh, 80, 526

*Order and Chaos in Stellar and Planetary Systems*
*ASP Conference Series, Vol. 316, 2004*
*G. Byrd, K. Kholshevnikov, A. Mylläri, I. Nikiforov, and V. Orlov, eds.*

# Properties of Stellar Trajectories in Numerical Dynamical Models of Open Stellar Clusters

V. M. Danilov and Ye. V. Leskov

*Astronomical Observatory, Ural State University, Ekaterinburg, Russia*

**Abstract.** Stellar trajectories in models of open star clusters (OCs) that are nonstationary in the regular field were investigated. Estimates of the maximum Lyapunov characteristic exponent $\lambda$ of stellar trajectories in OC models were performed. Averaged $\lambda$ values in considered cluster models are approximately equal to $\overline{\lambda} \simeq 1$ Myr$^{-1}$. Cluster cores are the regions of high stochasticity and haloes are the regions of more ordered motions. Values of $\overline{\lambda}$ and sizes of the region of high stochasticity in a cluster core increase with increasing of the density of cluster model. Fourier analysis of stellar trajectories in considered cluster models was performed. Distributions of stellar trajectories on periods of their motions with the most power spectrum value in a cluster were constructed. The "peaks" with periods corresponding (or close) to the periods commensurable with the period of the system regular field oscillations are displayed on these distributions.

## 1. Introduction

In the work by Danilov & Dorogavtseva (2003) the models of open star clusters (OCs), non-stationary in a regular field close to gravitational instability and driven on circular orbits in a galaxy field, are considered. During violent relaxation time, $\tau_{\rm vr}$, the radial fluctuations with practically constant amplitude and period are established, the radial motions of stars are appreciably synchronized.

The research of distributions of trajectories of stars in models of clusters on the periods with greatest value of spectral density for a trajectory is of interest. The analysis of such distributions and their comparison for OC models with a various degree nonstationary will allow us to establish the most significant of periodicity in movements of stars, and also will allow us to investigate influence of synchronization of radial movements of stars in clusters on movements of stars with frequencies incommensurable to frequency of radial oscillations of OC model. The purpose of this work is the research of properties of trajectories of motion of stars in OC models, non-stationary in a regular field.

## 2. Maximum Lyapunov Exponent Estimations

In our work the trajectories of movement of stars in six OC models are investigated. The parameters of models can be seen in Table 1 (Danilov & Dorogavtseva 2003). Estimations of the maximum Lyapunov characteristic exponent $\lambda$ of trajectories of movement of stars in OC models according to (Lihtenberg & Liberman 1984) are executed. Average values $\overline{\lambda}$ in OC models are found. Values of $\lambda$ can be put in conformity both initial, and final values of phase coordinates of stars. It allows one to look for dynamic evolution of a higher stochasticity zone with $\lambda > \overline{\lambda}$ in model of a cluster, see Figure 1. At the initial time moment $t = 0$ the area of significant stochasticity in OC models is located in cores and in the field of small and intermediate values of velocity $V$. During evolution of OC models the regions of significant stochasticity are propagated to regions of smaller values stellar distance from center $r$ and absolute stellar velocity $V$ (for OC model 1 see Figure 1). Trajectories of stars with great values $\lambda$ are

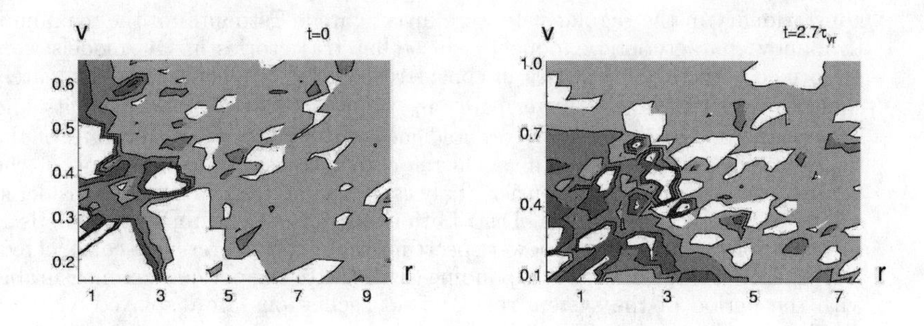


Figure 1. Dependencies $\lambda = \lambda(r, V)$ for $t = 0$ and $t = 2.7\tau_{\rm vr}$ (here $r$ in pc and $V$ in $\rm km\,s^{-1}$; $\lambda \in [0.08\ \rm Myr^{-1},\ 3.64\ Myr^{-1}]$). More dark areas correspond to the large $\lambda$ values.

located in cores of OC models, and trajectories with small $\lambda$ are located on the periphery of clusters. Thus, in the non-stationary OC core a region of the raised stochasticity is formed; in the halo a region of more regular motions of stars is formed. In more dense OC models the sizes of area of higher stochasticity in a core grow. In the space $(r, V)$ in nuclei of OC models extended areas high stochasticity are penetrated by narrow areas with small stochasticity, and on periphery of clusters in the fields of regular movement occasionally it is possible to meet small areas of higher stochasticity. Thus, the areas of regular and stochastic movement in space $(r, V)$ settle down beside with each other.

Distributions of trajectories of stars of magnitude $\lambda$ are constructed. The magnitude $\overline{\lambda}$ grows with increase in density and reduction of a degree of non-stationarity of considered OC models. Thus, the degree of stochasticity of motions of stars in OC models decreases at the large amplitudes of oscillations of a regular field. At account of $\lambda$ value for each trajectory this $\lambda$ very quickly grows and achieves the greatest value already at small values $t$, then the changes $\lambda$ become insignificant. Thus, "sticking" trajectory of a star in phase space to area with the given value $\lambda$ during the clusters dynamical evolution takes place.

## 3. Trajectory Distributions on Their Parameters

The oscillations of a regular field of a cluster lead to the oscillations of value of virial coefficient $\alpha(t) = 2E/W$, here $E$ and $W$ are the total and potential energies of cluster relatively the cluster center. For $\alpha(t)$ and for each trajectory of a star in the given work the amplitudes, periods $T$ and phases of first three harmonics (with the greatest amplitudes) variation of $r$ and $\dot{r} = \frac{dr}{dt}$ values with time were determined according to (Laskar 1988; Carpintero & Aguilar 1998). These three harmonics contain up to 95% of energy of spectra of dependencies $r = r(t)$ for trajectories of stars with small $\lambda$ values in our OC models. The

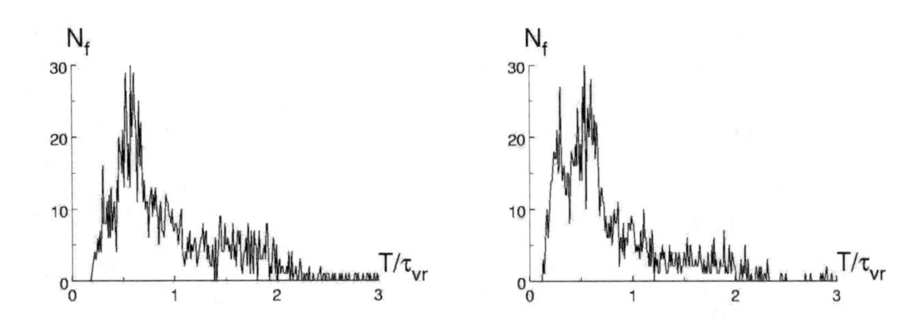


Figure 2. Trajectories distribution on the periods of oscillations $r$ values (left) and $\dot{r}$ values (right).

total distributions $N_f$ of these three basic harmonics of trajectories on the periods and phases of oscillations are constructed (for OC model 1 see Figure 2). Such distributions allow us to judge the frequencies and phases of most significant oscillatory motions of stars on radius $r$ in a cluster. On distributions of the most significant oscillations not casual "peaks" corresponding to or near to frequencies, commensurable with frequency of the first harmonic of oscillations of $\alpha(t)$ value are found. Prevalence of phases stars motions on $r$ close to a phase of the first harmonic of OC regular field (and the virial coefficient) oscillation is visible for the most non-stationary OC models.

## References


Danilov, V. M., & Dorogavtseva, L. V. 2003, AZh, 80, 526

Lihtenberg, A., & Liberman, M. 1984, Regular and stochastic dynamics (Moscow: Mir), 313 (in Russian)

Laskar, J. 1988, A&A, 198, 341

Carpintero, D. D., & Aguilar, L. A. 1998, MNRAS, 298, 1

*Order and Chaos in Stellar and Planetary Systems*
*ASP Conference Series, Vol. 316, 2004*
*G. Byrd, K. Kholshevnikov, A. Mylläri, I. Nikiforov, and V. Orlov, eds.*

# On the Cluster Dynamics in the Galactic Field

A. A. Davydenko and L. P. Ossipkov

*Department of Space Technologies and Applied Astrodynamics, St. Petersburg State University, Universitetskij pr. 28, Staryj Peterhof, St. Petersburg 198504, Russia*

**Abstract.** Two problems of the cluster dynamics in the Galactic tidal field are briefly discussed. We consider Hill's stability for clusters moving on circular orbits and the gross-dynamics of clusters moving along radial orbits.

## 1. Introduction

We consider the dynamics of a star cluster under the action of its own gravitation and Galactic tidal forces. After Bok's classical paper (1934) many authors studied the dynamics of clusters moving along a circular orbit (e.g., Rein 1936; Mineur 1939; Chandrasekhar 1942; van Wijk 1949; Cimino 1956; Kuzmin 1963; Jefferys 1976; Danilov 1982; Ossipkov 1993). Some authors considered vertical cluster oscillations in the Galactic field and a "shock" resulting from crossing an equatorial plane by a cluster (e.g., Spitzer 1987).

It seems plausible that galactic orbits of globular clusters are rather elongated, as well as orbits of some open clusters. There were several attempts to study the dynamics of clusters, whose orbits were non-circular (e.g., Kozhanov 2003).

In this work we discuss some steps towards the general analysis of a problem.

## 2. The Dynamics of Clusters on Circular Orbits

A model of cluster motion along a circular orbit in a stationary axisymmetric Galactic potential is the simplest. Unfortunately, there are some gaps in earlier works. Let $x$, $y$, $z$ be a rotating frame with an origin coincident with a cluster center and an axis $x$ directed to the Galactic anticenter. Under usual assumptions the equations of star motion under the joint action of cluster gravitation and Galactic tidal, Coriolis and centrifugal forces can be written as follows:

$$\ddot{\mathbf{r}} + 2\mathbf{\Omega}_0 \wedge \dot{\mathbf{r}} = \nabla \left[\Phi_g(\mathbf{r}) + \Phi_c(\mathbf{r})\right], \tag{1}$$

where $\Phi_g = \frac{1}{2}\left(\kappa_R^2 x^2 - \kappa_z^2 z^2\right)$, $\Phi_c$ is a (positive) potential of a cluster, $\mathbf{\Omega}_0$ is a vector of circular angular velocity at cluster distance from the Galactic center, $\kappa_R = \sqrt{4A(A-B)}$ ($A$ and $B$ are Oort's constants) and $\kappa_z$ are local parameters of the Galactic potential. We can put $\kappa_R \approx 42\,\mathrm{km\,s^{-1}kpc^{-1}}$, $\kappa_z \approx 85\,\mathrm{km\,s^{-1}kpc^{-1}}$ for the Solar neighborhood. It is known that Eq. (1) admits a Jacobi integral $H = \frac{1}{2}\dot{\mathbf{r}}^2 - [\Phi_g(\mathbf{r}) + \Phi_c(\mathbf{r})]$. Zero velocity surfaces can be found from an equation $H + \Phi_g + \Phi_c = 0$. So, Hill's stability of a cluster can be studied on the basis of the Jacobi integral. The problem is reduced to finding a maximal size of a cluster (of given mass and density law) for which its external surface touches a critical Hill's surface (i.e., a maximal closed surface of zero velocities). We have to solve the following equation for that surface:

$$2\varphi(\xi, \eta, \zeta) + \left[\xi^2 - \left(\frac{\kappa_z}{\kappa_R}\right)^2 \zeta^2\right] = C. \tag{2}$$

Here $C$ is a dimensionless Jacobi constant, $\varphi = \Phi_c(\mathbf{r})\left(G\mathcal{M}\kappa_R\right)^{-2/3}$ is a dimensionless potential of a cluster, $\mathcal{M}$ is its mass, $\xi$, $\eta$, $\zeta$ are dimensionless coordinates with $\left(G\mathcal{M}/\kappa_R^2\right)^{1/3}$ as a unit of length.

For spherical clusters $\Phi(\mathbf{r}) = G\mathcal{M}/r$ for external points. Then Eq. (2) can be solved easily (Antonov & Latyshev 1971). If $\kappa_R = 2\kappa_z$ then the Hill's critical surface crosses coordinate axes at $\xi_* = \pm 1$, $\eta_* = \pm 2/3$, $\zeta_* = 1/2$. If a cluster is uniform then its critical density $\varrho_* = \kappa_R^2/\pi G\beta$ with $\beta \approx 0.167$ in contrast with $\beta \approx 4/3$ according to Bok (1934)[1].

In the above analysis we considered a stationary cluster. Time-dependent models can be studied on the basis of gross-dynamic equations (Ossipkov 2001).

## 3. Clusters Crossing the Galactic Center

Now let us consider an opposite case when a cluster moves along a rectilinear orbit in the equatorial plane. We suppose that a central part of the Galaxy resembles a uniform spheroid, and its potential is equal to $P^2 - \frac{1}{2}Q^2R^2$. The case of a small massive core (a central black hole) must be studied with another technique. We take into account that a cluster size is much smaller than an orbital apocenter. Let $x$, $y$, $z$ be a frame moving together with a baricenter of a cluster, the $x$ axis being directed along a cylindrical radius. Then we find the following equations of star motion:

$$\ddot{x} = \frac{\partial \Phi_c}{\partial x} - Q^2 x\,, \qquad \ddot{y} = \frac{\partial \Phi_c}{\partial y} - Q^2 y\,, \qquad \ddot{z} = \frac{\partial \Phi_c}{\partial z}. \tag{3}$$

Now we are able to write down the collisionless Boltzmann equation for a cluster under consideration and find moment equations, generalizing Lagrange–Jacobi tensor equations, and equations for a kinetic energy tensor, as it was done for a cluster on a circular orbit (Ossipkov 1993, 2001). We omit that standard gross-dynamic analysis. The main results are the following:

- an equilibrium state does exist for that model;
- it is stable.

It is possible to generalize the theory taking into account a dynamical friction of clusters on stars of the Galactic background.

---

[1]A factor $\pi$ was erroneously omitted in f-la (1) for Bok's critical density by Ossipkov (2001).

## 4. Concluding Remarks

Intermediate cases of cluster orbits are of great interest, of course. The main difficulty lies in the non-integrability of equations of cluster motion in the Galaxy. If we restrict ourselves to planar orbits of clusters we need in analytical representation of coordinates of the cluster center as explicit functions of time. Using Lyapounov's method of finding periodic solutions provides a relevant tool for solving that problem. Preliminary calculations for some plausible Galactic models showed that this method yields good results.

## References

Antonov, V. A., & Latyshev, I. N. 1971, AZh, 48, 854 (in Russian)

Bok, B. J. 1934, Harvard Circular, 384, 1

Chandrasekhar, S. 1942, Principles of Stellar Dynamics (Chicago: Univ. Chicago Press)

Cimino, M. 1956, Atti Acad.Naz.Lincei. Rend.Sc.fis.mat.nat., 20, 217

Danilov, V. M. 1982, AZh, 59, 253 (in Russian)

Jefferys, W. H. 1976, AJ, 81, 983

Kozhanov, T. S. 2003, Gravitational Dynamics of Hierarchical Systems (Almaty: Kazak Univ.) (in Russian)

Kuzmin, G. G. 1963, Tartu Publ., 34, 10 (in Russian)

Mineur, H. 1939, Ann.d'Astrophys., 1939, 2, 1

Ossipkov, L. P. 1993, Vestnik SPb Univ.Ser.1, 1, 125 (in Russian)

Ossipkov, L. P. 2001, in ASP Conf. Ser., Vol. 228, Dynamics of Star Clusters and the Milky Way, ed. S. Dieters, B. Fuchs, A. Just, R. Spurzem & R. Wielen (San Francisco: ASP), 341

Rein, N. F. 1936, Trudy Sternberg Astron.inst., 9, 191 (in Russian)

Spitzer, L., Jr. 1987, Dynamical Evolution of Globular Clusters (Princeton: Princeton Univ. Press)

van Wijk, U. 1949, Ann.d'Astrophys., 12, 81

*Order and Chaos in Stellar and Planetary Systems*
*ASP Conference Series, Vol. 316, 2004*
*G. Byrd, K. Kholshevnikov, A. Mylläri, I. Nikiforov, and V. Orlov, eds.*

# Fractal Properties of Hierarchical Star Clusters

Vladimir Surdin
*Sternberg Astronomical Institute, Universitetskij pr. 13, Moscow 119992, Russia*

**Abstract.** It is shown that hierarchical star clusters (H-clusters) containing tens of components may exist in the Galaxy. The number of stars ($N$) in $\varepsilon$ Lyrae-type hierarchical systems is limited by the tidal action of regular gravitational field of the Galaxy and stochastic encounters with giant molecular clouds. In principle, the existence of H-clusters with $N = 256 \div 512$ is possible. But in fact, we can expect to find these systems with $N \leq 16$ in the recent astrometric surveis. Maximum fractal dimension of the simple stable H-cluster is $D = 0.3 \div 0.4$, which is much less than the fractal dimension of molecular clouds ($D = 2.3 \div 2.5$). This may be the reason of rare formation of high populated H-systems.

## 1. Chaotic and Hierarchical Stellar Systems

All stars are members of stellar systems—binary and multiple stars, open and globular clusters, and galaxies. In principle, each of these systems, except binary stars, may have one of two types of the internal organization: hierarchical or chaotic. Binary systems represent a degenerate case possessing an extreme degree of stability.

A multiple star of the chaotic type (C-system) is exemplified by the Trapezium of Orion, in which the mutual distances between its four stars are the same order of magnitude. These systems are extremely unstable and disintegrate in a time close to the dynamical time. There exist also multiple systems of the hierarchical type (H-systems), possessing asymptotic stability. Such systems are exemplified by multiple stars of the type $\varepsilon$ Lyr consisting of two relatively close pairs separated by an appreciable distance (in fact, there is fifth component in the system, but $\varepsilon$ Lyr is classical example of 4-star H-system). At each hierarchy level the dynamics of such a system is close to the dynamics of a binary star.

## 2. Structure of Hierarchical Stellar System

H-system (H-cluster) is characterized by the number of stars it contains ($N$) and by the maximum hierarchy level in their asociation ($H$). The relation between these parameters depends on the type of packing and varies within the range from $N = H + 1$ for systems with minimally dense packing of the type $n + 1$ to $N = 2^H$ for systems with maximally dense packing of the type $n + n$.

If all stars have the same mass $m$, as well as the same eccentricities of all orbits in the system ($e$) and the same ratios of the semimajor axes of orbits at the neighboring hierarchy levels ($k = a_{i+1}/a_i$), then the mean mass density in a

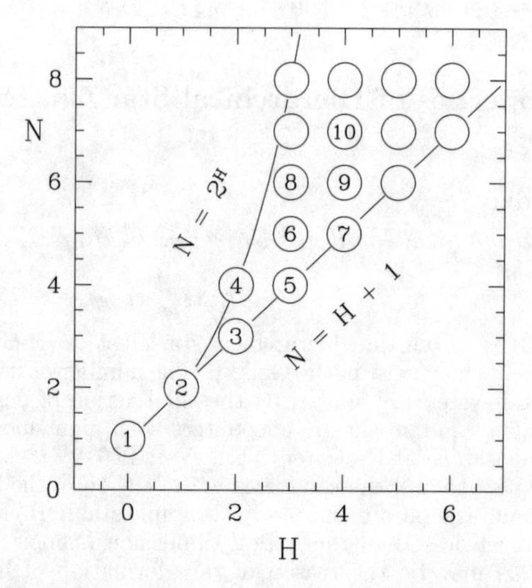


Figure 1. Number–hierarchy diagram. $N$ is multiplicity, i.e., the number of stars in a system. $H$ is the hierarchy level of the system. Examples of systems: single stars (1); binaries (2); triples (3); four components' $\varepsilon$ Lyr (4); $\varepsilon$ Boo (5); $\zeta$ UMa + 80 UMa—the system Mizar–Alcor, where Mizar ($\zeta$ UMa) itself is a system of $\varepsilon$ Lyr-type (6); DN UMa (7); Castor, or $\alpha$ Gem (8); $\beta$ Cap (9); $\nu$ Sco (10).

cluster of the type $n+1$ is

$$\rho_{n+1} = \frac{3mN^4k^6}{32\pi\,[a(1+e)(N-1)k^N]^3}; \tag{1}$$

and the mean mass density in a cluster of the type $n+n$ is

$$\rho_{n+n} = \frac{3mN(k-1)^3}{4\pi\,[a(1+e)N^{\log_2 k}]^3}, \tag{2}$$

where $a$ is the semiaxis of binary systems of the first hierarchy level (i.e., the minimum semiaxis of the system).

## 3. Stability of Hierarchical Stellar System and its Maximum Population

Disruption of a stellar cluster in the regular tidal field of the Galaxy occurs when Roche's condition is fulfilled

$$\rho \leq \frac{9M_{\mathrm{G}}(R)}{4\pi R^3}, \tag{3}$$

where $M_{\mathrm{G}}(R)$ is the mass of the Galaxy within the radius of the galactic orbit of the system ($R$). For the model of the Galaxy with constant circular velocity $V$, we have $M_{\rm G}(R) = RV^2/G$. At $V = 220$ km/s we obtain the critical density of the system:

$$\rho_c = 8\left(\frac{R}{1\ \mathrm{kpc}}\right)^{-1} M_{\odot}/\mathrm{pc}^3, \tag{4}$$

Comparing (1) and (2) with (4), we obtain the constraint on the number of stars in an H-cluster that is stable in the tidal field of the Galaxy. Since we are interested in the most populated and, consequently, maximally dense systems, we assume the values of the quantities appearing in (1) and (2) to be minimum. The lenght of the semiaxis of a close binary system is virtually limited to $a \approx 6R_{\odot}(m/M_{\odot})^{1/3}$. Since we assume that the masses of all stars are $m = 1M_{\odot}$, $a = 6R_{\odot}$. The time of desintegration of triple stars increases rapidly with the growth of the ratio $k = a_{i+1}/a_i$ and becomes virtually infinite at $k = 8$. Therefore we assume $k = 8$ and $e = 0.6$. Then in the range of galactocentric distances $2 < R < 30$ kpc, the maximum numbers of stars in H-systems are

$$N_{n+n}^{\max} = 512, \tag{5}$$

$$N_{n+1}^{\max} = 9. \tag{6}$$

For all intermediate types of packing the value of $N^{\max}$ lies between these two limit values, which are independent of the age of the system. However, for systems of age $>10^8$ years there is a high probability to encounter with giant molecular clouds (GMCs). Assuming the mean density of the cloud to be $\rho_{\rm GMC} \approx 30\ M_{\odot}/\mathrm{pc}^3$ we obtain maximum values for the population of elderly H-clusters:

$$N_{n+n}^{\max} = 256, \tag{7}$$

$$N_{n+1}^{\max} = 8. \tag{8}$$

## 4. Fractal Properties of the Ideal H-Clusters and their Paucity

The simplest hierarchical stellar system is fractal object. From (2) we obtain fractal dimension for H-cluster of the type $n + n$ with diameter $L$:

$$D_{n+n} = \frac{\log N}{\log L} = \frac{1}{\log_2 k} = \log_k 2. \tag{9}$$

Theoretically, the stable H-cluster has $k_{\min} = 8$; but real multiple systems has $k_{\min} = 5 \div 6$. Then maximum fractal dimension of stable H-cluster is $D = 0.3 \div 0.4$. We have to compare it with structure of protostellar medium to understand the formation of H-clusters.

All components of interstellar medium have fractal properties. In this paper we have concentrated on the molecular component of the ISM—an ensemble of molecular clouds. In the wide range of their masses—from $10^{-2}$ to $10^7\ M_{\odot}$—the volume fractal dimension of the clouds is terminated by the narrow range of $D = 2.4 \div 2.5$. This is much bigger of the maximum fractal dimension of stable H-cluster ($D = 0.3 \div 0.4$). This may be the reason of rare formation of high populated H-systems.

**Acknowledgments.** This work was partly supported by RFBR (grant 03-02-16288).

*Order and Chaos in Stellar and Planetary Systems*
*ASP Conference Series, Vol. 316, 2004*
*G. Byrd, K. Kholshevnikov, A. Mylläri, I. Nikiforov, and V. Orlov, eds.*

# On Problems of Probably Double and Multiple Open Star Clusters

R. F. Ziyakhanov, S. N. Nuritdinov, M. M. Muminov, and Yu. Ch. Muslimova

*National University of Uzbekistan and Astronomical Institute of Uzbek Academy of Sciences, Tashkent; nurit@astrin.uzsci.net*

## 1. Introduction

The existence of probable double open star clusters (OSCs) in Magellanic Clouds has been shown by Bhatia & Hatzidimitriou (1988). The possible existence of multiple and double OSCs in the Galaxy was mentioned by several authors (see, for example, references in Muminov et al. 2000) but nobody has done a concrete study of this problem by analysis of physical criteria so far. The conclusion about the existence of probably double and multiple OSCs is made on the basis of similarity of the basic astrophysical and astrometric characteristics of their components: ages, spectra, radial velocities, distances, proper motions of components and so on. Earlier we have listed probably double and multiple OSCs (Latipov et al. 1997; Ziyakhanov 2003). Below we discuss their physical connection problem as well as the criteria for physically double and multiple OSCs created.

## 2. Statement of the Problem

Studying the multiple OSCs, one can see the optically projected systems, which have very close angular distances between components but no physical connection between them at all. They are so-called optically double or multiple systems. After determination of distances between components of the systems, comparison of their physical and dynamic characteristics and developing appropriate criteria of duality or multiplicity only we can answer to the question on their physical connection. Selecting physically double or multiple systems it is necessary to pay attention not only to a nearness of coordinates, but also of ages, distances, proper motions, radial velocities, stellar contents and other parameters. The first lists of probably double and multiple OSCs were already published by us (Muminov et al. 2000; Ziyakhanov 2003). Is there any physically connected and enough stable OSC system? Yes, in case of double OSCs. $\chi$ and h Persei is a well-known example.

In contrast to the stable periodic motion in the problem of two bodies the system of three and more bodies has a chaotic internal dynamics which is fraught with the mutual closings of components and an energy exchange between them. As a result the multiple system should break up. It shows the necessity to solve the problem of probably multiple OSCs. Till now the OSC researchers did not pay enough attention to the multiple OSC existence problem. Our explorations show it is possible to reveal ~30 probably multiple OSCs distant up to ~3 kpc.

But this is most important that nobody has created physical criteria for double and multiple OSCs yet. In our opinion these criteria should take into account not only the nearness of the basic physical parameters but also their interaction with the gravitational field of the Galaxy. To solve the problem of the physical duality and multiplicity of OSCs we have studied the observational data, dividing these systems separately to double, triple or higher multiplicity OSCs.

## 3. Double OSCs

In contrast to stars and galaxies it is much more difficult to find out whether OSCs are physically double systems. It demands more careful study which is connected with location of the exact members of the clusters, determination of all physical and kinematic characteristics, and also creating the physical criteria of duality.

At first we listed the optically double OSCs to analyse the separate pairs. With this purpose the probable double OSCs of our Galaxy were revealed analysing some known catalogues of OSCs [Lynga 1987 (1992); Mermilliod 1995; Kharchenko et al. 2003; Loktin et al. 1994] based of the nearness of their individual characteristics (coordinate, distance, a relative spatial location, age, etc.). Then a preliminary list can be made based on the angular distance between components of double OSCs (see also Subramaniam et al. 1995), which were calculated by us for the mutual comparison of OSC pairs, knowing distances of components. However the OSC distances of different authors have a wide scatter that demands their specification and the analysis of the heterogeneity problem in the distances determination. We believe that the data on the spatial distances between components and the differences of the separate characteristics of double OSCs can characterize the physical connection within a pair if use the statistical parameters for their comparison.

Actually from the point of view of the physics the double OSCs can be evidently physically connected and stable if total energy of given system is negative. On the basis of this requirement we offer the following criterion of the duality:

$$\left(\vec{V}_1 - \vec{V}_2\right)^2 < 2(m_1 + m_2)\left\{\frac{G}{d} + \left[\frac{1}{m_2}\Phi(R_1, Z_1) + \frac{1}{m_1}\Phi(R_2, Z_2)\right]\right\}, \qquad (1)$$

where $V_i$ and $m_i$ are space velocities and masses of components of probably double OSC accordingly; $d$ is a distance between components; $\Phi(R, Z)$ is a gravitational potential of the Galaxy at the point $(R, Z)$

$$\Phi(R, Z) = U(R, Z) - U(R_0, Z_0) + \frac{\Omega^2(R^2 - R_0^2)}{2}, \qquad (2)$$

which has been written taking into account the centrifugal forces. Here $U(R, Z)$ is the rotationally symmetric potential of our Galaxy, $(R_0, Z_0)$ is the coordinate of the centre of mass of probably double cluster, and $\Omega$ is the rotation frequency of stars around of the Galaxy centre. This criterion is applied by us for a number of OSCs from our list of probable double clusters for which required values are known enough. Let us note, the inequality (1) is better the OSC mass is more. Therefore during the concrete calculations we utilized the value of mass which derived from analysis of membership probability of stars to the OSC. We calculated the rest values (the distances between double cluster components, relative velocities) based on known exact formulae. Their accuracy depends mainly on the distance determination accuracy. The prepared list of probably double OSCs is divided into three classes for statistical analysis of their physical characteristics.

## 4. Search for Multiple OSCs

Basing on the computer analysis of OSC catalogues we prepared a list of probably triple OSCs. The basic physical values have been credited from the electronic versions of a number of catalogues (Lynga 1987; Mermilliod 1995; Kharchenko et al. 2003; Loktin et al. 1994). Often the triple OSCs consist of one optical double and the third cluster located relatively close to them. Then we have found a physical criterion for the triple OSCs in the gravitational field of our Galaxy by analogy with (1). An elaboration of criterion for high multiplicity OSCs represents separate big question that will be given by us in the other paper. Basing on the OSC catalogues we prepared a list of OSCs with higher multiplicity. The selection was made using the sources mentioned above taking into account the nearness of coordinates, distances, ages, and so on. Often quadrupole and quintuple OSCs can be divided into two double or to double and triple systems.

## 5. Conclusions

One can see the problem of physical multiplicity of OSCs is rather difficult in comparison with the problem of double OSCs in means of selection, the statistical analysis and one also demands a preliminary discussion of various factors, in particular, not only a nearness of the main parameters of multiplicity, but also the physical conditions in their vicinity.

At present we have made the following within a subject:

1. Lists of 32 probably double and 28 probably multiple OSCs were prepared.
2. The statistical analysis of the list of double OSCs was carried out. The statistical criteria of the OSC duality have been created.
3. The exact criterion of physical duality and the stability of OSCs is created taking into account the gravitational field of the Galaxy.
4. The statistical analysis of the list of 28 probably multiple OSCs was carried out. The criterion for physically triple OSCs was created.

The problem of physical multiplicity of OSCs is rather difficult in comparison with the problem of double OSCs from the point of view of selection, the statistical analysis and one also demands a preliminary discussion of various factors, in particular, not only a nearness of the main parameters of multiplicity, but also physical conditions in their regions.

## References

Bhatia, R. K., & Hatzidimitriou, D. 1998, MNRAS, 230, 215

Muminov, M. M., Nuritdinov, S. N., Latypov, A. A., & Muslimova, Yu. Ch. 2000, A&A Trans., 18, 645

Latypov, A. A., Nuritdinov, S. N., Muslimova, Yu. Ch., & Malahov, D. 1997, Problems of Astrophysics and Applied Astronomy (Tashkent), 29

Subramaniam, A., Gorti, U., Sagar, R., & Bhatt, H. C. 1995, A&A, 302, 86

Mermilliod, J.-C. 1995, http://obswww.unige.ch/webda/navigation.html (a site devoted to stellar clusters, WEBDA)

Kharchenko, N. V. et al. 2003, Proper motions of open clusters, VizieR On-line Data Catalog: I/280A (originally published in Kin.Fiz.Neb.Tel, 17, 409), http://vizier.cfa.harvard.edu/viz-bin/VizieR?-source=J/AZh/80/291

Lynga, G. 1987 (VII/92), Open Cluster Data 5th Edition, http://vizier.cfa.harvard.edu/viz-bin/VizieR?-source=VII/92A

Loktin, A. V., Matkin, N. V., & Gerasimenko, T. P. 1994, Catalogue of open cluster parameters from *UBV*-data, Weighted mean parameters of open clusters, http://vizier.cfa.harvard.edu/viz-bin/VizieR?-source=V/96

Ziyakhanov, R. F. 2003, Uzbek Physical J., 4 (in press)

# Part V

# Structure and Kinematics of the Milky Way Galaxy

*Order and Chaos in Stellar and Planetary Systems*
*ASP Conference Series, Vol. 316, 2004*
*G. Byrd, K. Kholshevnikov, A. Mylläri, I. Nikiforov, and V. Orlov, eds.*

# The Distance to the Center of the Galaxy: the Current State-of-the-Art in Measuring $R_0$

Igor' I. Nikiforov

*Sobolev Astronomical Institute, St. Petersburg State University, Universitetskij pr. 28, Staryj Peterhof, St. Petersburg 198504, Russia*

**Abstract.** The current situation regarding the problem of the Sun–Galactic center distance, $R_0$, is discussed. A new three-dimensional classification of $R_0$ measurements is suggested. Despite the importance of certain of recent results reviewed, overall progress in solving the problem of $R_0$ remains slow. At this time a "best value" for $R_0$, based upon 65 original $R_0$ estimates from 57 works published since 1974, is $7.9 \pm 0.2$ kpc.

## 1. Introduction

The value of the Galactic center distance $R_0$ has a great impact on astronomy and astrophysics (see, e.g., Reid 1993, hereafter R93). Because of this, following the first measurement of $R_0$ by Harlow Shapley (1918) astronomers have consistently made considerable effort to determine this fundamental parameter. Besides, such studies are stimulated by an "inconstancy" of the constant $R_0$. Table 1 gives an idea of the "evolution" of $R_0$ value. Till the eighties of the XXth century, even the "adopted" ("standard", "mean", "best", "best estimate") value for $R_0$ was grossly changed. Later this variation became closer, and yet the accuracy of the adopted value remained low as did rather wide the scatter of individual estimates of $R_0$.

Table 1. Evolution of "best value" and individual estimates for $R_0$

| Source | | $R_0$ (kpc) | Covered range of years |
|---|---|---|---|
| Shapley | 1918 | $\approx$13 | |
| Baade | 1951 | 8.7 | |
| Oort-Baade value (see, e.g., Woolley 1963) | 1954 | 8.2 | |
| IAU standard | 1964 | 10 | |
| de Vaucouleurs | 1983 | $8.5 \pm 0.5$ | 1974–1982 |
| Kerr & Lynden-Bell | 1986 | $8.54 \pm 1.1$ | 1974–1986 |
| → IAU standard | 1986 | 8.5 | |
| Feast | 1987 | $7.8 \pm \sim 0.8$ | 1974–1986 |
| Reid | 1989 | $7.7 \pm 0.7$ | 1972–1986 |
| Reid | 1993 | $8.0 \pm 0.5$ | 1972–1993 |
| Individual estimates (Reid 1993) | | 6.2–10.4 | 1972–1993 |
| Individual estimates (this paper) | | 6.5–8.7 | 1994–2003 |

After the last comprehensive review of the problem by R93, a number of advances has been made: i) *new luminosity calibrations* of traditional distance indicators and suggestions of *new distance indicators*, both from the *Hipparcos* astrometric catalogue and additional data; ii) new *absolute* (i.e., not using luminosity calibrations) estimates of $R_0$; iii) efforts towards *perfecting methods* of deriving $R_0$; iv) updated relative (i.e., using calibrations) estimates of $R_0$ from *new data*, more numerous and/or more reliable. In this paper, we review the current status of research on the problem of $R_0$, emphasizing the results published since the R93 analysis. We suggest a new classification of $R_0$ measurements and, with it, compute a current "best value" for $R_0$.

## 2. Sources of Errors in Estimates of $R_0$

In almost all cases, measuring $R_0$ involves two constituent problems:

Problem 1. To obtain the heliocentric distances, which may be called the "*reference distances*" (RDs) in measuring $R_0$, and also other data (velocities, metallicities etc.) for some Galactic objects, or "reference objects" (ROs).

Problem 2. To derive *a value of $R_0$ from an analysis of reference distances* and other data.

Any estimate of $R_0$ is always expressed in the overall scale of accepted RDs; consequently, an error in the scale biases the $R_0$ estimate. Except that a *proper method of deriving* $R_0$ (i.e., a method of solving Problem 2) also introduces systematic errors of its own. If for no other reason than this, a distinction needs to be drawn between these two problems.

From this simple reasoning, *objective* sources of errors in an estimate of $R_0$ can be broken down into the following categories:

I. *Systematic errors*:

I.1. **Systematic errors in reference distances** owing to a) errors in the distance scale(s), or b) fallacies in the absolute method of estimating distances.

I.2. Systematic errors in other observational data.

I.3. **Fallacies in the proper method of deriving an estimate of $\mathbf{R_0}$** from accepted RDs and other input data.

II. *Random errors*:

II.1. Random errors in observational data.

II.2. A statistical uncertainty of the proper method of deriving an estimate of $R_0$ from accepted observational data.

In addition, Reid (1989, 1993) found clear evidence for a *subjective* (psychological) source of errors—so called the "**bandwagon effect**", i.e., a tendency for new published $R_0$ values to cluster around the "correct answer" (a current "accepted value" of $R_0$). A possibility for easy "correction" of a "wrong" $R_0$ estimate results from uncertainties in the distance scale(s), systematic errors in the proper method, and an arbitrariness in the procedure of analyzing data. Then the systematic variations of *rescaled* $R_0$ estimates with time in R93 actually give an evaluation of the impact of two factors last named—∼1.5 kpc! By this is meant that perfecting proper methods of deriving $R_0$ estimates, which has not yet received due attention, is essential for solving the problem of $R_0$ as are obtaining new observational data and new calibrations of distance scales, elaborating new proper methods, and studying the internal nature of ROs.

Besides, it is well bear in mind that the term "center of the Galaxy" implies different definitions in different methods: i) a *central singularity of the spatial distribution* of some luminous matter or a *"central" object* (Sgr A*, IRS16, Sgr B2, etc.); ii) a *dynamic center*, i.e., a barycenter; and iii) other, more peculiar, definitions. Conceivable discrepancies between the positions of different "centers" (Blitz 1994) can be another systematic factor in the problem of $R_0$.

## 3. Classification of $R_0$ Measurements

A classification of $R_0$ measurements, which correctly reflects at least the main specific features of *diverse* sources of errors in different approaches and a non-uniqueness of the term "Galactic center", would have to be multi-dimensional. Such a classification could be useful in revealing probable sources of errors in $R_0$ measurements of a given type and hence in perfecting these measurements, as well as in deriving a "best value" for $R_0$ with due regard for statistical and systematic uncertainties.

We classify measurements of $R_0$ according to *three* independent features:

I. According to **the type of proper method of measuring** $R_0$, i.e., the type of fundamental technique for analyzing the RDs and other input data:

I.1. *Spatial methods*: Shapley's and Baade's methods, cone of avoidance, drop in distribution along the $X$-axis (Harris 1980), central object (pioneered by Moran et al. 1987). In all these methods, by the Galactic center is implied a spatial singularity, the data analyzed are on the spatial distribution of ROs, the main troubles are selection and statistical effects.

I.2. *Kinematic methods.* Here, by the Galactic center is meant a barycenter (a singularity in the phase space), the data are on the phase distribution (spatial coordinates and velocities) of ROs, the main troubles are inadequacies in the accepted kinematic model and in fitting assumptions. These methods are widely diversified. They may be subdivided in respect to at least four independent features: the composition of the kinematic model in terms of considering noncircular motions, the functional forms representing the Galactic circular rotation law, the fitting technique for deriving $R_0$, and the way of allowing for an uncertainty in RDs (see details in Nikiforov 2003, hereafter N03).

I.3. *Dynamic methods* are based on comparing independent data for the space and phase distributions, i.e., for the local density and non-local rotation law (e.g., Olling & Merrifield 1998). Here, it is assumed that the barycenter coincides with the spatial singularity. These methods combine the troubles of both spatial and kinematic methods, and, again, have a specific source of systematics—a dependence of $R_0$ value on *numerous* added assumptions of varied Galactic characteristics. Hence dynamic methods are obviously exceeded in total reliability by spatial and kinematic ones, all other things being equal.

I.4. *Non-phase methods* by Belikov & Syrovoj (1977) and by Surdin (1980). The data are on the distribution of ROs in a non-phase space (i.e., which is not a subset of the phase one), by the Galactic center is meant a singularity in this space. Such a space combines the configuration one (spatial coordinates) with an RO characteristic different from the velocity (e.g., the metallicity). The main troubles are inadequacies in suppositions as to this characteristic's behaviour and in fitting assumptions; this reminds of systematics of kinematic methods.

II. According to **the way of finding the reference distances**:

II.1. *Empirical $R_0$ measurements* rely on empirical determinations of RDs: a) *relative* measurements, based on distance scales eventually using luminosity calibrations from local objects; b) *absolute* (geometrical) measurements, based on RDs found without luminosity calibrations: from a modeling systematic motions in a shell of RO (e.g., Moran et al. 1987), by the statistical parallax method (e.g., Huterer, Sasselov, & Schechter 1995; Genzel et al. 2000; Eisenhauer et al. 2003), by comparing the angular radius of RO shell with the absolute one found from the phase lag (e.g., Honma & Sofue 1996), from the "orbiting binary" technique for the star S2 and source Sgr A* (Eisenhauer et al. 2003).

II.2. *Theoretical $R_0$ measurements* are uncommon. They rely on determinations of RDs from theoretical constraints: the Eddington limit and a theoretical maximum luminosity of planetary nebulae (see refs in R93).

III. According to **the type of reference objects**, $R_0$ measurements are from: globular clusters; RR Lyrae variables; Mira variables; classical Cepheids; OB stars; H II regions and molecular clouds observed in CO; open clusters; molecular clouds observed in OH; stars at the main-sequence turn-off; masers at the periphery of newly-formed, massive stars; OH/IR stars; planetary nebulae; X-ray sources. Again, new RO types are recently suggested: the centroid of K and M giants in the Galactic bulge (Huterer et al. 1995); red clump stars (Paczyński & Stanek 1998); $\delta$ Scuti stars (Morgan, Simet, & Bargenquast 1998); red supergiants (Glushkova et al. 1998); the central star cluster (Genzel et al. 2000); the compact radio source Sgr A* (Eisenhauer et al. 2003).

For details of some classes of $R_0$ measurements and for corresponding refs see Feast (1987) and R93. More details of this classification as a whole and of individual classes are available in N03.

## 4. Recent Results

This section reviews, in terms introduced, only papers published since 1994. Most of earlier papers were covered by Feast (1987) and R93 (see also N03).

### 4.1. Perfecting Methods

Developing *Shapley's method*, Rastorguev et al. (1994) performed a modeling the spatial distribution of globular clusters by the *maximum likelihood method*, taking into account selection effects. Yet their technique can underestimate an $R_0$ value if the interstellar dust exhibits a prominent concentration toward the Galactic center, and not just toward the Galactic plane.

*Kinematic methods.* Pont, Mayor, & Burki (1994) applied the *two-dimensional $\chi^2$ fitting*, considering an uncertainty in distance moduli, to model the radial velocity field. This technique removes the bias of ordinary least squares, and yet it causes an $R_0$ estimate to depend on assumptions of velocity dispersion and of errors in distance moduli.

Nikiforov & Petrovskaya (1994) and Nikiforov (1999a,b) suggested and used a technique for constructing *optimally smooth rotation models*. Metzger, Caldwell, & Schechter (1998) tried to introduce an *elliptical distortion component* into the kinematic model. However, the problem of distinguishing between a tiny structure in the rotation law and perturbations due to streaming motions, with commonly used data on only a "local" galactocentric sector, remains unsolved. No $R_0$ estimate with regard to spiral wave perturbations appeared after Byl & Ovenden (1978). An attempt to take into account a *difference between HI and RO rotations curves*, likely caused by streaming motions, was done by Nikiforov (2000).

*Dynamic methods.* Olling & Merrifield (1998) considered the *gaseous component* in dynamic modeling, yet this refining does not cure the main troubles of dynamic methods (see section 3).

### 4.2. New Types of Reference Objects in Relative Measuring $R_0$

*Red clump stars.* Using them as new distance indicators was proposed by Paczyński & Stanek (1998). Large number of these stars both in the Solar neighborhood (~1000 in the *Hipparcos* catalogue) and in Baade's window (~10000) gives a high precision of the luminosity calibration ($\pm\sim 0^{\rm m}_{.}02$–$0^{\rm m}_{.}03$) as well as of $R_0$ estimates by Baade's method ($\pm\sim$0.1 kpc, see refs in Table 2). Problems with the metallicity dependence for red clump stars (see Reid 1999, hereafter R99, and refs within) appear not to be more difficult than similar problems for traditional distance indicators (Stanek et al. 2000). To re-normalize the $R_0$ estimates from such ROs, we adopt a calibration for the Baade's window clump, $M_I(\mathrm{RC}) = -0.24 \pm 0.15$ at $[\mathrm{Fe/H}]_{\mathrm{BW}} = -0.15$ (see refs in N03).

*δ Scuti stars* are fainter variables as against RR Lyrae stars. A possibility to take the high-amplitude δ Scuti stars as ROs in Baade's method arose only recently (Morgan et al. 1998; McNamara et al. 2000) from the publications of the OGLE dark matter survey and *Hipparcos* catalogue. There appears to be more efficiently to use only the metal-strong variables (McNamara et al. 2000). We adopt $M_V(\delta\ \mathrm{Sct}) = -3.725 \log P - 1.933$, $P$ is the variable's period, with an RDs' contribution of $\sigma_{\mathrm{RD}} = 0^{\rm m}_{.}15$ to the $R_0$ error budget (see N03).

*Red supergiants*, used by Glushkova et al. (1998) in a kinematic method, can be promising ROs due to a good accuracy of their distances and $\gamma$-velocities.

### 4.3. New Luminosity Calibrations for Traditional Reference Objects

The publication of the *Hipparcos* catalogue has greatly stimulated studies in the recalibration of distance indicators. The following in this subsection is a short summary of results from analyses of *Hipparcos* and other recent data.

*RR Lyrae variables.* Faint absolute magnitudes were confirmed at least for the local RR Lyraes: $\langle M_V(\mathrm{RR})\rangle \sim 0^{\rm m}_{.}75 \pm 0^{\rm m}_{.}15$ at $[\mathrm{Fe/H}] = -1.3$ (cf. R93 and R99, for more recent refs see N03). The problem of the metallicity dependence of $M_V(\mathrm{RR})$, like the suggested problem of a mismatch between the field and cluster calibrations, remains unresolved (R99).

*Globular clusters.* The MS-fitting to *Hipparcos* subdwarf data points to an average *increase* of $0^{\rm m}_{.}2$–$0^{\rm m}_{.}3$ in distance moduli as compared to Harris' (1996) $M_V(\mathrm{RR})$-based scale (R99; Carretta et al. 2000; N03). However, astrometric methods and the comparison with local white dwarfs lead to values of $M_V(\mathrm{HB})$ $0^{\rm m}_{.}1$–$0^{\rm m}_{.}2$ fainter than the subdwarf fitting (Carretta et al. 2000; N03). Therefore returning to the "canonical" (brighter) calibration, $M_V(\mathrm{HB}) = 0^{\rm m}_{.}6$ at $[\mathrm{Fe/H}] = -1.3$, seems to be most cautious at present. Here, the relation $M_V(\mathrm{HB}) = 0.2[\mathrm{Fe/H}] + 0.86$ and a value of $\sigma_{\mathrm{RD}} = 0^{\rm m}_{.}15$ were adopted (see N03).

Table 2. Recent relative estimates for $R_0$

| Source | Method(s)[a]: Reference Objects | $R_0$ (kpc) | Comments |
|---|---|---|---|
| | | *Spatial measurements:* | |
| Rastorguev et al. (1994) | Spatial modeling: globular clusters | $(7 \pm 0.5)$<br>$7.3 \pm 0.55$ | $M_V(\mathrm{RR}) = 0.38[\mathrm{m/H}] + 1.32$, $M_V(\mathrm{HB}) = 0.6$ |
| Layden et al. (1996) | B: RR Lyrae stars | $7.6 \pm 0.4$ | $M_V(\mathrm{RR}) = 0.15[\mathrm{Fe/H}] + 0.95$ |
| Morgan et al. (1998) | B: $\delta$ Scuti stars | $(7.6 \pm 0.35^b)$<br>$7.7 \pm 0.4$ | $M_V(\delta\ \mathrm{Sct}) = -3.73 \log P - 1.91$ |
| Paczyński & Stanek (1998) | B: red clump stars | $(8.4 \pm 0.2)$ | $M_I(\mathrm{RC}) = -0.279$. Revised by Stanek et al. (2000) |
| Udalski (1998) | B: RR Lyrae stars | $(8.1 \pm 0.25)$<br>$8.4 \pm 0.3$ | $M_V(\mathrm{RR}) = 0.18[\mathrm{Fe/H}] + 1.06$ |
| Alves (2000) | B: red clump stars | $(8.24 \pm 0.42)$<br>$8.28 \pm 0.42$ | $M_K(\mathrm{RC}) = -1.61$<br>$[M_I(\mathrm{RC}) = -0.23]$ |
| McNamara et al. (2000) | B: $\delta$ Scuti stars<br>B: RR Lyrae stars | $7.86 \pm 0.3^b$<br>$(7.79 \pm 0.08)$<br>$7.10 \pm 0.07$ | $M_V(\delta\ \mathrm{Sct}) = -3.725 \log P - 1.93$<br>$M_V(\mathrm{RR}) = 0.55$ at $[\mathrm{Fe/H}] = -1.46$ |
| Stanek et al. (2000) | B: red clump stars | $(8.67 \pm 0.1)$<br>$8.72 \pm 0.1$ | $M_I(\mathrm{RC}) = -0.227$ |
| Gould et al. (2001) | B: red clump stars | $(8.63 \pm 0.12)$<br>$8.60 \pm 0.12$ | $M_I(\mathrm{RC}) = -0.247$ at $[\mathrm{Fe/H}] = -0.15$ |
| | | *Kinematic measurements:* | |
| Nikiforov & Petrovskaya (1994) | HI: H II regions | $(7.5 \pm 1.0)$ | Revised by Nikiforov (2000) |
| Pont et al. (1994) | MR: clas. Cepheids | $(8.09 \pm 0.30)$<br>$8.20 \pm 0.30$ | $(m - M)_0^{\mathrm{LMC}} = 18\overset{\mathrm{m}}{.}47$ |
| Dambis et al. (1995) | MR: clas. Cepheids | $(7.1 \pm 0.5)$<br>$7.4 \pm 0.5$ | $(m - M)_0^{\mathrm{LMC}} = 18\overset{\mathrm{m}}{.}4^b$ |
| Glushkova et al. (1998) | MR: 3 types of ROs | $(7.3 \pm 0.3)$ | Kholopov's ZAMS |
| Metzger et al. (1998) | MR: clas. Cepheids | $7.66 \pm 0.32$ | $(m - M)_0^{\mathrm{LMC}} = 18\overset{\mathrm{m}}{.}50$ |
| Nikiforov (1999b, 2000) | MR, HI: CO clouds | $8.2 \pm 0.7$ | |
| | | *Dynamic measurements:* | |
| Olling & Merrifield (1998) | Comparing dynamic model with Galactic constants | $(7.1 \pm 0.4)$ | Assumptions of Oort's $A$ and $B$ constants, of Galactic dynamic model, of $-B/(A-B)$ etc. |
| | | *Non-phase measurements:* | |
| Surdin & Feoktistov (1999) | Surdin's (1980): globular clusters | $(8.6 \pm 1.0)$<br>$9.2 \pm 1.1$ | $M_V(\mathrm{RR}) = 0.2[\mathrm{Fe/H}] + 1.00$. See Surdin (1999) |

[a]B = Baade's, MR = modeling rotation of the Galaxy, HI = comparison with H I.
[b]See details in N03.

*Mira variables.* We adopt the relation $M_K = 0.84 - 3.47 \log P$ with the zero-point from the *Hipparcos* parallaxes of local Miras (Whitelock & Feast 2000), independent on other calibrations. It is only slightly brighter than the pre-Hipparcos PL($K$) relation (see N03). A value of $\sigma_{\mathrm{RD}}$ is also about $0\overset{\mathrm{m}}{.}15$.

*Classical Cepheids.* Analyses of *Hipparcos* and additional data yielded contradictory results—from an *increase* of distance scales, so that an LMC distance module of $(m - M)_0^{\mathrm{LMC}} \sim 18\overset{\mathrm{m}}{.}6 \div 18\overset{\mathrm{m}}{.}7$ is derived from Cepheids (Feast et al.), to a *decrease* of distance scales, so that $(m - M)_0^{\mathrm{LMC}} \lesssim 18\overset{\mathrm{m}}{.}4$ (Rastorguev et al.); see discussions and refs in R99 and N03. Thus the old calibration, $(m - M)_0^{\mathrm{LMC}} = 18\overset{\mathrm{m}}{.}5$, with $\sigma_{\mathrm{RD}} = 0\overset{\mathrm{m}}{.}2$, seems to be a good compromise today. (That is not adopting some LMC distance; a value of $(m - M)_0^{\mathrm{LMC}}$ is used here only to compare varied distance scales for Cepheids.)

*OB stars.* No significant correction was obtained for main-sequence stars, but a *decrease* of $\sim 0\overset{\mathrm{m}}{.}3$ in distance moduli was confirmed for *giants and supergiants* (see R99 and N03). Here, a correction of $-0\overset{\mathrm{m}}{.}15$ and a value of $\sigma_{\mathrm{RD}} = 0\overset{\mathrm{m}}{.}5$ were adopted for $R_0$ estimates from OB stars (see N03).

*Open clusters.* On the average, previous distances were confirmed—small corrections of opposite signs were obtained for different scales (see R99 and N03).

### 4.4. New Absolute Measurements of $R_0$

All *spatial* absolute $R_0$ estimates were obtained by the "central object method" in which a distance to a sole RO, assumed to be in the very center of the Galaxy, is taken as a value of $R_0$ without any further handling. In recent studies, the statistical parallax method was used to derive an absolute RD—by Huterer et al. (1995) for the centroid of K and M giants in the Galactic bulge, by Genzel et al. (2000) and Eisenhauer et al. (2003) for the central (within $10''$) star cluster. Unfortunately, this method is rather uncertain, even statistically (see Table 3).

A *kinematic* estimate of $R_0$ based on absolute RDs for OH/IR stars from comparing their shell angular radii with phase lags (Honma & Sofue 1996) is also not very accurate, mainly due to a limited number of ROs and systematic errors of the kinematic method applied.

Against this background, the very recent estimate (formally spatial) by Eisenhauer et al. (2003), $R_0 = 7.94 \pm 0.42$ kpc, with the absolute RD derived from a modeling the motion of the star S2 orbiting the massive black hole and compact radio source Sgr A*, is undoubtedly a major achievement.

Table 3. Recent absolute estimates for $R_0$

| Source | Method of Estimating $R_0$[a]: Reference Object(s) | Method of Estimating RD(s) | $R_0$ (kpc) |
|---|---|---|---|
| | *Spatial measurements:* | | |
| Huterer et al. (1995) | C: centroid of K and M giants | Statistical parallax | $8.21 \pm 0.98$ |
| Genzel et al. (2000), Eisenhauer et al. (2003) | C: central stellar cluster (within $10'' = 0.5$ pc) | Statistical parallax | $7.2 \pm 0.9$ |
| Eisenhauer et al. (2003) | C: S2/Sgr A* | Orbiting binary | $7.94 \pm 0.42$ |
| | *Kinematic measurements:* | | |
| Honma & Sofue (1996) | HI: OH/IR stars | Comparing angular radius with phase lag | $7.7 \pm 0.8$[b] |

[a]C = "central object method", HI = comparison with H I.
[b]See details in N03.

### 4.5. Traditional Reference Objects and Methods: Updated Results

A number of recent relative $R_0$ estimates was derived using *renewed observational data* on ROs, while by traditional methods and from ROs of traditional types.

In this class of results, among *spatial* methods the *Baade's* one alone was recently taken (Layden et al. 1996; Udalski 1998; McNamara et al. 2000); only the RR Lyrae variables, mainly in Baade's Window, were used as ROs. *Kinematic* methods were applied by Dambis, Mel'nik, & Rastorguev (1995) to classical Cepheids and by Glushkova et al. (1998) to a combined sample of classical

Table 4. "Best values" for $R_0$ from different groups of $R_0$ estimates

| Covered range of years | All estimates $R_0$ | $N_{est}$ $N_{pap}$ | Relative estimates $R_0$ | $N_{est}$ $N_{pap}$ | Absolute estimates $R_0$ | $N_{est}$ $N_{pap}$ |
|---|---|---|---|---|---|---|
| 1974–1993 | $7.85 \pm 0.24$ | 41<br>37 | $7.98 \pm 0.24$ | 35<br>31 | $7.61 \pm 0.68$ | 4<br>4 |
| 1994–2003 | $7.89 \pm 0.19$ | 24<br>20 | $7.93 \pm 0.23$ | 19<br>16 | $7.78 \pm 0.33$ | 5<br>4 |
| **1974–2003** | $\mathbf{7.90 \pm 0.17}$ | 65<br>57 | $8.03 \pm 0.20$ | 54<br>47 | $7.73 \pm 0.31$ | 9<br>8 |

Cepheids, open clusters, and red supergiants. Surdin & Feoktistov (see Surdin 1999) revised, using the catalogue of globular cluster by Harris (1996), a value of $R_0$ obtained by one of *non-phase* methods (Surdin 1980).

## 5. A Current "Best Value" for $R_0$

Tables 2 and 3 list recent $R_0$ estimates, relative and absolute respectively. Relative estimates of $R_0$ were rescaled according to the adopted calibrations (see sections 4.2 and 4.3); original $R_0$ values are given in parentheses. In Table 2, only statistical errors in $R_0$ are presented. In Table 3, $R_0$ errors reflect all statistical uncertainties and systematic errors in RDs (if given in papers).

To try to account for varied sources of errors and covariances among different $R_0$ estimates, the rescaled $R_0$ values were averaged with weights in four steps: over homogeneous groups of $R_0$ estimates (having the same classes of method, of RDs, and of ROs), over all calibration groups and over independent ones, and the final averaging (see N03). Some categories of unreliable $R_0$ estimates (e.g., dynamic), the estimates from combined samples of ROs, and the revised results were not used in averaging (N03). The $R_0$ values, taken for this procedure, are plotted versus publication year in Figure 1. The resulting "best $R_0$ values" for some groups of $R_0$ estimates ($N_{est}$ is the number of estimates, $N_{pap}$ is the number of papers) are presented in Table 4.

With the adopted calibrations, variations of $R_0$ with time (Figure 1) are not so high as in R93, due in part to rescaling for $R_0$ values from OB stars in this review, contrary to R93. Differences between $R_0$ estimates obtained before 1994 and more recent ones are, on average, nonsignificant (Table 4). Relative and absolute best values agree within their errors (Table 4); this implies the correctness of modern distance scales, at least on average. Combining all results obtained since 1974 yields a current "best value" of $R_0 = 7.9 \pm 0.2$ kpc.

Notice that Eisenhauer et al.'s (2003) estimate alone, $R_0 = 7.94 \pm 0.42$ kpc, does not solve entirely the problem of $R_0$ because of, at least, an insufficient accuracy—it does not reject values of $R_0 = 7.5$ or 8.5 kpc preferred by some authors (see N03). Besides, Eisenhauer et al. (2003) predict slow future improvements in the accuracy of this estimate. Thus to gain a further progress in the $R_0$ problem, more efforts in *different* lines of investigation are required.

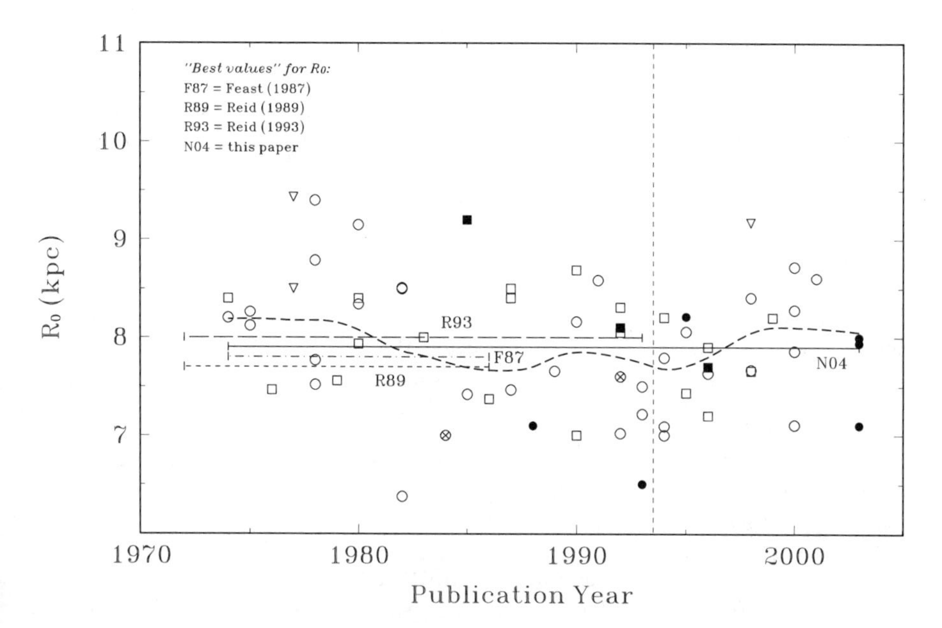


Figure 1. Rescaled estimates of $R_0$ (see text) versus publication year since 1974. The dashed curve is a smoothed dependence. Solid symbols, open ones, and crosses ($\times$) in symbols are $R_0$ estimates derived respectively from absolute, relative, and theoretical RDs by spatial (circles), kinematic (boxes), and non-phase methods (triangles).

**Acknowledgments.** This work was partly supported by the Leading Scientific School (Grant NSh-1078.2003.2).

## References

Alves, D. R. 2000, ApJ, 539, 732

Baade, W. 1951, Publ.Obs.Univ.Mich., 10, 7

Belikov V. T., & Syrovoj V. V. 1977, Ast.Tsirk., 968, 5 (in Russian)

Blitz, L. 1994, in ASP Conf. Ser., Vol. 66, Physics of the Gaseous and Stellar Disks of the Galaxy, ed. I. R. King (San Francisco: ASP), 1

Byl, J., & Ovenden, M. W. 1978, ApJ, 225, 496

Carretta, E., Gratton, R. G., Clementini, G., & Fusi Pecci, F. 2000, ApJ, 533, 215

Dambis, A. K., Mel'nik, A. M., & Rastorguev, A. S. 1995, Pis'ma AZh, 21, 331

de Vaucouleurs, G. 1983, AJ, 268, 451

Eisenhauer, F., Schödel, R., Genzel, R., Ott, T., Tecza, M., Abuter, R., Eckart, A., & Alexander T. 2003, ApJ, 597, L121

Feast, M. W. 1987, in Proc. of a NATO Advanced Study Institute, The Galaxy, ed. G. Gilmore & B. Carswell (Dordrecht: Reidel), 1

Harris, W. E. 1980, in IAU Symp. 85, Star Clusters, ed. J. E. Hesser (Dordrecht: Reidel), 81

Harris, W. E. 1996, AJ, 112, 1487

Honma, M., & Sofue Y. 1996, PASP, 48, L103

Huterer, D., Sasselov, D. D., & Schechter, P. L. 1995, AJ, 110, 2705

Genzel, R., Pichon, C., Eckart, A., Gerhard, O. E., & Ott, T. 2000, MNRAS, 317, 348

Glushkova, E. V., Dambis, A. K., Mel'nik, A. M., & Rastorguev, A. S. 1998, A&A, 329, 514

Gould, A., Stutz, A., & Frogel, J. A. 2001, ApJ, 547, 590

Kerr, F. J., & Lynden-Bell, D. 1986, MNRAS, 221, 1023

Layden, A.C., Hanson, R. B., Hawley, S. L., Klemola, A. R., & Hanley, C. J. 1996, AJ, 112, 2110

McNamara, D. H., Madsen, J. B., Barnes, J., & Ericksen B. F. 2000, PASP, 112, 202

Metzger, M. R., Caldwell, J. A. R., & Schechter, P. L. 1998, AJ, 115, 635

Moran, J. M., Reid, M. J., Schneps, M. H., Gwinn, C. R., Genzel, R., Downes, D., & Rönnäng, B. 1987, in AIP Conf. Proc., 155, The Galactic Center, ed. D. C. Backer (New York: AIP), 166

Morgan, S. M., Simet, M., & Bargenquast, S. 1998, Acta Astronomica, 48, 509

Nikiforov, I. I. 1999a, Astrofizika, 42, 399

Nikiforov, I. I. 1999b, Ast.Rep., 43, 345

Nikiforov, I. I. 2000, in ASP Conf. Ser., Vol. 209, Small Galaxy Groups, ed. M. J. Valtonen & C. Flynn (San Francisco: ASP), 403

Nikiforov, I. I. 2003, Ph. D. Thesis, Saint Petersburg State University (in Russian), http://www.astro.spbu.ru/astro/publications/disser/NikiforovII/

Nikiforov, I. I., & Petrovskaya, I. V. 1994, Ast.Rep., 38, 642

Olling, R. P., & Merrifield, M. R. 1998, MNRAS, 297, 943

Paczyński, B., & Stanek, K. Z. 1998, ApJ, 494, L219

Pont, F., Mayor, M., & Burki, G. 1994, A&A, 285, 415

Rastorguev, A. S., Pavlovskaja, E. D., Durlevich, O. V., & Filippova, A. A. 1994, Pis'ma AZh, 20, 688

Reid, I. N. 1999, ARA&A, 37, 191

Reid, M. J. 1989, in IAU Symp. 136, The Center of the Galaxy, ed. M. Morris (Dordrecht: Kluwer), 37

Reid, M. J. 1993, ARA&A, 31, 345

Shapley, H. 1918, AJ, 48, 154

Stanek, K. Z., Kaluzny, J., Wysocka, A., & Thompson, I. 2000, Acta Astronomica, 50, 191

Surdin, V. G. 1980, AZh, 57, 959

Surdin, V. G. 1999, A&A Trans., 18, 367

Udalski, A. 1998, Acta Astronomica, 48, 113

Whitelock, P. A., & Feast, M. W. 2000, MNRAS, 319, 759

Woolley, R. v. d. R. 1963, J.Brit.Astron.Assoc., 73, 131

*Order and Chaos in Stellar and Planetary Systems*
*ASP Conference Series, Vol. 316, 2004*
*G. Byrd, K. Kholshevnikov, A. Mylläri, I. Nikiforov, and V. Orlov, eds.*

# The Classification of Cepheids by Pulsation Modes

Marina V. Zabolotskikh, Alexey S. Rastorguev, and Ivan E. Egorov

*Sternberg Astronomical Institute, Universitetskij pr. 13, Moscow 119992, Russia*

**Abstract.** The neural network algorithm is used to classify the cepheids of our Galaxy by pulsation modes. The LMC and SMC cepheids with $I$-photometry have been used as a training samples. Infrared photometric data for cepheids of our Galaxy have been taken from Berdnikov's database. 13 cepheids from our sample of Galactic cepheids with periods less then 10 days were shown to have wrong classification in GCVS.

## 1. Introduction

Classical cepheids can be regarded as the most "important" stars that define the distance scale in the Universe, due to the presence of the "period–luminosity" relation for these variables. But as it was shown in our early work (Rastorguev et al. 1999) by the statistical parallax method, the distance scales of short-periodic and long-periodic cepheids can be different. From 76 cepheids with pulsation periods $P_{\rm pls} > 9^{\rm d}$ the distance scale factor $p \approx 0.93 \pm 0.09$, whereas from 155 cepheids with $P_{\rm pls} < 9^{\rm d}$ $p \approx 0.83 \pm 0.06$. (We note, that for $p < 1$ the distance scale should be made longer.) We concluded at that time, that considerable contamination of short-periodic sample by the unrecognized first-overtone pulsators could be one of the possible reasons of this difference.

The ratio of first-overtone to fundamental tone pulsation periods is ≈0.7 (Alcock et al. 1995). It can be easily shown, that for Berdnikov et al.'s (1996) "period–luminosity" relation and Dean et al.'s (1978) "period–intrinsic color" relation the distances of these unrecognized first-overtone cepheids should be multiplied by the factor of 1.35. Therefore, we estimated the fraction of such unrevealed first-overtone pulsators as ≈20–30%.

Only for cepheid with very well known distance we could in principle determine directly the pulsation mode. As for bulk cepheids, the problem seems to be more complex. One way to do this lies in photometric data, i.e., light curves. For example, the cepheids with small amplitude and almost symmetrical light curve were classified as the first-overtone pulsators in General Catalog of Variable Stars (GCVS; Kholopov et al. 1985).

In early works the attempts have been made to improve the classification. Commonly the cepheid's light curve is Fourier-decomposed as

$$m(t) = R_0 + \sum_{i=1}^{N} R_i \cos\left[2\pi i(t - T_0)/P_{\rm pls} + \phi_i\right],$$

and the ratios of Fourier-coefficients and the phases

$$R_{ij} = R_i/R_j, \quad \phi_{ij} = j\phi_i - i\phi_j,$$

have been used for classification.

First attempt to determine the localization of Large Magellanic Cloud (LMC) s-cepheids (with sinusoidal light curve) on the diagrams amplitude–$\log P_{\rm pls}$ and $\langle V\rangle$–$\log P_{\rm pls}$ has been made by Connolly (1980). In the papers of Antonello & Poretti (1986), Antonello et al. (1990) and Mantegazza & Poretti (1992) the possible ways to separate s-cepheids from normal cepheids in our Galaxy were searched for. Authors have shown, that s-cepheids form an almost separate group on the diagrams $R_{21}$–$P_{\rm pls}$ and $\phi_{21}$–$P_{\rm pls}$. S-cepheids with $P_{\rm pls} < 3^{\rm d}$ are certainly first-overtone pulsators. But in the first-overtone region of this diagrams normal cepheids were also found. Moreover, s-cepheids can be found between fundamental-tone pulsators. Buchler & Moskalik (1994) compared the 114 Small Magellanic Cloud (SMC) cepheids with $P_{\rm pls} < 7^{\rm d}$ and cepheids in our Galaxy. They supposed that a difference exists between the cepheids in these galaxies. We believe that this impression was due to the lack of observational data.

Zakrzewski et al. (2000) analyzed the Fourier-decomposition's components of the light curves of the Galactic cepheids. After OGLE (The Optical Gravitational Lensing Experiment) photometric survey has been appeared, with large number of LMC and SMC cepheids' light-curves in $I$-photometric band (Udalski et al. 1999a,b), new perspectives opened to continue the classification work.

## 2. The Data

The majority of OGLE photometric data was measured in the wide infrared band $I$. More than three thousands cepheids (1228 in LMC and 2028 in SMC) have good quality light curves. Because all cepheids in Magellanic Clouds are nearly at the same distance from the Sun, they can be classified by pulsation mode directly from $I$–$\log P_{\rm pls}$ diagram (at the same luminosity the cepheid pulsating in first overtone have smaller period). In Figure 1 such diagram for LMC cepheids is shown (for SMC cepheids all diagrams are very similar). Here we use the extinction insensitive index $W_I = I - 1.55(V - I)$ instead of $I$ to correct photometric data for differential extinction. The coefficient of 1.55 results from the standard interstellar extinction law, i.e, the relation between the extinction in $I$-band and the color excess $E(V - I)$. We would like to note two points. First, it can be seen that the fraction of first-overtone pulsators is very large. Second, there are no first-overtone cepheids with period greater than 9 days. Because we have choosen only most reliably classified pulsators in fundamental tone and first overtone, our Figure 1 is different from Figure 2 in paper Udalski et al. (1999a). Also fundamental tone and first-overtone pulsators are separated on $R_{21}$–$\log P_{\rm pls}$ and $\phi_{21}$–$\log P_{\rm pls}$ diagrams (see Figure 2).

Original observations of 213 Galactic short-periodic cepheids made by many authors have been taken from Berdnikov's data base. After preliminary processing we derived $I$-band light curves.

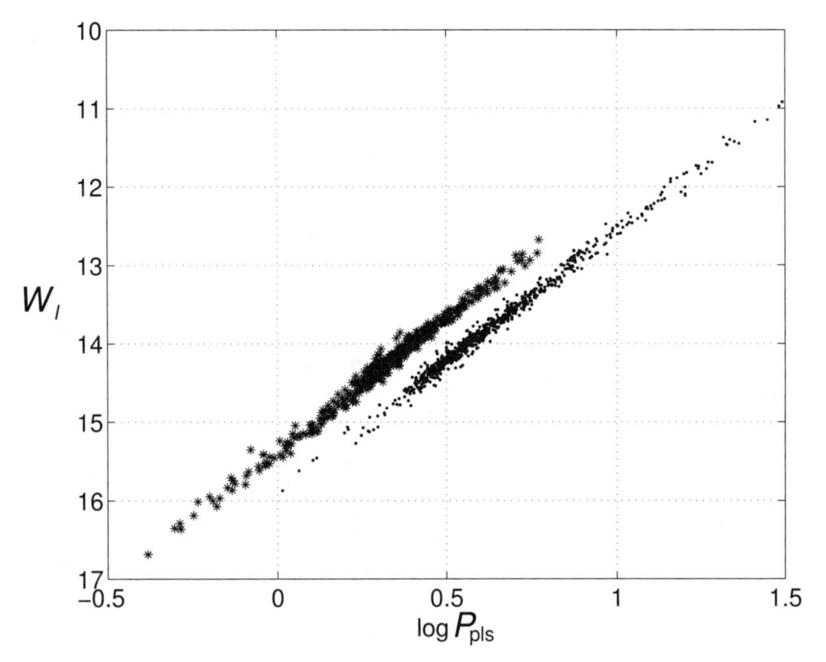


Figure 1. The $W_I$–log $P_{\rm pls}$ relation for cepheids in LMC. Dots and asterisks show fundamental tone and first-overtone pulsators respectively.

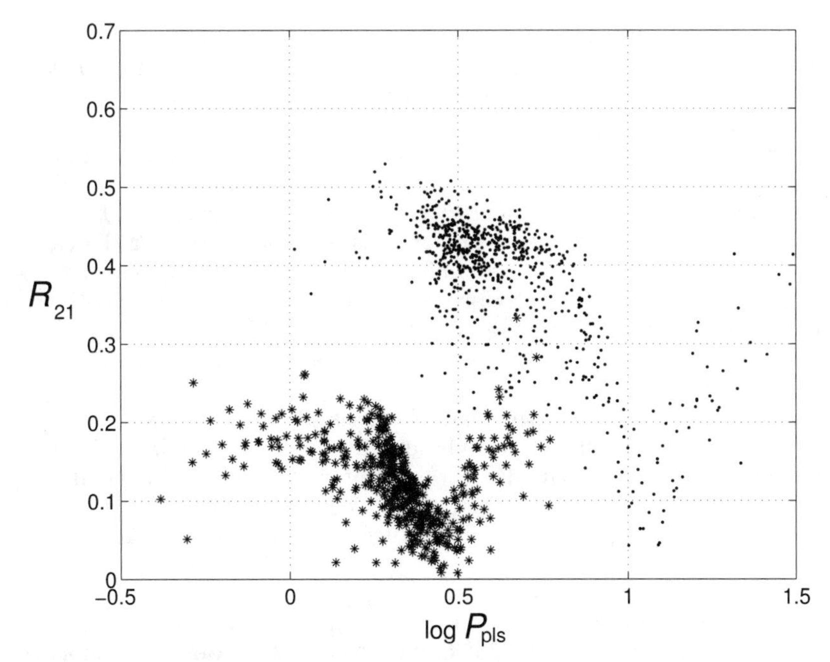


Figure 2. The $R_{21}$–log $P_{\rm pls}$ relation for LMC cepheids. Dots and asterisks show fundamental tone and first-overtone pulsators respectively.

## 3. The Method

To compare the properties of the light-curves of the cepheids in LMC, SMC and our Galaxy, we used the neural network technique. This technique applied in many areas of astronomy, but for the first time we use it to classify the cepheids by pulsation modes. The information about neural networks is accessible for example from ftp://ftp.sas.com/pub/neural/. In our case the multilayer perceptron neural network has been trained using back propagation algorithm. The training sample contains the set of input data (the parameters of the light curve) together with corresponding outputs (the classification of the cepheid as fundamental or first-overtone pulsator), and the network learns to infer the relationship between these two. We performed Fourier-decomposition of all light curves up to 7th order. The input set of parameters included the pulsation period, all amplitudes and phases and their ratios. LMC and SMC cepheids were used as the training samples.

The use of neural network have at least two advantages. Often we don't know which parameters are the best for classification. If we preset too many input parameters, the network itself can choose the best input subset for classification. For example, our network with LMC cepheids as the training sample, has found $R_{21}$ as the best input parameter. The possibility of estimation the classification errors is another advantage of neural network.

## 4. Results

The results of classification can be seen in Figure 3. For most stars that we used in our paper of 1999 the infrared data are present in Berdnikov's database. The neural network with LMC cepheids as training sample found 38 first-overtone pulsators, whereas one with SMC cepheids as training sample found 33 first-overtone pulsators. Among all stars 13 cepheids with $P_{\rm pls} < 10^{\rm d}$ (Table 1) have wrong classification in GCVS (they are marked by black circles in Figure 3). We would like to note that 6 of these cepheids have been used for the calculation of statistical parallaxes. So the number of unrecognized first-overtone pulastors is less then expected and therefore doesn't completely explain the difference in distance scales. Probably, as we noted earlier, the "period-luminosity" relation of Berdnikov & Efremov (1985) is more adequate than new one.

It is of special interest to analyze the pulsation modes of 9 cepheids—members of open clusters that have been used by Berdnikov et al. (1996) as the base of "period-luminosity" relation. Even only one cepheid of this sample was unresolved overtone/fundamental pulsator, it could completely destroy real "period-luminosity" relation. Fortunately, 5 of these cepheids have correct classification, but for 4 remaining cepheids there are yet no good $I$-band photometric data.

**Acknowledgments.** We are very grateful to L. N. Berdnikov for providing the photometric data on the galactic cepheids. This work was supported by the Russian Foundation for Basic Research (project nos 02-02-16677 and 03-02-06588) and President Grant NSH-389-2003-2 of Russian Federation.

Table 1. Cepheids with wrong classification in GCVS

| Cepheid | GCVS classification | Our Classification |
|---|---|---|
| FN Aql | First overtone | Fundamental tone |
| V1162 Aql | First overtone | Fundamental tone |
| V496 Aql | First overtone | Fundamental tone |
| V378 Cen | First overtone | Fundamental tone |
| V659 Cen | Fundamental tone | First overtone |
| IR Cep | Fundamental tone | First overtone |
| AV Cir | Fundamental tone | First overtone |
| BP Cir | Fundamental tone | First overtone |
| V465 Mon | Fundamental tone | First overtone |
| V504 Mon | Fundamental tone | First overtone |
| AU Peg | Fundamental tone | First overtone |
| SS Sct | Fundamental tone | First overtone |
| DK Vel | Fundamental tone | First overtone |

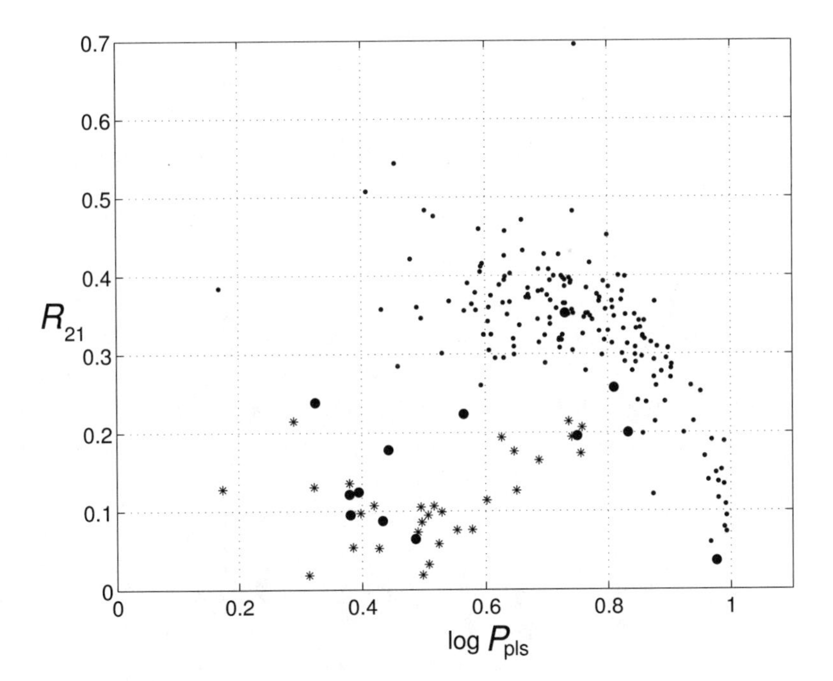


Figure 3. The $R_{21}$–$\log P_{\rm pls}$ relation for cepheids in our Galaxy. Dots and asterisks show fundamental tone and first-overtone pulsators respectively. Cepheids with wrong classification in GCVS marked by black circles. Training sample: LMC cepheids.

## References

Alcock, C., Allisman, R. A., Axelrod, T. S. et al. 1995, AJ, 109, 1653
Antonello E., & Poretti, E. 1986, A&A, 169, 149
Antonello, E., Poretti, E., & Reduzzi, L. 1990, A&A, 236, 138
Berdnikov, L. N., & Efremov, Y. N. 1985, Ast.Tsirk., 1388, 1
Berdnikov, L. N., Vozyakova, O. V., & Dambis, A. K. 1996, Ast.Lett., 22, 838
Buchler, J. R., & Moskalik, P. 1994, A&A, 292, 450
Connolly, L. P. 1980, PASP, 92, 165
Dean, J. F., Warren, P. R., & Cousins, A. W. J. 1978, MNRAS, 183, 569
Kholopov, P. N. et al., ed. 1985, General Catalog of Variable Stars, 4th edn. (Moscow: Nauka)
Mantegazza, L., & Poretti, E. 1992, A&A, 261, 137
Rastorguev, A. S., Glushkova, E. V., Dambis, A. K., & Zabolotskikh, M. V. 1999, Ast.Lett., 25, 689
Udalski, A., Soszynski, I., Szymanski, M. et al. 1999a, Acta Astronomica, 49, 223
Udalski, A., Soszynski, I., Szymanski, M. et al. 1999b, Acta Astronomica, 49, 437
Zakrzewski, B., Ogloza, W., & Moskalik, P. 2000, Acta Astronomica, 50, 387

*Order and Chaos in Stellar and Planetary Systems*
*ASP Conference Series, Vol. 316, 2004*
*G. Byrd, K. Kholshevnikov, A. Mylläri, I. Nikiforov, and V. Orlov, eds.*

# Local Motions in the Milky Way Galaxy from HIPPARCOS Data

Leonid Ossipkov

*Department of Space Technologies & Applied Astrodynamics, St. Petersburg State University, Universitetskij pr. 28, Staryj Peterhof, St. Petersburg 198504, Russia*

Aleksandr Mylläri

*Department of Information Technology, University of Turku, Lemminkaisenkatu 14-18 A, 20520, Turku, Finland*

Chris Flynn

*Tuorla Observatory, University of Turku, Väisäläntie 20, Piikkiö 21500, Finland*

**Abstract.** The *HIPPARCOS* data provide new possibilities for making good use of the following methods: a method of extrapolation to zero Heliocentric distances for determining the local values of stellar kinematic parameters, a method of extrapolating to stars moving along a rectilinear orbits to find the Solar Galactocentric velocity and a method of extrapolating to stars moving along circular orbits to find the local circular velocity. Algorithms for finding these parameters are discussed.

## 1. Introduction

Stellar Kinematics serves as a basis for Galactic Dynamics and is a fundamental method of analyzing the structure and dynamics of Galactic subsystems (e.g., Binney & Merrifield 1998; Binney & Ossipkov 1999). Many kinematic parameters, such as LSR (local standard of rest) velocity (defined in Chandrasekhar's 1942 sense) and the velocity ellipsoid must be considered as *local.* Chandrasekhar (1942) emphasized the rigorous definition of either assumes that the Galaxy is a continuous stellar fluid, whereas observations give us individual positions and velocities of stars. As a consequence the determination of the local kinematic parameters of the Galaxy is not trivial. We use some techniques here which have been suggested in the literature but have remained practically unused.

The *HIPPARCOS* Catalogue ESA 97 provides us with new opportunities to re-examine the fundamental data of local stellar kinematics. It seems timely to develop new methods of treating observational data and to clarify some basic concepts of Stellar Kinematics. In present study we make some steps in this direction.

One of main ideas developed in this work is an idea of *extrapolating* Stellar Kinematics to the Sun. We are interested mainly in local values of number star density, Solar velocity, velocity ellipsoid, etc., but not averaged over the whole volume available from observations. On one hand, using very nearby stars (say within a few parsecs) leads to too few stars to determine the local kinematic parameters. On the other hand, when using more distant star data, the samples are both incomplete and kinematically biased; such samples give results which are not local. Worth mentioning here is the "distance effect" (Ogrodnikoff 1936) that leads to systematic distortion of the velocity ellipsoid.

The problem was clearly understood by Chandrasekhar (1942). Later a concept of macroscopic volume element was developed in Galactic Dynamics (Chandrasekhar 1943; Ogorodnikov 1948, 1958), and the idea of extrapolation to zero Heliocentric distance for finding local values of Galactic parameters has appeared. Agekian (1962) first derived a local value of number density by this method. Later various modifications of the method were used for determining number density by Agekian & Ogorodnikov (1974), Orlov & Titov (1994). Then Ogorodnikov suggested that kinematic parameters could be found by a related procedure (Agekian & Ogorodnikov 1974), and Ogorodnikov & Ossipkov (1978) determined so the Solar motion from radial velocities of nearby stars. Nezhinskij, Ossipkov, & Kutuzov (1995) used a method of extrapolation to the zero distance to determine the local velocity ellipsoid for the open cluster subsystem.

In this study a method of extrapolation to zero distance, in combination with some other ideas discussed below, has been used to determine the Galactocentric Solar velocity (Section 2) and the circular velocity at the Solar Galactocentric distance (Section 3). Our preliminary results were published by Mylläri, Flynn, & Ossipkov (2001) and Ossipkov, Mylläri, & Flynn (2001).

## 2. Galactocentric Solar Velocity

In this and next sections we assume that Galactic potential is stationary and axisymmetric. Thus, circular orbits exist, and the angular momentum relative the rotation axis, $L$, is conserved along stellar orbits. We describe a method of finding the Galactocentric velocity of the Sun.

To find the Solar motion with respect to the Galactic centre we have to measure velocities of objects whose mean motion in the Galaxy is negligible. The angular momentum of such stars must be close to zero. High-velocity stars, metal-poor short-periodic variables and globular clusters are usually considered suitable objects; despite there being evidence of their slow net rotation.

The method we suggest is an elaboration of ideas by Ossipkov (1980), Carlberg & Innanen (1987), and Kinman (2000). If there are no stars crossing the rotation axis we find the mean Heliocentric velocity of stars passing nearby the axis and then extrapolate that obtained result to fictitious stars on rectilinear orbit. Also, the extrapolation to the zero Heliocentric distance is necessary as discussed above. So, our algorithm works as follows.

1. We choose the value of the Galactocentric Solar distance $R_0$ and the preliminary value of the Solar Galactocentric velocity $\mathbf{V}_\odot = (U_\odot, V_\odot, W_\odot)$. This allows one firstly to transform Heliocentric rectilinear coordinates $x$, $y$, $z$ and velocities $u = \dot{x}$, $v = \dot{y}$, $w = \dot{z}$ into Galactocentric cylindrical coordinates $R$, $\theta$, $z$ and the corresponding velocities $v_R$, $v_\theta$, $v_z$. Here $x$ is directed to the center of the Galaxy, $y$ in the direction of Galactic rotation. Secondly, we can calculate the angular momentum $L = Rv_\theta$ for all stars under consideration. Actually, we

do not need in $W_\odot$ for further calculations, and dependence of $L$ on $U_\odot$ is very slight for stars of the Solar neighborhood.

2. We choose a Solar neighborhood volume of radius $r$ and consider all stars within this volume. Then we take a subsample of stars with angular momentum less than some value $L$. We calculate an average azimuthal and, also, average radial and vertical velocities $\Theta(L; r; U_\odot, V_\odot; R_0)$, $\Pi(L; r; U_\odot, V_\odot; R_0)$, $Z(L; r; U_\odot, V_\odot; R_0)$ for the subsample. This procedure is repeated for a sequence of decreasing values of $L$. Then we extrapolate these velocities to the zero momentum and obtain estimates of the Solar Galactocentric velocity, denoted by $\mathbf{V}_e(r; U_\odot, V_\odot; R_0) = (\Pi_e, \Theta_e, Z_e)$,

$$\begin{aligned} \Theta_e(r; U_\odot, V_\odot; R_0) &= \lim_{L\to 0} \Theta(L; r; U_\odot, V_\odot; R_0), \\ \Pi_e(r; U_\odot, V_\odot; R_0) &= \lim_{L\to 0} \Pi(L; r; U_\odot, V_\odot; R_0), \\ Z_e(r; U_\odot, V_\odot; R_0) &= \lim_{L\to 0} Z(L; r; U_\odot, V_\odot; R_0). \end{aligned} \tag{1}$$

3. We repeat operations of the previous step for a set of decreasing values of $r$ and extrapolate $\mathbf{V}_e(r; U_\odot, V_\odot; R_0) = (\Pi_e, \Theta_e, Z_e)$ to the value $r = 0$ obtaining $\mathbf{V}_0(U_0, V_0; R_0) = (\Pi_0, \Theta_0, Z_0)$,

$$\begin{aligned} \Theta_0(U_\odot, V_\odot; R_0) &= \lim_{r\to 0} \Theta_e(r; U_\odot, V_\odot; R_0), \\ \Pi_0(U_\odot, V_\odot; R_0) &= \lim_{r\to 0} \Pi_e(r; U_\odot, V_\odot; R_0), \\ Z_0(U_\odot, V_\odot; R_0) &= \lim_{r\to 0} Z_e(r; U_\odot, V_\odot; R_0). \end{aligned} \tag{2}$$

In the ideal case output values $\Pi_0$, $\Theta_0$ coincide with input velocities $U_\odot$, $V_\odot$.

4. We repeat these steps for a set of input values of $U_\odot$, $V_\odot$ and compare $\Pi_0$, $\Theta_0$ with $U_\odot$, $V_\odot$. We calculate differences

$$\begin{aligned} \Delta_\Theta &= \Theta_0(U_\odot, V_\odot; R_0) - V_\odot, \\ \Delta_\Pi &= \Pi_0(U_\odot, V_\odot; R_0) - U_\odot \end{aligned} \tag{3}$$

and solve equations

$$\Delta_\Theta(U_\odot, V_\odot; R_0) = 0, \qquad \Delta_\Pi(U_\odot, V_\odot; R_0) = 0, \tag{4}$$

i.e., interpolate to vanishing values of differences $\Delta_\Theta$, $\Delta_\Pi$ to find Solar Galactocentric radial and azimuthal velocities $U_S(R_0)$, $V_S(R_0)$ at the chosen $R_0$. Substituting these values into the last of Eqs. (2) we find the Solar vertical velocity $W_S(R_0)$. Note that after extrapolation to the zero Heliocentric distance (2) the results should not depend on $U_\odot$, i.e., we did not interpolate $\Delta_\Pi \to 0$.

The question arises of how to fulfil extrapolations (1), (2) and interpolation (4) in practice? We have checked two variants of extrapolation (1), namely, linear $\Theta_e = \Theta + k_L^{(1)} L$ and parabolic $\Theta_e = \Theta + k_L^{(2)} L^2$. We concluded from our preliminary calculations that for all values of the parameters the function $\Theta(L; r; U_\odot, V_\odot; R_0)$ becomes almost constant for small $L$, and using the first approximate dependence leads to unreliable results. We therefore worked with the second approximation that seems preferable from physical considerations.

Ogorodnikov & Ossipkov (1978) have considered two variants of extrapolation (2) to the zero distance, linear $\Theta_0 = \Theta_e + k_r^{(1)} r$ and parabolic $\Theta_0 = \Theta_e + k_r^{(2)} r^2$. It follows from our Monte-Carlo simulations (which we do not describe here) that occasionally the parabolic extrapolation gives more precise results, but that as a rule linear extrapolation is preferable. Our calculations have shown that when we restrict ourselves to $r \leq 200$ pc the function $\Theta_e(r; \mathbf{V}_\odot; R_0)$ is almost constant. So, we used a linear dependence.

## 3. Circular Velocity

Once we know the Galactocentric Solar velocity $V_S$, the natural way to find the circular velocity, $V_c$, at the Solar Galactocentric distance is to measure the Solar velocity relative a point on the circular orbit. The latter technique involves extrapolating Stellar Kinematics to circular orbits, and was suggested by Loktin (1978). Loktin himself estimated orbital eccentricities of stars from Parenago's potential. Following this idea we use the theory of epicyclic orbits in the stationary Galactic potential (e.g., Chandrasekhar 1942; Ogorodnikov 1958). It is then straightforward to show that the orbital eccentricity of epicyclic orbits, $e$, is given by

$$e^2 = \frac{(R - R_0)^2}{R_0^2} + \frac{V_R^2}{R_0^2 \kappa^2}.$$

Here $R$ is the Galactocentric cylindrical distance, $V_R$ is the Galactocentric velocity component in the $R$-direction, $\kappa$ is the epicyclic frequency, $\kappa^2 = 4B(B-A) = -4BV_c/R_0$, where $A$, $B$ are Oort's dynamical constants. Then

$$e^2 = \left(1 - \frac{R}{R_0}\right)^2 + \frac{1+\lambda}{4}\left(\frac{V_R}{V_c}\right)^2 \tag{5}$$

with $\lambda = A/(-B)$.

The algorithm we suggest operates as follows.

1. We choose a preliminary value of the Solar Galactocentric distance $R_0$. We suppose we know components of the Solar Galactocentric velocity, $U_S(R_0)$, $V_S(R_0)$, found as described in the previous Section. Then we can calculate Galactocentric velocities $V_R$, $V_\theta$ for all stars in our sample.

2. We choose preliminary values of the circular velocity $V_c$ and for $\lambda$. Then we can estimate eccentricity of all stars $e = e(R, V_R; V_c; \lambda, R_0)$, according to (5). The derived orbital eccentricity may differ from true eccentricities for elongated orbits but give a very good approximation for the most of stars in the Solar neighborhood as they are predominantly on low eccentricities orbits close to circular.

3. We take values for $e$ and for a Heliocentric distance $r$, consider only stars with eccentricities smaller than $e$ and Heliocentric distances less than $r$. We find an average azimuthal velocity of such stars, $v_\theta = v_\theta(e; r; V_c; \lambda, R_0)$. We compute it for a sequence of decreasing values of $e$ and find

$$v_0 = \lim_{e \to 0} v_\theta(e; r; V_c; \lambda, R_0), \tag{6}$$

i.e., extrapolate $v_\theta$ to a circular orbit.

4. We repeat operations of the previous step for a set of decreasing values of $r$ and extrapolate to $r = 0$, i.e., find

$$v^*(V_c; \lambda, R_0) = \lim_{r \to 0} v_0(r; V_c; \lambda, R_0). \tag{7}$$

Then we calculate a difference $\Delta_c = v^*(V_c; \lambda, R_0) - V_c$ and interpolate to $\Delta_c = 0$. The solution of the equation $\Delta_c(V_c; \lambda, R_0) = 0$ is a circular velocity $V_c(\lambda, R_0)$ for given values of $\lambda$, $R_0$.

5. We repeat the above calculations for different values of $\lambda$. For a flat rotation curve $A = -B$, $\lambda = 1$. When $A = 14.8$ km s$^{-1}$ kpc$^{-1}$, $B = -12.4$ km s$^{-1}$ kpc$^{-1}$ (values found by Feast & Whitelock 1997 from *HIPPARCOS*) we obtain $\lambda = 1.19$. For the older estimates $A = 15$ km s$^{-1}$ kpc$^{-1}$, $B = -10$ km s$^{-1}$ kpc$^{-1}$ we obtain $\lambda = 1.5$. Then we repeat the procedure for various values of $R_0$.

## 4. Some Results

The above considerations were applied to *HIPPARCOS* stars with known space velocities (circa 22,000 stars). Our preliminary results were given by Ossipkov, Mylläri, & Flynn (2001). These studies were done under an erroneous assumption that stellar velocities used (we were working with precompiled catalogue of stellar velocities based on HIPPARCOS data) were not corrected for the Solar motion. We have corrected this error now.

We find the following values of the Solar azimuthal velocity $V_S(R_0)$ for various $R_0$:

| $R_0$ [kpc] | $V_S$ [km s$^{-1}$] |
|---|---|
| 7.0 | $236.5 \pm 0.4$ |
| 7.5 | $239.9 \pm 0.4$ |
| 8.0 | $243.1 \pm 0.4$ |
| 8.5 | $245.0 \pm 0.4$ |
| 9.0 | $246.5 \pm 0.4$ |
| 9.5 | $247.6 \pm 0.4$ |
| 10.0 | $247.7 \pm 0.4$ |

Note that the quoted errors are formal errors of the interpolation.

Extrapolating to circular orbits we have found that the result practically does not depend on $\lambda$ and $R_0$. The Solar velocity relative the circular velocity is equal to 17 km s$^{-1}$.

**Acknowledgments.** C.F. thanks for Academy of Finland for support through its funding of the ANTARES program for space research. The work of L.O. was partly supported by the Leading Scientific School Grant NSh-1078.2003.02.

## References

Agekian, T. A. 1962, in Course of Astrophysics and Stellar Astronomy, Vol. 2, ed. A. A. Mikhailov (Moscow: Fizmatgiz), 447 (in Russian)

Agekian, T. A., & Ogorodnikov, K. F. 1974, Highlights of Astronomy, 3, 451

Binney, J., & Merrifield, M. 1998, Galactic Astronomy (Princeton: Princeton Univ. Press)

Binney, J., & Ossipkov, L. P. 1999, Kin.Fiz.Neb.Tel, Suppl., no. 2, 43

Carlberg, R. G., & Innanen, K. A. 1987, AJ, 94, 666

Chandrasekhar, S. 1942, Principles of Stellar Dynamics (Chicago: Univ. Chicago Press)

Chandrasekhar, S. 1943, Ann.NY Acad.Sci., 45, 131

Feast, M. W., & Whitelock, P. A. 1997, MNRAS, 291, 683

Kinman, T. D. 2000, in Abstracts of Talks and Posters presented at the Intern. Conf. of the Astron. Gesellschaft at Heidelberg, March 20–24, 2000, ed. R. E. Schielicke (Hamburg: Astron. Gesellshaft), 35

Loktin, A. V. 1978, Ast.Tsirk., 1021, 1 (in Russian)

Mylläri, A., Flynn, C., & Ossipkov, L. P. 2001, in Stellar Dynamics: from Classic to Modern, ed. L. P. Ossipkov & I. I. Nikiforov (St. Petersburg: St. Petersburg Univ. Press), 51

Nezhinskij, E. M., Ossipkov, L. P., & Kutuzov, S. A. 1995, in IAU Symp. 164, Stellar Populations, ed. P. C. van der Kruit & G. Gilmore (Dordrecht: Kluwer), 374

Ogorodnikov, K. F. 1948, Uspekhi Ast.Nauk, 4, 4 (in Russian)

Ogorodnikov, K. F. 1958, Dynamics of Stellar Systems (Moscow: Fizmatgiz) (in Russian)

Ogorodnikov, K. F., & Ossipkov, L. P. 1978, Ast.Tsirk., 995, 1 (in Russian)

Ogrodnikoff, K. 1936, MNRAS, 96, 866

Orlov, V. V., & Titov, O. A. 1994, AZh, 71, 525 (in Russian)

Ossipkov, L. P. 1980, Ast.Tsirk., 1105, 1 (in Russian)

Ossipkov, L. P., Mylläri, A., & Flynn, C. 2001, in Stellar Dynamics: from Classic to Modern, ed. L. P. Ossipkov & I. I. Nikiforov (St. Petersburg: St. Petersburg Univ. Press), 48

*Order and Chaos in Stellar and Planetary Systems*
*ASP Conference Series, Vol. 316, 2004*
*G. Byrd, K. Kholshevnikov, A. Mylläri, I. Nikiforov, and V. Orlov, eds.*

# Distribution and Motion of Bright Stars within 500 pc

George Gontcharov

*Central Astronomical Observatory at Pulkovo, Pulkovskoe sh. 65/1, St. Petersburg 196140, Russia*

**Abstract.** We collected and analyzed various data of about 20000 best investigated stars to develop the Local Stellar System (LSS) database at http://www.astro.spbu.ru/. As the first use of the database we present the complete 6-parameter ($\alpha$, $\delta$, $\pi$, $\mu_\alpha$, $\mu_\delta$, $V_r$) distribution and motion of almost all stars with $M < -2^m$ within 500 pc for past 20 million years, i.e., for temporal interval comparable with the life-time of young massive stars. Their common motion as the Orion stream of the Orion local spiral arm is evident for the last 20 Myears. The Gould belt, OB associations, giant cloud complexes, young stellar clusters and all other large-scale structures of the local spiral arm have been developed with the evolution of the Orion stream. Two structures of these stars extend over 1 kpc in the Solar neighbourhood for a ten Myears: the Great rift and the Great tunnel.

To investigate distribution and motion of the stars in the Local Stellar System[1] we collected, analyzed and sometimes reprocessed various data for stars with precise parallaxes and radial velocities. The Hipparcos, WEB, Barbier-Brossat's, WDS, 6th catalog of visual orbits and dozens other catalogs and papers were used (full list will be published elsewhere). As a result, the LSS database contains $\alpha$, $\delta$, $\pi$, $\mu_\alpha$, $\mu_\delta$, $V_r$, multiplicity parameters, photometry in at least 3 bands, spectra and cross identification for all components of about 20000 stellar systems.

About 1000 stars with $M < -2^m$ (200 supergiants and 800 OB stars with mass $>10\ M\odot$) within 500 pc are in the LSS database. This sample is almost complete for this part of the space. Moreover, a half of these stars are visible by naked eye including Polar, Antares, Canopus, $\gamma$ Cas, Acrux, Spica and 7 in Orion asterizm. The median precision is $\pm 0.7$ mas for $\alpha$, $\delta$ and $\pi$; $\pm 0.7$ mas/year for $\mu_\alpha$ and $\mu_\delta$; and $\pm 1$ km/s for $V_r$. Therefore, the motion is known with the acceptable precision $\pm 20$ pc for 20 Myears, i.e., for temporal interval comparable with the life-time of these young massive stars.

We use standard galactic coordinates $X$, $Y$, $Z$ and standard velocity components $U$, $V$, $W$. The galactic rotation is found and taken into account according to Eggen (1996). The velocity of supergiants w.r.t. the Sun is $V = 0.009X - 11.3$ whereas the one of OB stars is $V = 0.001X - 13.3$, where $V$ in km/s, $X$ in pc. As the tracers of the spiral arm these stars show that the Sun rotates about 12 km/s faster than the spiral arm.

The stars under consideration show strong clusterization making OB associations, young clusters and subclusters of Big Deeper, Coma and others between

[1] Financially supported by the Russian Foundation for Basic Research grant #02-02-16570

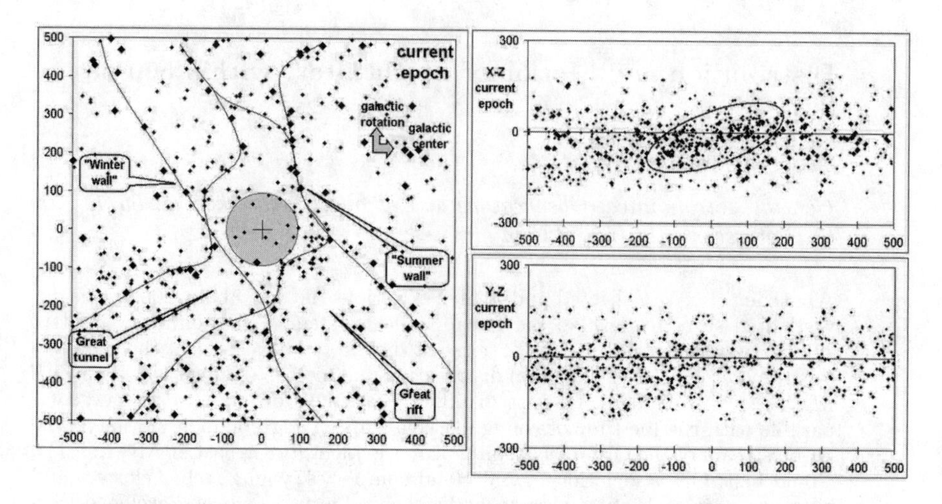


Figure 1. $XY$ (galactic plane), $XZ$ and $YZ$ distribution of bright stars for the current epoch: 200 supergiants and 800 OB stars shown as large and small diamonds. A clusterization of the stars is evident: the Great "walls" rich in bright stars visible on summer and winter sky, Great tunnel and Great rift poor in bright stars, Local bubble within 100 pc from the Sun (nearly the space well parallaxed before Hipparcos) and the Gould belt (the ellipse on $XZ$). Distances in pc, Sun in center.

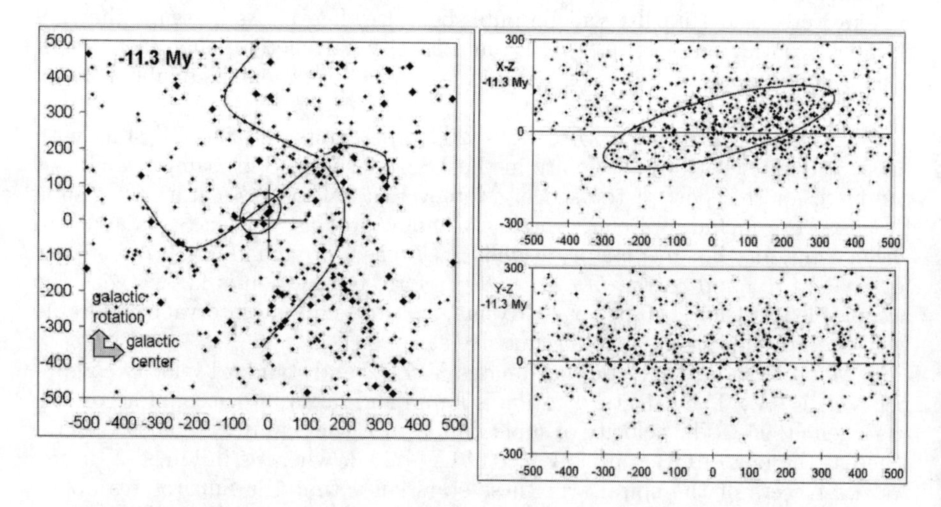


Figure 2. Same as Figure 1, but for −11.3 Myears. A leading Orion stream core with several supergiants (marked by the circle on $XY$) was just within 50 pc from the Sun. Rigel was $-5^m$, other 4 Orion supergiants were brighter than $-2^m$. A spiral structure appeared on $XY$. Some structure to be the Gould belt is shown by ellipse on $XZ$.

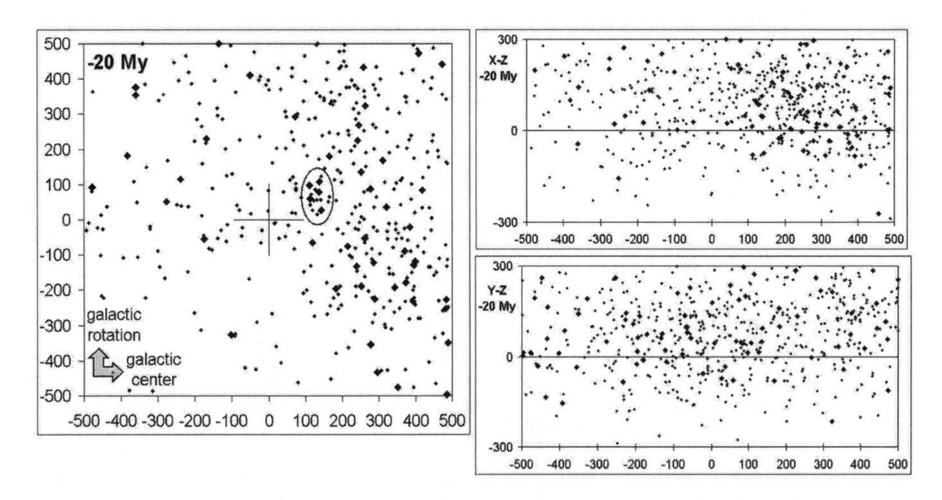


Figure 3. Same as Figures 1 and 2, but for $-20$ Myears. The Orion stream was a compact cloud closer to the galactic center and higher above the galactic plane. Probably that was its birth after a great series of supernova explosions. A leading Orion stream core (including all 7 bright stars of the Orion asterism) are marked by the circle on $XY$. On $XZ$ nothing resembled the Gould belt could be found.

giant cloud complexes. This makes 4 main local structures: 2 "walls" of bright stars visible on the summer and winter sky and 2 voids, the Great tunnel mentioned by Olano (2001) and Great dark rift on the summer Milky way. The voids contain some series of dark clouds: America, Ophiuchus, Coal sack, Gum nebula, Orion-Eridanus bubble, Taurus cloud and others. Many planetary nebulae such as the Ring, Dumbbell and Helix are along the "summer wall" whereas many clusters such as Pleiades, Hyades, $\alpha$ Per are along the "winter wall". The Gould belt appears to be a part of those greater structures. It seems no one structure survives 20 Myears.

Almost all considered bright stars show a common motion w.r.t. the Sun with negative $U$, $V$ and W velocity components. They are the Orion stream. Initially extended along $Y$ and $Z$ this structure becomes extended along $Y$ and $X$ for recent Myears. Also a rotation or deceleration of the stream is evident from a correlation of $U$ velocity component with Y coordinate.

The evolution of the Orion stream should be considered as the greatest event of the last 20 Myears. The transit of the leading Orion stream core near the Solar system 11.3 Myears ago could influence the Earth climate.

## References

Eggen, O. 1996, AJ, 112, 1595
Olano, C. A. 2001, AJ, 121, 295

*Order and Chaos in Stellar and Planetary Systems*
*ASP Conference Series, Vol. 316, 2004*
*G. Byrd, K. Kholshevnikov, A. Mylläri, I. Nikiforov, and V. Orlov, eds.*

# Negative *K*-Effect in Motion of the Gould Belt Stars

V. V. Bobylev

*Central Astronomical Observatory, Pulkovskoe ch. 65/1, St. Petersburg 196140, Russia*

**Abstract.** The kinematics of the OB-stars with heliocentric distances $0.2 < r < 0.6$ kpc has been investigated on the basis of Hipparcos data complemented with the radial velocities. It is established that the $K$-effect depends on the stellar ages, namely, $K\overline{r} = +4.1 \pm 0.8$ km/s for the youngest OB-stars and $K\overline{r} = -5.0 \pm 0.6$ km/s for the oldest OB-stars. The oldest giants and supergiants have negative $K$-effect, namely, $K\overline{r} = -4.0 \pm 0.9$ km/s for AF-stars and $K\overline{r} = -5.8 \pm 1.0$ km/s for GKM-stars. The analysis of the $K$-effects leads to the kinematic estimation of the Gould Belt age as $T \approx 60$ Myr. It is established that there exists an evolutionary connection between the Gould and Vaucouleur–Dolidze Belts (we also associate it with the Sirius supercluster), namely, the oldest stars are in stage of compression. Due to slow rotation of these stars, the youngest OB-stars form the Gould Belt and survive expansion state. Note that the compression the stage is changed into the expansion one ≈60 Myr ago.

## 1. Introduction

It is well known that there exists a positive $K$-effect in motion of the OB stars. In the paper by Torra et al. (2000) it was shown, on the basis of Hipparcos data complemented with the radial velocities, that the $K$-effect in motion of the OB stars depends on heliocentric distance and stellar age. For the youngest OB-stars with age <30 Myr and distances 0.1–0.6 kpc it was found that $K = +7.1 \pm 1.4$ km/s/kpc, but for distant stars with age <60 Myr and distance 0.6–2 kpc $K = -2.9 \pm 0.6$ km/s/kpc. For the oldest OB stars with distances <2 kpc it was established that $K = -6.0 \pm 0.6$ km/s/kpc (Bobylev 2004).

The aim of this paper is selection of those nearby stars which are characterized by the maximum $K$-effect (both positive and negative). Such an approach allows to obtain kinematic parameters of the hydrogen cloud from which the stars under consideration were formed at different time moments.

## 2. Method

The galactic rectangular coordinate system is the following: $x$-axis is directed towards the galactic center ($l = 0°$, $b = 0°$), $y$-axis coincides with the direction of the galactic rotation ($l = 90°$, $b = 0°$) and $z$-axis is directed towards the north galactic pole ($b = 90°$). In linear Ogorodnikov–Milne model we follow designations by Clube (1972) and those used by Mont (1977). The velocity of a star $\mathbf{V}$ with heliocentric radius vector $\mathbf{r}$ can be described by the following

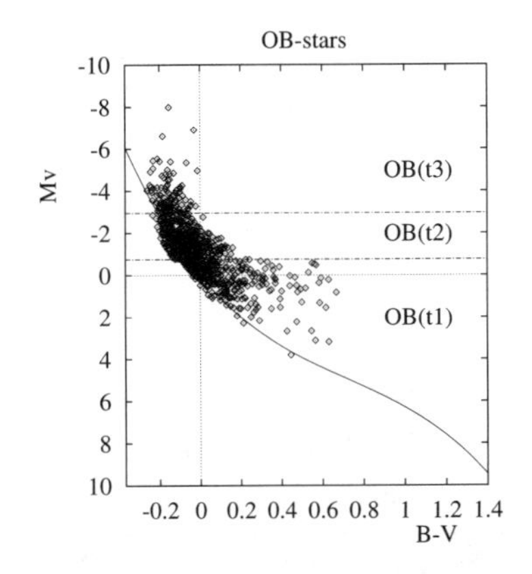


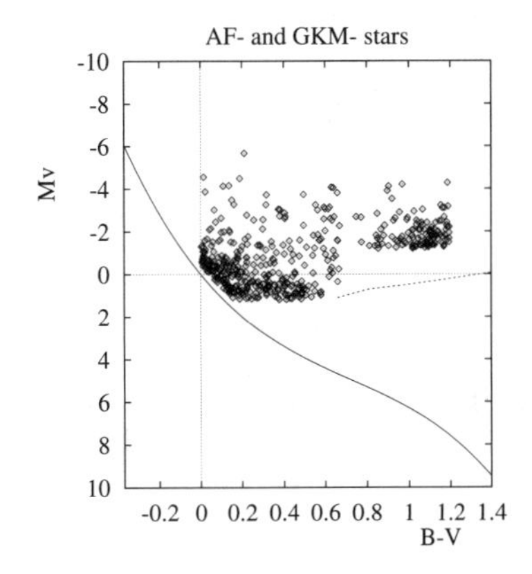


Figure 1. Hertzsprung–Russell diagram.

equation in vectorial form

$$\mathbf{V}(r) - \mathbf{V}_{GR} = \mathbf{u}_\circ + M\mathbf{r} + \mathbf{V}'. \qquad (1)$$

Here $\mathbf{u}_\circ(u_\circ, v_\circ, w_\circ) = -V_\odot(u_\odot, v_\odot, w_\odot)$ is the mean centroid velocity due to peculiar solar motion; $\mathbf{V}_{GR}$ is the velocity due to general galactic rotation; $\mathbf{V}'$ is the residual velocity and we assume that this velocity is of random character; $M$ is a deviation matrix and its components are partial derivatives $\mathbf{u}(u_1, u_2, u_3)$ with respect to $\mathbf{r}(r_1, r_2, r_3)$. Matrix $M$ can be divided into symmetric $M^+$ and asymmetric $M^-$ parts. In accordance with Ogorodnikov (1965) we shall call these parts as local deformation and local rotation tensors correspondingly:

$$M^+_{pq} = \frac{1}{2}\left(\frac{\partial u_p}{\partial r_q} + \frac{\partial u_q}{\partial r_p}\right)_\circ, \quad M^-_{pq} = \frac{1}{2}\left(\frac{\partial u_p}{\partial r_q} - \frac{\partial u_q}{\partial r_p}\right)_\circ \quad (p,\ q = 1, 2, 3). \qquad (2)$$

The following Oort constants $A = +13.7 \pm 0.6$ km/s/kpc and $B = -12.9 \pm 0.4$ km/s/kpc (Bobylev 2004) were used to take into account the general galactic rotation.

## 3. The Working Sample

The equatorial coordinates, parallaxes and proper motions were taken from Hipparcos (ESA 1997), radial velocities from the compilation of Barbier-Brossat & Figon (2000). We used single stars with parallaxes, radial velocities and proper motions. For all stars the following limitation on the residual spatial velocities is common: $|V_{UVW}| < 60$ km/s. Individual stellar ages were not determined. The stars were divided into the age groups using $B - V$, $M_V$ designated as OB(t3), OB(t2), OB(t1), AF and GKM. In order to obtain the extremal values of the $K$-effect this separation was made on the basis of kinematic characteristics—$M_{11}$ and $M_{22}$. In Figure 1 shows the position of the considered stellar groups on Hertzsprung–Russell diagram.

Table 1. Values $M$, $M^+$, $M^-$, $K_{xy}$ and $\lambda_{1,2,3}$ are given in km/s/kpc, $V_\odot(u_\odot, v_\odot, v_\odot)$, $K_{xy}r$ and $\sigma_{1,2,3}$ in km/s, $L_\odot$, $B_\odot$, $L_{1,2,3}$, $B_{1,2,3}$, $l_{1,2,3}$, $b_{1,2,3}$ in degrees; $N_\star$ is the number of stars

| | OB(t3) | OB(t2) | OB(t1) | GKM | AF |
|---|---|---|---|---|---|
| $N_\star$ | 71 | 449 | 322 | 124 | 336 |
| $\bar{r}$ (kpc) | 0.240 | 0.361 | 0.325 | 0.315 | 0.274 |
| $u_\odot$ | $12.5_{(0.9)}$ | $13.4_{(0.5)}$ | $10.0_{(0.7)}$ | $12.1_{(1.1)}$ | $13.0_{(1.0)}$ |
| $v_\odot$ | $14.9_{(0.9)}$ | $13.0_{(0.5)}$ | $9.2_{(0.7)}$ | $10.2_{(1.1)}$ | $11.3_{(1.0)}$ |
| $w_\odot$ | $7.7_{(0.9)}$ | $8.0_{(0.5)}$ | $5.3_{(0.7)}$ | $8.6_{(1.1)}$ | $5.9_{(1.0)}$ |
| $V_\odot$ | $20.9_{(0.9)}$ | $20.3_{(0.5)}$ | $14.6_{(0.7)}$ | $18.0_{(1.1)}$ | $18.2_{(1.0)}$ |
| $L_\odot$ | $50.0_{(2.6)}$ | $44.2_{(1.4)}$ | $42.9_{(2.9)}$ | $40.2_{(3.8)}$ | $41.1_{(3.3)}$ |
| $B_\odot$ | $21.6_{(2.5)}$ | $23.1_{(1.4)}$ | $21.3_{(2.9)}$ | $28.5_{(3.8)}$ | $19.1_{(3.3)}$ |
| $M_{11}$ | $21.8_{(5.4)}$ | $-8.8_{(1.8)}$ | $-14.2_{(2.5)}$ | $-33.5_{(5.1)}$ | $-36.0_{(5.2)}$ |
| $M_{12}$ | $-3.5_{(3.1)}$ | $5.4_{(1.5)}$ | $14.2_{(2.7)}$ | $4.2_{(4.1)}$ | $11.1_{(4.2)}$ |
| $M_{13}$ | $-19.9_{(10.8)}$ | $-15.1_{(4.6)}$ | $41.6_{(9.7)}$ | $-21.2_{(8.8)}$ | $-17.2_{(6.1)}$ |
| $M_{21}$ | $-18.9_{(5.4)}$ | $-2.7_{(1.8)}$ | $5.2_{(2.4)}$ | $-7.4_{(5.1)}$ | $-4.9_{(5.2)}$ |
| $M_{22}$ | $12.7_{(3.1)}$ | $-3.2_{(1.5)}$ | $-16.8_{(2.7)}$ | $-3.4_{(4.1)}$ | $6.8_{(4.2)}$ |
| $M_{23}$ | $-0.9_{(10.2)}$ | $-9.9_{(4.6)}$ | $-15.5_{(9.7)}$ | $-13.7_{(8.7)}$ | $-3.0_{(6.0)}$ |
| $M_{31}$ | $-1.8_{(5.3)}$ | $-0.1_{(1.8)}$ | $-0.8_{(2.4)}$ | $3.9_{(5.1)}$ | $7.2_{(5.2)}$ |
| $M_{32}$ | $1.9_{(3.1)}$ | $1.2_{(1.5)}$ | $-0.5_{(2.7)}$ | $2.1_{(4.1)}$ | $-0.5_{(4.2)}$ |
| $M_{33}$ | $11.8_{(10.1)}$ | $4.0_{(4.5)}$ | $11.6_{(9.7)}$ | $11.0_{(8.7)}$ | $-12.2_{(6.0)}$ |
| $M^+_{12}$ | $-11.2_{(3.1)}$ | $1.3_{(1.2)}$ | $9.7_{(1.8)}$ | $-1.6_{(3.3)}$ | $3.1_{(3.3)}$ |
| $M^+_{23}$ | $0.5_{(5.3)}$ | $-4.4_{(2.4)}$ | $-8.0_{(5.0)}$ | $-5.8_{(4.8)}$ | $-1.7_{(3.7)}$ |
| $M^+_{13}$ | $-10.8_{(6.0)}$ | $-7.6_{(2.5)}$ | $20.4_{(5.0)}$ | $-8.6_{(5.1)}$ | $-5.0_{(4.0)}$ |
| $M^-_{21}$ $(\Omega_z)$ | $-7.7_{(3.1)}$ | $-4.0_{(1.2)}$ | $-4.5_{(1.8)}$ | $-5.8_{(3.3)}$ | $-8.0_{(3.3)}$ |
| $M^-_{32}$ $(\Omega_x)$ | $1.4_{(5.3)}$ | $5.5_{(2.4)}$ | $7.5_{(5.0)}$ | $7.9_{(4.8)}$ | $1.3_{(3.7)}$ |
| $M^-_{13}$ $(\Omega_y)$ | $-9.1_{(6.0)}$ | $-7.5_{(2.5)}$ | $21.2_{(5.0)}$ | $-12.6_{(5.1)}$ | $-12.2_{(4.0)}$ |
| $K_{xy}$ | $+17.3_{(3.1)}$ | $-6.0_{(1.2)}$ | $-15.5_{(1.8)}$ | $-18.5_{(3.3)}$ | $-14.6_{(3.3)}$ |
| $K_{xy}\bar{r}$ | $+4.1_{(0.8)}$ | — | $-5.0_{(0.6)}$ | $-5.8_{(1.0)}$ | $-4.0_{(0.9)}$ |
| $T$ (Myr) | $57_{(10)}$ | — | $63_{(7)}$ | $53_{(9)}$ | $67_{(22)}$ |
| $L_1, B_1$ | 151, 24 | 185, −12 | 320, −25 | 5, 11 | 356, 11 |
| $L_2, B_2$ | 268, −45 | 283, −16 | 57, −13 | 98, 16 | 146, −13 |
| $L_3, B_3$ | 223, 55 | 222, 59 | 352, 62 | 241, 71 | 85, 84 |
| $\lambda_1$ | 33.6 | −12.3 | −34.3 | −35.3 | −37.2 |
| $\lambda_2$ | 11.7 | −4.8 | −8.1 | −4.8 | −11.5 |
| $\lambda_3$ | 1.0 | 9.2 | 23.0 | 14.3 | 7.2 |

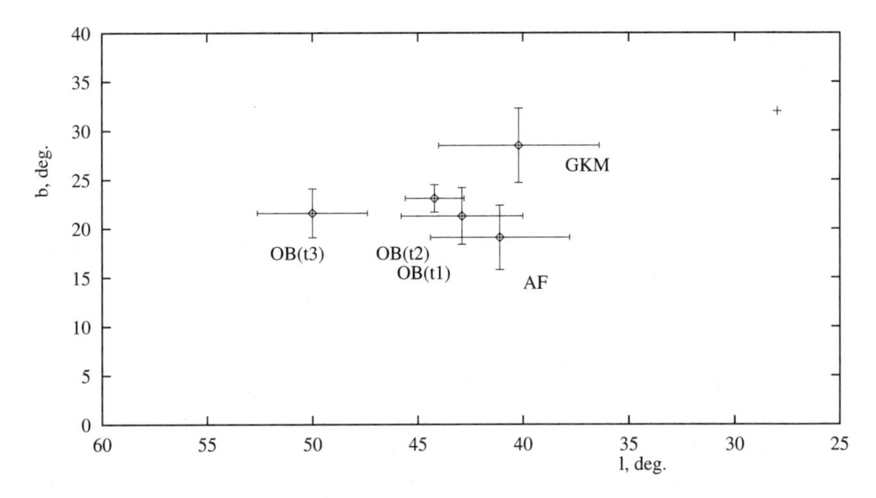


Figure 2. Coordinates of the solar apex.

## 4. Results

Table 1 the following components are given: components of the solar motion $u_\odot$, $v_\odot$, $w_\odot$ and $V_\odot$, $L_\odot$, $B_\odot$; nine components of the deviation matrix $M$ and three components of the deformation tensor $M^+$ calculated using the latter and three components of the rotation tensor $M^-$ ($\Omega_x, \Omega_y, \Omega_z$ are the projections of the rigid-body rotation vector); eigen-values of the deformation tensor $\lambda_{1,2,3}$ and its eigen-vectors $L_{1,2,3}$ and $B_{1,2,3}$ which are placed so that $\lambda_{1,2,3}$ are defined by $x$-, $y$-, $z$-axes respectively.

## 5. Discussion

### 5.1. Linear Motion of the Stars with Respect to LSR

In Figure 2 the coordinates of the solar apex are plotted. Here the apex calculated using solar motion by Dehnen & Binney (1998) is marked by "+" in the upper right-hand corner. The youngest OB(t3) stars have the maximal velocity with respect to LSR (Dehnen & Binney 1998). This motion takes place practically along one coordinate $y$ ($v_\odot$). We can conclude that the motion of the OB(t3) stars is directed towards $l = 270°$ with linear velocity $V = 9.6 \pm 1.1$ km/s. For the OB(t1) stars this velocity is $V = 5.7 \pm 1.2$ km/s in the same direction. The linear motion of the OB stars obtained and its direction are in good agreement with the velocity ≈8 km/s towards $l \approx 270°$ found by Zabolotskikh et al. (2002) for OB associations, and are also in good agreement with the results of Bobylev (2004) which show that practically all the nearest and distant OB stars have linear velocity ≈10 km/s towards $l \approx 270°$.

### 5.2. $K$-effect

We can see that $K$-effect is a function of stellar ages. Note that for the OB(t3) and OB(t1) stars both $M_{11}$ and $M_{22}$ values $[K_{xy} \stackrel{\text{def}}{=} (M_{11} + M_{22})/2]$ are well determined. Significant value $K$ obtained for the GKM and AF stars are only determined by $M_{11}$ only. The positive $K$-effect of the young stars means that they are in the expansion state while the oldest stars are characterized by the negative $K$-effect which testifies to their contraction state. Does the spatial expansion of the OB(t3) stars exist? This question arises because, from the formal point of view, the invariant $K_{xyz} = (M_{11} + M_{22} + M_{33})/3 = 15.5 \pm 4.0$ km/s/kpc significantly differs from zero. This estimation is only true for spherically distributed velocity vectors only. Thus, for the flat Belt [OB(t3) stars] it does not hold. On the other hand, the OB(t3) stars have three positive eigen values $\lambda_{1,2,3} = (33.6, 11.7, 1.0)$ km/s/kpc, which means that $\lambda_3 = 1.0$ km/s/kpc can be considered as a coefficient of the spatial expansion of the considered stars (like the Hubble constant). But because the value $M_{33} = 11.8 \pm 10.1$ km/s/kpc is determined with of very large error we can conclude that motion in $xy$-plane, i.e., practically a planar case, takes place.

In Table 1 the so called "life time" of the stellar system $T$ is given, which was calculated using formulae by analogy with the Hubble constant, but in our case for $xy$-plane: $T = 0.9775 \times 10^9/K$ yr, here $K$ in km/s/kpc (Murrey 1986). As seen from Table 1, value $T$ is of the order of 60 Myr. This value is in good agreement with the age of the Gould Belt of 30–60 Myr determined by Torra et al. (2000).

### 5.3. Self-Rotation of the Stars

As seen from Table 1, only two values of vector $(\Omega_x, \Omega_y, \Omega_z)$ are well determined: $\Omega_y = -21.2 \pm 5.0$ km/s/kpc for the OB(t1) stars and $\Omega_z = -4.0 \pm 1.2$ km/s/kpc for the OB(t2) stars. It is very important that the rotational vector (or the rotational axes) of the OB(t1) stars lies in the galactic $xy$-plane. We encounter some difficulties in differentiation the deformations due to rotation and those due to expansion-contraction effect and as well as in determination of characteristics of the rotation center. In paper by Bobylev (2004) it was shown on the basis of the Bottlinger formulae that practically all the nearest OB stars rotated around $z$-axis with angular velocity $\approx -10$ km/s/kpc with the center of rotation $r_\circ =$ 150 pc in direction $l_\circ = 160°$. We adopt these values as rotational characteristics for all the stars considered.

### 5.4. Orientation of the Deformation Tensor Axes

Following Torra et al. (2000) for the Gould Belt, we adopt the inclination $i_G =$ 16–20° and direction of ascending node $\Omega_G = 275$–295° (coordinates of the pole are $l_G = 185$–205° and $b_G = 74$–70°). Following Dolidze (1980) for Vaucouleur–Dolidze Belt, we adopt the inclination $i_{VD} = 40°$ and direction of ascending node $\Omega_{VD} = 30°$ (coordinates of the pole are $l_{VD} = 300°$ and $b_{VD} = 50°$). Note that Vaucouleur–Dolidze Belt is also associated with the oldest subsystem in the solar neighborhood, the so-called Sirius supercluster (Olano 2001), whose age is 500 Myr. Analysis of values $L_i$, $B_i$ (Table 1) shows that the orientation of the deformation tensor for OB(t1) stars is close to the orientation of the

Vaucouleur–Dolidze Belt. Orientation of the deformation tensor for OB(t2) stars practically coincides with the orientation of the Gould Belt. Taking into account the direction of self-rotation (clockwise motion as seen from the north galactic pole) of the OB-stars we can conclude that the transformation of the deformation tensor axes from the oldest stars [OB(t1)] to the youngest stars [OB(t3)] takes place due to rotation.

## 6. Conclusions

The hydrogen cloud from which the Gould Belt stars were formed $\approx$60 Myr ago was in the state of contraction with linear velocity $\approx$4 km/s at $r \approx 0.3$ kpc [negative $K$-effect with respect to the old giants OB(t1), AF and GKM]. At that time the outflow of matter took place from the inner part of the cloud which continues up today with linear velocity $\approx$4–5 km/s at $r \approx 0.3$ kpc [positive $K$-effect of young stars OB(t3)]. Early state of the cloud could be connected with the Vaucouleur–Dolidze Belt. This cloud shows slow self-rotation around $z$-axes with the angular velocity $\approx -10$ km/s/kpc in the same direction as the general galactic rotation. The present state of the cloud can be connected with the Gould Belt. This kinematic scenario does not contradict the dynamical Olano's (2001) model as well as the 3D dynamical model of interstellar gas by Perrot & Grenier (2003).

**Acknowledgments.** This research has been supported by Russian Foundation for Basic Research under grant No. 02-02-16570.

## References

Barbier-Brossat, M., & Figon, P. 2000, A&AS, 142, 217
Bobylev, V. V. 2004, Ast.Lett., 30, 159
Clube, S. V. M. 1972, MNRAS, 159, 289
Dehnen, W., & Binney, J. J. 1998, MNRAS, 298, 387
Dolidze, M. B. 1980, Ast.Lett., 6, 92
du Mont, B. 1977, A&A, 61, 127
ESA SP-1200, 1997, The Hipparcos and Tycho Catalogue
Murrey C. A. 1983, Vectorial Astrometry (Bristol: Adam Hilger)
Olano, C. A. 2001, AJ, 121, 295
Ogorodnikov, K. F. 1965, Dynamics of Stellar Systems (L.: Pergamon Press)
Perrot, C. A., & Grenier, I. A. 2003, A&A, 404, 519
Torra, J., Fernández, D., & Figueras, F. 2000, A&A, 359, 82
Zabolotskikh, M. B., Rastorguev, A. S., & Dambis, A. K. 2002, Ast.Lett., 28, 454

*Order and Chaos in Stellar and Planetary Systems*
*ASP Conference Series, Vol. 316, 2004*
*G. Byrd, K. Kholshevnikov, A. Mylläri, I. Nikiforov, and V. Orlov, eds.*

# Stellar Kinematics by Vectorial Harmonics

Veniamin Vityazev and Alexey Shuksto

*The Sobolev Astronomical Institute, St. Petersburg State University, Universitetskij pr. 28, Staryj Peterhof, St. Petersburg 198904, Russia; e-mails: vityazev@venvi.usr.pu.ru, simply@pisem.net*

**Abstract.** To study the stellar kinematics, we propose a new method based on decomposing of proper motions on a set of orthogonal vectorial spherical functions. The first degree harmonics of the decomposition are identical with the terms of the Ogorodnikov–Milne model, whereas the high order harmonics are able to detect the effects that are beyond the model. The method is applied to examine the stellar velocity field for various samples from the HIPPARCOS catalogue. Besides the classical contributions to the proper motions (Solar motion, rotation of the Galaxy, etc.) our method revealed some harmonics the physics of which requires further study.

## 1. Outline of the Method

Let the proper motions in galactic coordinate system be $\mu_l(l,b)\cos b$ and $\mu_b(l,b)$. Introduce the vectorial function as

$$\vec{f}(l,b) = \mu_l(l,b)\cos b\,\vec{e}_l + \mu_b(l,b)\,\vec{e}_b, \tag{1}$$

where $\vec{e}_l$ and $\vec{e}_b$ are the unit vectors in the directions of longitude and latitude. We are looking for decomposition of the proper motions in such a way that

$$\vec{f}(l,b) = \sum_{j=1}^{\infty}\left[t_j\vec{T}_j(l,b) + s_j\vec{S}_j(l,b)\right], \tag{2}$$

where the functions $\vec{T}_j(l,b)$ and $\vec{S}_j(l,b)$ are given by equations in Mignard & Morando (1990):

$$\vec{T}_j(l,b) = R_j\left[\frac{\partial V_j(l,b)}{\partial b}\,\vec{e}_l - \frac{1}{\cos b}\frac{\partial V_j(l,b)}{\partial l}\,\vec{e}_b\right], \tag{3}$$

$$\vec{S}_j(l,b) = R_j\left[\frac{1}{\cos b}\frac{\partial V_j(l,b)}{\partial l}\,\vec{e}_l + \frac{\partial V_j(l,b)}{\partial b}\,\vec{e}_b\right], \tag{4}$$

$$j = 1, 2, \ldots,$$

here $R_j$ is the normalizing coefficient, and $V_j(l,b)$ is the spherical function. The set of functions $\vec{T}_j(l,b)$ and $\vec{S}_j(l,b)$ forms the basis, i.e., a complete orthonormal system of vectorial harmonics.

The coefficients $t_j$ and $s_j$ may be derived from

$$t_j = (\vec{f}, \vec{T}_j) = \int_0^{2\pi} \int_{-\frac{\pi}{2}}^{+\frac{\pi}{2}} \left[ f_l(l,b) T_{j_l}(l,b) + f_b(l,b) T_{j_b}(l,b) \right] \cos b \, dl \, db, \quad (5)$$

$$s_j = (\vec{f}, \vec{S}_j) = \int_0^{2\pi} \int_{-\frac{\pi}{2}}^{+\frac{\pi}{2}} \left[ f_l(l,b) S_{j_l}(l,b) + f_b(l,b) S_{j_b}(l,b) \right] \cos b \, dl \, db. \quad (6)$$

In case of the Ogorodnikov–Milne model (du Mont 1977) the stellar velocity field is given by expression

$$\vec{V} = \vec{V}_0 + M^+ \vec{r} + M^- \vec{r}, \quad (7)$$

where the following notations are used:

- $\vec{V}_0$ is the reflex of the Solar motion with respect to given centroid of stars. This velocity is defined by components $U$, $V$, $W$ in the directions of the principal galactic axes $x$, $y$, $z$;
- $M^+$ is the diverging matrix with the dilation coefficients $M^+_{11}$, $M^+_{22}$, $M^+_{33}$, and $M^+_{12}$, $M^+_{13}$ $M^+_{23}$ standing for shears in the galactic planes $(x, y)$, $(x, z)$, $(y, z)$. Since proper motions reflect tangential motions only, we set $M^+_{22} = 0$. In this case the unknowns $M^+_{11}$ and $M^+_{33}$ are replaced with $M^*_{11} = M^+_{11} - M^+_{22}$ and $M^*_{33} = M^+_{33} - M^+_{22}$ respectively (Clube 1972);
- $M^-$ is the rigid rotation matrix whose elements are defined over the components $\omega_1$, $\omega_2$, $\omega_3$ of the angular vector rotation of the stellar system about axes $x$, $y$, $z$.

In our method, prior to vectorial decomposing the Solar motion components $U$, $V$, $W$ must be eliminated from the proper motions.

The crucial point of our method is that the kinematical parameters of the Ogorodnikov–Milne model and the low-order coefficients of the decomposition (2) are connected by the following equations:

$$t_1 = \frac{\omega_3}{R_1}, \qquad t_2 = \frac{\omega_2}{R_2}, \qquad t_3 = \frac{\omega_1}{R_3}, \quad (8)$$

$$s_4 = \frac{M^*_{33} - \frac{1}{2} M^*_{11}}{2R_4}, \quad (9)$$

$$s_5 = \frac{M^+_{23}}{R_5}, \qquad s_6 = \frac{M^+_{13}}{R_6}, \quad (10)$$

$$s_7 = \frac{M^+_{12}}{2R_7}, \qquad s_8 = \frac{M^*_{11}}{4R_8}, \quad (11)$$

whereas the rest of harmonics do not belong to the Ogorodnikov–Milne model and may be used to study the effects that are beyond the model.

## 2. Application to HIPPARCOS

This study is essentially based on the HIPPARCOS catalogue (ESA 1997) with its unprecedented accuracy of astrometric data—positons, proper motions and parallaxes.

First of all, it is desirable to see what kinematics can be derived from all stars listed in the catalogue. To do this we used 113646 stars (the rest were discarded as having no full astrometric information). Two solutions have been made. In the first one we employed the classical LSM routine for solving the Ogorodnikov–Milne equations, in the second one—we used the decomposing of proper motions on vectorial harmonics.

Both solutions gave practically the same parameters of the Ogorodnikov–Milne model. Besides the Solar motion, only the Oort's coefficients turned out to be reliable.

Below, we show the velocity of the Sun $V_\odot$ and the coordinates $L_\odot$, $B_\odot$ of the Solar apex, together with the values of the Oort's coefficients $A$ and $B$ obtained by us in comparison with the values recommended by IAU:

$$\begin{array}{llll} \text{IAU:} & V_\odot = 19.7 \pm 0.01\,\frac{\text{km}}{\text{s}}, & L_\odot = 55\overset{\circ}{.}86, & B_\odot = 23\overset{\circ}{.}55. \\ \text{HIP:} & V_\odot = 23.3 \pm 0.01\,\frac{\text{km}}{\text{s}}, & L_\odot = 60\overset{\circ}{.}1 \pm 0\overset{\circ}{.}3, & B_\odot = 11\overset{\circ}{.}2 \pm 0\overset{\circ}{.}3. \end{array}$$

$$\begin{array}{ll} \text{IAU:} & A = 15.0 \pm 0.9\,\frac{\text{km}}{\text{s}\cdot\text{kpc}}. \\ \text{HIP:} & A = 13.5 \pm 2.0\,\frac{\text{km}}{\text{s}\cdot\text{kpc}}. \end{array}$$

$$\begin{array}{ll} \text{IAU:} & B = -10.0 \pm 1.4\,\frac{\text{km}}{\text{s}\cdot\text{kpc}}. \\ \text{HIP:} & B = -12.6 \pm 1.6\,\frac{\text{km}}{\text{s}\cdot\text{kpc}}. \end{array}$$

From these we estimated the local angular velocity of the Galaxy to be $\Omega = A - B = \omega_3 - M^+_{12} = 26.1 \pm 2.5\,\frac{\text{km}}{\text{s}\cdot\text{kpc}}$, corresponding to period of rotation as $P = 236$ Myr.

Thus we see that in the frame of the Ogorodnikov–Milne model the rotation of stars is flat. This conclusion follows from both solutions.

## 3. "Extra-Model" Components of the Proper Motions

When applied to stellar kinematics, the main advantage of the vectorial harmonics over traditional approach is a chance to detect the motions which are not included in the Ogorodnikov–Milne model. Indeed, in the global solution the method of vectorial functions detected the terms $(-12.9 \pm 4.6) \times \vec{S}_{10}$, $(12.2 \pm 4.4) \times \vec{S}_{14}$, $(-12.7 \pm 4.6) \times \vec{S}_{20}$, $(11.1 \pm 4.3) \times \vec{S}_{34}$ (all in km s$^{-1}$ kpc$^{-1}$). Besides the global solution we applied our method to several samples of stars with different distances and spectral classes. The "extra-model" terms specified by the functions $\vec{T}_4$, $\vec{T}_6$, $\vec{S}_{10}$ and $\vec{S}_{14}$ were found to be common to all examined samples including the global solution. The analytical expressions of these terms look as follows:

$$\vec{T}_4(l,b) = R_4 \sin 2b\, \vec{e}_l, \tag{12}$$

$$\vec{T}_6(l,b) = R_6 \left(\cos 2b \cos l\, \vec{e}_l + \sin b \sin l\, \vec{e}_b\right), \tag{13}$$

$$\vec{S}_{10}(l,b) = R_{10} \left[\left(2\cos^2 b \sin b \cos l - \sin^2 b \cos b \cos l + \frac{1}{5}\cos b \cos l\right) \vec{e}_b - \left(\sin^2 b \sin l - \frac{1}{5}\sin l\right) \vec{e}_l\right], \tag{14}$$

$$\vec{S}_{14}(l,b) = -R_{14} \left(3\cos^2 b \sin b \cos 3l\, \vec{e}_b + 3\cos^2 b \sin 3l\, \vec{e}_l\right). \tag{15}$$

In conclusion, we state that contribution to the proper motions of the "extra-model" components may be very high. It is comparable and sometimes even exceeding the contribution of "classical" terms (see Figure 1). The next paper will be devoted to the physical properties of the "extra-model" terms found here.

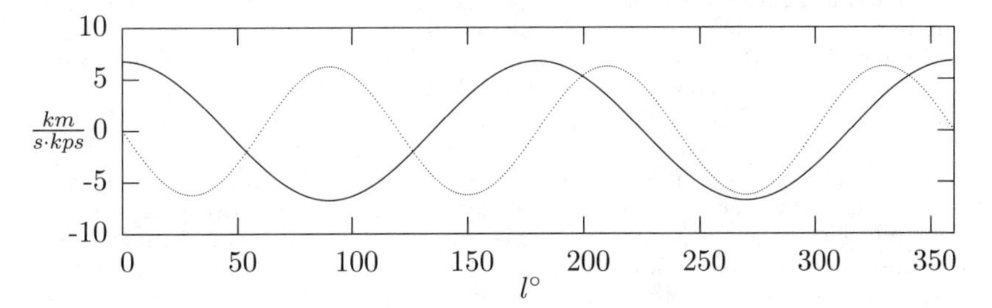


Figure 1. Contribution to the proper motions in longitude from the model harmonic $\vec{S}_7$ (Oort's coefficient $A = M_{12}^{+}$, solid line) in comparison to the significant "extra-model" harmonic $\vec{S}_{14}$ (dashed line).

**Acknowledgments.** The authors appreciate the support of this work by the grant 02-02-16570 of the Russian Fund of Fundamental Research and by the grant of the Leading Scientific School 00-15-96775.

## References

Clube, S. V. M. 1972, MNRAS, 159, 289

du Mont, B. 1977, A&A, 61, 127

ESA 1997, The Hipparcos and Tycho Catalogues

Mignard, F., & Morando, B. 1990, Motion of Celestial bodies, Astrometry and Astronomical Reference Frames, Proc. of the JOURNEES 1990 (Paris), 151

*Order and Chaos in Stellar and Planetary Systems*
*ASP Conference Series, Vol. 316, 2004*
*G. Byrd, K. Kholshevnikov, A. Mylläri, I. Nikiforov, and V. Orlov, eds.*

# Evidences of Axial Symmetry Rupture in the Solar Neighborhood

Santiago Alcobé

*Group of Astronomy and Geomatics, Univ. Politecnica de Catalunya, Gran Via, 421, 08015 Barcelona, Spain*

Rafael Cubarsi

*Dept. Matematica Aplicada IV, Campus Nord, Univ. Politecnica de Catalunya, Jordi Girona, 1-3, 08034 Barcelona, Spain*

**Abstract.** It is well known that population moments are good indicators of some features—such as symmetries—of the velocity distribution. The trivariate velocity distribution moments of representative local stellar samples from CNS3 and HIPPARCOS catalogs are evaluated, where a superposition of stellar populations is assumed. The samples have been selected looking for the maximum entropy of the mixture probability and a numerical-statistical algorithm has been applied for computing the partial kinematic parameters. In order to give an interpretation to the obtained results, some dynamic models based on Chandrasekhar's approximation are reviewed. According to those models, values of specific moments are different as expected for some of the stellar populations. In particular the moment $\mu_{\varpi\theta}$, and also several odd-order moments, are clearly not null. The same statistical analysis is carried out for both star samples. For the closer one, derived from CNS3, a stellar component, associated with old disk stars, could fit in an axially symmetric model. Nevertheless, for the sample selected from HIPPARCOS, which takes a much larger scale height as CNS3, even in the case of a Galactic disk with three stellar components, all partial populations require at least point-axial symmetry in order to explain their kinematic behavior.

## Symmetries of the Mixture Components

The Galactic system is interpreted as a coexistence of several galactic components, like bulge, thin and thick disk, stellar halo, corona, etc. These parts are also composed of one or more stellar populations or population components. They are stellar groups with common astrophysical properties and, in particular, from a kinematical viewpoint, each group has reached the statistical equilibrium so that their velocity probability density function, $f$, can be assumed of normal type in the peculiar velocities (Cubarsi 1990). The behavior of the Galactic system is described in nearly all the regions according to a conservative dynamic system, satisfying the Liouville equation, leading to the well known Chandrasekhar system of equations. Since this is a linear differential equation for $f$, the superposition principle can be applied in order to obtain the shape of the potential and the kinematical parameters of each stellar population. From the above assumptions several dynamic models have been developed in order to predict values for the covariance matrix—in particular vertex deviation—and mean velocities for each population component (Cubarsi et al. 1997 and references therein). Hereafter a galactocentric cylindrical coordinates system is used.

Table 1. Moments, means and vertex deviation depending on the symmetry assumptions for a population component. The indices 1, 2 and 3 relate to radial, tangential and vertical directions

| **Distribution Symmetry** | **Time Dep.** | **Sym. Plane** | **Mean Velocity** | **Second-Order Moments** | **Vertex Dev.** |
|---|---|---|---|---|---|
| $\frac{\partial}{\partial\theta}$ | $\frac{\partial}{\partial t}$ | z=0 | $v_1, v_2, v_3$ | $\mu_{ij}$ | $\varepsilon$ |
| | | | Axial symmetry | | |
| $\frac{\partial}{\partial\theta}=0$ | $\frac{\partial}{\partial t}=0$ | Yes | $0, v_2, 0$ | $\mu_{11}, \mu_{22}, \mu_{33}, \mu_{13} \neq 0$ | $\varepsilon=0$ |
| $\frac{\partial}{\partial\theta}=0$ | $\frac{\partial}{\partial t}\neq 0$ | Yes | $v_1, v_2, v_3$ | $\mu_{11}, \mu_{22}, \mu_{33}, \mu_{13} \neq 0$ | $\varepsilon=0$ |
| $\frac{\partial}{\partial\theta}=0$ | $\frac{\partial}{\partial t}\neq 0$ | No | $v_1, v_2, v_3$ | $\mu_{11}, \mu_{22}, \mu_{33}, \mu_{12}, \mu_{13} \neq 0$ | $\varepsilon\neq 0$ |
| | | | Point axial symmetry | | |
| $f(\theta)=f(\theta+\pi)$ | $\frac{\partial}{\partial t}=0$ | Yes | $v_1, v_2, v_3$ | $\mu_{11}, \mu_{22}, \mu_{33}$, $\mu_{12}, \mu_{13}, \mu_{23} \neq 0$ | $\varepsilon=\varepsilon(\theta)$ |
| $f(\theta)=f(\theta+\pi)$ | $\frac{\partial}{\partial t}\neq 0$ | Yes | $v_1, v_2, v_3$ | $\mu_{11}, \mu_{22}, \mu_{33}$, $\mu_{12}, \mu_{13}, \mu_{23} \neq 0$ | $\varepsilon=\varepsilon(\theta)$ |

The overall distribution belongs to a sub-class which can be defined from a set of cumulant constraint equations (Cubarsi 1992; Alcobe 2001), and we are interested in some particular cases of the constraints, attending to specific symmetry hypotheses, in order to test underlying distribution symmetries, working from the total $k$-statistics of the input sample. The following situations are studied: *symmetry plane*, *non-differential movement*, and *axial symmetry*.

## Application to Stellar Samples

A statistical analysis has been applied to two different samples. The first one is composed by all stars with known space velocities from the Third Catalogue of Nearby Stars CNS3 (Gliese & Jahreiß 1991). The second one has been selected from HIPPARCOS catalogue (ESA 1997). It contains all stars with radial velocity data present in Hipparcos Input Catalogue (HINCA)(Turon et al. 1992) and limited to a heliocentric distance of 300 pc. Samples are supposed to be composed by several subsamples with gaussian distribution in the peculiar velocities. The criteria for selecting a sample is looking for the maximum entropy of the mixture probability. Each maximum defines a stellar sample whose moments and cumulants up to fourth order are evaluated. For both catalogues three maximums of entropy are found.

The selected sample is that which leads to the best $\chi^2$ value. For both samples it is found at $\chi^2$=15 for 15 degrees of freedom what shows that the approximation fits very well with the data. For CNS3, this is the maximum corresponding to $\|\mathbf{v}\| = 180\ \mathrm{km\,s^{-1}}$ and $\|\mathbf{v}\| = 210\ \mathrm{km\,s^{-1}}$ for HIPPARCOS. After applying the segregation method to each sample, two partial gaussian populations are obtained.

Expressing the results in terms of standard deviations and vertex deviation $\varepsilon$ that moments for CNS3 are:
Component I: $\sigma_1 : \sigma_2 : \sigma_3 = (34 \pm 2) : (19 \pm 6) : (17 \pm 2)$; $\varepsilon = 9 \pm 7$.
Component II: $\sigma_1 : \sigma_2 : \sigma_3 = (64 \pm 5) : (46 \pm 14) : (39 \pm 4)$; $\varepsilon = 10 \pm 16$.

And for HIPPARCOS:

Component I: $\sigma_1 : \sigma_2 : \sigma_3 = (28 \pm 1) : (15 \pm 2) : (12 \pm 1)$; $\varepsilon = 9 \pm 3$.
Component II: $\sigma_1 : \sigma_2 : \sigma_3 = (65 \pm 2) : (40 \pm 9) : (41 \pm 2)$; $\varepsilon = 7 \pm 5$.
The following aspects about symmetries are remarked:

Odd-moments in the $z$-coordinate are clearly null, not only for the partial components but also for the global samples. We conclude that it exists a symmetry plane in the solar neighborhood.

Even though for both global samples the moment $\mu_{111}$ is clearly null, the $\mu_{122}$ instead seems to be slightly different from zero (more clearly seen for Hipparcos due to lower errors). All other odd-moments in the radial component are null too. This tells us that there is a small but not negligible differential movement. This result is later found when the two components are segregated from the global sample.

Concerning the moments $\mu_{12}$ and $\mu_{23}$, the second one is null in all cases. For the moment $\mu_{12}$, only that corresponding to the Galactic component associated with the thick disk segregated from CNS3 is null while others are clearly not negligible. This result agrees with authors who find not negligible vertex deviation for all stellar populations (Dehnen 1998).

It is deduced then that there is no axial symmetry either for the global samples nor for the partial ones. Nevertheless, it is seen that the thick disk component shows axial symmetry if the scale heigh is small enough as for CNS3.

This results are similar to those found in a previous paper (Alcobé & Cubarsi 2001). The same procedure was applied to a sample selected from CNS3 but in that case it was limited not by maximum entropy but by the classical assumption *mean* $\pm 2\sigma$. The fitting was also quite good, $\chi^2 = 26$ for 15 degrees of freedom.

## References

Alcobé, S. 2001, PhD Thesis, Universitat de Barcelona, Barcelona

Alcobé, S., & Cubarsi, R. 2001, in ASP Conf. Ser., Vol. 230, 43

Cubarsi, R. 1990, AJ, 99, 1558

Cubarsi, R. 1992, AJ, 103, 1608

Cubarsi, R., Ninkovic, S., Sanz-Subirana, J., & Alcobé, S. 1997, in Structure and Evolution of Stellar Systems, ed. T. A. Agekian, A. A. Myllari & V. V. Orlov (St. Petersburg: St. Petersburg Univ. Press), 36

Dehnen, W. 1998, AJ, 115, 2384

ESA, 1997, The Hipparcos Catalogue, ESA SP-1200 (Vol. I–XVII)

Gliese, W., & Jahreiß, H. 1991, Third Catalogue of Nearby Stars, Astronomisches Rechen-Institut, Heidelberg

Turon et al. 1992, The Hipparcos Input Catalogue, ESA SP-1136

*Order and Chaos in Stellar and Planetary Systems*
*ASP Conference Series, Vol. 316, 2004*
*G. Byrd, K. Kholshevnikov, A. Mylläri, I. Nikiforov, and V. Orlov, eds.*

# On the Form of the Velocity Ellipsoid for Flat Subsystems. Theory and Observations

Leonid P. Ossipkov

*Department of Space Technologies & Applied Astrodynamics,*
*St. Petersburg State University, Universitetskij pr. 28, Staryj Peterhof,*
*St. Petersburg 198504, Russia*

**Abstract.** Kuzmin's theoretical formula $(\sigma_R/\sigma_z)^2 = 1+(\sigma_R/\sigma_\phi)^2$ is recalled. It is compatible with new data for stars of various ages.

## 1. Introduction

A correlation between kinematics and spectral types of stars was discovered in the 1950s by P. P. Parenago, N. Roman, T. Rootsmäe and others. Later it was transformed into the age dependence of the velocity ellipsoid (Mayor 1974; Wielen 1977). Now that dependence is usually considered as a well established fact of Galactic Kinematics (Lacey 1991; Binney & Merrifield 1998). It was confirmed by *HIPPARCOS* data (Dehnen & Binney 1998). However some authors conclude that the velocity dispersions are almost constant for stars of various ages (Palouš & Piskunov 1985; Strömgren 1987; Quillen & Garnett 2000). Two main explanations of the age increase of the velocity dispersions were suggested (Binney 2001). According to the first one that dependence reflects the conditions of star formation, e.g., it results from a secular decay of turbulent motions in the interstellar medium (Larson 1979). An alternative explanation (supported now by the most investigators) considers it as a statistical effect of accelerations acting on stars after they form. According to Spitzer & Schwarzschild (1951), Lebedynsky (1951) and others the increase of the velocity dispersions may be due to star scattering from initially circular orbits by massive perturbers, such as Giant Molecular Clouds (GMCs) (e.g., Icke 1982; Lacey 1984; Kamahori & Fujimoto 1986). Some authors considered a stochastic accelerations by black holes of the Galactic halo (Lacey & Ostriker 1985; Fuchs, Dettbarn, & Wielen 1994) or by dark massive clusters (Carr & Lacey 1987). Spitzer & Schwarzschild assumed that the velocity distribution was isotropic, but Lebedynsky stressed it must be triaxial.

## 2. Kuzmin's Theory

Following Lebedynsky's ideas, Gurevich (1953) derived an approximate formula for a form of the velocity ellipsoid resulting from interactions with clumps of gravitating matter. According to Gurevich, a quasi-equilibrium value of $\sigma_z^2/\sigma_R^2 = {}^1/_2\left[1+(\sigma_\phi^2/\sigma_R^2)\right](1+\tau/\tau')^{-1}$, where $\tau$ is the relaxation time, and

$\tau' \approx \tau$ is the time of increasing a flattening of a subsystem under consideration. Then Kuzmin (1961, 1963, 1964, 1973) developed a more detailed theory. He assumed that unperturbed stellar orbits were epicyclic and distinguished direct changes of orbital elements due to random pushes (a diffusion in the space of integrals of motion) and indirect changes caused by a secular evolution of the smoothed Galactic potential. Kuzmin (1961) averaged the equations of motion and derived the general equations for time variations of the velocity dispersions and mean amplitudes of star oscillations relative a circular orbit. To develop an analytical theory he made some assumptions of mathematical nature. It seems that they are plausible in the case of star scattering by GMCs. Note that an embryo of the Coulomb logarithm was not constant in Kuzmin's theory. He made an additional assumption that a form of the velocity ellipsoid remains time-independent. Then Kuzmin concluded that the Lindblad's relation $\sigma_R/\sigma_\phi = 2\omega/\kappa_R$ (where $\omega$, $\kappa_R$ are the circular and the epicyclic frequencies) is valid, and $(\sigma_R/\sigma_z)^2 = 1 + (\sigma_R/\sigma_\phi)^2$. Later Kuzmin tried to generalize the theory. In his last article on the subject (Kuzmin 1973) the dynamical friction was taken into account and more general formulae for the time dependence of the velocity dispersions were derived (see also a paper by Wielen 1977). Kuzmin found that if a dimensionless time $t\kappa$ ($\kappa$ is a coefficient of the dynamical friction) is small, then his formula for $\sigma_R/\sigma_z$ remains valid. So, we can expect that Kuzmin's formula must be valid for stars of intermediate age, as the universal first approximation.

## 3. Other Theories

For decades Kuzmin's theory remained practically unknown, and when GMCs were discovered the problem was revisited. Lacey (1984) has carefully analyzed scattering by GMCs but found that his theory conflicts with observations predicting an excessive ratio $\sigma_z^2/\sigma_R^2$. The latter seems surprising for standpoints of theories by Kuzmin and Lacey coincide. So, theories that describe the heating of a stellar disk under the combined influence of a stochastic spiral structure and GMCs were developed (Sellwood & Carlberg 1984; Jenkins & Binney 1990), the role of GMCs being responsible for $\sigma_z$, for spiral irregularities are incapable of increasing vertical velocities (Binney 2001). Alternative theories were published by Kokubo & Ida (1992), Ida, Kokubo, & Makino (1993), and others. These authors noticed gaps in the Lacey's analysis. Ida et al. (1992) found that Lacey's excessive value of $\sigma_z$ is a consequence of his assumption on the constancy of the Coulomb logarithm. Unfortunately they do not give an alternative analytical formula.

## 4. Observations

To check Kuzmin's formula, we used recent velocity ellipsoid determinations by various authors (Palouš & Piskunov 1985; Strömgren 1987; Mignard 2000; Quillen & Garnett 2000; Brosche, Schwan, & Schwarz 2001; Zabolotskikh, Rastorguev, & Dambis 2002). Following a standard statistical analysis we found that Kuzmin's formula is statistically confirmed for all stars excluding very young ones. Details of calculations will be published later.

**Acknowledgments.** I am indebted to Dr I. Nikiforov for valuable advice.

## References

Binney, J. J. 2001, in ASP Conf. Ser., Vol. 230, Galaxy Disks and Disk Galaxies, ed. J. G. Funes & E. M. Corsini (San Francisco: ASP), 63

Binney, J., & Merrifield, M. 1997, Galactic Astronomy (Princeton: Princeton Univ. Press)

Brosche, P., Schwan, H., & Schwarz, O. 2001, Ast.Nachr., 322, 15

Carr, B. J., & Lacey, C. G. 1987, ApJ, 316, 23

Dehnen, W., & Binney, J. J. 1998, MNRAS, 298, 387

Fuchs, B., Dettbarn, C., & Wielen, R. 1994, in Ergodic Concepts in Stellar Dynamics, ed. V. G. Gurzadyan & D. Pfenniger (Berlin: Springer), 34

Gurevich, L. E. 1954, Voprosy Cosmogonii, 2, 150 (in Russian)

Icke, V. 1982, ApJ, 254, 517

Ida, S., Kokubo, E., & Makino, J. 1993, MNRAS, 263, 835

Jenkins, A., & Binney, J. 1990, MNRAS, 245, 305

Kamahori, H., & Fujimoto, M. 1986, PASJ, 38, 77

Kokubo, E, & Ida, S. 1992, PASJ, 44, 601

Kuzmin, G. G. 1961, Tartu Publ., 33, 351 (in Russian)

Kuzmin, G. G. 1963, Tartu Teated, 6, 19 (in Russian)

Kuzmin, G. G. 1964, Tartu Publ., 34, 10 (in Russian)

Kuzmin, G. G. 1973, in Dynamics of Galaxies and Star Clusters, ed. T. B. Omarov (Alma-Ata: Nauka), 76 (in Russian)

Lacey, C. 1984, MNRAS, 208, 687

Lacey, C. 1991, in Dynamics of Disc Galaxies, ed. B. Sundelius (Göteborg: Göteborgs Univ. and Chalmers Univ. Technology), 257

Lacey, C., & Ostriker, J. P. 1985, ApJ, 299, 633

Larson, R. 1979, MNRAS, 186, 470

Lebedynsky, A. I. 1951, Dokl.AN SSSR, 79, 41 (in Russian)

Mayor, M. 1974, A&A, 32, 321

Mignard, F. 2000, A&A, 354, 522

Palouš, J., & Piskunov, A. E., 1985, A&A, 145, 102

Quillen, A. C., & Garnett, D. R. 2000, astro-ph/0004210

Sellwood, J. A., & Carlberg, R. G. 1984, ApJ, 282, 61

Spitzer, L. Jr., & Schwarzschild, M. 1951, ApJ, 114, 385

Strömgren, B. 1987, in The Galaxy, ed. G. Gilmore & R. Carswell (Dordrecht: Reidel), 229

Wielen, R. 1977, A&A, 60, 263

Zabolotskikh, M. V., Rastorguev, A. S., & Dambis, A. K. 2002, Pis'ma AZh, 28, 516 (in Russian)

*Order and Chaos in Stellar and Planetary Systems*
*ASP Conference Series, Vol. 316, 2004*
*G. Byrd, K. Kholshevnikov, A. Mylläri, I. Nikiforov, and V. Orlov, eds.*

# Investigation of the Possible Links between Different Characteristics of Galactic Disk Dwarf Stars from the Spectral Classes F and G

G. A. Malasidze and R. M. Dzigvashvili

*Tbilisi State University, University st. 2, Tbilisi 0143, Georgia*

**Abstract.** The dependence between the elements of the galactic orbits on the ages of the dwarf stars from spectral classes F and G are examined based on a specific form of the gravitational potential. Results are obtained mainly for the stars situated in the solar neighborhood. The possibility that these dependencies are different in other areas of the galaxy is not excluded.

We intend to study the age variation of several most interesting invariants of motion of dwarf stars from spectral classes F and G in the neighborhood of the Sun. For this purpose we employ the theory for the determination of flat stellar orbits (developed by authors in Kuzmin & Malasidze 1970 and Malasidze 1971) and the important form of the isochronic potential

$$\Phi = 2\Phi_0 \left(1 + \sqrt{1 + 8\frac{R^2}{R_0^2}}\right)^{-1}, \tag{1}$$

which permits the solution of the flat orbit problem in elementary functions:

$$\theta - \theta_1 = \pm \left[\arctan\left(\sqrt{\frac{1-u_2}{1-u_1}}\tan\varphi\right) + \sqrt{\frac{2y}{1+2y}}\varphi\right], \tag{2}$$

$$\omega_c(t - t_1) = x^{-\frac{3}{2}} \arctan\left(\sqrt{\frac{u_2}{u_1}}\tan\varphi\right) + \frac{\sqrt{1+2y}}{2x}\frac{p\sin 2\varphi}{1 - p\sin^2\varphi}. \tag{3}$$

We use the data of space motion components from the paper by Edvardsson et al. (1993) (see their Table 11). Here $(R, \theta)$ denote polar coordinates in the galactic plane, $\Phi_0 = \Phi(0)$ is the gravitational potential in the center of the system, $R_0$ and $t$ are length and time parameters. $\theta_1$ and $t_1$ are arbitrary constant of the integration of equations of motions and variables $x$, $y$ and $\omega_c$ are defined as follows:

$$\omega_c^2 = 4\Phi_0 R_0^{-2}, \quad x = -I_1(2\Phi_0)^{-1}, \quad y = I_2^2(2\Phi_0)^{-1}R_0^{-2}, \tag{4}$$

where $I_1$ is doubled total energy and $I_2$ is the integral of the kinetic moment of the star. Here $u_1$ and $u_2$ are roots of the following equation:

$$(1 + 2y)u^2 - (1 + x)u + x = 0. \tag{5}$$

Parametrical equation of the orbit with parameter $0 \leq \varphi \leq \pi/2$ may be written as follows:

$$R = R_1(1 - p\sin^2\varphi)^{-1}\sqrt{1 + m\sin^2\varphi}, \tag{6}$$

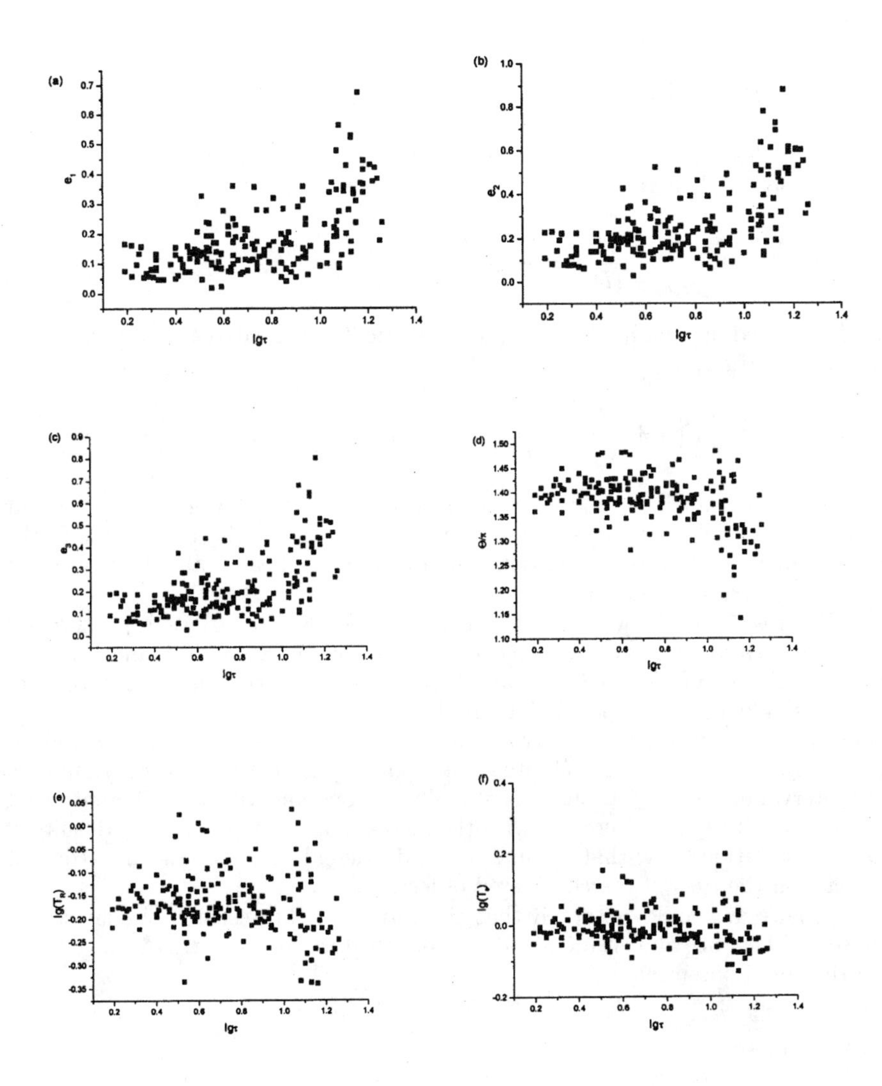


Figure 1. Stellar orbital elements vs the age logarithm in $10^9$ years: (a)–(c) the eccentricities according the different definitions; (d) doubled apsidal angle; (e), (f) anomalistic and sidereal periods in $10^9$ years, respectively.

where $R_1$ is the pericentric distance to the star and $p$ and $m$ are:

$$p = (u_1 - u_2)u_1^{-1}, \quad m = (u_1 - u_2)(1 - u_1)^{-1}. \tag{7}$$

We consider eccentricities defined by the following equations:

$$e_1 = (R_2 - R_1)(R_1 + R_2)^{-1}, \quad e_2 = 1 - 8xy(1-x)^{-2}, \quad e_3 = 1 - (1 + P_R|P_\theta|^{-1})^{-2}. \tag{8}$$

Variables $P_R$ and $P_\theta$ denote adiabatic invariants of the motion (Landau & Lifshiz 1988). Their ratio may be calculated using Eq. (1):

$$\frac{P_R}{|P_\theta|} = \frac{1 + \sqrt{2xy} - \sqrt{x(1+2y)}}{2\sqrt{2xy}}, \tag{9}$$

while doubled apsidal angle $\Theta$ and anomalistic $T_R$ and sidereal $T_\theta$ periods may be calculated as:

$$\Theta = \pi\left(1 + \sqrt{\frac{2y}{1+2y}}\right), \quad \omega_c T_R = \pi x^{-3/2}, \quad T_\theta = 2\pi\frac{T_R}{\Theta}. \tag{10}$$

In the calculations of the $x$ we have used the total velocity of the star instead of the velocity projection on the galactic plane. This corresponds to the enhancement of $|V_R|$ for the chosen stars by an insignificant value that does not affect results of the calculations.

Figures 1a–1c show eccentricities calculated by Eq. (8) vs logarithms of the star ages for over two hundred dwarf stars in the galactic plane from the spectral classes of F and G. In all three cases the variation of eccentricities is almost identical. In particular, in the interval $0 \leq \log\tau \leq 1$ it is almost constant, while at $\log\tau > 1$ the rapid increase of the eccentricity is occurred for older objects. The similar picture is plotted on Figure 1d for apsidal angle in the interval $0 \leq \log\tau \leq 1$, i.e., it is also almost constant with the difference that at $\log\tau > 1$ it rapidly decreases with the increase of the object age $\tau$. Results of our investigation show that anomalistic and sidereal periods on average do not depend on the ages of the considered objects.

Finally, it should be noted that the results of the research similar to the presented here are almost independent from the form of the form of the galactic gravitational potential.

## References

Kuzmin, G. G., & Malasidze, G. A. 1970, Publ.Tartusk.Astrofiz.Observ., 38, 181

Malasidze, G. A. 1971, Bull.Abastum.Astrophys.Observ., 40, 123

Edvardsson, B., Andersen, J., Gustafsson, B., Lambert, D. L., Nissen, P. E., & Tomkin, J. 1993, A&A, 275, 101

Landau, L. D., & Lifshiz, E. M. 1988, Vol. I, Mechanics, 49, 199

*Order and Chaos in Stellar and Planetary Systems*
*ASP Conference Series, Vol. 316, 2004*
*G. Byrd, K. Kholshevnikov, A. Mylläri, I. Nikiforov, and V. Orlov, eds.*

# TYCHO2: Search for Stellar Groups Using Wavelet Transform

Elena Kazakevich, Veniamin Vityazev, and Victor Orlov

*Sobolev Astronomical Institute, St. Petersburg State University, Universitetskij pr. 28, Staryj Peterhof, St. Petersburg 198504, Russia; e-mails: elen@ek3286.spb.edu, VITYAZEV@venvi.usr.pu.ru, vor@astro.spbu.ru*

**Abstract.** The wavelet transform was used to identify the inhomogeneities in the distribution on the celestial sphere of stars from Tycho2 Catalogue. The method was tested by simulations. The wavelet coefficients maps (for the samples of stars from different visual magnitude intervals) around the whole celestial sphere were constructed at different restrictions for visual magnitude values. Statistically significant groups of stars in the region $l \in [0°, 180°]$, $b \in [-30°, -10°]$ were identified.

## 1. Introduction

The study of star distribution plays important role in galactic astronomy. In particular, it is used for star clusters detection. Among a big variety of astrometric catalogues, the combination of rather high accuracy (average error in proper motions is about 2.5 mas/yr) and large amount of stars take an advantage for Tycho2 Catalogue. This catalogue contains 2539913 stars and more than 95% of them are fainter than $9^{\mathrm{m}}$. *So for the first time we can investigate the distribution of faint stars on celestial sphere* (there are no parallaxes in this catalogue).

## 2. Description of the Method

We applied the wavelet transform technique that was used in cosmology (see, e.g., Slezak et al. 1990) in our investigation. The so-called Mexican hat (MHAT) wavelet was used to detect the inhomogeneities in the distribution on the celestial sphere of Tycho2 stars. For a given zone of celestial sphere we define pixels $(i, j)$ in 1° grid in $l$- and $b$-directions. The wavelet transform for the scale $\sigma$ (thereafter it is given in degrees) of the star sample was then computed in each pixel $(i, j)$ according to:

$$W(i,j,\sigma) = \sum_n \left[2 - \frac{(i-l)^2 + (j-b)^2}{\sigma^2}\right] e^{-\frac{(i-l)^2+(j-b)^2}{2\sigma^2}}, \tag{1}$$

where $l$, $b$ are galactic coordinates of the star.

### 2.1. Testing of the Method

The method was tested by simulations. The uniform distribution of 9000 stars on celestial sphere in the strip with a length of 180° and a width of 10° with three clusters at $l \in [16°, 19°]$, $b \in [3°, 5°]$ of 60 stars, $l \in [85°, 88°]$, $b \in [5°, 7°]$ of 90 stars and $l \in [150°, 154°]$, $b \in [4°, 6°]$ of 80 stars had been simulated. The map of wavelet coefficients obtained with $\sigma = 2$ can be seen in Figure 1. Thus, as we can see from Figure 1, our wavelet analysis is able to reveal the stellar groups with angular sizes corrsponding to the scale parameter $\sigma$.

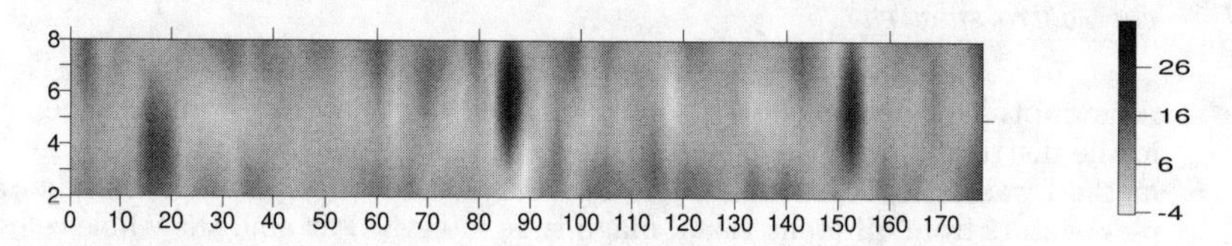


Figure 1. Map of the wavelet coefficients obtained for the stars from the simulated catalogue with $\sigma = 2$.

### 2.2. Criteria for Group Identification

The statistical significance of the revealed groups was estimated using simulated uniform distribution of the same number of events. The distribution of the wavelet coefficients in simulated catalogue was fitted with Gaussian function. Using parameters of the Gaussian we set the statistical significance level for the groups identification. We considered the threshold for stellar groups detection equals to $3\sigma'$, where $\sigma'$ is a parameter of the Gaussian that corresponds to a probability 0.003 that wavelet coefficient value is due to random fluctuations.

## 3. Results

We choose for our study the belt region with coordinates $l \in [0°, 180°]$, $b \in [-30°, -10°]$. Making the restriction on visual magnitude value we divided all stars into six samples: sample 1 contains stars brighter than $8^{\mathrm{m}}$, samples 2–5 are within $[8^{\mathrm{m}}, 12^{\mathrm{m}}]$ with magnitude interval of $1^{\mathrm{m}}$ and sample 6 consists of the stars fainter than $12^{\mathrm{m}}$. Star clusters (black) and voids (gray) detected in this region at $\sigma = 2$ among the stars from samples 1, 4, 5 are shown in Figure 2, clusters and voids detected in the same region at $\sigma = 1$ among the faint stars from sample 6 are shown in Figure 3. The coordinates of the star clusters are cited in first two columns of the Table 1. The coordinate intervals covers pixels where wavelet coeficients overlapped the threshold. The star sample where the cluster was revealed are noted in the third column.

Following from bright to faint stars we can trace how the distribution of the inhomogeneities on celestial sphere is changing. Clustering becomes more intensive among stars fainter than $11^{\mathrm{m}}$. These groups are predominantly formed by distant stars. However, the selection effects may play an important role for the faint stars because the Tycho2 catalogue is complete till $11\overset{\mathrm{m}}{.}5$ only.

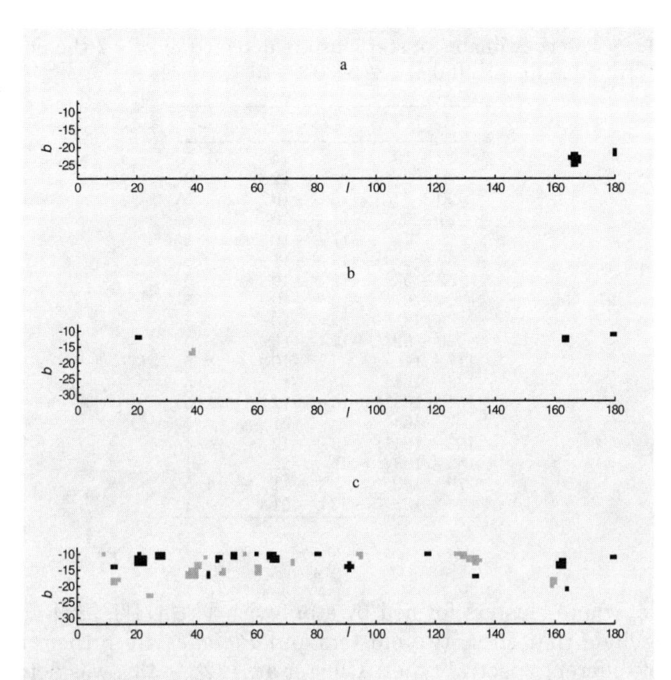


Figure 2. Star clusters detected with $\sigma = 2$ among the stars: a) brighter than $8^{\rm m}$, b) with $V \in [10^{\rm m}, 11^{\rm m}]$, c) with $V \in [11^{\rm m}, 12^{\rm m}]$.

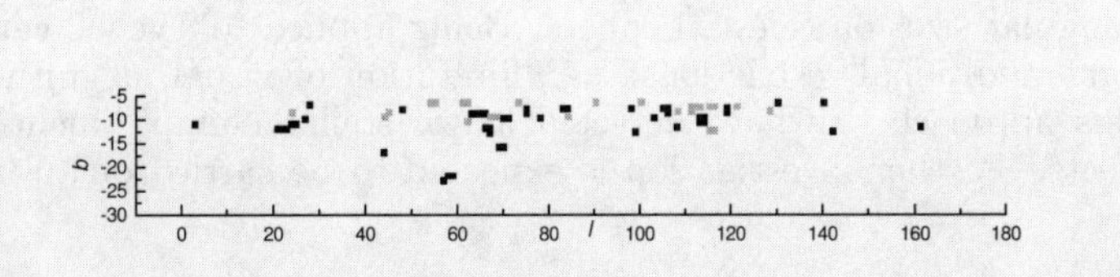


Figure 3. Star clusters (black) and voids (gray) detected with $\sigma = 1$ among the stars fainter than $12^{\rm m}$.

Two known open clusters (*Hyades* and *Pleiades*) were found with probability 0.997 at $\sigma = 2$ among the bright stars ($V$ less than $8^{\rm m}$) from the region. With $\sigma = 1$ we have identified *Pleiades* and one more cluster at $[148°, -7°]$ corresponding to *Melotte 20* (see Lynga 1987). One new cluster at $[142°, -10°]$ was detected with $\sigma = 1$ among stars of sample 2. *NGC 752* was revealed at this scale analysing sample 3. There are no statistically significant groups of stars revealed with $\sigma = 2$ among the stars with $V \in [8^{\rm m}, 10^{\rm m}]$.

Among the stars of sample 4 with $V \in [10^{\rm m}, 11^{\rm m}]$ three clusters were identified with $\sigma = 2$. One of them at $[21°, -12°]$ corresponds to *Ruprecht 147* (see Lynga 1987), cluster at $[180°, -11°]$ can be associated with two known clusters (*NGC 1746* and *NGC 1758*). These three clusters were detected also among stars with $V \in [11^{\rm m}, 12^{\rm m}]$, two of them (at $[21°, -12°]$ and $[163°, -13°]$) were identified among faint stars with $V$ more than $12^{\rm m}$. *Pleiades* can also be seen

Table 1. The coordinates of star clusters detected at $\sigma = 2$ the belt region with coordinates $l \in [0°, 180°]$, $b \in [-30°, -10°]$

| $l$ | $b$ | sample |
|---|---|---|
| $12 \div 13$ | $-14$ | 5 |
| $20 \div 21$ | $-12$ | 4 |
| $20 \div 23$ | $-13 \div -10$ | 5 |
| $27 \div 29$ | $-11 \div -10$ | 5 |
| 44 | $-17 \div -16$ | 5 |
| $47 \div 48$ | $-12 \div -11$ | 5 |
| $52 \div 53$ | $-11 \div -10$ | 5 |
| $64 \div 67$ | $-12 \div -10$ | 5 |
| $80 \div 81$ | $-10$ | 5 |
| $90 \div 92$ | $-15 \div -13$ | 5 |
| $117 \div 118$ | $-10$ | 5 |
| $133 \div 134$ | $-17$ | 5 |
| $161 \div 163$ | $-14 \div -12$ | 5 |
| 164 | $-21$ | 5 |
| $163 \div 164$ | $-13 \div -12$ | 4 |
| $165 \div 168$ | $-25 \div -22$ | 1 |
| $179 \div 180$ | $-11$ | 4, 5 |
| 180 | $-22 \div -21$ | 1 |

in Figure 2c, where clusters formed by stars with $V \in [11^{\mathrm{m}}, 12^{\mathrm{m}}]$ are presented. We can conclude that these two clusters and *Pleiades* are rather rich and contain stars of different spectral types. Cluster at $[142°, -13°]$ was detected among stars from sample 6 at both scales. The other clusters drawn in Figures 2c and 3 are previously unknown.

Thus, our method gives an opportunity to identify the stellar groups of various angular sizes on celestial sphere. Being applied to Tycho2 catalogue, it needs some additional restrictions, e.g., limitation of visual magnitude values. Using this approach, we have detected a few stellar clumps, including some new objects. Further investigation is expected to be carried out using proper motions.

## References

Slezak, E. et al., 1990, A&A, 227, 301

Lynga, G. 1987, Catalogue of open star clusters, 5-th edition (Uppsala Obs., Sweden)

*Order and Chaos in Stellar and Planetary Systems*
*ASP Conference Series, Vol. 316, 2004*
*G. Byrd, K. Kholshevnikov, A. Mylläri, I. Nikiforov, and V. Orlov, eds.*

# Statistics of Double Stars

George Gontcharov

*Central Astronomical Observatory at Pulkovo, Pulkovskoe sh. 65/1, St. Petersburg 196140, Russia*

**Abstract.** The statistics of visual (CPM, fixed, linear and orbital), interferometric, astrometric (delta-mu, VIM and orbital), spectroscopic (long-period, SB1 and SB2) and photometric (eclipsing, occultation and close) stellar pairs are given based on Hipparcos, WDS, TDSC, CCDM, ARIHIP, $6^{th}$ catalog of visual orbits, $9^{th}$ catalog of spectroscopic orbits, $4^{th}$ catalog of interferometric measurements and others. The data are collected into a database in order to make calibrations[1].

A complete and robust classification of stellar pairs by method:

- Close pairs on the sky with both components visible, i.e., **visual pairs**. They could be divided by further study into: optical pairs, common proper motion pairs (usually with separation >20000 AU), fixed pairs with separation <20000 AU, linear pairs with slightly changing separation and/or positional angle and orbital pairs with elliptical motion of the both components.

- **Interferometric pairs** observed exclusively by interferometry.

- **Astrometric pairs** with variation of photocenter's proper motion. They should be divided into: delta-mu pairs with proper motion significantly varies from one catalog to another, astrometric pairs with detected elliptical motion of photocenter and variability-induced movers (VIM) with a specific motion and variation of photocenter.

- **Spectroscopic pairs** with periodic variations of radial velocity due to elliptical motion. They should be divided into: long-period SB (spectral pairs) with fixed composite spectrum, SB1 with a component's lines (including exoplanets) and SB2 with motion of 2 sets of lines.

- **Photometric pairs** with specific variability. They include: eclipsing pairs, occultation pairs with a photometric variation in lunar or other occultation and close pairs such as cataclysmic variables and so on.

The main sources used are $6^{th}$ Catalog of Orbits of Visual Binary Stars by Hartkopf & Mason (2003) with 4865 orbits for 1633 stellar systems (in addition to visual it contains about 300 orbits obtained almost exclusively from

---

[1]Financially supported by the Russian Foundation for Basic Research grant #02-02-16570

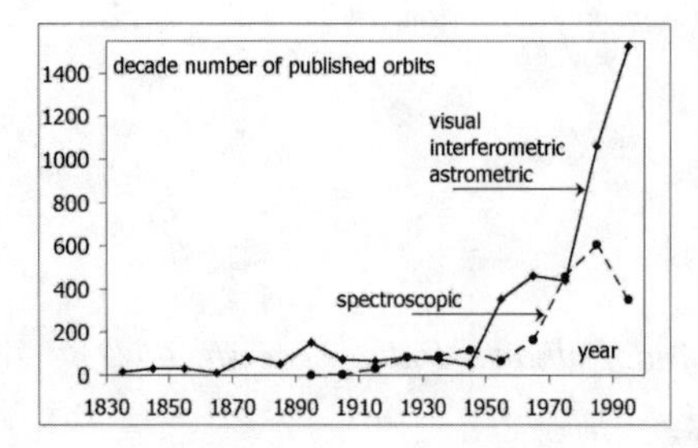


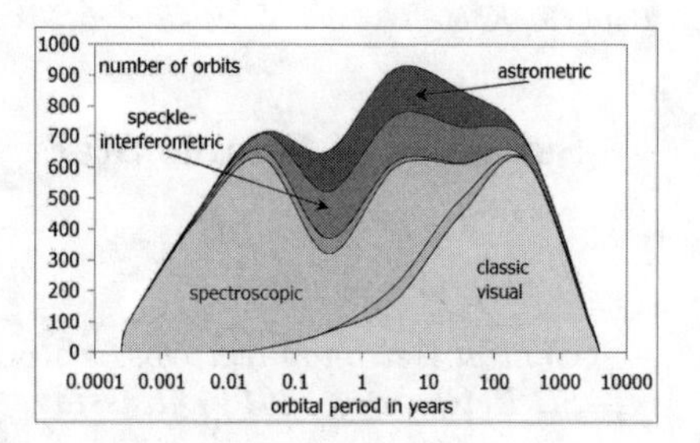


Figure 1. Left: Post-Hipparcos revolution in discovery and study of visual, interferometric and astrometric binaries is seen in the number of published orbits (similarly for observations). No such a rise for spectroscopic binaries. Right: As a result, the distribution of known orbits along with orbital period (consequently with semi-major axis) becomes almost normal. Some gap at fractions of year can be explained by observational selection due to weather and seasonal periodicity. Some spectroscopic orbits also known as visual or speckle ones are marked as 2 darker borders of the spectroscopic domain.

speckle-interferometric observations and about 300 exclusively astrometric orbits); 9$^{\text{th}}$ Catalogue of Spectroscopic Binary Orbits by Pourbaix & Tokovinin (2003) with 1895 spectroscopic orbits; 4$^{\text{th}}$ Catalog of Interferometric Measurements of Binary Stars by Hartkopf et al. (2003) with 47000 pairs resolved by interferometry and occultations; WDS by Mason, Wycoff, & Hartkopf (2003) with over 500000 means of about 100000 visual and astrometric pairs; TDSC by Fabricius et al. (2002) with results for 103259 binaries including 13251 new visual binaries; CCDM by Dommanget & Nys (2002) with 105838 components of 49325 visual binaries; ARIHIP catalog by Wielen et al. (2002) with data for over 7000 delta-mu binaries among Hipparcos stars.

Unrecognized optical pairs, CPMs and fixed pairs are about 96% of nearly 100000 currently known pairs. About 30% of them have a few observations for the only epoch. 92% are simple pairs and 5% are non-hierarchical triples. 32 of 40 nearest stars are known or suspected members of double and multiple systems. Non-single stars show almost the same HR diagram as for single ones: the majority of components are main sequence and slightly evolved stars with masses from $0.5\,M_{\odot}$ to $5\,M_{\odot}$. 70% of known pairs have magnitude difference of components $<3^m$. 70% of known pairs are in the northern celestial hemisphere where more observers. The known orbits demonstrate expected distribution along orbital inclination and eccentricity (less orbits with very small and very large eccentricity). Co-planar orbits in multiple systems are not usual. Also there is no statistical evidence of a relation between orientation of the orbits and any plane in the Solar system.

We use the USNO B1.0 catalog by Monet et al. (2003) containing about 1 billion stars with positions, proper motions (for about 20% of stars) and $B$, $R$, $I$ magnitudes together with the Aladin Interactive Sky Atlas at the CDS in order to estimate the number of unrecognized faint visual pairs in the digitized sky surveys. It shows that at least several millions ($>1\%$) images in the surveys are undetected visual pairs with separation of 2–10 arcsec.

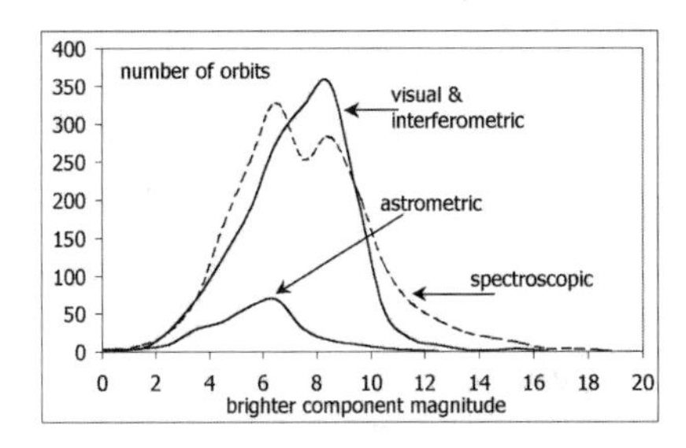


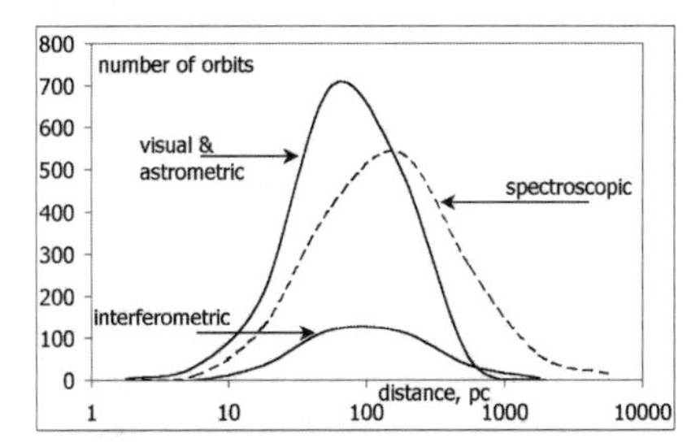


Figure 2. Left: Distribution of known orbits with primary's magnitude. The distribution of observations looks similar. This illustrates the fact that the primaries of 95% known pairs and orbits are Hipparcos stars. The distribution of astrometric binaries is determined by the fact that they are discovered in decade-long series of observations earlier engaging only bright stars. Right: The most targets of all methods lie between 20 and 400 pc (looks similar for observations and orbits).

A comparison of positions and proper motions of Tycho2, AC2000.2, TAC2, CPC2, M2000, CMC12, UCAC and other catalogs could be a new source of astrometric binaries in the range $8^m$–$12^m$.

Complete information on position, proper motion, parallax, radial velocity, multiplicity, photometry in at least 3 bands, spectra and cross identification of about 20000 best studied singles and pairs with a Hipparcos component is selected by us into a database at http://www.astro.spbu.ru/ in order to make calibrations of various parameters of single and double stars.

## References

Dommanget, J., & Nys, O. 2002, Catalog of Components of Double and Multiple Stars, Obs. Royal Belg., http://cdsweb.u-strasbg.fr/viz-bin/Cat?I/274

Fabricius, C. et al. 2002, Tycho Double Star Catalogue, A&A, 384, 180

Hartkopf, W. I., & Mason, B. D. 2003, 6$^{\text{th}}$ Catalog of Orbits of Visual Binary Stars, USNO, http://ad.usno.navy.mil/wds/orb6/orb6text.html

Hartkopf, W. I. et al. 2003, 4$^{\text{th}}$ Catalog of Interferometric Measurements of Binary Stars, USNO & CHARA, http://ad.usno.navy.mil/wds/int4.html

Mason, B. D., Wycoff, G. L., & Hartkopf W. I. 2003, Washington Double Star Catalog, USNO, http://ad.usno.navy.mil/wds/

Monet, D. G. et al. 2003, USNO B1.0, http://www.nofs.navy.mil/data/fchpix/

Pourbaix, D., & Tokovinin, A. 2003, 9$^{\text{th}}$ Catalogue of Spectroscopic Binary Orbits, http://sb9.astro.ulb.ac.be/

Wielen, R. et al. 2002, Astrometric Catalogue ARIHIP, Astron. Rechen-Institut Heidelberg, http://www.ari.uni-heidelberg.de/arihip

*Order and Chaos in Stellar and Planetary Systems*
*ASP Conference Series, Vol. 316, 2004*
*G. Byrd, K. Kholshevnikov, A. Mylläri, I. Nikiforov, and V. Orlov, eds.*

# Orientation of AMP-Orbits of Pulkovo Programme Binary Stars in the Galaxy Coordinate Frame

A. A. Kisselev and L. G. Romanenko

*Central Astronomical Observatory at Pulkovo, Pulkovskoe sh. 65/1, St. Petersburg 196140 Russia*

**Abstract.** The orientation of orbits of 30 binaries obtained at Pulkovo by the apparent motion parameters (AMP) method is determined. The distributions of orbital plane poles and periastron directions in the Galaxy coordinate frame are analyzed. It's shown that generally the distributions obtained do not contradict the accidental case. At the same time the orbit's planes for the wide binaries (with orbital period over 1000 years) mostly tend to have a steep inclination to the Galactic plane and their periastron axes are directed towards the Galactic Centre. This tendency is more evident for the wide double star orbits with the eccentricity exceeding 0.7.

## 1. Introduction

The photographic observations of visual double stars at Pulkovo were renewed in 1960 with the 26-inch Zeiss refractor ($F = 10.4$ m). The programme initiated by Professor A. N. Deich has included nearby double stars with suspected dark companion and other stellar pairs with large common proper motions. The main goal of the observations was to determine the orbits and masses of the stars. Later astrometric data obtained for most of the stars were extended with HIPPARCOS parallaxes. First results of the observations were relative positions and motions of 200 double stars gathered in Catalogue (Kisselev et al. 1988). Methods of observations and reduction were also described in the issue in detail.

## 2. Data and Technique

Here we consider some results of astrometric investigation of 30 binaries based on Pulkovo photographic observations during 1970–2000. The observational basis consists of about 2000 photographic plates, each containing 10–20 exposures. Observational time was evenly distributed between the objects: 3–4 plates every 2–3 years for each binary. The plates were measured manually with the help of semiautomatic Zeiss "Ascorecord" measuring machine. The seasonal errors of relative positions of the components were $\pm(0.005'' \div 0.010'')$ for most binaries. The observed arc of the apparent orbits did not exceed 3–5 degrees. For determination of orbit elements in these conditions we had to use the apparent motion parameters (AMP) method (Kisselev & Kiyaeva 1980; Kisselev et al. 1997) which allows to solve the problem on the basis of short arc observations provided with relative radial velocities of the components are known. We were kindly provided with these data by A. Tokovinin (Tokovinin & Smekhov 2002;

Tokovinin 1994). In this way we were able to determine the spatial vectors $r\vec{R}$ and $v\vec{V}$ of position and velocity of component $B$ relative to component $A$ corresponding to a time moment in the middle of observational period.

We use $\vec{R}$ and $\vec{V}$ for unit vectors which are expressed in astrocentric coordinate system as follows:

$$\vec{R} = \begin{bmatrix} \sin\theta\cos\beta \\ \cos\theta\cos\beta \\ \pm\sin\beta \end{bmatrix}, \tag{1}$$

$$\vec{V} = \begin{bmatrix} \sin\psi\cos\gamma \\ \cos\psi\cos\gamma \\ \sin\gamma \end{bmatrix}. \tag{2}$$

Here $\theta$, $\psi$ are position angles of $\vec{R}$ and $\vec{V}$ in the tangential plane; $\beta$, $\gamma$ are inclination angles of $\vec{R}$ and $\vec{V}$ with respect to the tangential plane. The values of $r$ and $v$ are determined by AMP-method unequivocally, but the inclination sign of $\vec{R}$ (i.e., the vector direction) is ambiguous, since:

$$\cos\beta = \rho/\pi_t r\ , \tag{3}$$

$$\sin\beta = \pm\sqrt{1-\cos^2\beta}\ . \tag{4}$$

The AMP orbital elements of the double stars under study were published elsewhere (Kisselev et al. 1997; Kisselev 2000). Here we examine the spatial geometry of these orbits trying to ascertain the peculiarity of their orientations in the Galaxy coordinate frame. The AMP-method gives us an opportunity to analyze this problem using the vector products as follows:

$$\vec{Q} = (\vec{R}\times\vec{V})/|\vec{R}\times\vec{V}|, \tag{5}$$

$$\vec{\Omega} = (\vec{Z}\times\vec{Q})/\sin i, \tag{6}$$

$$\vec{\Pi} = \vec{\Omega}\cos\omega + (\vec{Q}\times\vec{\Omega})\sin\omega. \tag{7}$$

Here the unit vector $\vec{Q}$ determines the direction to orbital pole, $\vec{\Omega}$ is the direction of nodal axis, $\vec{\Pi}$ is one of periastron axis, $i$ is the angle corresponding to inclination of orbital plane to the tangential plane, $\omega$ is the angular separation between $\vec{\Omega}$ and $\vec{\Pi}$ in orbital plane. Unit vectors $\vec{Q}$, $\vec{\Omega}$, $\vec{\Pi}$ obtained by this manner are determined in astrocentric tangential system $T(\xi,\eta,\zeta)$, where $Z(0,0,1)$ coincides with $\zeta$.

To transpose the vectors $\vec{Q}$ and $\vec{\Pi}$ into the galactic coordinate frame we use the matrix product as follows:

$$\vec{R}_G = \mathbf{G}\times\mathbf{T}^T\times\vec{R}_T. \tag{8}$$

Here the matrix

$$\mathbf{T}^T = \begin{bmatrix} -\sin\alpha & -\cos\alpha\sin\delta & \cos\alpha\cos\delta \\ \cos\alpha & -\sin\alpha\sin\delta & \sin\alpha\cos\delta \\ 0 & \cos\delta & \sin\delta \end{bmatrix} \tag{9}$$

converts the vector $\vec{R}$ from astrocentric system to the equatorial system and the matrix **G** converts it from equatorial system to the galactic system.

To determine the matrix **G** we have used the galactic orientation angles in the equatorial system: inclination $62\overset{\circ}{.}9$, ascending node $\alpha = 18^{\rm h}51\overset{\rm m}{.}4$, $\delta = 0°$, $l = 32\overset{\circ}{.}93, b = 0°$ (equinox being 2000.0).

## 3. Results

The results of converting vectors $\vec{Q}$ and $\vec{\Pi}$ are given in Table 1 where the basic data essential for dynamics of double stars' systems are shown, namely: parallax, orbital period and eccentricity.

The orbit data (see Table 1) are sorted by the period in ascending order. There are two equally possible orbits, which are dynamically identical, but different in orientation. Each possibility is estimated with weight "1" (last column of Table 1), but in some cases (if the binary has a long-term history of observations) we were able to determine the orbit unequivocally, these orbits are estimated with weight "2". The galactic longitude and latitude $(l, b)$ of $\vec{Q}$ and $\vec{\Pi}$ as derived by formulae (5) and (7) are also given in Table 1.

We have analyzed the data from Table 1 to find the most common directions of the vectors $\vec{Q}$ and $\vec{\Pi}$ for the wide and close binaries in the vicinity of the Sun. The results of this analysis are presented in Table 2.

## 4. Conclusions

The obtained data allow us to conclude that:

1. If we consider all the obtained orbits, the observed distributions of poles $\vec{Q}$ (selection with weight 60) and periastron directions $\vec{\Pi}$ do not contradict to the case of random orientations of orbital planes.

2. If we consider only the wide binaries with orbital period over 1000 years (selection with weight 36), we notice that orbital planes are mostly steeply inclined to the galactic plane; and periastron axes are directed to the Galaxy center. This tendency is more evident for orbits of wide binaries with the eccentricity exceeding 0.7 (selection with weight 18).

Table 1. Orientation of AMP-orbits of the Pulkovo VBS in the galactic frame

| $N$ | $N_{ADS}$ | $\pi_t$ | $P$ | $e$ | $l_Q$ | $b_Q$ | $l_{\Pi}$ | $b_{\Pi}$ | $\beta$ | $W$ |
|---|---|---|---|---|---|---|---|---|---|---|
| 1 | 9031 | .075 | 157 | .48 | 242 | -33 | 142 | -16 | 0 | 2 |
| 2 | 8242 | .028 | 277 | .41 | 14 | -25 | 275 | -19 | +47 | 2 |
| 3 | 9090 | .089 | 327 | .98 | 310 | -6 | 352 | +81 | -64 | 1 |
| | | | | .98 | 146 | -40 | 283 | -40 | +64 | 1 |
| 4 | 48 | .087 | 512 | .22 | 245 | -9 | 331 | +24 | +26 | 2 |
| 5 | 15229 | .040 | 531 | .74 | 205 | +72 | 122 | -2 | -52 | 2 |
| 6 | 2427 | .074 | 540 | .49 | 148 | +47 | 169 | -41 | -40 | 1 |
| | | | | .57 | 155 | -42 | 281 | -34 | +40 | 1 |
| 7 | 5983 | .061 | 590 | .73 | 356 | +42 | 167 | +48 | -53 | 2 |
| 8 | 11632 | .280 | 590 | .32 | 322 | +24 | 247 | -31 | +27 | 2 |
| 9 | 14636 | .287 | 680 | .46 | 307 | +37 | 123 | +53 | -34 | 2 |
| 10 | 10345 | .037 | 690 | .46 | 118 | +20 | 72 | -63 | -35 | 2 |
| 11 | 15600 | .032 | 710 | .85 | 128 | +47 | 188 | -25 | -18 | 1 |
| | | | | .92 | 126 | -7 | 220 | -25 | +18 | 1 |
| 12 | 8002 | .061 | 730 | .36 | 147 | -69 | 170 | +20 | -39 | 1 |
| | | | | .32 | 24 | -25 | 171 | -60 | +39 | 1 |
| 13 | 2757 | .041 | 1100 | .89 | 313 | +1 | 325 | -89 | -3 | 1 |
| | | | | .87 | 304 | +4 | 334 | -86 | +3 | 1 |
| 14 | 7251 | .166 | 1100 | .24 | 321 | -59 | 239 | +5 | -21 | 2 |
| 15 | 10329 | .040 | 1100 | .71 | 63 | +19 | 318 | +39 | 0 | 2 |
| 16 | 9167 | .026 | 1400 | .27 | 51 | -59 | 332 | +7 | +46 | 2 |
| 17 | 8236 | .023 | 1800 | .38 | 206 | +8 | 125 | -47 | 0 | 2 |
| 18 | 497 | .019 | 1900 | .53 | 148 | -41 | 303 | -46 | 0 | 2 |
| 19 | 5436 | .021 | 2000 | .88 | 239 | +13 | 151 | -9 | -38 | 1 |
| | | | | .36 | 239 | -7 | 335 | -44 | +38 | 1 |
| 20 | 8861 | .076 | 3900 | .33 | 228 | -30 | 278 | +47 | -42 | 1 |
| | | | | .50 | 315 | -35 | 96 | -48 | +42 | 1 |
| 21 | 8100AC | .068 | 4400 | .57 | 330 | +17 | 61 | +2 | -62 | 1 |
| | | | | .43 | 88 | -69 | 218 | -14 | +62 | 1 |
| 22 | 8250 | .043 | 4800 | .48 | 139 | +31 | 266 | +45 | -7 | 1 |
| | | | | .38 | 126 | +31 | 246 | +40 | +7 | 1 |
| 23 | 12169 | .040 | 7400 | .62 | 44 | -22 | 243 | -67 | -38 | 1 |
| | | | | .74 | 128 | +33 | 117 | -57 | +38 | 1 |
| 24 | 10386 | .049 | 15000 | .38 | 92 | -20 | 21 | +43 | -38 | 1 |
| | | | | .56 | 136 | +15 | 204 | -55 | +38 | 1 |
| 25 | 6646 | .014 | 18000 | .88 | 271 | -25 | 192 | +21 | -38 | 1 |
| | | | | .87 | 352 | -10 | 259 | -17 | +38 | 1 |
| 26 | 12815 | .047 | 38000 | .87 | 36 | -8 | 343 | +76 | -38 | 1 |
| | | | | .55 | 153 | +6 | 53 | +60 | +38 | 1 |
| 27 | 10044 | .010 | 50000 | .48 | 136 | +68 | 232 | +3 | -38 | 1 |
| | | | | .59 | 27 | +16 | 271 | +57 | +38 | 1 |
| 28 | 3593 | .029 | 53000 | .88 | 23 | +2 | 293 | -19 | -46 | 1 |
| | | | | .89 | 292 | -15 | 203 | +3 | +46 | 1 |
| 29 | 10759 | .045 | 56000 | .62 | 141 | -86 | 189 | +3 | +38 | 2 |
| 30 | 11061 | .019 | 450000 | .94 | 347 | -64 | 232 | -12 | -31 | 1 |
| | | | | .95 | 273 | +15 | 180 | +14 | +31 | 1 |

Table 2. Orientation of AMP-Orbits of Wide Double Stars

| | | Pole latitude | | | Periastron longitude | | | |
|---|---|---|---|---|---|---|---|---|
| Selection | $N$ % | $< -30°$ | $\|\beta\| < +30°$ | $> +30°$ | 0° ±45° | 180° ±45° | 90° ±45° | 270° ±45° |
| All orbits | 60 | 16 | 32 | 12 | 11 | 17 | 13 | 14 |
| actual dis. | 100 | 28 | 53 | 20 | 18 | 28 | 22 | 32 |
| $P > 1000$ | 36 | 11 | 21 | 4 | 9 | 8 | 6 | 13 |
| actual dis. | 100 | 31 | 58 | 11 | 25 | 22 | 17 | 36 |
| $e > 0.70$ | 18 | 2 | 12 | 4 | 5 | 7 | 2 | 4 |
| actual dis. | 100 | 11 | 67 | 22 | 28 | 39 | 11 | 22 |
| random dis. | 100 | 25 | 50 | 25 | 25 | 25 | 25 | 25 |

## References

Kisselev A. A. et al. 1988, Catalogue of relative positions and motions of 200 visual double stars as observed at Pulkovo by 26-inch refractor in 1960–1986 (Leningrad: Nauka), 1 (in Russian).

Kisselev A. A., & Kiyaeva O. V. 1980, AZh, 57, 1227 (in Russian).

Kisselev A. A. et al. 1997, in Visual Double Stars: Formation, Dynamics, Evolutionary Tracks, ed. J. A. Docobo et al. (Kluwer), 377

Tokovinin A. A., & Smekhov M. G. 2002, A&A, 382, 118

Tokovinin A. A. 1994, Ast.Rep., 38, 258

Kisselev A. A. et al. 2000, Izvestia GAO at Pulkovo, N214, 239 (in Russian).

*Order and Chaos in Stellar and Planetary Systems*
*ASP Conference Series, Vol. 316, 2004*
*G. Byrd, K. Kholshevnikov, A. Mylläri, I. Nikiforov, and V. Orlov, eds.*

# One Stochastic Model of Stellar Population

A. A. Vyuga

*Sobolev Astronomical Institute, St. Petersburg State University, Universitetskij pr. 28, Staryj Peterhof, St. Petersburg 198504, Russia*

**Abstract.** Special kind of Kolmogorov equations, where transitive probabilities linearly take into account partial mass of object and restrictions on it capacitance is used in model. Mean spectrum of stellar population masses, generated by these equations recurrently, is two-modal.

The role of gravitational factor in fragmentation of proto-stellar clouds is determining (Surdin 2001). On the other hand, at collapse of a fragment we can see the second order fragmentation, caused by redistribution of movement moment to the outer parts and energy generation into internal parts. Also, already existing massive objects by their interactions make catalytic effect. As result the picture of matter streams and stars birth is very complicated. We assume here the next probabilistic simplification:

1) at its appearance every object has the mass $\mu$;

2) already existing objects are non-destructive and matter absorption realizes by portions, which values are equal to $\mu$;

3) maximum possible mass of object is $\mu m$.

Thus we have the uniform grid with step $\mu$ for consideration of mass distribution. So that if $j = 1, \ldots, m$ denotes dimensionless mass of object and $k_j$ is the amount of objects with mass $j$ into the population, then the vector $(k_1, k_2, \ldots, k_m)$ characterizes the mass spectrum. Total amount of objects into population is $N = k_1 + k_2 + \ldots + k_m$ and total dimensionless mass of population is $M = k_1 + 2k_2 + \ldots + mk_m$. We suppose also what the growth of $M$ by 1 may be realized at one of the next incoincident events:

a) with the probability $Q_0(k_1, k_2, \ldots, k_m)$, the new object of mass 1 appears;

b) with the probability $Q_j(k_1, k_2, \ldots, k_m)$, the one and only one object of mass $1 \leq j \leq m-1$ absorbs the unit portion of mass.

The places of the matter portions resulting in transformation of a cloud into stellar population may be presented in Bernoulli scheme by Kolmogorov equations for $M = 2, 3, \ldots$

$$P(k_1, k_2, \ldots, k_m) = P(k_1 - 1, k_2, \ldots, k_m)\, Q_0(k_1 - 1, k_2, \ldots, k_m) + \sum_{j=1}^{m-1} P(\ldots, k_j + 1, k_{j+1} - 1, \ldots)\, Q_j(\ldots, k_j + 1, k_{j+1} - 1, \ldots), \quad (1)$$

where the transitive probabilities are connected by normalizing condition

$$\sum_{j=0}^{m-1} Q_j(k_1, k_2, \ldots, k_m) = 1 \tag{2}$$

and probability $P(k_1, k_2, \ldots, k_m)$ may be interpreted as a functional, allowing to obtain the most probable (or mean expected) mass spectrum (in the form of histogram) inside the population under consideration. In particular, if $m = M$ and $Q_j(k_1, k_2, \ldots, k_m) = k_j/n$, that is we have equiprobable matter placing among $n$ cells of unrestricted capacitance, from (1) the well known distribution for equiprobable placing (Kolchin at al. 1976) follows.

Here we consider the elementary linear model at assumption

$$Q_j(k_1, k_2, \ldots, k_m) = \frac{jk_j}{M}(1 - j/m), \quad 1 \le j \le m. \tag{3}$$

Thus, we assume that absorbing ability of a fragment is proportional to its partial mass $jk_j$, but its capacitance decreases linearly with approaching to maximal possible mass $m$. In this case from condition (2) we find

$$Q_0(k_1, k_2, \ldots, k_m) = 1 - \sum_{j=1}^{m} \frac{jk_j}{M}(1 - j/m) = \frac{1}{mM}\sum_{j=1}^{m} j^2 k_j, \tag{4}$$

that is, the probability of a new condensation appearance (with mass $\mu$) is proportional to mean square masses of condensations already existing in a cloud. Under assumptions (3), equation (1) becomes for $M = 2, 3, \ldots$

$$P(k_1, k_2, \ldots, k_m) = P(k_1 - 1, k_2, \ldots, k_m)\frac{1}{m(M-1)}\left(\sum_{j=1}^{m} j^2 k_j - 1\right) +$$
$$\sum_{j=1}^{m-1} P(\ldots, k_j + 1, k_{j+1} - 1, \ldots)\frac{j(k_j + 1)}{M - 1}(1 - j/m). \tag{5}$$

It is evident that for all $M$

$$\sum_{k_1 + 2k_2 + \ldots + mk_m = M} P(k_1, k_2, \ldots, k_m) = 1. \tag{6}$$

Now, if $P(1) = 1$, then for $M = 2$ from (5) follows $P(2) = 1/m$, $P(0,1) = 1 - 1/m$ and for $M = 3$ we have $P(3) = 1/m^2$, $P(1,1) = 3(1 - 1/m)/m$, $P(0,0,1) = (1 - 1/m)(1 - 2/m)$.

We can find by induction from (5) what

$$P(k_1, k_2, \ldots, k_m) = \frac{S(k_1, k_2, \ldots, k_M)}{(M-1)!}\prod_{j=1}^{M-1}(1 - j/m)^{\sum_{i=j+1}^{m} k_i}, \tag{7}$$

the values of $S$ are determined by equations

$$\begin{aligned} S(k_1, k_2, \ldots, k_M) &= S(k_1 - 1, k_2, \ldots, k_M) \left( \sum_{j=1}^{m} j^2 k_j - 1 \right) + \\ & \sum_{j=1}^{M-1} S(\ldots, k_j + 1, k_{j+1} - 1, \ldots)(k_j + 1)j, \qquad (8) \\ & S(1) = 1, \quad M = 2, 3, \ldots \end{aligned}$$

One can see that the expressions (7), (8) perform a separation of variables $m$ and $M$. By induction from (8) one finds that

$$S(M) = S(0, \ldots, 1) = (M-1)!, \quad S(k_1, k_2) = C_M^{k_1} (M-1)!. \qquad (9)$$

As to expected average spectrum of mass in terms of partial quantities of objects, it is presented by vector

$$(\overline{k}_1, \overline{k}_2, \ldots, \overline{k}_m) = \sum_{k_1 + 2k_2 + \ldots + mk_m = M} (k_1, k_2, \ldots, k_m)\, P(k_1, k_2, \ldots, k_m) \qquad (10)$$

and in terms of partial masses it is presented by vector

$$(\overline{k}_1, 2\overline{k}_2, \ldots, m\overline{k}_m). \qquad (11)$$

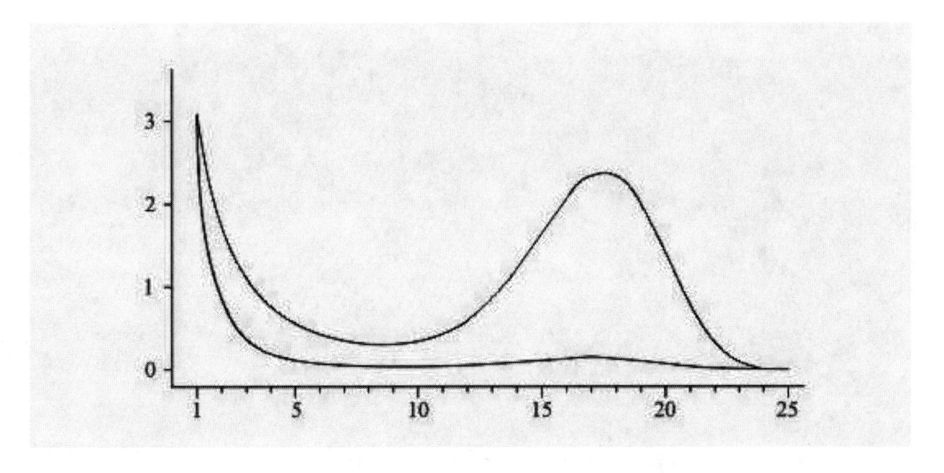


Figure 1. Mean expected distribution of objects of the mass (10), (11) into population with $M = 25$ and $m = 40$.

Thus, in this model the mass spectrum depends on two parameters—a whole population mass $M$ and the maximum possible object mass $m$. For example, if $M = 7$, we have the following numerical expressions for the spectrum (10):

$$m = 5: \quad (1.88, 0.663, 0.495, 0.410, 0.134, 0, 0),$$

$$m = 10: \quad (1.16, 0.345, 0.222, 0.265, 0.322, 0.232, 0.0605).$$

It is obvious what the width of spectrum (10), (11) is equal to $\min(m, M)$.

One can see what the distribution of stars by mass has two tops (see Figure 1). The luminosity function observed for bright stars is in agreement with this picture as a whole. Thus, it seems hopeful for next more accurate considerations.

## References

Surdin, V. G. 2001, Stars birth, 261

Kolchin, V. F., Sevastianov, B. A., & Chistiakov, V. P. 1976, Probable placing, 223

# Part VI

# Outside the Milky Way

*Order and Chaos in Stellar and Planetary Systems*
*ASP Conference Series, Vol. 316, 2004*
*G. Byrd, K. Kholshevnikov, A. Mylläri, I. Nikiforov, and V. Orlov, eds.*

# Once Again about the Origin of the System of the Giant Stellar Arcs in the Large Magellanic Cloud

Yu. N. Efremov

*Sternberg Astronomical Institute, Universitetskij pr. 13, Moscow 119992, Russia*

**Abstract.** The origin of the arc-shaped stellar complexes in the LMC4 region is still unknown. These perfect arcs could not have been formed by O stars and SNe in their centers: strong arguments exist also against the possibility of their formation from infalling gas clouds. The origin from microquasar/GRB jets is not excluded, because there is a strong concentration of X-ray binaries in the same region and the massive old cluster NGC 1978, a probable site of formation of binaries with compact components, is there also. The last possibility is that the source of energy for formation of the stellar arcs and the LMC4 supershell might be the the giant jet from the nucleus of the Milky Way, which might be active a dozen Myr ago.

## 1. Introduction

In the field of the supershell LMC4 in north-east LMC a few huge arcs of young stars and clusters have long been known. The goal of this paper is to turn attention to this unique system and the lasting enigma of its origin. The brightest of these arcs was first noted by Westerlund & Mathewson (1966), who wrongly identified it with Shapley's "Constellation III"; nowadays it is known as association LH77 or "Quadrant", whereas the near-by smaller arc was called "Sextant" (Efremov & Elmegreen 1998). These arcs are parts of exact circles, they are the most perfect formations of a half of a dozen more or less similar structures found inside other galaxies (Efremov 2001a). So the LMC arcs are the best example of the ordered stellar systems, whose regular appearance is surely not due to gravitation. They are surely dynamically unstable.

The Quadrant arc is inside the H I superbubble LMC4, and Westerlund & Mathewson (1966) suggested that both features were formed in result of a Super-Supernova outburst. The whole system of the arcs in this region was first noted by Hodge (1967). All in all, this area may host five arcs of young stars and clusters (see Figures 1–4 in Efremov 2001a).

The problem of origin of this system of arcs was first considered by Efremov & Elmegreen (1998). They found that age and radius of Quadrant are about 16 Myr and 280 pc, and of Sextant are 7 Myr and 170 pc, and suggested the formation of the arcs from the gas shells, swept up by the sources of central pressure with energy about $10^{52}$ ergs.

The H I superbubbles considered to have been formed by multiple supernovae and O stars or by the impact of the high velocity clouds (Tenorio-Tagle & Bodenheimer 1988). The supershells of swept up gas should break up to star clusters located on an arc of a circle. However a lot of superbubbles are known but only a few stellar arcs (Efremov 2001a; 2002). Only in IC 2574 galaxy the H I supershell is known inside which there is an older cluster surrounded by younger ones (Stewart & Walter 2000). However, even in this sample the younger clusters are located in a random way, by no means in the regular arc structure. Also, only 6 of 44 H I superbubbles in the Ho II galaxy contain clusters suitable to be their progenitors (Rhode et al. 1999), yet even these supershells are not surrounded by younger cluster arcs as well. It is possible that the regular arcs of clusters were formed by other processes and because of this they are sush rare objects.

## 2. The LMC Arcs were Formed in a Special Way

Apart from the general difficulties with explanation of the supershell origin, the understanding of the LMC arcs faces specific problems. Efremov (1999) and Efremov & Elmegreen (1999) have noted that it is not clear why around the clusters whose very existence is doubtful, there could be such exotic structures, missing around all known clusters elsewhere. Later on, Braun et al. (2000) found that stellar population inside the LMC4, including the center of Quadrant arc, have an age close to the arc age. These authors disproved also the conclusion of Dopita et al. (1985) about the age gradient from the LMC4 center. It is not clear at all, why the giant arcs, such rare structures, are available in the LMC in an number not less than two (and probably five) and, moreover, why they are in the neighborhood of each other.

The Quadrant and Sextant arcs are parts of exact circles, instead of being the ellipses appropriate to inclination of the LMC plane to the sky plane. The latter should be the case provided they are rings in the LMC disk plane. The most plausible explanation of this is both arcs are cup-shaped, they are segments of spherical surfaces, seen in projection (see Figure 3 in Efremov 2001a).

If the sources of the pressure were the SNe or O stars and if the gas density was uniform, the arcs must have shape of the rings, not arcs or segments of spheres, because the young clusters, as well as the highest gas density, are observed in the plane of the symmetry of the galactic disks. For the case of the isotropic pressure, the model experiments (Efremov et al. 1999) demonstrated that the shape of the partial sphere may arise in two events: the center of pressure is at outskirts of an isolated dense cloud, or at some height outside the galactic plane. The latter position, however, leads to the special orientation of the arc relative to the line of intersection of the LMC and sky planes. This orientation contradicts the observed LMC arcs (Efremov et al. 1999, Figure 6), apart from the Fifth arc only.

Since the special positions of progenitors of all arcs in dense isolated clouds are improbable, they might be formed by an anisotropic pressure, sources of which may be located anywhere inside or outside the disk of the LMC. There exist however other possibilities.

## 3. High Velocity Clouds' Impact

The hypothesis about oblique infalling of group of high velocity clouds (HVCs) could probably explain a neighborhood of arcs, as well as the similar orientation of Quadrant and Sextant arcs. However it is not the case for the Third and Fifth arcs. The different ages of arcs are hardly compatible with this hypothesis, also. Only the region near the SNR N49 (see Dopita et al. 1985 and references therein) demonstrates the H I velocity perturbations, which are barely seen also in the Kim et al. (2003) data. The issue is how long the peculiarities in H I distribution and velocities might exist after a HVC impact.

Anyway, this hypothesis is the most natural explanation of absence of counterpart arcs. Their regular shape recalls the bow shock appearance of the leading edge of some galaxies (say DDO 165 and Ho II in M81 group) due to the ram pressure of the intergalactic medium. Within framework of the HVC hypothesis one could suggest that these clouds were dense enough and the stellar arcs were not formed from the swept up gas shell, but are relicts of the bow shock at the interface of the LMC gas and the impacting HVCs (Efremov 2002). Earlier Braun et al. (2000) have proposed ram pressure from the Milky Way halo gas to explain the LMC4 supershell, yet they have not suggested the mechanism of formation for the stellar arcs.

## 4. The GRB Hypothesis

As a source supplying $10^{52}$–$10^{53}$ ergs to the ISM and suitable to swept up even the largest H I supershells, the Gamma-ray bursts (GRBs) were suggested; later on, this idea was applied to explanation of the LMC stellar arcs (Efremov 1999, 2001b and references therein). Their concentration in the same region of the LMC was explained by ejection of the GRB progenitors as s result of the dynamic interaction of stars in the dense core of a star cluster. The unique massive 1 Gyr old cluster is indeed there, it is NGC 1978, famous for its highly elliptical shape. The pairs of compact objects, NS+NS or NS+BH might merge in a GRB event rather soon after ejection, at few hundred parsec distance from the cluster.

This hypothesis implies an origin of binary stars with compact components as a result of dynamic interactions in dense cores of rather old star clusters, which has been supported now by many simulations of the dynamical evolution of clusters. However, there are also data pointing to the connection of the GRBs with some type of the powerful Supernovae (Hypernovae). Their energy is sufficiently large to form the stellar arcs, but Hypernovae are suggested to have be originated from massive stars, which should not exit in the old cluster, and escape of which with rather high velocity is improbable.

The main argument for the Hypernovae origin of GRBs is that their afterglow is observed mostly in regions of star formation. However, the GRBs, which progenitors have been ejected from the same clusters, might trigger the star formation (like the LMC arcs). There is also independent evidence for the origin of the short GRBs from NS+NS or NS+BH merging (Perna & Belczynski 2002). Also, Bulik & Beclczynski (2003) found that NS+NS progenitors of GRBs must be short lived and traced star formation. In our opinion, the distribution of GRBs afterglows in a summary galaxy is close to that of the classical old globular clusters and not to the regions of star formation. From the distribution of GRBs in redshift it follows that their average age was around a few Gyrs, just compatible with the NGC 1978 age (Efremov 2001b).

However, the prevailing opinion now is that GRB energy is emitted within the narrow superrelativistic jets whose energy is about only $10^{51}$ erg. At any rate, the only known LMC Soft repeating Gamma-ray source is also in the LMC4 area, within the SNR N49. The perturbations of H I velocities just in this region might indicate the origin of this source in a Hypernovae event.

## 5. The Microquasar Hypothesis

There are other sources of relativistic jets which may be long standing and whose accumulated energy is high enough. Microquasars (MQSs) are X-ray binaries which are able to emit variable relativistic jets (Mirabel 2002); some 70% of the X-ray binaries in our Galaxy may be MQSs (Fender & Maccarone 2003). Their jets interact with the interstellar medium and can therefore initiate star formation (Mirabel 2003). Also, for some of them, high space velocities are observed, the examples being GRO J1655-40, Cyg X-1, LS 5039 and Sco X-1; for the latter the origin in a globular cluster was suggested (Mirabel & Rodriques 2003).

A puzzling concentration of X-ray binaries (XRBs) is known in northern part of the LMC4 supershell, i.e., near NGC 1978. Half of XRBs, known in the LMC, are there, within 3 square degrees of total 80 square degree of the LMC area. The natural explanation is again their formation in NGC 1978 and then ejection from this cluster (Efremov 1999). Now many examples of connection between XRBs and the massive old clusters are known and many simulations published, confirmed the possibility of XRB formation in and ejection from dense old clusters.

There are little doubts that jets from MQSs may trigger star formation, providing they act long enough. The kinetic energy of jet ouflow observed now in SS433, the most famous MQS in our Galaxy, is $10^{39}$ erg and during the life-time of about 20 000 years, the jets have put in the interstellar medium $10^{51}$ ergs (Dubner et al. 1998); if the stage of the jet outflow, similar observed now in SS433 will be prolonged for $10^5$ years, they could create arcs similar observed in the LMC. The opening angle of the arc might be determined by the precession of jets, like known in SS433, and also by its multiprecession, like suggested by Fargion for GRBs (see Efremov & Fargion 1999 and references therein).

However, the initially narrow jets may form arcs with a large opening angle in any event, owing to instabilities arising around a relativistic jet. Moiseev et al. (2000) have put forward a hypothesis that the cones of ionizations describing morphology of regions, emanating narrow lines in a number of Seyfert galaxies, are coupled to hydrodynamic instability caused by a velocity break between the interstellar medium and a jet; the axes of cones coincide with axes of jets, and the central angles at centre of cones can make 50–60 degrees. We may accept that the pressure of the hot gas in such broad cone around of a narrow jet is capable of initiating star formation in a cup-shaped shell of swept up interstellar gas, and that the similar situation may exist for jets of GRBs and/or of MQSs.

## 6. The Jet from the Milky Way Core?

At any rate, the star formation in wider regions near tips of relativistic jets is well known in galaxies Cen A and NGC 4258. The sheet of contact of a jet working surface with interstellar matter itself contains the hot spots which can be a power source for creation of the expanding shell of the swept up gas, like supergiant bubbles known around tips of very long jets in some radiogalaxies.

The very tip of the giant jet of radiogalaxy 3C 445 has the appearance of the arc 9 kpc long, compared with 450 kpc long jet, and the synchrotron emission of the arc is seen also in optical wavelength (Prieto et al. 2002). If a similar ratio is fair for the LMC case, it could mean, that the sources of jets generated the arc-shaped stellar complexes might be at dozens kiloparsecs from the LMC.

The hypothesis may be advanced that the the source of the jet was the nucleus of our Galaxy. It is not active now, yet various signs of its past activity are long suspected. Long jets at rather small angles to the plane are known in some spiral galaxies; Gopal-Krishna & Irwin (2000) believed that such jets may give origin to the expanding shells, especially in the region of the fast decrease of the galaxy gas density. It is known also that the precessing long jet of NGC 4258 galaxy was sometime ago in the plane of the galaxy (Cecil et al. 2000).

All in all, existing data do not contradict the suggestion that some 15 Myr ago the Galaxy emitted the long relativistic jet which might be directed to the LMC. It might form the LMC4 H I superbubble whereas the hot spots within it might give rise to smaller shells, relics of which are observed now as the giant stellar arcs. The ratio between the jet length and the size of the arc at its tip is 50 for the 3C 445 case, and it is about the same between the distance to the LMC and size of the LMC4 supershell. Note the exceptional nature of the LMC4 superbubble which is the largest in the LMC and is about completely empty from the gas, unlike any other supershells in the LMC. The position of LMC4 at outskirts of the LMC H I disk consists with mechanism of supershell formation by a galactic jet flaring at the gas density gradient, suggested by Gopal-Krishna & Irwin (2000).

Anyway, the angle between the MW center and the plane of the LMC, seen from the center of the LMC, is about 63 degrees (van der Marel 2001). Note also that the direction of the LMC proper motion to ENE is compatible with the relative ages and positions of the stellar arcs, the oldest Fourth arc being at the East and youngest Sextant being at the West. The slow precession motion of the suggested jet might be also a reason for this correlation of ages and positions. Then the LMC4 superbubble might be a result of the pressure from the Quadrant stars, as Efremov & Elmegreen (1998) have suggested.

The hypothesis of the origin of the arcs under action of the past jet from the Milky Way nucleus does not explain the presence in the same region of the LMC the excess density of X-ray binaries and the unusual cluster, but does connect the stellar arcs and the LMC4 superbubble origin. It may be crazy enough to be correct. More data on the past history of the Milky Way nuclear activity and surely on the LMC4 region are needed to check it.

An almost indentical version of this text with 8 figures added is published as astro-ph/0310720.

**Acknowledgments.** The support from the RFBR (project 03-02-16288) and Scientific School Foundation (project NSh 389.2003.2) is appreciated. Many thanks to the unknown referee!

## References

Braun, J., de Boer, K. S., & Altmann M. 2000, astro-ph/0006060

Bulik, T., & Belczynski, K. 2003, astro-ph/0310550

Cecil, G., Greenhill L. J., DePreee, C. G. et al. 2000, ApJ, 536, 675

Dopita, M., Mathewson D. S. & Ford, V. L. 1985, ApJ, 297, 599

Dubner, G. M., Holdaway, G. M., Goss, W. M., & Mirabel, I. F. 1998, AJ, 116, 1842

Efremov, Yu. N. 1999, Ast.Lett., 25, 74

Efremov, Yu. N. 2001a, Ast.Rep., 45, 769

Efremov, Yu. N. 2001b, in Gamma-ray bursts in the afterglow era, ed. E. Costa, F. Frontera & J. Hjorth (Springer), 243; astro-ph/0102161

Efremov, Yu. N. 2002, Ast.Rep., 46, 791

Efremov, Yu. N., & Elmegreen, B. E. 1998, MNRAS, 299, 643

Efremov, Yu. N., & Elmegreen B. G. 1999, in New Views of the Magellanic Clouds, IAU Symposium No. 190, ed. You-Hua Chu et al. (Chelsea: Sheridan Books), 422

Efremov, Yu. N., & Fargion, D. 1999, astro-ph/9912562

Efremov, Yu. N., Ehlerova S., & Palous J. 1999, A&A, 350, 457

Fender, R. & Maccarone, T. 2003, astro-ph/0310538

Gopal-Krishna & Irwin J. A. 2000, A&A, 361, 888

Hodge P. W. 1967, PASP, 79, 29

Kim S., Staveley-Smit, L., Dopita, M. A. et al. 2003, ApJS, 148, 473

Mirabel, I. F. 2002, astro-ph/0211085

Mirabel, I. F. 2003, astro-ph/0302195

Mirabel, I. F. & Rodrigues, I. 2003, astro-ph/0301580

Moiseev, A. V., Afanasiev, V. I, Dodonov, S. N., Mustevoi, V. V., & Khrapov, S. S. 2000, astro-ph/0006323

Perna R., & Beclczynski K. 2002, ApJ, 570, 252

Prieto, M. A., Brunetti, G., & Mack K.-H. 2002, astro-ph/0210304

Rhode, K., Salzer, J. J., Westphal, D. J., & Radice, L. A. 1999, AJ, 118, 323

Stewart, S. G., & Walter, F. 2000, AJ, 120, 1794

Tenorio-Tagle G., & Bodenheimer P. 1988, ARA&A, 26, 145

van der Marel, R. P. 2001, astro-ph/015340

Westerlund, B. E., & Mathewson, D. S. 1966, MNRAS, 131, 371

*Order and Chaos in Stellar and Planetary Systems*
*ASP Conference Series, Vol. 316, 2004*
*G. Byrd, K. Kholshevnikov, A. Mylläri, I. Nikiforov, and V. Orlov, eds.*

# Properties of Globular Cluster Candidates in the Local Volume Dwarf Galaxies from the HST

M. E. Sharina

*Special Astrophysical Observatory, Russian Academy of Sciences, N. Arkhyz, KChR 369167, Russia*

**Abstract.** Forty five nearby dwarf galaxies were searched for globular cluster candidates (GCCs) by using Hubble Space Telescope WFPC2 imaging in $V$ and $I$. The sample consists of 8 dwarf spheroidal (dSph), 28 irregular (dIrr) and 7 "transition" type (dIrr/dSph) galaxies with angular dimensions less then $2'$ situated at the distance 1.9–6.3 Mpc in the field and in the nearby groups: M81, Centaurus A, Sculptor, Canes Venatici I cloud. Additionally, I found GCCs in KK084, dSph companion of NGC3115, situated at the distance of about 9.7 Mpc (Tonry et al. 2001), and Holmberg IX, the tidal dwarf companion of M81 (Boyce et al. 2001). dSph and dIrr/dSph galaxies comprise the weakest part of our sample with mean surface brightness $\mu_V > 24\overset{\mathrm{m}}{.}6/\square''$ and $\mu_V > 24\overset{\mathrm{m}}{.}0/\square''$ respectively. Four dSphs—KK211, KK221, Scu22, and KK084—were found to host 16 globular cluster candidates with integral colors $0.9 < V - I < 1.4$. There are no GCCs in dIrr/dSph galaxies. Eleven dIrr galaxies have low surface brightness $\mu_V > 24\overset{\mathrm{m}}{.}0/\square''$. Only two of them, DDO53 and KK200, contain GCCs. Eighteen dIrr galaxies have mean surface brightnesses falling in the range of $22\overset{\mathrm{m}}{.}4/\square''$–$24\overset{\mathrm{m}}{.}0/\square''$. Nine of them contain 67 GCCs including 31 GCCs of UGC3755 and 22 GCCs of Holmberg IX. The majority of GCCs have absolute visual magnitudes $-9.0 \div -5.1$, and linear half-light radii 3–11 pc, which are within values typical for Galactic globular clusters. Almost all GCCs (90%) show pronounced concentration toward the galaxy centers with linear projected separation $0.6 \pm 0.3$ kpc. The histogram of color distribution for all GCCs in dIrrs shows obvious bimodality with the peaks near $V-I = 0.6$ and $V-I = 0.95$.

## 1. Introduction

Before discussing the properties of globular cluster candidates in the Local Volume dwarf galaxies, I briefly summarize the main characteristics for the objects of study. Globular clusters are centrally concentrated, spherical systems with masses of $10^4 \leq M\ [M_\odot] \leq 10^{6.6}$, luminosities from $-10\overset{\mathrm{m}}{.}55$ (Mayall II in M31) to $0\overset{\mathrm{m}}{.}2$ (AM-4), and tidal radii ranging from $\sim$10 to $\sim$100 pc (Grebel 2002). They are bound, long-lived objects, whose lifetimes may extend beyond a Hubble time. In 12 of the 36 Local Group (LG) galaxies globular clusters have been detected. It is interesting to search for globular clusters in dwarf galaxies of different morphology situated beyond the LG and to study properties of GCCs.

## 2. Data and Results

The images used were taken from the HST archives. The objects were selected by applying FIND task of the DAOPHOT-II (Stetson 1987) package implemented in

MIDAS. The detection threshold was set at 4 sigma above the background. The minimum full width at half maximum input parameter (FWHM) used was $\sim 0.''2$. (The stellar FWHM is $\sim 0.''15$.) The measured WFPC2 magnitudes F606w and F814w were converted into Johnson–Cousins $V$ and $I$ magnitudes using equations derived by Holtzman et al. (1995). Finally, the magnitudes of GCCs were corrected for Galactic extinction, and absolute magnitudes were computed by applying the accurate distance moduli. Half-light radii and linear projected separations of GCCs from the centers of parent galaxies were reduced to linear measure in kpc to make obtained parameters to be compared. After the procedure of primary selection of the objects and photometry I made visual inspection of GCCs on CCD images.

The properties of globular cluster candidates in the Local Volume dwarf galaxies are summarized in Figure 1. Four dSphs were found to host 16 globular cluster candidates with integral colors $0.9 < V - I < 1.4$ (Figure 1a). The $V - I$ color distribution for GCCs in all dIrrs reveals two peaks at $V - I = 0.6$ and 0.95 (see Figure 1a). GCC luminosity functions of dSph galaxy KK084, dwarf irregular galaxies UGC 3755 and Holmberg IX are shown in Figures 1b, 1c, and 1d, correspondingly. In the case of Holmberg IX, GCC absolute magnitude distribution has a power law and not a Gaussian form. Half-light radii of the most dSphs, including KK084, and of dIrrs correlate with their linear projected separations from the centers of parent galaxies and central surface brightness (Figures 1e and 1f). The presence of such relationship follows from modeling the processes of dissipation and dynamical friction effects on the number of star clusters in a galaxy (Surdin & Arkhipova 1998). It is interesting, that similar relationship between half-light radii, central surface brightness and galactocentric distances is not observed for GCCs associated with the isolated galaxy, UGC 3755, and the young galaxy, Holmberg IX (Figures 1g and 1h). This suggests that these galaxies are less affected by dynamical evolution.

**Acknowledgments.** I wish to thank G. Byrd, K. Kholshevnikov, A. Myllari, V. Orlov for organizing a such interesting conference.

## References

Boyce, P. J., Minchin, R. F., Kilborn, V. A. et al. 2001, ApJ, 560, L127

Grebel, E. K. 2002, in Extragalactic Star Clusters, IAU Symp. 207, ed. D. Geisler, E. K. Grebel & D. Minniti (San Francisco: Astronomical Society of the Pacific), 132

Holtzman, J. A., Burrows, C. J., Casertano, S. et al. 1995, PASP, 107, 1065

Stetson, P. B. 1987, PASP, 99, 191

Surdin, V. G., & Arkhipova, N. A. 1998, Ast.Lett., 24, 343

Tonry, J. L., Dressler, A., Blakeslee, J. P. et al. 2001, ApJ, 546, 681

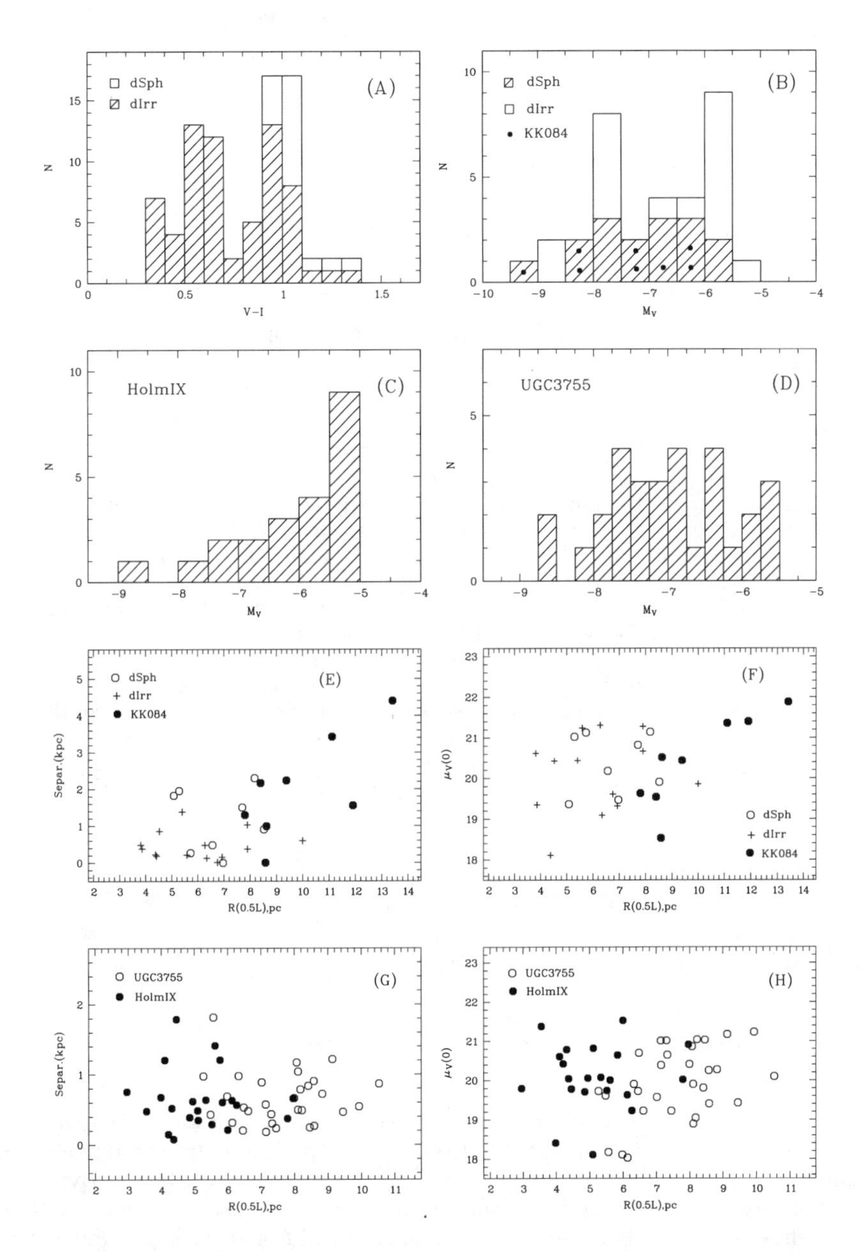


Figure 1. (a) Color distribution for all GCCs in the studied galaxies. (b) GCC luminosity function of dwarf spheroidal galaxy KK084, compared with GCC population of other sample dSphs and dIrrs (except UGC3755 and Holmberg IX). (c), (d) GCC luminosity functions of dwarf irregular galaxies Holmberg IX and UGC3755. (e), (g) GCC half-light radii as a function of linear projected separation of GCCs from the center of corresponding galaxy. (f), (h) GCC half-light radii as a function of GCC central surface brightness.

*Order and Chaos in Stellar and Planetary Systems*
*ASP Conference Series, Vol. 316, 2004*
*G. Byrd, K. Kholshevnikov, A. Mylläri, I. Nikiforov, and V. Orlov, eds.*

# Does the Galaxy NGC4622 Have a Pair of Leading Arms?

Gene Byrd

*Univ. of Alabama, Tuscaloosa, AL (USA)*

Tarsh Freeman

*Bevill State, Brewer Campus, Fayette AL (USA)*

Ronald Buta

*Univ. of Alabama, Tuscaloosa, AL (USA)*

**Abstract.** Combined with ground-based Doppler shift observations, our Hubble Space Telescope observations of gas cloud silhouettes and bulge color asymmetries in the two-way arm galaxy, NGC4622, have indicated that it has a pair of arms which wind outward in the same clockwise (CW) sense on the sky as the disk orbital motion. This is against current conceptions of arm sense in spiral galaxies. We would like to verify these conclusions. The orbital motion of disk material can be ascertained by using the time delay between favorable points for gas cloud aggregation (where a peak of $I$ emission occurs), the resulting star formation, and then the subsequent ionization and dispersal of the gas as the star complex lights up in the $B$ band. We have observationally verified this $IB$ time delay via the resonance ring in NGC3081. We use our HST color observations to verify the CW orbit angular sense in NGC4622, the same as the direction its pair of leading spiral arms wind outward.

## 1. Introduction

Do spiral arms lead or trail outward relative to orbital motion? This was the subject of controversy until G. de Vaucouleurs (1958) found all to trail in a small sample of highly inclined spirals. Trailing is accepted as the rule today. Byrd and collaborators (1989) pointed out NGC4622's two-way arms. Our Figure 1 HST photo shows the arms in the stellar disk and young blue associations. Going from outside inward, there is a pair of arms which wind out clockwise (CW). Then there is a lop-sided ring. Finally, there is an inner arm which winds outward CCW. NGC4622 must have one (or two) leading arm(s)! Byrd, Freeman, & Howard (1993) found that by assuming a flat rotation curve, two trailing outer arms and a single inner leading arm could be produced a plunging, disk plane passage by a small perturber. For this pattern to exist in NGC4622 its disk must turn CCW.

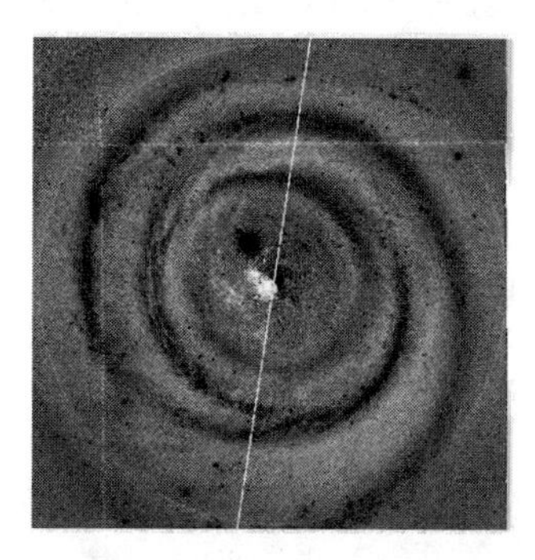

Figure 1. HST $B - I$ color index image of NGC4622. Note the white line of nodes (NNE, top, and SSW, bottom). Blue color is dark and red white. Note dark stellar associations and light colored dust silhouettes.

## 2. Our First HST Determination of Which Arms Lead or Trail

We needed to observationally find which sense (CW, CCW) NGC 4622's disk turns on the sky. We used ground-based radial velocity observations and our HST image determine this. First, we needed the direction of the line of nodes, the direction of the intersection of the disk with the imaginary plane of the sky, 90 degrees to the line of sight, through the galaxy center. We used the elongation direction of the outer isophotes from a circular shape to determine this line. The line of nodes is shown as a white line in Figure 1. It extends NNE (top) toward SSW (bottom).

We also used a ground-based H$\alpha$ emission velocity field for NGC4622. This also double checked the line of nodes in that the maximum speed away is observed away from the observer is toward the NNE side of the disk and the maximum toward on the SSW side, as indicated by the isophotes. This is sketched in Figure 2.

We used two methods on the HST images to determine disk orientation. First we used dust silhouettes seen against the bulge starlight near the center of our HST image of NGC4622 to determine the nearer edge of the disk. These can be seen in Figure 1 near the center and to the left (east) of the line of nodes as white filaments on the disk. Figure 2 shows the logic of our determination. A "coin" in Figure 2 shows the disk orientation with the silhouettes indicated as dark areas on the nearer edge of the coin.

Determining the disk turning sense on the sky determines the sense of the arms. Via silhouettes the east edge is nearer. The NNE line of nodes recedes. The disk thus orbits CW on the sky. Thus the inner single CCW arm trails. The outer CW pair thus must lead or wind outward into the direction the disk material orbits!

Near to the center ($<12''$) we also looked at the color of the stellar bulge of NGC4622 ninety degrees to the line of nodes. This also indicates the the left (east) side is nearer (p. 650 in Buta, Byrd, & Freeman 2003).

Our conclusion is against current conceptions of arm sense for pairs of arms in spiral galaxies. Our conclusion from HST images that the two outer NGC4622 arms lead was simply not acceptable to some. One well-known astronomer said, "You're the backward astronomers who found the backward galaxy." Aside from these general objections, because of the low 26° inclination of NGC4622, we want to substantiate this radical conclusion in every way possible. We do this here via the *IB* age time sequence which we established using the galaxy NGC3081 in Freeman, Byrd, & Buta (2004), in this conference proceedings.

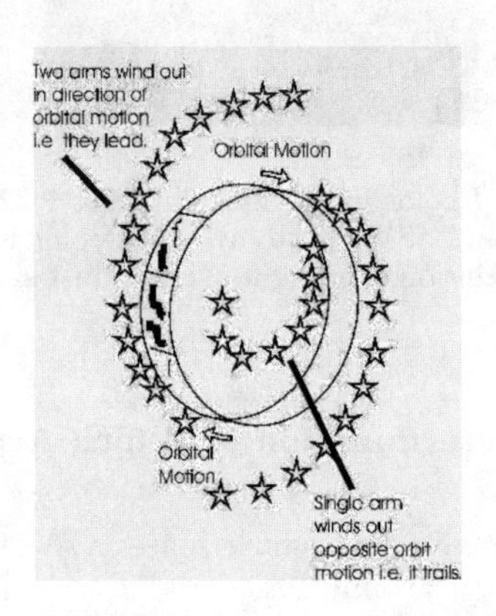


Figure 2. Schematic of disk orientation and rotation sense on the sky. Dust lanes are dark. Stars are stellar arms and associations. "Coin" shape shows near and far edges with north edge having red doppler shift, away, and south blue, toward, as indicated.

## 3. The *IB* Time Delay Method for One-Fold Peturbations

Here is a summary of this method. Our discussion and notation are similar to those for periodic orbits under a bar ($m = 2$) perturbation in the text, Binney & Tremaine (1987). Freeman, Byrd, & Buta (2004) discuss bar periodic orbits very near CR which are qualitatively similar to the following equations except here $m = 1$. $\Omega_0$ is the undisturbed circular orbit angular rate and $\Omega_p$ is the constant angular rate at which a one-sided ($m = 1$) perturbing potential turns. The circular orbital angular rate can vary with radius while the perturbation potential angular rate is taken to be constant with radius. Disk objects in undisturbed circular motion have an angular coordinate time derivative relative to the bar of

$$\frac{d\phi}{dt} = (\Omega_0 - \Omega_p). \tag{1}$$

Disk objects are perturbed from a circular orbit radius, $r_0$, radially and azimuthally by

$$r_1 = r - r_0 \quad \text{and} \quad \phi_1 = \phi - \phi_0, \tag{2}$$

where $\phi_0$ is a reference angle for undisturbed motion and $\phi_1$ is the angular perturbation from undisturbed circular motion. Denote $\Phi$ as the net original plus perturbation potential and $\Phi_0$ as the circularly symmetric, unperturbed potential evaluated at $r_0$. $\Phi^A_{p,m}$ is the amplitude (<0) of the azimuthal component of the disturbing potential where $m$ is the multiplicity of the disturbing potential. A "bar" corresponds to a two-fold ($m = 2$) perturbation. A one-sided perturbation corresponds to $m = 1$. The pattern rate is presumed to be the same for all components.

The epicyclic frequency is defined as usual

$$\kappa = \left( 4\,\Omega_0^2 + r_0 \frac{d\Omega_0{}^2}{dr} \right)^{1/2}_{r_0} = \sqrt{2}\Omega_0, \tag{3}$$

where the right hand side is for a flat rotation curve of constant circular speed, $V_0$.

We define co-rotation as the radius, $r_{CR}$, where $\Omega_p = \Omega_0$. At resonances with the perturbing potential, $r_1$ or $\phi_1$ can become large. In either of these regions gas clouds would collide strongly and periodic cloud orbits would not occur. The colliding clouds and any stars formed as a result of the collision would be thrown into non-circulating non-periodic orbits.

In general, we are interested in regions somewhat away from $CR$ where perturbations, surface density enhancements, cloud collisions, and star formation occur gently so that the clouds and stars are still in lop-sided periodic orbits. The radial perturbation is for small damping

$$r_1 \approx -\left[ 2\Omega_0 \frac{\Phi^A_{p,m}}{r_0(\Omega_0 - \Omega_p)\,\Delta} \right] \cos\left[m(\Omega_0 - \Omega_p)t\right], \tag{4}$$

where the amplitude

$$C_{r,m} = -\left[\ldots\right].$$

Here

$$\Delta = |\kappa^2 - m^2(\Omega_0 - \Omega_p)^2|, \tag{5}$$

where $\Delta$ can be zero at $m = 2$ Lindblad resonances. In our region of interest near co-rotation, we have a non-zero $\Delta \approx \kappa^2$.

Taking the orbital angular rate to smoothly decline with radius, $(\Omega_0 - \Omega_p) > 0$ inside CR and the periodic orbits are elongated to one side of the center. Outside CR the orbits are elongated to one side but with the long dimension 180° opposite the orbits inside CR. The same is true for the angular velocity perturbations from circular motion.

By way of an example of the above qualitative logic, Figure 3 shows an $n$-body simulation demonstration of this 180° difference. We apply a turning one fold sinusoidal perturbation to a flat rotation curve disk. This mimics the region of NGC4622 where the single arm and ring occur where the rotation curve is approximately flat. The perturbation is only a few percent of the disk plane potential and the disk itself is halo dominated with very little self gravity. Note the above described 180° asymmetry of the maximum radial perturbations.

## 4. Application of the *IB* Age Sequence Method to NGC4622

We use the *IB* sequence to determine how the disk of NGC4622 turns. We identify the gap at 22 arc sec between the inner single arm and the lop-sided

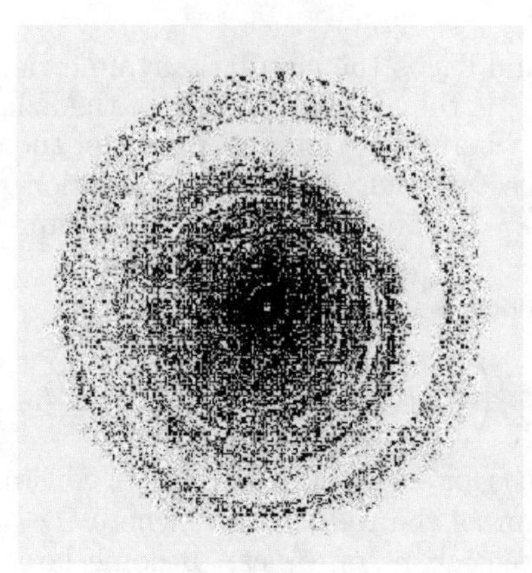

Figure 3. An $n$-body simulation of the flat rotation curve portion of NGC4622's disk perturbed by a turning sinusoidal one-fold perturbation. Note the similarities to equation (4) inside and outside CR. There is little disk self-gravity. Empty region is CR radius.

ring as the co-rotational radius for a one sided perturbation. Assume as from the previous results that the disk and one sided pattern turn CW. The gas clouds interior to CR orbit faster than CR, those exterior more slowly. Clouds crowd together in a manner to promote star formation in $I$ band maxima $\sim 180°$ apart on the single arm and larger lop-sided ring. After the stars in associations reach the main sequence and disperse the gas surrounding them, the associations "light up" in the blue. There should thus be an $IB$ color age sequence. If our previous conclusions are correct for the way the disk on NGC4622 turns on the sky, this $IB$ sequence will be CW for the inner arm and CCW for the outer lop-sided ring.

We verify this prediction observationally in Figure 4 below which is a Fourier analysis of our HST images in the $I$, $V$, and $B$ color bands with $I$ dotted, $V$ dashed, and $B$ solid. The CCW position angle from north of the $m = 1$ (one fold) color intensity maximum is plotted as a function of radius. The inner single arm extends 17 to 21 arc sec and the lop-sided ring from 22 to 28 arc sec. Note that the $IB$ sequence of the inner single arm ($<22''$) is downward (CW) as expected and the $IB$ color sequence around the lop-sided ring ($>22''$) is upward (CCW) as expected. The disk material thus orbits CW on the sky and the outer pair of arms' CW unwinding is a leading pattern!

## 5. Conclusions

We had verified the leading nature of the outer pair of arms in NGC4622 by two methods: dust silhouettes, and color asymmetry both across the the line of nodes of the disk. Here we add a third determination using the $IB$ color delay on the interior and also on the exterior of the co-rotation radius between the inner single arm and the lop-sided ring. We thus find further substantiation of exceptional nature of this galaxy.

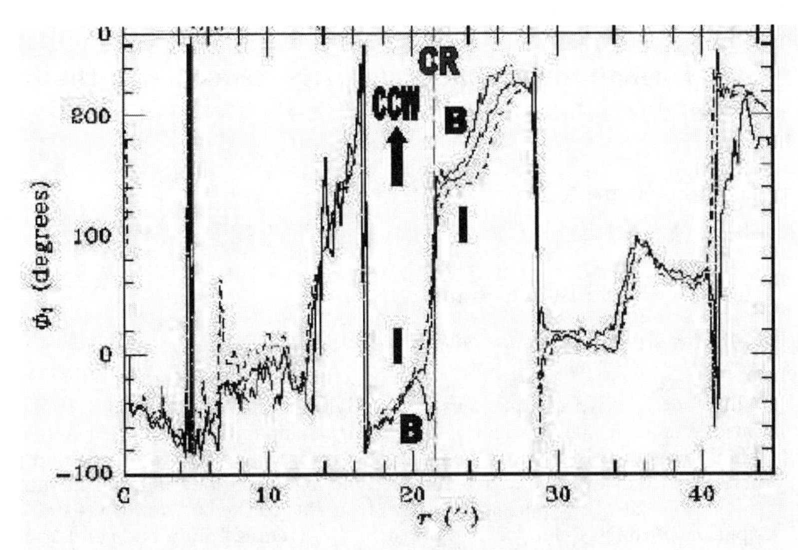


Figure 4. A Fourier analysis of our HST images in the $I$, $V$, and $B$ color bands with $I$ dotted, $V$ dashed, and $B$ solid in position angle versus radius.

Buta, Byrd, & Freeman (2003) discuss other examples of "two-way" galaxies. The somewhat weak stucture in the "Blackeye Galaxy" NGC4826 has also been found to have a single inner trailing arm and winding oppositely from an outer leading pair. R. Buta and graduate student R. Grouchy in recent observations have identified another two-way galaxy, ESO297-27. It has a single inner arm and two oppositely winding outer arms. NGC 3124 has oppositely winding multi-armed spiral structure in its inner and outer parts.

**Acknowledgments.** This work was supported by grants NASA/STScI GO 8707 and NSF AST-0206177 (TF and GB).

## References

Binney, J., & Tremaine, S. 1987, Galactic Dynamics (Princeton, New Jersey: Princeton Univ. Press)

Buta, R., Byrd, G., & Freeman, T. 2003, AJ, 125, 634

Byrd, G. G., Freeman, T., & Howard, S. 1993, AJ, 105, 477

Byrd, G. G., Thomasson, M., Donner, K. J., Sundelius, B., Huang, T.-Y., & Valtonen, M. J. 1989, CeMDA, 45, 31

de Vaucouleurs, G. 1958, ApJ, 127, 487

Freeman, T., Byrd, G., & Buta, R. 2004, Order and Chaos in Dynamical Systems, in press

*Order and Chaos in Stellar and Planetary Systems*
*ASP Conference Series, Vol. 316, 2004*
*G. Byrd, K. Kholshevnikov, A. Mylläri, I. Nikiforov, and V. Orlov, eds.*

# Star Formation and the Color–Age Sequence in the Inner Resonance Ring of the Galaxy NGC3081

Tarsh Freeman

*Bevill State, Brewer Campus, Fayette AL (USA)*

Gene Byrd, and Ronald Buta

*Univ. of Alabama, Tuscaloosa, AL (USA)*

**Abstract.** We complement Buta, Byrd, & Freeman's (2004) HST observations of NGC3081's inner $r$ ring using an analytical approach and n-body simulations of this galaxy. We find: A long-lived ($5 \times 10^9$ yr) gas cloud ring of periodic orbits forms under a rotating bar perturbation. The misalignment of the major axis of the nuclear ring ($nr$) from 90° to the bar can be created by dissipation in the gas clouds. The perturbation creates pointy ends and transverse crowding in the $r$ ring at the major axis, both favorable to stellar association formation. Ring elongation increases with perturbation strength. The azimuthal surface mass density variation and the fractional long wavelength light variation give a stellar disk surface density of roughly 50 $M_{\odot}/\mathrm{pc}^2$ at the $r$ ring, 1/5 that of a 100% disk sufficient to explain the rotation curve, i.e., the galaxy is halo dominated. NGC3081's disk and the bar pattern turn counter clockwise (CCW) on the sky with a faster orbital angular rate than the pattern inside the ring edge. Thus the $r$ ring's CCW displacement between the $m = 2$ Fourier $I$ and $B$ peaks along the $r$ ring from the bar inside co-rotation (CR) must result from the time delay before "lighting up" in the blue by stellar associations formed near the bar. Further observations are needed to verify the NGC3081 "formation to light up" delay. The outer edge of the ring at the bar orbits with the pattern speed, supported by the superposition of the $I$ and $B$ peaks and the wide range of association ages there. The ring edge is smaller than CR. Near the theoretically expected halo value, our simulations form gas cloud "associations" near the ends of the bar as observed. Grant support NASA/STScI GO 8707 and NSF AST-0206177 (TF, GB).

Figures for the Abstract follow.

## References

Buta, R., Byrd, G., & Freeman, T. 2004, AJ (in press)

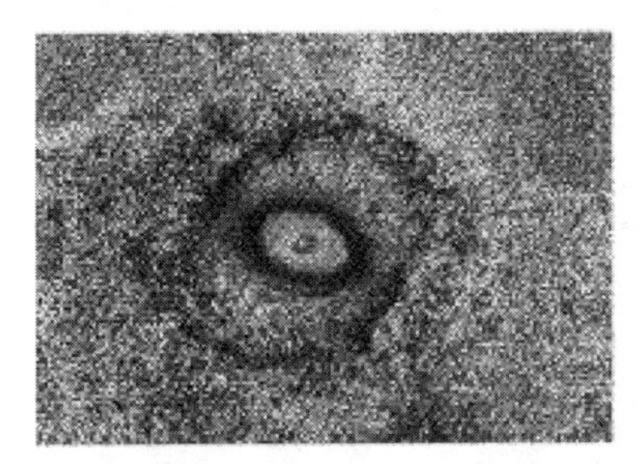

Figure 1. $B - I$ color index map of NGC3081. Blue is dark gray and red lighter. Note small nuclear ($nr$, larger inner $r$, and largest outer $R_1$, $R_2$ rings). Disk turns CCW on sky. Outer long dimension of $r$ ring is about 38.5″ (7 kpc). Galaxy is tilted about 34° with a 97° position angle line of nodes.

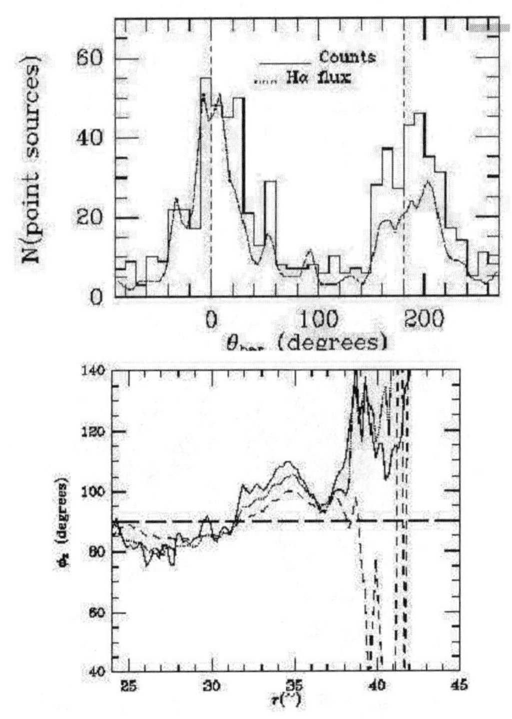


Figure 2. Top: HST association number histogram vs angle around the ring in the galaxy plane (Buta, Byrd, & Freeman 2004). Note CCW asymmetry of peaks of number and Hα young association emission (dotted curve) at bar ends (dashed). Bottom: Fourier analysis around/along the ring. CCW angles from 90° to bar of $m = 2$ $IVB$ color peaks over the 34″ to 37″ ring are plotted vs. radius. Bar is 90° line. $B$ solid, $V$ dotted, and $I$ dashed. Fourier amplitude shows smooth change over the 34″ to 37″ radius range with CCW separation of $I$ and $B$.

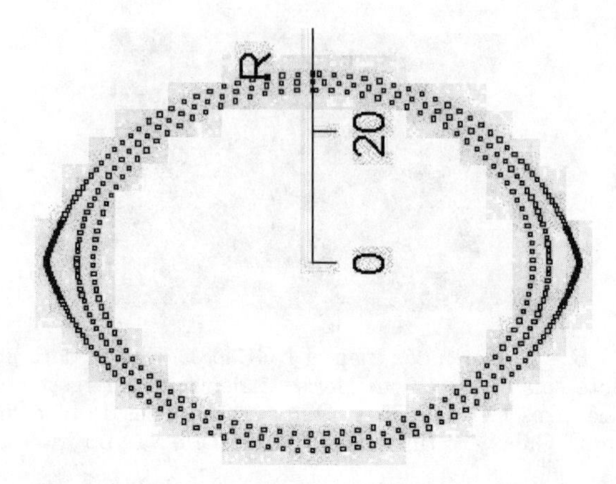


Figure 3. Analytic model of $r$ ring. CR = 54″. Flat rotation curve bar potential perturbation is 0.03. Dots are at equal intervals of time. Note transverse crowding at major axis favoring formation of associations there. Extreme crowding and slow motion relative to pattern speed occurs at the pointy ends so that associations form and grow older there. At smaller radii the extremely young assocations should form near the major axis then move to progressively larger CCW angles as they age.

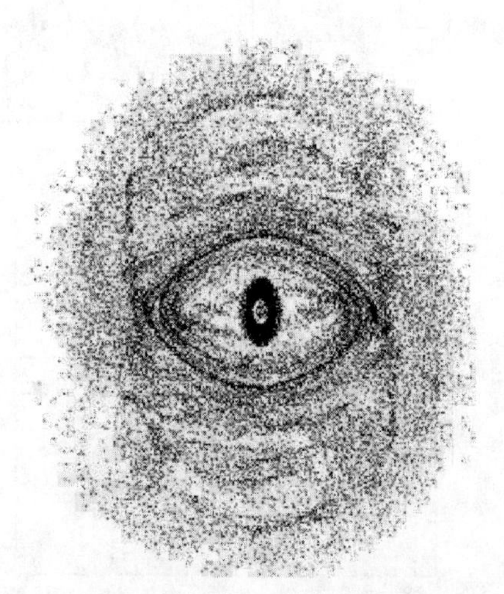

Figure 4. $N$-body gas cloud disk simulation about $4 \times 10^9$ yr after onset of bar perturbation (Byrd, Freeman, & Buta 2004). Comparing to Figure 1, note $nr$, $r$ and $R_1$ rings. The disk gas cloud surface density is 1/15 that needed to create the flat rotation curve (halo dominated). Clumps (associations) form near the analytically predicted pointy ends of the rings as predicted. At lower gas cloud surface densities, the clumps do not form (as expected for the program "softening").

*Order and Chaos in Stellar and Planetary Systems*
*ASP Conference Series, Vol. 316, 2004*
*G. Byrd, K. Kholshevnikov, A. Mylläri, I. Nikiforov, and V. Orlov, eds.*

# The Ordered Matter Outflow from AGNs

E. Y. Vilkoviskij and S. N. Yefimov

*Fesenkov Astrophysical Institute, 480020 Almaty, Kazakhstan*

**Abstract.** Remarkable ordered structures, the so-called "line-locking" effect are seen in the UV spectra of some active galactic nuclei. The theory and a numerical model of the phenomena are briefly discussed.

## 1. Introduction

The evidences of the outflowing streams in the inner regions of active galactic nuclei (AGNs) are observed as absorption details in the UV and X-ray bands. Sometimes they show regular structures (e.g., Q1303+308, see Foltz et al. 1987). The theoretical interpretation of this phenomenon is presented.

## 2. The Model

Our theoretical model of the outflows is based on the unification scheme of AGNs shown in Figure 1. We suppose the two-phase outflow (cold clouds in the hot gas) is at the internal surface of the obscuring torus.

### 2.1. Numerical Calculations

Our theory includes the equation of dynamics of the hot gas (taking into account both the gravitation forces of the central black hole and the compact stellar cluster); the equations of motion of cold clouds; the equations of the radiation transfer in the clouded media; and the ionization balance equations (Vilkoviskij & Karpova 1996, Vilkoviskij et al. 1999). As the radiation pressure forces (both in spectral lines and continuum) are included in the equations of motion, the task is non-linear, and we solve it numerically. An example of the solution is presented in Figure 2.

One can see that the line-locking is provided by the ladder-like structure of the velocity-distance dependence, due to the non-linear effects of the radiation pressure and radiation transfer connection. The same physics can explain the stable narrow details in the UV-spectra of the Seyfert galaxies.

The conditions which promote the line-locking effect are: the acceleration due to the action of the line-scattered radiation must exceed the continuum absorption acceleration, and the mass flow in the clouds should be limited. The total calculated spectrum from optical to X-ray bands is presented in Figure 3.

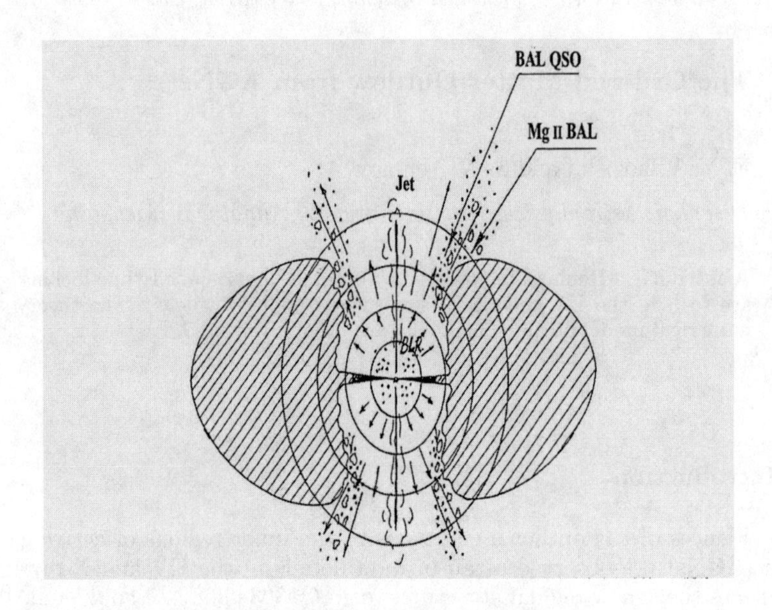


Figure 1. The unification scheme of AGN outflows.

## References

Foltz, C. D., Weymann, R. Y., Morris, S. L., & Turnshek, D. A. 1987, ApJ, 317, 450
Vilkoviskij, E. Y., & Karpova, O. G. 1996, Pis'ma AZh, 22, 148
Vilkoviskij, E. Y., Efimov, S. N., Karpova, O. G., & Pavlova, L. 1999, MNRAS, 309, 80

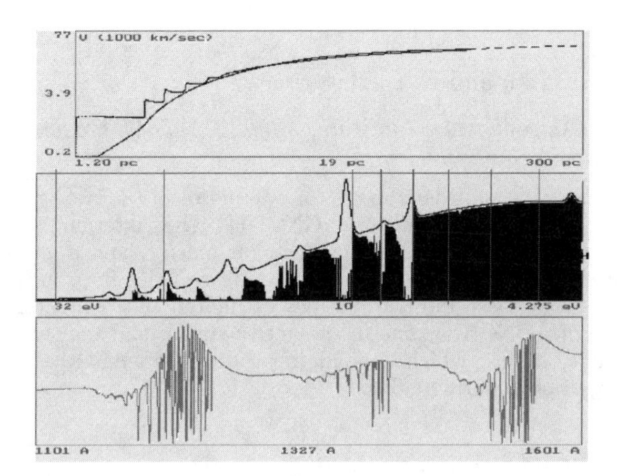


Figure 2. The result of model calculation. The 3 panels in the picture show (up to down): 1) the velocity distribution of the cold clouds ($\log R$–$\log V$), 2) the resulting spectrum in 300–3000 Å band, and 3) a smaller part of the spectrum with a better resolution.

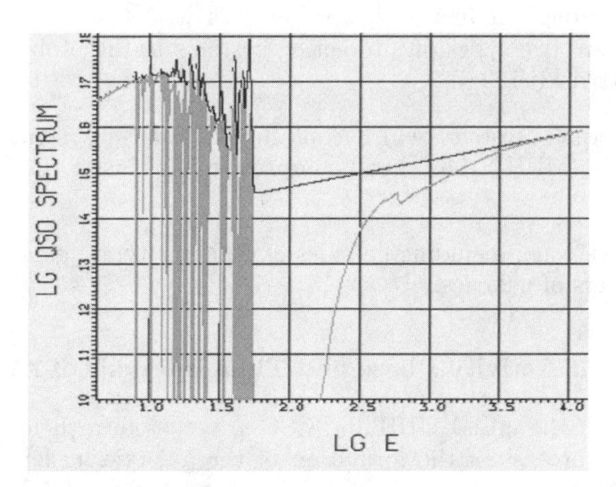


Figure 3. The total absorption spectrum (in gray, $E$ in eV).

*Order and Chaos in Stellar and Planetary Systems*
*ASP Conference Series, Vol. 316, 2004*
*G. Byrd, K. Kholshevnikov, A. Mylläri, I. Nikiforov, and V. Orlov, eds.*

# Interacting Subsystems in the Evolution of AGNs

E. Y. Vilkoviskij and E. B. Baturina

*Fesenkov Astrophysical Institute, 480020 Almaty, Kazakhstan*

**Abstract.** We consider the role of interactios of the two types of subsystems related to active galactic nuclei (AGNs): (1) The "internal" subsystems, i.e., a supermassive black hole, a compact stellar cluster, and a gas subsystem. (2) The "external" subsystems, i.e., interacting galaxies in groups and clusters of galaxies. From these interactions, we estimated two characteristic times: the duty time of AGNs (the time in the active state), and the time in between the high activity states, which depends on the redshift and the environment (the clustering of neighboring galaxies).

## 1. Introduction

One of the main tasks of the active galactic nucleus (AGN) theory is working out of the theoretical models of the AGN internal structure and evolution. After discovering of the unification scheme of the AGNs, the task of creation of the general physical model of AGN became real. The complication of the task in comparison to, say, the models of stellar structure and evolution, consist, in particular, the complex structure of AGNs. Being in the centers of galactic nuclei, the AGNs have compound structures including several physically different, but strongly interacting subsystems. On the other hand, the interaction of the galaxies itself strongly influence the evolution of AGNs.

So, there are two types of interacting systems in the problem of evolution of AGNs (Vilkoviskij 1998):

i) the "internal" system, which contains 3 main subsystems, the massive black hole (MBH), the central compact stellar cluster (CSC) and the gas subsystem;

ii) the "outer" one, including galaxies of different types, galactic groups and the clusters of galaxies.

## 2. The Bright Activity Phase and Time Intervals of AGNs

We suppose that the initial MBH and CSC are created in the centers of young galaxies in the process of the formation of the galactic nuclei. The "bright" phase of the AGNs (the phase of the Eddingtone accretion rate) is short, about

$$T_1 \sim 5 \times 10^7 \, \varepsilon_{0.1} \text{ years,} \tag{1}$$

where $\varepsilon_{0.1}$ is the energy output per unit mass in $0.1c^2$ units.

The "next generation" AGNs are produced by merging of the galaxy containing MBH, with the neighbor galaxies. We suppose that the merging produced the new CSC, which provides the mass source for the next activity cycle (Vilkoviskij & Czerny 2002). Every activity cycle includes the 3 phases: the "gas cocoon" (observed as the supraluminal IRAS sources), the bright AGN (the quasar surrounded with the obscuring torus) and the fading AGN with slowly diminished brightness (the Seyfert galaxies and radio-sources).

The intensity of interaction and merging of galaxies in galactic groups and clusters presents the important factor of the galactic evolution and consequently provides the second characteristic time, the time between the "duty cycles":

$$T_2(t) \approx 1.4 \times 10^{10}\, F(M, f, t)(N_3 \Sigma_2 V_2)^{-1} \text{ years}, \tag{2}$$

where $M$ is the mass of the galaxy, harbouring MBH, $f$ is a function, depending on kinematics of the collision, $t$ is the cosmological time, $N_3$ is the number of the neighboring galaxies in the units $10^3$, $\Sigma_2$ is the collision cross-section in 100 kpc$^2$ and $V$ is the velocity in 100 km/s. Averaging $T_2(t)$ for different groups and clusters, and taking into account also the "internal" galactic evolution due to stellar evolution, one can obtain both galactic and AGN luminosity function evolution. As a first step we intend to investigate the evolution of the brightest quasars, using the $N$-body simulations for the evolution of the AGNs in the central galaxies of the rich clusters, which are most strongly influenced by merging (Kontorovich & Krivitskii 1995).

## References

Kontorovich, V. M., & Krivitskii, D. S. 1995, Pis'ma AZh, 22, 643

Vilkoviskij, E. Y. 1998, Izvestija RAN Ser.phys., 62, 1721

Vilkoviskij, E. Y., & Czerny, B. 2002, A&A, 387, 804

*Order and Chaos in Stellar and Planetary Systems*
*ASP Conference Series, Vol. 316, 2004*
*G. Byrd, K. Kholshevnikov, A. Mylläri, I. Nikiforov, and V. Orlov, eds.*

# On the Dynamics of Clusters of Galaxies in the Expanding Universe

T. S. Kozhanov

*Kazakh National Agrarian University, Republic of Kazakhstan, 480100, Almaty, ave. Abai 8; e-mail: koblandy@mail.ru*

**Abstract.** We investigated dynamical evolution of gravitationally bound system on the background of smoothed and isotropically expanding unified two-component cosmological substratum. The equation of state of the system is similar to that of radiation (neutrinos) with pressure $p = \rho_r c^2/3$ and dark matter with zero pressure. The equation describing the behavior of bound systems has been obtained. It is shown that the possible expansion of clusters or superclusters of galaxies depends on the relation between the density of dark matter and the density of radiation. It was also revealed that the expansion of these systems depends also on the ratio of the density of dark matter at zero pressure and the density of central condensation. The results can be applied to the models of early evolution of inhomogeneous structure of the Universe.

## 1. Introduction

One of the most important problems of modern gravitational physics is investigation of behavior of ensemble of points and evolution of extended objects in the universe. The influence of gravitational field of continuously distributed and isotropically expanding unified two-component cosmological substratum in the form of dark matter without pressure and radiation is taken into account. The candidates for such type of matter are neutrino sea, gravitational waves, and high temperature gas.

The flat rotation curves of galaxies and large virial masses of galaxy clusters show that a dominant component of masses in the Universe is dark matter with cosmological density parameter $\Omega_{\text{dark}} \geq 0.1$ (Rubin 1983; Bosma 1981a,b; Faber & Gallaghar 1979). The first candidate to be such dark matter is intergalactic ionized gas (Fild 1973). According to Fild, the rich clusters (for example, Coma cluster) are X-ray sources with temperature $10^5$–$10^7$K and for such gas $\Omega \leq 0.05$. Observations do not exclude the presence of significant amount of gas with specified temperature in groups of so-called de Vaucouleurs galaxies. Moreover, there are numerous arguments in favor of existence of various forms of dark matter such as huge crown of galaxies, consisting of faint stars, dwarfs or dead galaxies and "black holes" in the space between galaxies, etc. (Peebles 1971). Probably, taking into account of these exotic objects of the most different nature will essentially change evaluation of the average matter density in the Universe.

Thus, dark matter shows itself only in gravitational interaction and plays a role in formation of large-scale structure of the Universe.

The main purpose of development of dynamical theory of already existing gravitationally bound systems on different scales and sizes is an explanation of their further evolution on the background of a unified cosmological two-component system. One has to take into account any assumption about the the contribution of invisible form of matter with respect to other kinds of matter.

## 2. Equation of Motion and Its Quantitative Analysis

Let us consider $n$ discrete and extended space objects in the regular two-component gravitational field of a homogeneous and isotropically expanding supercluster substratum. One may select inside the background a sphere of sufficiently large radius including a definite energy of both components of the field: radiation with density $\rho_{\rm r}$ and dark matter with density $\rho_{\rm dark}$ (the dark matter behaves like dust matter)

$$\rho_{\rm r} = A/R^4, \quad \rho_{\rm dark} = B/R^3, \tag{1}$$

where $A$ and $B$ are constants.

We may take as the center of this region the inertial reference frame origin, and let at the initial moment of time the center of gravitating medium coincides with the center of masses $m_i$, $i = 1, 2, \ldots, n$.

Write the equation of motion for the gravitational $n$-body problem taking into account the additional forces from the two-component medium (Kozhanov & Omarov 1974)

$$m_i \frac{d^2 \vec{r}_i}{dt^2} = -G \sum_{i \neq j}^{n} \frac{m_i m_j}{r_{ij}^3} \vec{r_{ij}} - \frac{4}{3} \pi G (2\rho_{\rm r} + \rho_{\rm d}) m_i \vec{r_i}, \tag{2}$$

where $\rho_{\rm r}$ is density of the radiating medium, $\rho_{\rm d} = \rho_{\rm dark}$ is density of the dark matter background, $G$ is gravitaty constant.

Using additive properties of mass $M = \sum m_i$ and radius-vector of centre of masses of the system, we obtain

$$\vec{r} = \frac{\sum_i m_i \vec{r_i}}{\sum m_i} = \frac{\sum m_i \vec{r_i}}{M}. \tag{3}$$

From equation (2), we find

$$\frac{d^2 \vec{r}}{dt^2} = -\frac{GM}{r^3} \vec{r} - \frac{4}{3} \pi G (2\rho_{\rm r} + \rho_{\rm d}) \vec{r}. \tag{4}$$

Equation (4) may be interpreted as the equation for test particle in the field of central condensation with mass $M$. It has the area integral, i.e., the law of angular momentum conservation for a unit mass. Then equation (4) can be written in the form

$$\frac{d^2 r}{dt^2} + \frac{GM}{r^2} - \frac{L^2}{r^3} = -\frac{4}{3} \pi G \left(2\rho_{\rm r} + \rho_{\rm d}\right) r, \tag{5}$$

$$r^2 \frac{d\varphi}{dt} = \text{const} = L.$$

Here we included the effect of pressure caused by massless particles in the form of ultra-relativistic gas radiation. Although the extra terms $\rho_{\rm r}(t)$ and $\rho_{\rm d}(t)$ due to substratum produce a force having spatial dependence of a harmonic oscillator, this does not lead to sinusoidal or cosinusoidal oscillations. If substratum density is small in comparison with other terms in equation (5), it does not lead to a simple harmonic motion. If it is large as compared with the other terms, then we have simple Keplerian motion. The radius $r$ in equation (4) has variations on the same time scale $t_c = \sqrt{r_0^3/GM}$ ($r_0$ is the radius of central condensation with mass $M$) as the one of the substratum densities. We study equation (5) in approximation that the densities of substratum change slowly in comparison with mean time $t_c = \sqrt{r_0^3/GM}$ for motions under the influence of central condensation mass $M$. It is clear that for early circular orbits, $r$ varies almost sinusoidally around its mean value $\overline{r}$, making approximately one cycle during one orbital period. According to Noerdlinger & Petrosian (1971), for sufficiently small amplitude we have

$$GM\overline{r} - L^2 + \frac{4}{3}\pi G\left(2\rho_{\rm r} + \rho_{\rm d}\right)\overline{r}^4 = 0. \tag{6}$$

Let us consider this equation in a few special cases. Differentiating this equation on time, we obtain

$$\frac{\dot{\overline{r}}}{\overline{r}} = -\frac{\alpha(1+8/3Z)}{1+4\alpha(1+2/Z)}\frac{\dot{\rho_{\rm d}}}{\rho_{\rm d}}, \tag{7}$$

where

$$\alpha = \frac{4\pi}{3M}\overline{r}^3\rho_{\rm d} = \frac{\rho_{\rm d}}{\rho_M}, \quad \rho_M = \frac{M}{\frac{4}{3}\pi G\overline{r}^3}, \quad z = \frac{\rho_{\rm d}}{\rho_{\rm r}}, \tag{8}$$

$\rho_M$ is the density of central condensation with radius $\overline{r}$.

In expanding universe $\dot{\rho_{\rm d}}$ is negative. In dust-filled universe without matter streaming toward $M$, $\dot{\rho_{\rm d}}/\rho_{\rm d} = -3\dot{R}/R = -3H$. Then expression (7) can be re-written in the form

$$\frac{\dot{\overline{r}}}{\overline{r}} = \frac{3\alpha(1+8/3z)}{1+4\alpha(1+2/z)}H. \tag{9}$$

Equation (9) depends on two parameters $\alpha$ and $z$. Then we consider different cases for these values:

I. Let $z \ll 1$, $\rho_{\rm d} \ll \rho_{\rm r}$, i.e., the Universe is radiation (or neutrinos) dominated, we obtain from equation (9)

$$\frac{\dot{\overline{r}}}{\overline{r}} \approx \frac{8\alpha/z}{1+8\alpha/z}H, \tag{10}$$

(a) $\alpha/z = \rho_{\rm r}/\rho_M \ll 1$, $\rho_{\rm r} \ll \rho_M$, then we have

$$\frac{\dot{\overline{r}}}{\overline{r}} \approx 8\frac{\alpha}{z}H. \tag{11}$$

In this case gravitationally bound superclusters are expanding slowly as compared with usual Hubble expansion. (b) $\alpha/z = \rho_{\rm r}/\rho_M \gg 1$, $\rho_{\rm r} \gg \rho_M$, then

$$\frac{\dot{\overline{r}}}{\overline{r}} \approx H. \tag{12}$$

In these cosmologies all bound systems will expand with the Universe according to Hubble-like expansion.

II. Let $z = 1$, $\rho_{\rm d} = \rho_{\rm r}$. In this case we have the Universe with equal portion of matter. Then from equation (9) we obtain

$$\frac{\dot{\bar{r}}}{\bar{r}} \approx \frac{11\alpha}{1+12\alpha}H. \qquad (13)$$

(a) $\alpha \gg 1$, $\rho_{\rm d} \gg \rho_M$. In this case expansion of bound systems will also expand with the universe of an equivalent Hubble-like expansion

$$\frac{\dot{\bar{r}}}{\bar{r}} \approx \frac{11}{12}H. \qquad (14)$$

(b) $\alpha \ll 1$, $\rho_{\rm d} \ll \rho_M$. Then gravitationally bound systems will expand slower with respect to the Hubble expansion

$$\frac{\dot{\bar{r}}}{\bar{r}} \approx 11\alpha H. \qquad (15)$$

III. Let $z \gg 1$, $\rho_{\rm d} \gg \rho_{\rm r}$. In this case we have dark matter dominating in the universe. From equation (9) we find

$$\frac{\dot{\bar{r}}}{\bar{r}} \approx \frac{3\alpha}{1+4\alpha}H. \qquad (16)$$

(a) $\alpha \gg 1$, $\rho_{\rm d} \gg \rho_M$, i.e., dark matter density is large as compared with the density of central condensation mass $M$. Then the expansion will take place

$$\frac{\dot{\bar{r}}}{\bar{r}} \approx \frac{3}{4}H. \qquad (17)$$

(b) $\alpha \ll 1$, $\rho_{\rm d} \ll \rho_M$ is the same as in (a).

$$\frac{\dot{\bar{r}}}{\bar{r}} \approx 3\alpha H. \qquad (18)$$

In this case dynamical behaviour of galaxy cluster almost coincides with case II.

## 3. Solution of the Equation of Motion

In future we will try to obtain the solution of the equation (4). For this purpose we shall introduce new variables (Mazaros 1974):

$$r(t) = R(t)[1 - x(t)], \qquad x(t) \ll 1. \qquad (19)$$

where $x(t)$ is a small variable, $R(t)$ is the usual expansion scale of the Universe, which satisfies to Fridman's cosmological equation for a unified two-component background matter

$$\frac{d^2R}{dt^2} = -\frac{4}{3}\pi G\left(2\rho_{\rm r} + \rho_{\rm d}\right). \qquad (20)$$

Substituting (19) in equation (5) and only up to the term $dx/dt$ we get

$$\frac{d^2x}{dt^2} + 2\frac{\dot{R}}{R}\frac{dx}{dt} - \frac{GM}{R^3} + \frac{L^2}{R^4} = 0. \tag{21}$$

Let us rewrite equation (21) in the form

$$\frac{H}{R}\frac{d}{dR}\left(\frac{dx}{dt}R^2\right) = \frac{GM}{R^3} - \frac{L^2}{R^4}, \tag{22}$$

where $H = \dot{R}/R$. It is found from cosmological equation (20), taking into account expansion (1),

$$\dot{R}^2 = \frac{8\pi G}{3R^2}(A + BR), \quad H = \sqrt{\kappa}\frac{\sqrt{A+BR}}{R^2}, \quad \kappa = \frac{8\pi G}{3}. \tag{23}$$

Substituting equation (23) into equation (22) we have

$$\frac{d}{dR}\left(\frac{dx}{dt}R^2\right) = \frac{GM}{R^2H} - \frac{L^2}{R^3H} = \frac{GM}{\sqrt{\kappa}\sqrt{A+BR}} - \frac{L^2}{\sqrt{\kappa}\sqrt{A+BR}R}. \tag{24}$$

Now integrating the equation (24) we get

$$\frac{dx}{dt}R^2 = \frac{2GM}{\sqrt{\kappa}B}\sqrt{A+BR} - \frac{L^2}{\sqrt{\kappa}\sqrt{A}}\ln\frac{\sqrt{A+BR}-\sqrt{A}}{\sqrt{A+BR}+\sqrt{A}} + C_1, \tag{25}$$

where $C_1$ is the constant of integration. If the peculiar velocity is not zero, we obtain a decay of the peculiar velocity $\sim R^{-2}$. However for the case of peculiar velocity, taking into account central inhomogeneity, we get

$$\frac{dx}{dt} = \frac{C_1}{R^2} + \frac{2GM}{\sqrt{\kappa}B}\frac{\sqrt{A+BR}}{R^2} - \frac{L^2}{\sqrt{\kappa A}R^2}\ln\frac{\sqrt{A+BR}-\sqrt{A}}{\sqrt{A+BR}+\sqrt{A}}. \tag{26}$$

Thus, the peculiar velocity consists of two parts. The first term describes the decay of the peculiar velocity due to the general expansion of the Universe and decreases more rapidly $(\sim R^{-2})$, than the second part which decreases as $R^{-3/2}$ due to the central condensation of mass $M$ and $L$. The formula (26) is written using relation $\rho_\mathrm{d}$ to $\rho_\mathrm{r}$ or $z = \rho_\mathrm{d}/\rho_\mathrm{r} = BR/A$ in form

$$\frac{dx}{dt} = \frac{\overline{C_1}}{z^2} + \frac{2GMB}{\sqrt{\kappa}A^{3/2}}\frac{\sqrt{1+z}}{z^2} - \frac{L^2B^2}{\sqrt{\kappa}A^{5/2}}\frac{1}{z^2}\ln\frac{\sqrt{1+z}-1}{\sqrt{1+z}+1}. \tag{27}$$

We consider the following cases: (a) $z \gg 1$, $\rho_\mathrm{d} \gg \rho_\mathrm{r}$, then peculiar velocity of the decay is proportional $\sim R^{-3/2}$ which is due to central condensation with mass $M$. (b) $z \ll 1$, $\rho_\mathrm{d} \ll \rho_\mathrm{r}$, then peculiar velocity decreases as $\sim R^{-2}$.

Now define $x$ from equation (26) taking into account expression (23)

$$x = \frac{3M}{4\pi B}\ln z - \frac{3L^2}{16\pi GA}\ln^2\frac{\sqrt{z+1}-1}{\sqrt{z+1}+1} + \tilde{C_1}\ln\frac{\sqrt{z+1}-1}{\sqrt{z+1}+1} + \tilde{C_2}, \tag{28}$$

where $\tilde{C}_2$ is the second constant of integration.

Using expression (19) we find

$$\frac{\dot{r}}{r} = \left[1 - \frac{\dot{x}}{(1-x)\,H}\right] H. \tag{29}$$

In this case equation (27) gives $\dot{x} \sim H/\alpha$. Thus we have

$$\frac{\dot{r}}{r} \approx H\left[1 - \frac{3}{4+3\alpha}\right] \text{ for } z \gg 1, \quad \text{or} \quad \frac{\dot{r}}{r} \approx H\left[1 + \frac{3}{4-3\alpha}\right] \text{ for } z \ll 1, \tag{30}$$

which is the result stated in equations (10)–(18). In general case this expression differs from the results of equations (10)–(18). An effect is as large as that shown in equation (30) and would seriously affect the evolution of clusters or superclusters of galaxies. This effect would apparently inhibit their formation as relics of high-density regions which are lagged behind the cosmological expansion.

Substituting equation (19) in equation (5) we expand the left-hand side of the equation (5) in power series of $x$ upto the term $x$. Then the equation (5) reduces to the form

$$z^2\,(z-1)\,\frac{d^2x}{dz^2} + z\left(\frac{3}{2}z - 1\right)\frac{dx}{dz} - (2az + 3b)\,x = az + b. \tag{31}$$

Here we use the following values

$$2a = \frac{1}{\alpha} = \frac{3M}{4\pi B}, \qquad b = \frac{9L^2}{8\pi GA}.$$

Formula (31) is an inhomogeneous hypergeometrical equation.

Solutions of (31) are Gauss hypergeometrical functions $F(\alpha, \beta, \gamma, z)$. The equation has a singular point at $z = 0$, 1, and $\infty$. Series expansions for the solutions exist around each of these singular points. In our case, in view of definition (31), the negative $z$ region is applicable, that is we can use the series expansions around $z = 0$.

On the basis of the equation mentioned above it is also possible to investigate dynamical behavior of gravitationally bound systems in the background of unified two-component model of the Universe. The results depend on whether the dark matter or ultra relativistic gas with pressure dominates.

## 4. Conclusions

In this work we developed the theory of gravitational dynamics of galaxy clusters evolving under the influence of expanding cosmological two-component system. Since the two-component nature of cosmic matter has recently become more important, especially in connection with the galaxy formation and the problem of their evolution, we expect that our investigation could play an important role in future studies.

For $z \ll 1$, particles of galaxy clusters are frozen into the substratum. This means that the gravitational fields of point mass particle and extended cosmic objects cause small reduction of expansion.

## References

Bosma, A., 1981a, AJ, 86, 1791
Bosma, A., 1981b, AJ, 86, 1825
Faber, S., & Gallaghar, J. 1979, ARA&A, 17, 135
Fild, G. B. 1973, Proc. IAU Symp. 63
Kozhanov, T. S., & Omarov, T. B. 1974, Trudy AFI of AN KazSSR, 26, 16 (in Russian)
Meszaros, P., 1974, A&A, 37, 225
Noerdlinger, P., & Petrosian, V. 1971, ApJ, 168, 1
Peebles, P. 1971, A&A, 11, 377
Rubin, V. C. 1983, Science, 220, 1139

*Order and Chaos in Stellar and Planetary Systems*
*ASP Conference Series, Vol. 316, 2004*
*G. Byrd, K. Kholshevnikov, A. Mylläri, I. Nikiforov, and V. Orlov, eds.*

# Influence of the Vacuum on the Galaxy Groups Dynamics

A. Minz and V. Orlov

*Sobolev Astronomical Institute, St. Petersburg State University, Universitetskij pr. 28, Staryj Peterhof, St. Petersburg 198504, Russia*

**Abstract.** Influence of cosmological vacuum on the galaxy groups dynamical evolution is studied. Numerical model built shows a good agreement with last observational data, especially with the Local Group and compact groups.

## 1. Introduction

Cosmological vacuum appeared in observational cosmology just a few years ago as a result of Supernovae Ia and CMB spectra observations. It has an unusual equation of state $P = -\rho c^2$ ($P$ is pressure, $\rho$ is density), and therefore acts as a repulsion force. It's estimated density is about 2 times higher, than that for the matter.

In the early 1980s Sandage mentioned that the Hubble Law was discovered on scales about 10 Mpc, while the inhomogeneity cell is much bigger (≈300 Mpc, according to the latest theoretical and observational works). "Classical" theory is useful only for the homogeneous spatial matter distribution, and thus nearly exact Hubble Law on such small scales cannot be explained.

Baryshev et al. (2001) assumed that cosmological vacuum can produce the Hubble Law on scales about 1–2 Mpc, with a very small velocity dispersion. The aim of the work was to build a simple model for testing this idea.

The main assumption made was that galaxies are moving under the modified Newtonian gravitation law, as velocities and potentials are non-relativistic:

$$\ddot{\vec{r}}_i = -\sum_{\substack{j=1 \\ i \neq j}}^{20} \frac{GM_j \vec{r}_{ij}}{\left(r_{ij}^2 + \epsilon_i^2 + \epsilon_j^2\right)^{\frac{3}{2}}} + \alpha^2 \vec{r}_i, \quad i = 1, 2, \ldots, 20, \tag{1}$$

where $\vec{r}_i$ is the $i$-th body radius-vector, $\vec{r}_{ij}$ is the separation vector between $i$-th and $j$-th bodies, $m_j$ are the masses, $\alpha = \sqrt{8\pi G\rho_V/3}$ is parameter (here $\rho_V$ is the vacuum density). As the assumed value of $\rho_V$ is $6.25 \times 10^{-30}$ g/cm$^3$ (according to WMAP data) we have $\alpha = 1.9 \times 10^{-18}$ sec$^{-1}$ or 57.5 km/sec/Mpc in traditional units.

The main task was to estimate the measured Hubble constant and velocity dispersion with respect to the Hubble law within a group of 20 galaxies. But we have only one good object to check our results: Local Group (LG). There's some information about other galaxy groups, but we can still say nothing about radial motions in their outer parts. Thus it seems to be useful to compare known group parameters, such as mean harmonic radius, l.o.s. velocity dispersion and mass-to-light ratios with the simulated ones. But these data show extremly high scattering. This is due to different galaxy catalogs used, methods applied and parameters choosen. Interlopers may play singnificant role also, in old catalogs especially (as no redshift data were used there). Thus one must be very careful in comparing results.

## 2. Measurment of the Hubble Constant

As it can be shown in semi-analytic analysis of one-dimensional case, systematic deviations from the linear Hubble law are significant. Therefore we need some criteria to choose galaxies useful for measuring Hubble constant. The easy way is to find selection is based on the distance from the center of mass of the group. The lower limit was taken empirically twice mean harmonic radius ($R_h$) of the group, while the upper limit was fixed at 3 Mpc (mean distance between groups near LG). Inscription of the linear Hubble law into selected points was made using least-squares method. The line has to pass through zero point due to physical reasons. Inclination of this line was taken as a measured Hubble constant for this group.

## 3. Model and Results

A group of twenty galaxies with a total mass of $10^{12} M_{\odot}$ was taken as a model object. Two galaxies were heavier than the others (like in the LG). Initial positions were randomly distributed within the sphere of a radius that varied from 400 to 1200 kpc. Velocities were randomized within a sphere and normalized to give a fixed initial virial ratio (its values were taken as 0, 0.5, 1, 1.5). Dark matter had isothermal distribution with a radius of 200 kpc and mass of about 80% of the total one. One thousand runs were calculated for cases each with and without vacuum. The main result is that the model without vacuum rarely gives an agreement with observations (just for compact and expanding groups), while the model with vacuum has a good agreement for a wide range of parameters. In both cases free parameters are the worthest ones. The scatter is extremely high for the compact groups and the value of the Hubble constant is too small for the loose ones.

The velocity dispersion is surprisingly small for most of cases. Generally, no difference can be made between models with and without vacuum. Mean values are about 20–80 km/sec with a high scatter.

Group catalogs were built from pseudo-observational catalogs of galaxies using classical friend-of-friend algorithm. There are less groups in models with vacuum, and they are wider. Line-of-sight velocity dispersions are about 30–110 km/sec and 20–70 km/sec for models with and without vacuum correspondently. Vacuum model is preferable, as this value for actual observational catalogs is 50–200 km/sec. Measured velocity dispersions for Maffei and NGC 6946 groups are 86 and 95 km/sec correspondently, with a good agreement with vacuum model.

Other parameters have too wide distributions to make strict conclusions, but model with vacuum seems to be more preferable.

In future some more work on comparing results with observations is planned, with more attention for mass-to-light ratio, which is one of the most important parameters.

## 4. Conclusions

The main result is that model with vacuum is useful for describing group evolution and Hubble law existing. Dispersions are not too high for both models. Group parameters are similar for models with and without vacuum. The dynamical evolution of the groups (excepting the closest ones) is not finished and there's no strict group borders.

## References

Baryshev, Yu. V., Chernin, A. D., & Teerikorpi, P. 2001, A&A, 378, 729

*Order and Chaos in Stellar and Planetary Systems*
*ASP Conference Series, Vol. 316, 2004*
*G. Byrd, K. Kholshevnikov, A. Mylläri, I. Nikiforov, and V. Orlov, eds.*

# Method of Determination of Spatial Luminosity Distribution Using 2D-Data

M. A. Mardanova
*Faculty of Applied Mathematics & Control Processes, St. Petersburg State University, Universitetskij pr. 28, Staryj Peterhof, St. Petersburg 198504, Russia*

**Abstract.** In this paper a method of solution of a deprojection problem is suggested. We find a spatial distribution of luminosity by using information on surface brightness distribution. A numerical algorithm is worked out for a disk model of a galaxy, consisting of three components, namely a bulge, a disk and a dust layer. For solving the problem and finding model parameters, a combination of two minimization methods was used. We have carried out an analysis of numerical errors and have estimated an efficiency of the offered method.

## 1. Introduction

Clearly, our task is pertained to the class of inverse problems. According to its definition, it is an ill-posed problem, complicated enough to require an approximate calculus technique. But a group of algorithms is known for a search of solutions which are stable to small changes of initial data. For instance, in a work presented by Tihonov & Arsenin (1986), the question of finding solutions is considered in a general view. We use a regularization method and remove an ill-posedness via construction of a smoothing functional. So, the problem of parameters determination of model distribution leads to a search of minimum of this functional. Another way of looking the problem is described in previous paper (Mardanova 2003).

## 2. Model

As an investigated model we took a model for disk galaxies (Hamabe 1982) and included to it some additions. So, instead its original two components three are considered. The third component is a dust layer. The luminosities of bulge and disk are described by the following formulas:

$$\begin{cases} l_b(R,z) = l_{0b} a_0^{\gamma+1} (a_0^2 + q^2)^{-(\gamma+1)}, \\ l_d(R,z) = l_{0d} e^{-\alpha R - \beta |z|}, \\ q^2 = R^2 + z^2/c^2, \end{cases} \tag{1}$$

where $l_{0b}$ and $l_{0d}$ are densities of luminosity at the center of bulge and disk respectively; $R$, $z$ are the cylindrical coordinates. By means of insertion of parameter $a_0$, we describe a central density of bulge; $c$ is the sphericity of the bulge. We note that the luminosity decreases with fixed $q$, if $\gamma$ increases and it increases as this parameter decreases. The structural parameters $\alpha$ and $\beta$ ($\alpha > \beta$) characterize an extension of our model in horizontal and vertical directions. For description of a third component is introduced the optical thickness (see Kodaira & Ohta 1994):

$$\tau = k_0 \int_{-\infty}^{+\infty} \exp\{-(R/R_a - z/z_a)\}\, d\zeta, \tag{2}$$

where $k_0$ is a coefficient of absorption at the dust layer center; it's value is considered as a constant. Parameters $R_a$ and $z_a$ are structural too, and $R_a > z_a$; $\zeta$ is the integration variable along the line of sight. So, we have a model with ten parameters and the vector of model parameters looks as:

$$p = (a_0, \alpha, \beta, \gamma, c, k_0, l_{0b}, l_{0d}, R_a, z_a). \tag{3}$$

Six of them are dimensional, they are: $a_0, k_0, l_{0b}, l_{0d}, R_a, z_a$. All calculations are made with non-dimensional values. Thus, our problem reduces to the determination of vector $p$ with four components, and now $p = (\alpha, \beta, \gamma, c)$.

## 3. Algorithm

As the posed problem is incomplete, formally its solution is the root of the following integral equation:

$$B_{\rm mod}(\xi, \eta; p) = \int_{-\infty}^{+\infty} \left[l_b(R, z; p) + l_d(R, z; p)\right] e^{-\tau(R,z;p)} d\zeta. \tag{4}$$

It must be solved with regard to the unknown vector of parameters $p$ by using photometric data, $B_{\rm obs}$. But such an approach to the problem does not look rational to us, evidently because of the bulky integration element (4) the volume of computations will be too big and take much time. That's why we use an idea of regularization method, just as for instance in (Kutuzov & Sergeev 1978) and introduce the smoothing functional:

$$\mathcal{F}(p) = \sum_{i=1}^{N} \omega_i \left[B_{\rm mod}(\xi_i, \eta_j; p) - B_{\rm obs}(\xi_i, \eta_j)\right]^2. \tag{5}$$

So, we replace the search process of the solution of an integral equation (4) by the solving of the constrained minimization problem for functional (5) with given linear constraints-inequalities for components $p_i$ of unknown vector $p$.

In formula (5) $B_{\rm obs}$ are the generated observational data; $N$ is a number of grid knots $\{\Gamma = (\xi_m, \eta_j)\}$ on the projection plane $O\xi\eta$ and $(\xi_m, \eta_j)$ are the coordinates of an observer. We project on this plane the spatial luminosity distribution $l$. Weights $\omega_i$ are assumed to be unity.

In previous paper (Mardanova 2003) we investigated spheroidal model (Kutuzov 1971) and considered an elliptical grid of knots. Here we take a grid from triangles. Their vertexes are placed at the points of intersection of elliptic arcs with lines outgoing from the origin of coordinates. The elliptic arcs are drawn in such way to obtain triangles with about equal areas.

From the geometrical point of view we calculate the sum of volumes of right prisms (they are 72). There are triangles in their bases and the volume of an each prism is found so:

$$V = (z_1 + z_2 + z_3)\, S/3, \tag{6}$$

where $S$ is an area of a corresponding triangle; verges $z_i$ (or applicates) are computed as

$$z_i = (B_{\mathrm{mod}} - B_{\mathrm{obs}})^2. \tag{7}$$

After this sum is found, we find its minimum value. So, the functional minimization for one cycle is realized. For the minimization, the method of the "direction search" was chosen, worked out by Hook and Jeevs (Himmelblaw 1975). In the calculation program, it was transformed to the method of "coordinatewise descent". In order to take into account all constraints to parameters, this method is modified and further it's considered in combination with method of penal functions. So, now we solve the problem of minimization of "expanded" functional (Aoki 1977):

$$F(p, \overline{k}) = \mathcal{F}(p) - \overline{k} \sum_{i=1}^{m} \frac{1}{g_i(p)}, \quad \overline{k} > 0, \tag{8}$$

where $g_i(p) \geq 0$, $i = \overline{1, m}$, are mentioned before linear constraints-inequalities and $\overline{k}$ is a penalty multiplier. Generation of observable data is produced by a way of "polar coordinates" (Knut 1977). According to it, we take two samples of random variables distributed evenly on an interval (0,1) as a source data for the computational procedure. As a result of execution of this procedure, we obtain two samples of random variables, distributed accordingly to the normal law on the same interval. These values we use as noise ill-wresting "true" values of observable data. In other words, at first we calculate the values of $B_{\mathrm{mod}}(\xi_i, \eta_j; p)$ at the points of grid $(\xi_m, \eta_j)$ for the selected vector of parameters $p$. After that we add noise to these values, so we obtain generated photometric data $B_{\mathrm{obs}}(\xi_i, \eta_j)$. Here the minimization of functional (8) is fulfilled by means of this algorithm as in previous work (Mardanova 2003). But a general minimum is found as a minimum of the sum of prism volumes with a corresponding choice of $p_i$.

For a completion of the numerical process we took two criteria. First of them is a test of modifications of error values of the approximate vector $\tilde{p}$ obtained. They are an absolute error, $\delta^i_{\mathrm{abs}}$ and a relative one, $\delta^i_{\mathrm{cor}}$. It is necessary that these errors decreased in each cycle, with the decreasing of a penalty multiplier $\overline{k}$. And the second condition is a fulfillment of the inequality:

$$|F(\tilde{p}, \overline{k}) - \mathcal{F}(\tilde{p})| < \overline{\varepsilon}, \tag{9}$$

for this algorithm $\overline{\varepsilon} = 0.001$. A variable $\tilde{p}$ is a value of a zero approach to vector parameters $p$. One can say we calculate difference between values of an initial functional, $\mathcal{F}(\tilde{p})$, and an "expanded" one, $F(\tilde{p}, \overline{k})$ for a given $k$. Further we check whether the difference (9) decreased in comparision with its value in preceding cycle or not.

## 4. Results

Let us go to the results of numerical experiments directly. As is said above the dimensionless vector of model parameters is: $p = (\alpha, \beta, \gamma, c)$. For the generation of observable data we take vector $p_{\text{obs}} = (0.64, 0.43, 2.9, 0.32)$; the same components $p_i$ are considered for an initial estimate, $\tilde{p}_0$. Functions $g_i(p)$ from (8), are the linear constraints-inequalities, look here in the following way:

$$\begin{cases} g_0(p_i) = -p_i \leq 0, \\ g_1(p_1) = p_1 - 1.5 \leq 0, \\ g_2(p_2) = p_2 - 1 \leq 0, \\ g_3(p_3) = p_3 - p_2 \leq 0. \end{cases} \tag{10}$$

The problem of minimum search (8) was solved for one vector $p$ and three samples of observable data $B_{\text{obs}}$. Furthermore, our casts realize for two variants of inclination $i$: for $i = 0°$ and $i = 60°$. This angle $i$ is the angle between the line of observer sight and equatorial plane of a galaxy.

Part of our results without taking subproducts into consideration is noted in Tables 1–5. In the third of them we demostrate results of calculations for $i = 60°$ and $\sigma = 0.1$, 0.2, 0.3. The parameter $\sigma$ is a dispersion and describes normal law of random value distribution. It's used when we generate photometric data.

Table 1. $i = 60°$, $\sigma = 0.1$, $\tilde{p} = (0.638, 0.430, 2.896, 0.321)$

| $\overline{k}$ | $F(\tilde{p}, \overline{k})$ | $\mathcal{F}(\tilde{p})$ | $G(\tilde{p})$ |
|---|---|---|---|
| $1 \times 10^{-1}$ | 2.518 | 2.063 | $4.55 \times 10^{-1}$ |
| $1 \times 10^{-2}$ | 2.082 | 2.036 | $4.55 \times 10^{-2}$ |
| $1 \times 10^{-3}$ | 2.041 | 2.036 | $4.56 \times 10^{-3}$ |
| $1 \times 10^{-4}$ | 2.0377 | 2.0373 | $4.59 \times 10^{-4}$ |
| $1 \times 10^{-5}$ | 2.03738 | 2.03734 | $4.59 \times 10^{-5}$ |

Table 2. $i = 60°$, $\sigma = 0.2$, $\tilde{p} = (0.600, 0.412, 2.848, 0.325)$

| $\overline{k}$ | $F(\tilde{p}, \overline{k})$ | $\mathcal{F}(\tilde{p})$ | $G(\tilde{p})$ |
|---|---|---|---|
| $1 \times 10^{-1}$ | 7.114 | 6.686 | $4.28 \times 10^{-1}$ |
| $1 \times 10^{-2}$ | 6.702 | 6.656 | $4.55 \times 10^{-2}$ |
| $1 \times 10^{-3}$ | 6.667 | 6.663 | $4.56 \times 10^{-3}$ |
| $1 \times 10^{-4}$ | 6.6678 | 6.6674 | $4.55 \times 10^{-4}$ |
| $1 \times 10^{-5}$ | 6.66759 | 6.66754 | $4.55 \times 10^{-5}$ |
| $1 \times 10^{-6}$ | 6.66761 | 6.66760 | $4.55 \times 10^{-6}$ |

Function $G(\tilde{p})$ from the tables is a penalty function, by using it we control acceptable bounds of parameters $p_i$:

$$G(\tilde{p}) = -\frac{1}{g_i(\tilde{p})}, \quad i = \overline{0, 4}. \tag{11}$$

Evidently, its value decreases on each step, hence a module of difference, $|F(\tilde{p}, \overline{k}) - \mathcal{F}(\tilde{p})|$ decreases too. It came out from formula (8). So, one can

Table 3. $i = 60°$, $\sigma = 0.3$, $\tilde{p} = (0.594, 0.425, 2.848, 0.325)$

| $\overline{k}$ | $F(\tilde{p}, \overline{k})$ | $\mathcal{F}(\tilde{p})$ | $G(\tilde{p})$ |
|---|---|---|---|
| $1 \times 10^{-1}$ | 15.916 | 15.488 | $4.28 \times 10^{-1}$ |
| $1 \times 10^{-2}$ | 15.643 | 15.598 | $4.55 \times 10^{-2}$ |
| $1 \times 10^{-3}$ | 15.635 | 15.630 | $4.56 \times 10^{-3}$ |
| $1 \times 10^{-4}$ | 15.6349 | 15.6344 | $4.55 \times 10^{-4}$ |
| $1 \times 10^{-5}$ | 15.6345 | 15.6344 | $4.55 \times 10^{-5}$ |

judge a possible completion of iterations by the values of function $G(\tilde{p})$ in an each cycle. There were estimated relative, $\delta_{\rm cor}$, and absolute, $\delta_{\rm abs}$, errors of integration in all points of a grid. We also calculated relative errors, $\delta_{\tilde{p}}$, of vector $p$. The value of its root-mean-square deviation, $\tilde{\sigma}$, is found by the formula:

$$\tilde{\sigma} = \sqrt{\sum_{i=1}^{m} (p^i_{\rm obs} - \tilde{p}_i)^2}. \tag{12}$$

Results of $\delta_{\tilde{p}}$ calculations for $i = 0°$, $60°$ and different $\sigma$ are shown in Tables 4–5:

Table 4. $i = 0°$

| $\sigma$ | m=1 | m=2 | m=3 | m=4 |
|---|---|---|---|---|
| 0.1 | $8.04 \times 10^{-3}$ | $-1.19 \times 10^{-2}$ | $1.77 \times 10^{-3}$ | $6.43 \times 10^{-3}$ |
| 0.2 | $2.30 \times 10^{-6}$ | $-1.01 \times 10^{-2}$ | $-8.88 \times 10^{-3}$ | $8.04 \times 10^{-3}$ |

Table 5. $i = 60°$

| $\sigma$ | m=1 | m=2 | m=3 | m=4 |
|---|---|---|---|---|
| 0.1 | $3.43 \times 10^{-3}$ | $2.33 \times 10^{-4}$ | $1.42 \times 10^{-3}$ | $-1.61 \times 10^{-3}$ |
| 0.2 | $6.27 \times 10^{-2}$ | $4.27 \times 10^{-2}$ | $1.77 \times 10^{-2}$ | $-1.60 \times 10^{-2}$ |
| 0.3 | $7.16 \times 10^{-2}$ | $1.07 \times 10^{-2}$ | $1.77 \times 10^{-2}$ | $-1.60 \times 10^{-2}$ |

By the analysis of the calculation's errors, we cleared up that as the $\sigma$ increased (both $i = 0°$ and $i = 60°$), the relative and absolute errors increased also. It is possible to explain the growth of the errors in the following way. It is known that probability of distribution of a some random variable $x$ by the normal law is equal to

$$P(x) = \frac{1}{\sigma\sqrt{2\pi}} e^{-\frac{x^2}{2\sigma^2}}. \tag{13}$$

And by the way it's affirmed that if the value of $\sigma$ less the curve of probability distribution have a bigger value of maximum (Gnedenko 1961). So, probability to get in a some interval $(x_1, x_2)$ will be more for that $x$, which has a small value of $\sigma$. From this it follows that in our case we obtain the very little noise for the generation observable data with $\sigma = 0.1$. As a consequence of this we obtain also the very little values of errors with the same value of $\sigma$. As may be seen from the tables when the value of $\sigma$ changes from 0.1 to 0.3, noise increases and hence errors increase too.

## 5. Conclusions

This paper is the continuation of our previous elaborations and it's considered as their alternative. In particular, here we set out oneself to estimate the efficacy of the proposed method as an object by the comparing it with debugged earlier numerical programs. The rate of convergence in the algorithm is not too large. The cause of this is the big thickening of current points, that's why the process of minimization fulfills slowly, as the frontiers of an admissible set approach. However, our analysis of errors shows that in the whole the chosen model and the results found one can be regarded as successful. Furthermore we can obtain more exact values of approach $\tilde{p}$ to $p$ by the decrease of $\sigma$ in procedure of generation of observable data. This is one of the ways to avoid the solution instability.

In this model we allowed for the effects of light absorption only. Further plans are to consider a model taking into account dispersion and making calculations with dimensional variables.

## References


Aoki, M. 1977, Introduction to optimization methods (Moscow: Nauka)

Gnedenko, B. 1961, Course of theory of chances (Moscow: Fizmatgiz)

Hamabe, M. 1982, PASJ, 34, 423

Himmelblaw, D. 1975, Applied Nonlinear Programming (Moscow: Mir)

Knut, D. 1977, Art of computer programming, 2 (Moscow: Nauka)

Kodaira, K., & Ohta, K. 1994, PASJ, 46, 155

Kutuzov, S. 1971, Talk at the Workshop on Extragalactic Astrophysics, Tartu astrophys. observ. (private communication)

Kutuzov, S., & Sergeev, V. 1978, ApJ, 14, 473

Mardanova, M. 2003, Vestnik SPb Univ.Ser.1, 132

Tihonov, A., & Arsenin, V. 1986, Methods for a decision of ill-posed problems (Moscow: Nauka)

# Part VII

# Models of Gravitating Systems

*Order and Chaos in Stellar and Planetary Systems*
*ASP Conference Series, Vol. 316, 2004*
*G. Byrd, K. Kholshevnikov, A. Mylläri, I. Nikiforov, and V. Orlov, eds.*

# To Recent 150th Anniversary of Sofia Kovalevskaya (1850–1891): Her Scientific Legacy in Celestial Mechanics of Equilibrium Figures of Fluid Mass in Axial Rotation

E. N. Polyakhova

*Dept. of Celestial Mechanics, Faculty of Mathematics and Mechanics, St. Petersburg State University, Universitetskij pr. 28, Staryj Peterhof, St. Petersburg 198504, Russia; e-mail: pol@astro.spbu.ru*

## 1. Primary Events and Dates in S. Kovalevskaya's Life and Scientific Work

In January 2000, the world's scientific community celebrated the 150th Centennial Jubilee Anniversary of Sofia Wasil'evna Kovalevskaya—outstanding Russian scientist-mathematician, writer, publicist, the world's first female professor, and elected Correspondent-Member of Academy of Sciences in St. Petersburg, Russia. Professionals in the areas of Mathematics, Mechanics, Physics and Astronomy, History and Literature acknowledge her heritage. As for her remarkable scientific legacy, S. Kovalevskaya had published just 9 works: 3 of pure mathematical nature and 6 on the mathematical applications to the problems of Celestial Mechanics (1 article), Physics of crystals (2 articles) and Classical Mechanics of rigid body rotational motion (3 articles).

With connection to this memorable date in the history of science, let us recall major events and dates of S. Kovalevskaya's (nee Korvin-Krukovskaya) life and scientific activity:

- 1850: 3(15) of January, born in Moscow.
- 1853–1858: lived with her parents in Kaluga.
- 1858: moved to country estate of Palibino, not far from the city of Velikiye Luki, now Pskov region of Russia.
- 1866: began to study mathematics with N. A. Strannolyubski during the family travels to St. Petersburg.
- 1867: traveled with family to Germany and Switzerland.
- 1868: 15(27) of September, marriage with Wladimir Kovalevsky (1842–1883); moved to St. Petersburg to study science.
- 1869: trip to Heidelberg to study with G. Kirchhoff (1824–1887) and H. Helmholtz (1821–1894); her trip to London.
- 1870: moved to Berlin and began private mathematical studies with Karl Weierstrass (1815–1897), the greatest analyst of his time.

- 1871: stayed in Paris during the city siege and the Paris Commune Days; helped her brother-in-law Victor Jaclaire to escape from prison.
- 1872–1874: worked in Berlin on Abelian integrals (Mathematics) and Saturn's rings stability (Mechanics).
- 1874: in her absence, the University of Gettingen awarded Kovalevskaya a Ph.D. in Mathematics and Master of Fine Arts with support of Karl Weierstrass for the cycle of three works over the period of 1870–1874: I. "To the theory of partial differential equations" (her first work on pure Mathematics first published in German as her dissertation "Zur Theorie der partiellen Differentialgleichungen," Inaugural Dissertation, 1874, and then republished with the same title in Journal fuer die Reine und Angewandte Mathematik, Berlin, 1875, Bd. 80, S. 1–32), II. "About reduction of same class third rank abelian integrals to elliptic integrals" (her second work on pure Mathematics, will be published only in 1884), III. "Addendum and remarks to Laplace's Saturn's rings shape research" (her first and only single work in Celestial Mechanics will be published later, only in 1885); returned back to Russia.
- 1874–1877: literary-publicist work in St. Petersburg; worked as a journalist in Suvorin's newspaper "New Time"; meetings with Mendeleev, Setchenov, Botkin, Butlerov, Chebyshev, Stoletov, Turgenev, Dostoevskiy, etc.
- 1878: birth of her daughter Sofia Wladimirovna Kovalevskaya (1878–1952).
- 1879: returned back to scientific work, with support from mathematician P. L. Chebyshev (1821–1894) gave a talk at the 6th Russian Naturalists' Convention in St. Petersburg on properties of Abelian integrals.
- 1880: moved to Moscow and was elected to be a member of Moscow Mathematical Society.
- 1881–1883: lead scientific work in Berlin and Paris on mathematical description of refraction of light in crystals.
- 1883: death of her husband, W. O. Kovalevsky; returned to Russia; gave a talk at the 7th Russian Naturalists' Convention in Odessa on refraction of light in crystals; moved to Stockholm as private assistant professor in Stockholm University by invitation of Professor G. M. Mittag-Leffler (1846–1927); published article IV entitled "About light refraction in crystals" (her first work in Physics, published in German as "Ueber die Brechung des Lichtes in kristallinischen Mitteln," Acta Mathematica, Stockholm, 1883, Bd. 6, S. 249–304).
- 1884: 30th of January (11th of February), gave her first lecture in Stockholm University; was appointed ordinary Professor of Stockholm University and invited to join editorial board of the journal "Acta Mathematica," founded by G. M. Mittag-Leffler and edited in Sweden (as it is known, Mittag-Leffler began discussions about the journal in 1881, the first issue of Acta was published in 1882); published article II (see above) in

German as "Ueber die Reduktion einer bestimmten Klasse Abel'schen Integrale 3-en Ranges auf elliptische Integrale," Acta Mathematica, Stockholm, 1884, Bd. 4, S. 393–414); published article V "About propagation of light in crystalline medium" (her second work on Physics of crystals, published in French as "Sur la propagation de la lumiere dans un milieu cristallisé," Comptes Rendus de Seances de l'Academie des Sciences, Paris, 1884, v. 98, p. 356–357, and then published again in Swedish as "Om ljusets fortplantning uti ett kristalliniskt medium," Ofversigt uf Konigl. Vetenskapsakademiens Foerhandlingar, Stockholm, 1884, Bd. 41, S. 119–121).

- 1885: published article III (see above) in German as "Zusaetze und Bemerkungen zu Laplace's Untersuchung ueber die Gestalt der Saturnringe" (Astronomische Nachrichten, Kiel, 1885, Bd. 111 (3), N. 2643, S. 37–48).
- 1886: published essay "Remembrances of George Elliott" in journal "Russian Thought" (1886, n. 6), based on her visits with the English writer in 1869 in London and on their long-term correspondence.
- 1887: worked on drama "Fight for Happiness," co-written with Swedish writer Anne-Charlotte Leffler-Edgren (later the play was staged in Russia).
- 1888: worked on the topic of rigid body rotation; was awarded Borden's Prize for the work VI that made her famous: "Problem of rigid body rotation around fixed point" (her first work on Classical Mechanics) by the Paris Academy of Sciences.
- 1889: published the article in French as "Sur la probleme de la rotation d'un corps solide autour d'un point fixe" (Acta Mathematica, Stockholm, 1889, Bd. 12, S. 177–232); published literary essay about M. E. Saltykov-Shchedrin for a French newspaper; after resolution of the issue of female acceptance in "Academia" by St. Petersburg Academy of Sciences on 14(26) of December 1890 and with active support of russian Academicians Chebyshev, Imshenetsky and Bunyakovsky, on 7(19) of November 1889, Kovalevskaya was elected to be a Correspondent-Member of Mathematics and Physics Division of St. Petersburg Academy of Sciences.
- 1890: published literary essay "Three days in peasants' university of Sweden"; worked on the book "Reminiscence of childhood," short story "Nihilist" (unfinished), and on the novel "Nihilstka" (female-nihilist), this last publication was not allowed in Russia (the novel only saw light in Russia in 1928), when it was published in Sweden very soon under the title "Vera Voronzoff"; Kovalevskaya was awarded the Sweden mathematical prize (King Oskar's Prize) for her work VII "About the property of differential equations system, defining the rotation of rigid body around fixed point" (her second work on Classical Mechanics, which was the development and continuation of the work VI); article VII was published in French as "Sur la proprieté du systeme d'equations differentielles qui definit la rotation d'un corps solide autour d'un point fixe" (Acta Mathematica, Stockholm, 1890, Bd. 14, S. 81–93); worked on article VIII about one special case of problem of rigid body rotation around fixed point when

integration is done by the means of ultra-elliptical functions of time (her third work on Classical Mechanics; the continuation and development of works VI and VII); article VIII was published in French as "Memoire sur un cas particulier du probleme de la rotation d'un corps pesant autour d'un point fixe, ou l'integration s'effectue a l'aide de fonctions ultraelliptique du temps" (Memoirs presentes par divers savants a l'Academie des Sciences de l'Institut National de France, Paris, 1890, v. 31, p. 1–62).

- 1891: book "Reminiscence of childhood" was published in Russian journal "European Bulletin" and in Swedish journal entitled as "Systrarna Rajevsky"; Kovalevskaya's work X about one Brunce's theorem from theory of potential (her third work on pure mathematics) was published in French as "Sur un theoreme de M. Bruns" (Acta Mathematica, Stockholm, 1891, Bd. 15, S. 45–52) soon after her death.

- 1891: death of Sofia Kovalevskaya on January 29th (February 10th) 1891 in Stockholm where she was buried.

Sofia Kovalevskaya is honoured by names of lunar crater and minor planet (asteroid): minor planet (NMP) "1859 Kovalevskaya" discovered in 1972, September, 4th, in Nautchny-Observatory in Crimea (former USSR), and preliminary numbered as 1972 RS2, is named by its discoverer L. V. Zhuravleva after this famous russian woman.

## 2. About Kovalevskaya's Scientific Legacy in Celestial Mechanics (Her Results in Celestial Mechanics of Equilibrium Figures of Fluid Masses in Axial Rotation)

As is known, S. Kovalevskaya got the fundamental results in the theory of partial differential equations (the classical Cauchy–Kovalevskaya's theorem). These her results were the first in the series of works which were completed in 20th century only. As usually, the result of Kovalevskaya in Classical Mechanics is widely known as concerned to her research on the rotation of a rigid body about a fixed point. She carried out this research after her appointment to a professorship in Stockholm in 1880s.

Her first article on it (see above article VII) won the Bordin Prize's of Paris Academy of Sciences in 1888, on which occasion the Prize Commitee doubled the usual prize money, in recognition of an unusual achievment. She became the leading woman mathematician, "Princess of Science," earning the respect and confidence of many scientists.

Unsimilarly to these usually and widely known remarkable and famous results of Kovalevskaya, it might be interesting to remember to one of her early articles.

In her single astronomical paper published in 1885 only (see above article III) but prepared long before, in 1874, Kovalevskaya has considered (according to P. S. Laplace's theory of Saturn rings structure), the profile shape and equilibrium stability of a single gravitating ring (fluid or monolite) around an attractive centre. She presented the development and essential expansion of

Laplace's results up to second order approximation in the determination of ring gravitational potential.

This her paper is mentioned and reffered very seldom and commented else seldomer. These her results were not applied by astronomers at all because since 1859 J. C. Maxwell's theory of Saturn rings with account of their particles differential keplerian orbiting was already successfully applied in Planetary Celestial Mechanics dealing with Saturn rings observations and researches.

Indeed, this article dealing with Saturn rings fluid or solid structure (after P. S. Laplace) contradicted obviously to Maxwell' theory. Nevertheless, this Sonya's single astronomical paper is now not only of historical-memorial interest but of great importance for the modern theory of rotating celestial bodies potentials.

To investigate equilibrium figures of fluid mass, was proposed to Sonya by russian mathematician P. L. Chebyshev in 1860s in St. Petersburg, long before her depart to abroad but she had shown no interest in this problem at that time. As is known, Chebyshev widely discussed mathematical problems with K. Weierstrass in many letters as his letter-correspondent. Very possible to suppose that the problem of Saturn ring shape was advised by Chebyshev again in 1870s.

Several years later, in 1881, Chebyshev advised to A. M. Lyapunov (1857–1918) to research this problem in order to develop the former results of both Maclaurin and Jacobi about equilibre ellipsoids. The so-called Chebyshev's problem was formulated as follows: to prove the existance of some new figures of equilibrium near to two kinds of Maclaurin's ellipsoids. Lyapunov liked this problem and accomplished its first approximation in 1884. Simultaneiusly, in 1880s, H. Poincaré began to solve this problem too. Poincaré has found some new figures one year later, in 1885, but not especially exactly. Lyapunov proved this result exactly in 1900s only.

## 3. Short History of Saturn's Rings Discovery and Early Primary Researches

- 1610–1616: Galileo Galilei (1564–1642) observes two faint objects to the left and right of Saturn.

- 1655: Christian Huygens (1629–1695) observes the single bright ring around Saturn.

- 1675: Gean Domenico Cassini (1625–1715) observes the gap (division) in this single ring (the so-called Cassini's gap).

- XVII and XVIII centuries: Development of theories of fluid or monolite solid single ring with one gap in it.

- 1755: Immanuel Kant (1724–1804) proposes the first theory of the differential rotation of particles in viscosity-resistant disc-like ring of Saturn in framework of his remarkable cosmogonic theory concept (Kant, 1755). Being written as a some philosophical thesis Kant's book was practically unknown to astronomers till beginning of XIX century when in 1805

Alexander Humboldt mentioned this book as an outstanding achievement of natural science.

- 1789–1798: Pierre Simon Laplace (1749–1827) publishes his theory of Saturn rings system structure (Laplace 1789, 1796) in framework of his famous cosmogonic theory concept.

    1. Laplace proposes the model of rings structure as the system of many concentric narrow thin rings, fluid (iced) or monolite (solid), every ring being of its own rotational velocity "from ring to ring."
    2. I. Kant's cosmogonic theory is unknown to Laplace at all.
    3. Laplace studies the stability of a single narrow monolite ring and proves both its instability at its constant thickness and alternatively its stability at variable thickness (cross-section). Laplace insists on the model of Saturn rings as a system of distinct monolite narrow rings, every ring being of variable cross-section.
    4. Then, Laplace derives the expression for the gravitational potential of a fluid circumplanetary ring to an outer point of its surface. His result obtained analytically shows that the fluid rotating ring cross-section by a plane containing the planet rotation axis is of elliptical shape, the ellipse major axis being located in the planet equatorial plane. The cross-section of this torus-like ring was named later after P. S. Laplace as "Laplace's ellipse."

- 1859: James Clerk Maxwell (1831–1879) writes in Cambridge University his manuscript "On the stability of the motion of Saturn's ring" published as the book and then later as the article (Maxwell 1859).

    1. Maxwell uses and continues Laplace's ideas of rings system but he considers every narrow ring as a system of discrete particles, each of them moving according to Kepler's Laws. It was the first idea of the ring differential rotation "from particle to particle."
    2. Maxwell proves that every narrow ring can be stable only when contains the heavy satellite inside it, mass of this satellite being 4.5 times as total mass of the ring.
    3. Maxwell's publication can be regarded as the first work on collective processes physics in history of science.

- 1870–1874: Sofia Kovalevskaya (1850–1891) finds more general solution of Laplace's problem of stability of a three-dimensional torus-like ring in rotation.

    1. She studies during her staying in Berlin the famous Laplace's book (in five volumes) "Traité de Mecanique Celeste" (Laplace 1796).
    2. Kovalevskaya is interesting in this branch of Celestial Mechanics for which solutions of the hyperelliptic functions theory may be successfully applied.

3. Kovalevskaya studies Laplace's results in planetary rings stability and finds out that his conclusion about the elliptic shape of a rotating torus-like ring cross-section is not enough exact and can be consider as the first order solution only.

4. She tries to resolve this problem in more general form: the rotating massive torus gravitational potential is calculated by use of hyper-elliptic functions, the general stability condition is taken as a "non-escape" staying of a particle on the rotating ring surface. Central planetary gravitation, ring gravitation and rotation frame inertia are taken into account.

5. Kovalevskaya finds that a particle stays stable on the three-dimensional ring surface when this ring cross-section has a not elliptic (after Laplace) but a some oval shape. The equation of a boundary curve of this cross-section was derived analytically and exposed as some infinite convergent series expansion. The first order term in Kovalevskaya's solution turned out to be "Laplace's ellipse" again, the terms of higher orders can be determined up to arbitrary desired order.

- 1885: eleven years later after her result of 1874, Kovalevskaya decides to publish her manuscript (see above discussion). She does it under strong and friendly insisting of her colleague Hugo Gylden (1841–1896), Director of Stockholm Royal Observatory, Fellow Editor of "Acta Mathematica," famous astronomer and mathematician. Gylden advises to publish her manuscript in german journal "Astronomische Nachrichten." He writes himself the preface to her article informing the readers that he considers the presented result to be very interesting and important for astronomy and mathematics.

The paper was finally edited in 1885 (see above article III in Section 1). The result had no especially influence and importance for Planetary Physics because Maxwell's prediction about rings structure and nature was compatible with rings observations made in XIXth century. Nevertheless this paper turned out very soon to be some remarkable and important stage in Celestial Mechanics in framework of celestial bodies potentials theory. It contains both elliptic and hyperelliptic functions application for the integration in stability theory of uniform fluid mass.

Kovalevskaya's paper was immediately high estimated by both Henri Poincaré and Francois Tisserand. The later exposed Kovalevskaya's successful results (in range of other significant and independent results of this problem) in his book (Tisserand 1891). Very soon her method attained the attention of specialists in hydrodynamics as a useful hydrodynamical analogy (Lamb 1932).

As it is known, this problem development is connected with many other famous names of scientists. The glorious line of their names during XVII–XX centuries can be divided by three main stages of development (see below), according to Paul Appell's classification (Appell 1921, 1932).

## 4. Main Stages of Equilibrium Figures Theory Development

*From Newton to Laplace.* Equilibrium figures of Solar System bodies: ellipsoids of rotation and ring-like figures (Newton 1687, in "Philosophiae Naturalis Principia Mathematica," considers the Earth as an uniformly rotating slightly oblate spheroid; Huygens 1732, in "Discours de la Cause de la Pesanteur," proves that the relative equilibrium of rotating fluid achieves when the sum of gravitation and inertia forces in any point of free surface directs normally to the surface of the isotropic body; Maupertui 1732; Clairaut 1737, 1743, in "Theorie de la Figure de la Terre," considers non-isotropic bodies equilibrium; Maclaurin 1740, 1742 generalizes Clairaut's results and proves that any oblate spheroid can be an equilibrium figure finding the relationship between angular velocity and meridional section eccentricity, Maclaurin's ellipsoids being varying from sphere to infinitely thin disc; Simpson 1743; Laplace 1789, 1796, 1798, 1825 proves that an uniformly rotating figure has level surfaces coinciding with surfaces of equal pressure and those of equal density; Legendre 1789, 1800, 1830).

*From Jacobi to Poincaré.* Three-axis ellipsoids stability (Jacobi 1834 investigates the three-axis uniform ellipsoid; Liouville 1834, 1843, 1846 proves that Jacobi's ellipsoids can be in equilibrium only at some specified values of their angular momentum, being varying from axisymmetric configuration to an infinitely thin needle; Smith 1838; Meyer 1842; Plana 1853; Matthiessen 1859; Lejeune-Dirichlet 1837, 1861; Poincaré 1892, 1895; Jeans 1903, 1916).

*Poincaré and Lyapunov.* Equilibrium stability of ellipsoidal and non-ellipsoidal bodies. Figures of Poincaré [Tompson and Tait 1883; Poincaré, 1885a–c, 1886, 1887, 1888, 1892, 1901, 1902, 1911, 1913, 1914; Kovalevskaya 1885; Schwarzschild 1897; Hadamard 1897; Darwin 1902, 1908, 1910; Liapunov (Liapunoff) 1903, 1904a,b, 1905, 1906, 1907, 1908, 1909, 1912, 1914, 1916a,b, 1925, 1927; Steklow 1908; Crudeli 1909, 1910; Amber 1918; Lichtenstein 1923, 1933; Nicliborc 1929, 1931, 1932; Wavre, 1926, 1932].

Some of papers from this Appell's classification are mentioned in References (see below) together with some other conventional papers and books on the discussed topic. According to this formal classification, Sofia Kovalevskaya's paper of 1885 has importance in the last branch of presented three stages of theory development. The theory concerns to stability figures of equilibrium of uniform fluid structures in rotation connected mainly with results of Poincaré and Lyapunov.

**Acknowledgments.** The work was supported by the RFBR and Leading Scientific School (Grants 00-15-96775 and 1078.2003.2).

## References

Antonov, V. A., Timoshkova, E. I., & Kholshevnikov, K. V. 1988, Introduction to Newtonian Potential Theory (Moscow: Nauka Publ.) (in Russian)

Appell, P. 1921, 1932, Traité de Mecanique Rationelle, T. 4 (Figures d'equilibre d'une masse liquide homogene en rotation) (Paris: Gauthier-Villars)

Appell, P. 1925, Henri Poincaré (Paris)

Bell, E. T. 1937, Men of Mathematics, Chap. The Last Universalist (N.Y.: Simon and Schuster)

Chandrasekhar, S. 1969, Ellipsoidal Figures of Equilibrium (New Haven, London: Yale Univ. Press)

Darwin, G. H. 1910, Scientific Papers, Vol. 3 (Cambridge), 436

Diacu, F., & Holmes, P. 1996, Celestial Encounter: the Origin of Chaos and Stability (Princeton University Press)

Esposito, L. W. et al. 1984, Saturn's rings: structure, dynamics, and particles properties, in Saturn, ed. Gehrels & Matthews (Univ. of Arizona Press)

Garten, K. 1932, Mathem.Zeitschr., 35, 684

Garten, K., & Maruhn, K. 1932, Mathem.Zeitschr., 35, 154

Jeans, J. 1903, Philosophycal Transactions of the Royal Society of London, 200, Ser. A

Jeans, J. 1919, Problems of Cosmogony and Stellar Dynamics (Cambridge)

Kant, I. 1755, Allgemeine Naturgeschichte und Theorie des Himmels

Lamb, H. 1932, Hydrodynamics, Chap. 12, 6th edn. (Cambridge)

Laplace, P. S. 1789, Memoire sur la Theorie de l'Anneau de Saturne (Paris: Memoires de l'Academie des Sciences) [ou dans: Traité de Mecanique Celeste, 1798, Pt. 1, Livre 3 (T. 2) (Sur les Figures de Corps Celestes), Chap. 6]

Laplace, P. S. 1796, Exposition du Systeme du Monde (Paris)

Liapunoff, A. M. 1903, Memoires de l'Academie Imperiale des Sciences de St. Petersburg, 8-me Serie, 14, N 7, 1

Liapunoff, A. M. 1904a, Memoires de l'Academie Imperiale des Sciences de St. Petersburg, 8-ieme Serie, 15, N 10, 1

Liapunoff, A. M. 1904b, Annales de la Faculté des Sciences de l'Université de Toulouse, 2-ieme Serie, 6, 5

Liapunoff, A. M. 1905/1906, Memoires de l'Academie Imperiale des Sciences de St. Petersburg, 8-ieme Serie, 17, N 3, 1

Liapunoff, A. M. 1906, 1909, 1912, 1914, Memoires de l'Academie Imperiale des Sciences de St. Petersburg, 8-me Serie, 1-ere Partie, Etudes generale du probleme; 2-ieme Partie, Figures d'equilibre derivees des ellipsoides de Maclaurin; 3-ieme Partie, Figures d'equilibre derivees des ellipsoides de Jacobi; 4-ieme Partie, Nouvelle formules pour la recherches des figures d'equilibre

Liapunoff, A. M. 1907/1908, Memoires de l'Academie Imperiale des Sciences de St. Petersburg, 8-ieme Serie, 22, N 5, 1

Liapunoff, A. M. 1908, Annales de la Faculté des Sciences de l'Université de Toulouse, 2-ieme Serie, 9

Liapunoff, A. M. 1909, Annales Scientifiques de l'Ecole Normale Superieure, 3-ieme Serie, 26 (Paris)

Liapunoff, A. M. 1916a, Bulletin de l'Academie Imperiale de Sciences, 6-ieme Serie, N 3, 139

Liapunoff, A. M. 1916b, Bulletin de l'Academie Imperiale de Sciences, 6-ieme Serie, 1-iere Partie, 470; 2-ieme Partie, 589

Liapunoff, A. M. 1925, 1927, Academie des Sciences de l'Union des RSS, 1-iere Partie, 1; 2-ieme Partie, 225

Lichtenstein, L. 1922, Mathem.Zeitschr., 13, 82

Lichtenstein, L. 1933, Gleichgewichtsfiguren Rotierender Fluessigkeit (Leipzig)

Lyttleton, R. A. 1953, The Stability of Rotating Liquid Masses (Cambridge)

Maruhn, K. 1932, Mathem.Zeitschr., 36, 122

Maruhn, K. 1933, Mathem.Zeitschr., 37, 463

Matthiessen, K. 1859, Neue Untersuchungen, Schriften d. Univ. zu Kiel, VI

Maxwell, J. C. 1859, Astronomical Society of London, Monthly Notices, 19, 297

Poincaré, H. 1885a, Bulletin Astronomique, 2, 109, 117, 405
Poincaré, H. 1885b, Comptes Rendus de l'Academie des Sciences, 100, 1068; 101, 307
Poincaré, H. 1885c, Acta Mathematica, 7, 259
Poincaré, H. 1886, Comptes Rendus de l'Academie des Sciences, April
Poincaré, H. 1887, Comptes Rendus de l'Academie des Sciences, Mars
Poincaré, H. 1888, Comptes Rendus de l'Academie des Sciences, Juin
Poincaré, H. 1892, Revue Generale des Sciences, Paris, Dec
Poincaré, H. 1901/1902, Philosophical Transactions of the Royal Society of London, 198, Ser. A, 333; 200, Ser. A, 67
Poincaré, H. 1902, Lecons professees a la Sorbonne en 1900 (Paris: Gauthier-Villard)
Poincaré, H. 1911, 1913, Lecons sur les Hypotheses Cosmogoniques (Paris: Gauthier-Villars)
Poincaré, H. 1911, Bulletin Astronomique, Juillet
Poincaré, Henri 1914, Paris. Alcan. (by Vito Volterra et al.)
Polyakhova, E. N. 2003, in Order and Chaos in Stellar and Planetary Systems, Abstracts of Intern. Conf. (St. Petersburg State University), 49
Tisserand, F. 1891, Traité de Mecanique Celeste, T. 2 (Paris: Gauthier-Villars)
Thomson, W., & Tait, P. G. 1883, Treatise on Natural Philosophy (Cambridge Univ. Press)
Schwarzschild, K. 1898, Annales d. Kgl. Sternwarte, 3, 238
Wavre, R. 1932, 1934, Figures Planetaires et Geodesie, Cahiers Scientifique, Fascicule 12, N 8, (Paris: Gauthier-Villard)

*Order and Chaos in Stellar and Planetary Systems*
*ASP Conference Series, Vol. 316, 2004*
*G. Byrd, K. Kholshevnikov, A. Mylläri, I. Nikiforov, and V. Orlov, eds.*

# Some Historical Notes on the Dynamics of Spherical Star Systems

Leonid P. Ossipkov

*Department of Space Technologies & Applied Astrodynamics, St. Petersburg State University, Universitetskij pr. 28, Staryj Peterhof, St. Petersburg 198504, Russia*

**Abstract.** A history of the Dynamics of Spherical Stellar Systems is briefly reviewed with an emphasis on contributions by T. A. Agekian and V. A. Antonov.

## 1. Introduction

The dynamics of spherically symmetrical star systems is much simpler than that of axisymmetric or triaxial ones. But it can be applied to such real astronomical objects as globular clusters and spherical galaxies. In these notes we will briefly review an instructive history of the theory and recall some seminal works.

## 2. Henri Poincaré and Sir Arthur Eddington

Dynamics of star clusters has appeared as a part of Stellar Dynamics (in its modern sense) in 1914–16 in papers by A. S. Eddington and J. H. Jeans. But we must remember that H. Poincaré was their precursor. In an article "La Voie Lactée et la théorie de gaz" (containing no formulae) Poincaré (1908) critically discussed a concept of "star gas" by Lord Kelvin and pointed out that ensembles of gravitating point masses resemble gas in Crookes' tube but not the ideal gas of classical kinetic theory. Among other interesting points he noted that star evaporation is possible, that gravitating systems are collisionless and suggested a model in which all stars reach an exterior boundary of a system. Some years later Poincaré (1911) continued to analyze properties of gravitating systems in his "Leçons sur les hypothéses cosmogoniques" containing the virial theorem and a statement known now as "Jeans' Theorem" (that is fundamental for Galactic Dynamics and trivial for a mathematician).

Works of J. H. Jeans were recently discussed by Dr. L. Ciotti and we note here only one of his papers (Jeans 1916). It was one of the first articles dealing with spherical systems only. In that paper he considered models of finite mass and infinite size and found asymptotic expansions for a potential and a density (his density law was much later criticized by I. King) and an equilibrium distribution function with isotropic velocity distribution. Generalizations of such expansions were studied only in 1960s and 1970s (e.g., Ossipkov 1978).

One of the first works on Stellar Dynamics was Eddington's (1916) great paper. It is now mostly cited as a work in which self-gravitating collisionless models with spherical velocity distribution were first constructed. An integral equation for a distribution function of such models was deduced. It was found to be of generalized Abel type. As examples collisionless analogues of Emden's polytropes were constructed. But other sections ot the paper are also of interest. At first, Eddington discussed attempts of physical justification of the Schuster–Plummer's density law that was used as an approximation to the density runs in some globular clusters (e.g., von Zeipel 1913). He showed that justifications by Plummer and von Zeipel were erroneous. Later other authors tried to deduce that law from various considerations, e.g., Idlis (1953) (a tidal truncation in the Galactic field and field of other clusters) and Minin (1952) (star evaporation).

Eddington tried to find an entropy maximum for spherical systems, but failed. It was clear for him (as well as for Poincaré) that stars can evaporate from gravitating systems and their statistical equilibrium is impossible but he has done no numerical estimates. Eddington was the first who has noted that spheres with Maxwellian velocity distribution (i.e., isothermal spheres), as well as spheres with Schwarzschild's one (studied by himself in a previous paper) are not compatible with Newtonian gravitation, and a simple velocity truncation is not enough. Unfortunately, that was ignored for decades.

Five years later Eddington (1921) discussed problems of Stellar Dynamics once more. He was close to constructing self-consistent homogeneous steady models with anisotropic velocity distribution, but could not do it correctly for Lindblad's diagram (the space of isolating integrals of motion) was not known at that time. The same can be written about K. Schwarzschild's fragment, published posthumously at that time.

Note that the first paper on Stellar Dynamics in Russian was Kostitsyn's (1922) one reviewing Eddington's work. Kostitsyn has also criticized the Schuster–Plummer's law from an observational point of view. By the way, V. A. Kostitsyn was a supervisor of K. F. Ogorodnikov.

## 3. Vyacheslav Stepanov, Albert Einstein and Others

For decades after the brilliant work of Jeans and Eddington there were only few publications on the subject we discuss. The most interesting of them was a brief paper by Stepanov (1928). Professor of Moscow University V. V. Stepanov was a prominent mathematician, an expert in the theory of differential equations. He was an author of an excellent textbook and (joint with Nemytsky) a monograph "Qualitative Theory of Differential Equations," translated into English. In 1920s he worked also in Dynamical Astronomy and in so-called Dynamical Cosmogony. At the beginning of his paper Stepanov noted that systems with circular orbits are possible for any spherical gravitational potential. Ten years later Einstein (1939) also wrote about this from the point of view of general relativity. But till the 1970s it seemed that such models have no real interest, though Agekian (1962a) discussed them briefly in his manual. Stepanov also discussed models with purely radial motions. He studied an equation for the density law of such models. When a radial velocity distribution is Maxwellian, the asymptotical form of the density is $\varrho \propto r^{-2}$, as for the isothermal model. At the center again $\varrho \propto r^{-2}$ as for a singular isothermal sphere. Such models were also discussed by Agekian (1962a) but their detailed study was undertaken only in the 1970s. So,

Stepanov was the first to show that models with anisotropic velocity distribution are possible.

We have to mention also a very pleasant paper by Shiveshwarkar (1936). Analyzing Boltzmann's equation written in spherical coordinates he proved that if a distribution function (not only a mass density) is spherical and a system is stationary then it depends only on three (not six) phase space coordinates, namely $r$, $v_r$, $v_t$. In accordance with the Jeans' theorem one can consider now two integrals of motion, namely, integrals of energy and total momentum as independent arguments of a distribution function. Ogorodnikov (1958) called this statement "Shiveshwarkar's theorem." What if a density law is spherical but a distribution function does not possess spherical symmetry? The problem was solved about 1960, when Lynden-Bell (1960) and Idlis (1961) independently showed how to construct spherical rotating systems (by using a kind of "Maxwell's demon"). The distribution function of such models is obviously a function of three integrals. At last we pay attention to a brief note by Dicke (1939) in which for the first time evaporation and gravothermal catastrophe were discussed as alternatives.

## 4. Cluster Evaporation

In the late 1930s prominent achievements were reached in the study of collisional star clusters. We refer to the seminal papers of Ambartsumian (1938), who was the first to estimate the rate of star evaporation from a system and showed that the process has a great significance for evolution of open clusters. We will not write more about that work for it is well known. It is interesting that some years later Spitzer (1940) independently obtained similar results. Ambartsumian's work was immediately appreciated and developed further by S. Chandrasekhar and many others.

A classical paper by Gurevich & Levin (1950a) was a natural continuation of Ambartsumian's work (note that at All-Union Cosmogonic Workshops of the 1950s Ambartsumian was a severe opponent of Gurevich). They developed a simple analytical theory of the final evolution of evaporating clusters (under assumptions that the virial theorem is valid and cluster energy is conserved). That work was known enough by Russians (e.g., Agekian 1962a) but was rediscovered several times (by I. King, S. von Hoerner, R. Miller & E. Parker).

A further development of the theory of star evaporation from clusters, mainly by M. Hénon, I. King, R. W. Michie, and L. Spitzer is out of the scope of the present paper. There were excellent reviews by Michie (1965), Petrovskaya (1968), Hénon (1973) and a book by Spitzer (1987). But we mention a hydrodynamic model by Agekian (1963) who concluded that the evolution of star clusters due to evaporation is approximately homological. At the same time a homological model was constructed by Hénon (1961). In another work Hénon (1960) has found that stars will not escape from clusters due to weak encounters (described by the Fokker–Planck equation) if star motion in the smoothed gravitational field is taken into account. But close (strong though rare) encounters must be treated as a purely discontinuous random process described by the Kolmogorov–Feller equation, as it was stressed by Agekian (1962a).

Agekian (1959a) found the a probability of a star encounter with a given velocity change and then derived and approximately solved a balance equation for a volume element (Agekian 1959b). The latter gave him the possibility to find a more precise expression for a rate of star evaporation. A detailed theory was developed by Agekian's students V. S. Kaliberda and I. V. Petrovskaya and others. Petrovskaya (1969) formulated a general problem and ten years later reviewed the main achievements (Pietrowskaja 1979). A very clear treatment of the theory was given by Kandrup (1980). In another paper Petrovskaya (1988) has found a generalization of Holtzmark's distribution of the random force from star encounters.

Agekian (1962b, 2001) has also studied star evaporation in spherical models with radial orbits. He showed that it can be sizable, but soon transverse velocities of stars will result from stellar encounters, and collisional evolution will be slow. It seems plausible that radial motions dominate at early evolutionary stages of star clusters (Agekian & Petrovskaya 1962). So, it follows from Agekian's analysis that globular clusters may loose a significant part of their mass at the beginning of their evolution.

## 5. From Statistical Mechanics to Gravothermal Catastrophe

If it is possible to neglect with star evaporation and formation of hard binaries (which was first studied by Gurevich & Levin 1950b) then it was believed the system will reach a state of statistical equilibrium that can be found by methods of Equilibrium Statistical Mechanics. The principal differences between a "stellar gas" (graviplasmas) and the ideal gas of the classical theory of gases were known, of course (e.g., Poincaré 1908; Ogorodnikov 1948). According to Kurth (1955c, 1957) statistical mechanics of gravitating systems cannot be constructed in principle, and discussion of the problem continues even now (e.g., Kiessling 1989; Padmanabkhan 1990).

Ogorodnikov (1958) has used a method of phase space cells for finding the most probable state of spherical gravitating systems. Of course, it was an isothermal sphere with a Maxwellian velocity distribution, as it can be expected[1]. The next step was taken by Antonov (1962a) in his third seminal paper. As it is now well known now, Antonov established that there is no state of absolute maximum of entropy for gravitational systems and its relative maximum can be reached under some conditions only (Binney & Tremaine 1987). It is difficult to read and understand Antonov's original treatment (Antonov 1962a, 1963) for it practically contains no physics but cumbersome calculations of the second variation of Boltzmann's $H$-function. Antonov's results became known after a lengthy paper by Lynden-Bell & Wood (1968). They explained the physical sense of the phenomenon discovered by Antonov and suggested it be called the "gravothermal catastrophe". The deep reason of that catastrophe lies in the fact that heat capacity of gravitating systems is negative as it follows immedi-

[1]Pay attention to the following paradox. An isothermal sphere is characterized with infinite values of mass and energy. But it is assumed in seeking for the most probable state that values of a mass and an energy of a system are fixed and finite. To avoid the contradiction Kurth (1955c, 1957) wrote the gravitating sphere must be of infinitely low density.

ately from the virial theorem[2]. J. Katz, R. and D. Lynden-Bell, W. Thirring were among further investigators of the problem. A recent discussion of the gravothermal catastrophe and related topics is by Lynden-Bell (2001).

## 6. Studies in Collisionless Equilibrium

After works by Eddington, Jeans, and Stepanov there were no new studies in constructing equilibrium distribution functions until a seminal paper by Camm (1952), who discovered generalizations of collisionless polytropes with anisotropic velocity distributions. He suggested models of two kinds. For the first kind of models star velocities $v = (v_r^2 + v_t^2)^{1/2}$ must vanish at the boundary of a model. For the second kind of models only $v_r$ vanishes. The star density can be found as a solution of a nonlinear Poisson's equation. Some limiting cases were discussed by Hénon (1959). Twenty years later similar models were independently suggested by Bisnovatyi-Kogan & Zel'dovich (1969b). In another paper Bisnovatyi-Kogan & Zel'dovich (1969a) found an anisotropic distribution function for a singular isothermal sphere ($\varrho \propto r^{-2}$) (see also Antonov, Ossipkov, & Chernin 1975).

Kurth (1955a, 1957) tried to establish general properties of steady distribution functions for spherical systems. Bouvier (1962 and some later publications) briefly discussed anisotropic distribution functions (including models with an ellipsoidal velocity distributions). He found some new models for the Schuster–Plummer's density law. But it was Veltmann (1961) who explicitly wrote down an integral relation between a density and a distribution function as a function of two classical integrals of motion. To find a distribution function for a given density law some additional assumptions must be done, and examples were given in papers by Kuzmin & Veltmann (1967a, 1967b, 1972, 1973 and others). At first they studied Schuster–Plummer's spheres, then Hénon's isochrones, and, at last, generalized isochrones (including a "limiting" model rediscovered later by Hernquist). Later achievements in constructing self-consistent spherical models are too recent to be described here (e.g., Binney & Tremaine 1987). We mention here only Gerhard's models (1991) that can be considered as a generalization of models with circular orbits, for a deviation from Lindblad's envelope (a curve of circular orbits on the energy–total momentum diagram) was considered by him to be an argument of the distribution function.

General properties of star orbits in spherical gravitating systems were studied by Contopoulos (1954), and also by Kurth (1955b, 1957). Among other subjects they proved the stability of circular orbits. Hénon (1959) has discovered the isochrone potential for which the period of radial star oscillations does not depend on the value of the total angular momentum integral (as for the Keplerian motion and for the harmonic oscillator).

Unfortunately, many distribution functions found for a given density law seems to be unphysical (e.g., they involve a delta function). Sometimes it seems preferable to construct macroscopic models as solutions of the equation of hydro-

---

[2]It seems that it was discovered for gravitating gaseous systems by H. Lane in 1869 but his paper is not available for me.

static equilibrium. That equation (i.e., the Jeans' equation for steady spherical systems) was firstly found by Coutrez (1950) (who also assumed that a radial streaming velocity may not vanish) and independently by Ogorodnikov (1957). Then Agekian & Petrovskaya (1962) has constructed a generalized isothermal model with $\sigma_r^2(r) = \text{const}$, $\sigma_t^2(r) = \text{const}$, but $2\sigma_r^2(r) \neq \sigma_t^2(r)$, i.e., anisotropic velocity distribution (Coutrez 1950 has also mentioned about this model but very briefly). Many authors calculated the radius-dependence of velocity dispersions for a given density law under various suppositions on the form of velocity ellipsoid. E.g., Binney (1980) has considered a sphere that follows de Vaucouleurs' profile for surface brightness.

## 7. Vadim Antonov and Others: Pioneering Studies in Collisionless Stability

Until the 1960s some authors discussed the stability of stellar systems, but mainly qualitatively and by analogies with classical equilibrium figures (e.g., Ogorodnikov 1958). Advances in studying equilibrium models of collisionless graviplasmas provided the birth of a new branch of Galactic Dynamics, namely the theory of stability of collisionless gravitating systems. It is well known that the first Antonov's paper (published in 1960) marked the appearance of that theory. In his fourth paper Antonov (1962b) (also Antonov 1963) has applied his variational method to spherical systems with isotropic velocity distribution. His main results were formulated by Binney & Tremaine (1987) as the four Antonov's Laws. They include the following fundamental statements: for spherical systems with an unperturbed distribution function $f_0 = f_0(E)$ and $f_0'(E) < 0$ all non-radial oscillations are stable; if the third derivative of a density with respect to a (positive) potential is positive then a model is stable (Binney & Tremaine 1987; Fridman & Polyachenko 1984). Antonov's paper was published in a non-astronomical journal that was not translated from Russian. But it was almost immediately appreciated and a lot of responses followed it. We mention here only two articles, by Lynden-Bell (1966) and by Kandrup & Sygnet (1985).

Antonov's Laws give almost exhaustive analysis of the stability problem for spheres with isotropic velocity distribution. But we may expect that velocity distribution in real spherical galaxies and star clusters, both collisional and collisionless is not spherical. Are such systems stable? Studying the stability of spherical systems with almost circular orbits was initiated by Academician Ya. B. Zel'dovich. Initially it was related to a problem of constructing models with a large gravitational redshift (Bisnovatyi-Kogan & Zel'dovich 1969a). At first, Bisnovatyi-Kogan, Zel'dovich, & Fridman (1968) has proved a stability of spheres with circular orbits to radial perturbations. That result seems evident, for if spherical envelopes do not intersect one another, any particle will move under gravity of the time-independent mass. Further extensive investigations (by Fridman's team) are out of our scope (see Fridman & Polyachenko 1984). It seems surprising that even rotating uniform spheres with circular orbits were found to be stable. But are all spheres with an intermediate form of the velocity ellipsoid (i.e., with radially flattened velocity distribution) stable (as it seems natural)? For a special case of a homogeneous sphere a positive answer was given by Antonov & Nuritdinov (1990).

The opposite case of spheres with radial orbits is of more interest for astronomical applications, as it was stressed by Agekian (Agekian 1962b; Agekian & Petrovskaya 1962). An instability of such models was proved by Antonov (1973). His work was fulfilled in summer 1972, and it was presented at the All-Union Meeting on Stellar Dynamics (Alma-Ata, October 1972) and published in its Proceedings. In the course of his work Antonov maintained contact with Fridman's team (that worked in Irkutsk at that time). At that meeting Zel'dovich, Polyachenko, Fridman, & Shukhman (1972) published a review[3] in which they discussed equilibrium models and the stability of collisionless gravitating systems. They wrote that "it is evident from quantitative considerations that clusters with radial star motion ... are unstable for there is no stabilizing influence of a stellar velocity dispersion in a tangential direction."[4] They added the following footnote: "Recently Antonov informed us that he succeeded in rigorous demonstrating instability of a gravitating system with radial mass motion." Antonov's proof (see Fridman & Polyachenko 1984) is based on using the quasi-classical approximation method that left room for doubt whether it is rigorous. Antonov's (2003) recent comment on that subject (I translate below) is of interest: "Immediately after a talk A. M. Fridman noted that intuitively the result looks simply as an application of Jeans' criterion to a model with zero dispersion of transverse velocities. But actually only one of the conditions for validity of the local analysis is fulfilled, ..., a wavelength of a perturbation is small in comparison with sizes of a system. Full locality requires that the growth time of a perturbation is much smaller than a characteristic time for particle movement. But both times are of the same order $(G\varrho)^{-1/2}$. So, the result was obtained out of limits of ability of the local analysis, and, as we believe, it keeps its significance."

The above discussion dealt with linear stability, of course. But using a space–time transformation (known as Kummer–Liouville's transformation in Mathematical Physics, as Meshchersky's one in Mechanics of Variable Masses, and Schürer's transformation in Stellar Dynamics) opens some possibilities for studying non-linear oscillations of homogeneous gravitating spheres. That was clear to Kurth (1957), but in the 1950s he did not succeed in discovering concrete models, and he constructed a model of an oscillating stellar sphere only much later (Kurth 1978) when similar results had already been reached by others. A. Kalnajs was the first who studied in 1973 non-linear oscillations of a homogeneous slab of stars, and V. A. Antonov has shown in 1973 how a circular cylinder transforms into a rotating board. Then Antonov & Nuritdinov (1975) used a method of Lagrangian displacements in a phase space and studied non-linear radial oscillations of a homogeneous sphere, a cylinder and a disk. Their model was rediscovered later by others. Then Nuritdinov (1977) showed that non-radial non-linear oscillations of spherical systems are impossible. The problem of stability of such non-linear oscillations arises naturally. It was studied in

[3]The review was distributed as one of preprints of Siberian Institute of Terrestrial Magnetism, Ionosphere, and Radio Wave Propagation (SibIZMIR) and was never published in refereed journals.

[4]Translated from Russian by the author.

works by Nuritdinov in the 1980s and later. Worth mentioning here is one of his review articles (Nuritdinov 1987).

## 8. Concluding Remarks

We restricted ourselves to Dynamics of Spherical Systems only and briefly considered its development to the 1980s. The tomography problem for spherical systems and the observational foundations of modelling real systems were not discussed nor were computer simulations. Of course, it is desirable to write a more extensive historical analysis of Galactic Dynamics. The author hopes that the present paper may be useful for this purpose.

**Acknowledgments.** It is a pleasure for me to thank A. D. Chernin for encouraging discussions and J. Binney for improvements in the original text.

## References

Agekian, T. A. 1959a, AZh, 36, 41 (in Russian)

Agekian, T. A. 1959b, AZh, 36, 283 (in Russian)

Agekian, T. A. 1962a, in Course of Astrophysics and Stellar Astronomy, Vol. 2, ed. A. A. Mikhailov (Moscow: Fizmatgiz), 528 (in Russian)

Agekian, T. A. 1962b, Vestn.Leningr.Univ., Ser.math.,mekh.,astron., 1, 152 (in Russian)

Agekian, T. A. 1963, AZh, 40, 318 (in Russian)

Agekian, T. A. 2001, in Stellar Dynamics: from Classic to Modern, ed. L. P. Ossipkov & I. I. Nikiforov (St. Petersburg: Sobolev Astronomical Institute), 440

Agekian, T. A., & Petrovskaya, I. V. 1962, Uchen.zap.Leningr.Univ., 307, 187 (in Russian)

Ambartsumian, V. A. 1938, Uchen.zap.Leningr.Univ., 22, 19 (in Russian)

Antonov, V. A. 1962a, Vestn.Leningr.Univ., Ser.math.,mekh.,astron., 7, 135 (in Russian)

Antonov, V. A. 1962b, Vestn.Leningr.Univ., Ser.math.,mekh.,astron., 19, 96 (in Russian)

Antonov, V. A. 1963, Applications of the Variational Method to Stellar Dynamics and Some Other Problems (PhD Thesis) (Leningrad Univ.) (in Russian, unpublished)

Antonov, V. A. 2003, Astronomical School's Report (Kyiv), 4, 35 (in Russian)

Antonov, V. A., & Nuritdinov, S. N. 1975, Vestn.Leningr.Univ., Ser.math., mekh.,astron., 7, 133 (in Russian)

Antonov, V. A., & Nuritdinov, S. N. 1990, Ast.Tsirk., 1545, 3 (in Russian)

Antonov, V. A., Ossipkov, L. P., & Chernin, A. D. 1975, in Dynamics and Evolution of Stellar Systems, ed. K. F. Ogorodnikov (Moscow; Leningrad: VAGO, GAO), 289 (in Russian)

Binney, J. 1980, MNRAS, 180, 873

Binney, J., & Tremaine, S. 1987, Galactic Dynamics (Princeton: Princeton Univ. Press)

Bisnovatyi-Kogan, G. S., & Zel'dovich, Ya. B. 1969a, Astrofizika, 5, 223 (in Russian)

Bisnovatyi-Kogan, G. S., & Zel'dovich, Ya. B. 1969b, Astrofizika, 5, 425 (in Russian)

Bisnovatyi-Kogan, G. S., Zel'dovich, Ya. B., & Fridman A. M. 1968, Dokl.AN SSSR, 182, 794 (in Russian)

Bouvier, P. 1962, Arch.Sci. (Genéve), 15, 162

Camm, G. L. 1952, MNRAS, 112, 155

Contopoulos, G. 1954, ZAp, 35, 67
Coutrez, R. 1950, Comm.Observ.Roy.Belgique, 15, 1
Dicke, R. H. 1939, AJ, 48, 108
Eddington, A. S. 1916, MNRAS, 76, 572
Eddington, A. S. 1921, Astron.Nachr., Jubiläumsnummer, 9
Einstein, A. 1939, Ann.Math., 40, 922
Fridman, A. M., & Polyachenko, V. L. 1984, Physics of Gravitating Systems, 2 vols (New York: Springer)
Gerhard, O. E. 1991, MNRAS, 250, 812
Gurevich, L. E., & Levin, B. Yu. 1950a, Dokl.AN SSSR, 70, 781 (in Russian)
Gurevich, L. E., & Levin, B. Yu. 1950b, AZh, 27, 273 (in Russian)
Hénon, M. 1959, Ann.d'Astrophys., 22, 126
Hénon, M. 1960, Ann.d'Astrophys., 23, 668
Hénon, M. 1961, Ann.d'Astrophys., 24, 369
Hénon, M. 1973, in Dynamical Structure and Evolution of Stellar Systems. Third Advanced Course of Swiss Society of Astron. Astrophys., ed. L. Martinet, M. Mayor (Geneva: Geneva Observ.), 183
Idlis, G. M. 1953, Dokl.AN SSSR, 41, 1305 (in Russian)
Idlis, G. M. 1961, Trudy Astrofiz.inst. (Alma-Ata), 1 (in Russian)
Jeans, J. H. 1916, MNRAS, 76, 567
Kandrup, H. E. 1980, Physics Reports, 63, 1
Kandrup, H. E., & Sygnet, J. F. 1985, ApJ, 298, 27
Kiessling, M. K.-H. 1989, J.Stat.Phys., 55, 203
Kostitsyn, V. A. 1922, Trudy Glavn.Rossijsk.Astrofiz.observ., 1, 28 (in Russian)
Kurth, R. 1955a, Astron.Nachr., 282, 97
Kurth, R. 1955b, Astron.Nachr., 282, 241
Kurth, R. 1955c, ZAMP, 6, 115
Kurth, R. 1957, Introduction to the Mechanics of Stellar Systems (London: Pergamon Press)
Kurth, R. 1978, Quart.Appl.Math., 36, 325
Kuzmin, G. G., & Veltmann, Ü.-I. K. 1967a, Tartu Publ., 36, 3 (in Russian)
Kuzmin, G. G., & Veltmann, Ü.-I. K. 1967b, Tartu Publ., 36, 470 (in Russian)
Kuzmin, G. G., & Veltmann, Ü.-I. K. 1972, Tartu Publ., 40, 281 (in Russian)
Kuzmin, G. G., & Veltmann, Ü.-I. K. 1973, in Dynamics of Galaxies and Star Clusters, ed. T. B. Omarov (Alma-Ata: Nauka), 82 (in Russian)
Lynden-Bell, D. 1960, MNRAS, 120, 204
Lynden-Bell, D. 1966, in IAU Symp. 25, The Theory of Orbits in the Solar System and in Stellar Systems, ed. G. Contopoulos (London; New York: Academic Press), 78
Lynden-Bell, D. 2001, in Stellar Dynamics: from Classic to Modern, ed. L. P. Ossipkov & I. I. Nikiforov (St. Petersburg: Sobolev Astronomical Institute), 408
Lynden-Bell, D., & Wood, R. 1968, MNRAS, 138, 495
Michie, R. W. 1965, ARA&A, 2, 49
Minin, I. N. 1952, Uchen.zap.Leningr.Univ., 153, 60 (in Russian)
Nuritdinov, S. N. 1977, Vestn.Leningr.Univ., Ser.math.,mekh.,astron., 1, 151 (in Russian)
Nuritdinov, S. N. 1987, in Star Clusters, ed. K. A. Barkhatova (Sverdlovsk: Urals Univ. Press), 121 (in Russian)

Ogorodnikov, K. F. 1948, Uspekhi Astron.Nauk, 4, 3 (in Russian)
Ogorodnikov, K. F. 1957, Dokl.AN SSSR, 116, 200 (in Russian)
Ogorodnikov, K. F. 1958, Dynamics of Stellar Systems (Moscow: Fizmatgiz) (in Russian)
Ossipkov, L. P. 1978, Astrofizika, 14, 225 (in Russian)
Padmanabkhan, T. 1990, Physics Reports, 188, 285
Petrovskaya, I. V. 1968, in Astronomy 1966, Kinematics and Dynamics of Stellar Systems, ed. K. F. Ogorodnikov, T. A. Agekian (Moscow: VINITI), 132 (in Russian)
Petrovskaya, I. V. 1969, AZh, 46, 824 (in Russian)
Petrovskaya, I. V. 1988, in IAU Colloq. 96, The Few Body Problem, ed. M. Valtonen (Dordrecht: Kluwer), 275
Pietrowskaja, I. W. 1979, Postepy Astronomii, 27, 257 (in Polish)
Poincaré, H. 1906, Bull.Soc.astron.France, 20, 153
Poincaré, H. 1911, Leçons sur les Hypothèses cosmogoniques (Paris: A. Hermann et fils)
Shiveshwarkar, S. W. 1936, MNRAS, 98, 749
Spitzer, L. Jr. 1940, MNRAS, 100, 396
Spitzer, L. Jr. 1987, Dynamical Evolution of Globular Clusters (Princeton: Princeton Univ. Press)
Stepanov, V. V. 1928, AZh, 5, 132 (in Russian)
Veltmann, Ü.-I. K. 1961, Tartu Publ., 33, 387 (in Russian)
von Zeipel, H. 1913, Kungl.Svenska Vetenskap.Handl., 51, 3
Zel'dovich, Ya. B., Polyachenko, V. L., Fridman, A. M., & Shukhman, I. G. 1972, Preprint SibIZMIR SO AN SSSR, 7 (in Russian)

*Order and Chaos in Stellar and Planetary Systems*
*ASP Conference Series, Vol. 316, 2004*
*G. Byrd, K. Kholshevnikov, A. Mylläri, I. Nikiforov, and V. Orlov, eds.*

# On Potential Formulae in Stellar Dynamics

Slobodan Ninković

*Astronomical Observatory, Volgina 7, 11160 Beograd 74, Serbia and Montenegro; e-mail: sninkovic@aob.bg.ac.yu*

**Abstract.** Analytical formulae for the gravitational potential developed for the case of spherical symmetry by Kuzmin and his disciples are generalised. The possibility of additional generalisation towards comprising the axial symmetry is also analysed.

## Introduction

There are many dynamical tasks in stellar astronomy requiring the gravitational potential to be (analytically) known, for example the calculation of star orbits in a steady and axially symmetric field (e.g., Ninković et al. 2002). Realistic formulae, suitable for use, are always "welcome" in describing stellar systems. This paper is a continuation of an earlier one by the present author (Ninković 2001).

Such a suitable potential formula was proposed by Tartu astronomers (e.g., Kuzmin & Malasidze 1970; Kuzmin & Veltmann 1973). The present contibution deals with cases of generalising this formula.

## Case of Spherical Symmetry

At first one presents the generalised formula for the most simple case—spherical symmetry. Its form is

$$\Pi = \frac{G\kappa\mathcal{M}_{\rm eq}}{(\kappa^2 r^2 + r_c^2)^{1/2} \pm r_0}; \tag{1}$$

the designations used are as follows: $\Pi$ is the potential; $G$ is the gravitation constant; $\kappa$ is a dimensionless positive constant; $\mathcal{M}_{\rm eq}$ is a constant of mass dimension; $r_c$ is the "core radius", or the main scale length of the stellar system under study; and $r_0$ is the "auxiliary" scale length.

In the original form developed by the Tartu astronomers only the positive sign was in the denominator. Then the mass constant in the numerator is equal to the total mass of the system. If the sign in the denominator is negative, then the density at some distance to the centre vanishes, to become negative afterwards. This circumstance was noticed by the Tartu astronomers, but, most likely, since a negative density is physically meaningless, no attention was paid.

However, there is no need to study always a mass-distribution model in which the volume covered by the system is infinite. Some models, where the matter of the system fills a finite volume, are well known, say King's (1962) model, also the models considered in the earlier paper of the present author (Ninković 2001). Recently, the present author (Ninković 2003) considered the case of formula (1) with the negative sign in the denominator. There a formula for the limiting radius (distance at which the density vanishes) is obtained. Also, in this case the mass constant in the numerator of formula (1) will not be equal to the total mass of the system. As shown by the present author (Ninković 2003), the ratio of this mass constant to the total mass depends on the ratio $r_0/r_c$. In the same paper the formula describing the density corresponding to potential (1) with the negative sign is also obtained. Clearly, it is applied inside the limiting radius only, beyond which the density is zero and, consequently, the potential is no longer described by means of (1), but by using the point-mass formula.

As for the dimensionless constant $\kappa$ [formula (1)], it should be said that it is not essential, since, as easily seen, by dividing by it both the numerator and the denominator one would have a "new stellar system" with the scale parameters $r_c/\kappa$ and $r_0/\kappa$ instead of the "old ones" $r_c$ and $r_0$. This parameter can be useful when formula (1), or any other based on it, is fitted to a particular model of a stellar system based observationally (e.g., Ninković 2002). Therefore, in theoretical analyses it can be assumed $\kappa = 1$. Then, if the sign in the denominator is positive, potential (1), as easily seen, is reduced to the so-called generalised isochrone one (Kuzmin & Veltmann 1973), i.e., to the very well-known Plummer–Schuster case ($r_0 = 0$). It should be noted that formula (1) comprises two more special cases (in both positive sign and $\kappa = 1$), but for which the density is infinite at the centre:

i) $r_c = 0$, $r_0$ different from zero;

ii) both $r_c$ and $r_0$ equal to zero-point mass.

The case i) was indicated by Kuzmin & Veltmann (1973), to be rediscovered by Hernquist (1990).

## Axial Symmetry

The general formula (1) derived for the case of spherical symmetry can be subjected to a further generalisation, towards the axial symmetry. This time the present author will follow the approach of Miyamoto & Nagai (1975). The term $r^2 + r_c^2$ in the denominator, in view of $r^2 = R^2 + z^2$, is replaced by $R^2+[h+(z^2+r_c^2)^{1/2}]^2$, where $h$ is a positive constant. The case $h = 0$ corresponds to the spherical symmetry.

It is well known that the Miyamoto–Nagai formula has been sufficiently frequently used for the purpose of fitting the potentials of real galaxies (e.g., Ninković et al. 2002). However, as seen from the given example, the use of this formula for the case of fitting the exponential disc leads to a high total mass of the disc, its ratio to the mass in the exponential model is about 1.5. On the other hand, if the values of the square in the contribution of the exponential disc to the circular velocity, obtained numerically, were expressed through a product of the gravitation constant and an "equivalent mass" within the given distance divided by that distance, the run of that equivalent mass would not be monotonous, at some distance a maximum would occur, after which the equivalent mass becomes decreasing. This does not hold for the Miyamoto–Nagai formula where the run of the equivalent mass is monotonous, just as in the case of the spherical symmetry (then the Miyamoto–Nagai formula is reduced to the Plummer–Schuster one). This fact indicates that an alternative approach is necessary. For instance, a relatively good fit to the exponential disc with the same total mass can be obtained by applying (1) after its generalisation towards the axial symmetry in the way of Miyamoto & Nagai. Then in the denominator one has the negative sign, whereas the auxiliary distance is equal to $0.48(r_c + h)$ and $\kappa$ is equal to 0.47.

## Conclusion

As can be seen from the text presented above, both the generalisation of the formulae proposed by the Tartu School [formula (1)] and a further generalisation towards the axial symmetry can be useful in describing of real stellar systems, i.e., subsystems of galaxies.

**Acknowledgments.** This work is a part of the project "Structure, Kinematics and Dynamics of the Milky Way", No 1468, supported by the Ministry of Science, Technology and Development of Serbia.

## References

Hernquist, L. 1990, ApJ, 356, 359

King, I. 1962, AJ, 71, 64

Kuzmin, G. G., & Malasidze, G. A. 1970, Publ.Tart.Astrof.Obs.im. V. Struve, 38, 181

Kuzmin, G. G., & Veltmann, Ü.-K. 1973, Publ.Tart.Astrof.Obs.im. V. Struve, 40, 281

Miyamoto, M. & Nagai, R. 1975, PASJ, 27, 533

Ninković, S. 2001, in Stellar Dynamics: from Classic to Modern, ed. L. P. Ossipkov & I. I. Nikiforov (St. Petersburg: St. Petersburg Univ. Press), 314

Ninković, S. 2002, JENAM 2002, Abstracts, No 1.13, 516

Ninković, S., Orlov, V. V., & Petrova, A. V. 2002, Pis'ma AZh, 28, 189

Ninković, S. 2003, Serb.Ast.Journal, 166, 39

*Order and Chaos in Stellar and Planetary Systems*
*ASP Conference Series, Vol. 316, 2004*
*G. Byrd, K. Kholshevnikov, A. Mylläri, I. Nikiforov, and V. Orlov, eds.*

# Gravitational Potential of Material Wide Ring, Filled by Rosette Orbit

B. P. Kondratyev, E. S. Mukhametshina, and N. G. Trubitsina

**Abstract.** We consider the broad material ring or disk, which is a result of two-dimensional averaging of mass of moving body by rosette orbit. Such disk may be also consists of the identical elliptic orbits uniformly portioned on azimuth angle. The surface density and space Newtonian potential of the disk are obtained.

Even Gauss (see, e.g., Subbotin 1968) used the method of averaging of mass of moving body along its orbit to study of secular perturbations in celestial mechanics. As a result of this averaging, he obtained a one-dimensional material elliptical ring; the linear density of the disk is inversely proportional to velocity of the moving body in given point of the orbit.

Here we consider a broad material ring or disk, which has been obtained as a result of *two-dimensional averaging of mass of moving body by rosette orbit.* Such disk may be consists of the set of identical elliptic orbits uniformly distributed on azimuth angle.

Let come from the energy integral of the two body problem. If the central mass $M$ is very big, the energy integral in the Keplerian problem is equal

$$v^2 = GM\left(\frac{2}{r} - \frac{1}{a}\right). \tag{1}$$

The velocity squared of the second body has two components $v^2 = v_r^2 + v_\theta^2$, moreover the azimuth velocity component is found from the law of conservation of the angular moment

$$r v_\theta = \kappa\sqrt{a\left(1-e^2\right)} = \sqrt{GM}\sqrt{a\left(1-e^2\right)}. \tag{2}$$

Then, we find the radial velocity component

$$v_r = \frac{\kappa\sqrt{\left(r_a - r\right)\left(r - r_p\right)}}{r\sqrt{a}}, \quad \left\{ \begin{array}{l} r_a = a\left(1+e\right), \\ r_p = a\left(1-e\right). \end{array} \right. \tag{3}$$

In the narrow ring at $(r, r+dr)$ testing body has the response time

$$dt \sim \mathrm{const} \times \frac{r\,dr}{\sqrt{\left(r_a - r\right)\left(r - r_p\right)}}. \tag{4}$$

We shall get surface density function of the ring, if divide $dt$ on $2\pi r\,dr$

$$\sigma\left(r\right) = \frac{\widetilde{C}}{\sqrt{\left(r_a - r\right)\left(r - r_p\right)}}. \tag{5}$$

The total mass of the ring is equal

$$M_{\text{ring}} = 2\pi\widetilde{C}\int\limits_{r_p}^{r_a}\frac{r\,dr}{\sqrt{(r_a - r)\,(r - r_p)}} = \pi^2\widetilde{C}\,(r_a + r_p)\,. \tag{6}$$

The relation allows us to find the constant $\widetilde{C}$.

By definition, the potential of ring with the surface density (5) is expressed by the double integral on its area

$$\varphi\,(r, x_3) = G\widetilde{C}\int\limits_{r_p}^{r_a}\frac{r'\,dr'}{\sqrt{(r_a - r')\,(r' - r_p)}}\int\limits_0^{2\pi}\frac{d\theta'}{\sqrt{r'^2 + r^2 + x_3^2 - 2r'x_1\cos\theta}}. \tag{7}$$

The internal integral here is equal

$$2\int\limits_0^{\pi}\frac{d\theta'}{\sqrt{r'^2 + r^2 + x_3^2 - 2r'r\cos\theta}} = \frac{4}{\sqrt{r'^2 + r^2 + x_3^2 + 2r'r}}K\left(\sqrt{\frac{4r'r}{(r' + r)^2 + x_3^2}}\right), \tag{8}$$

where $K\,(...)$ is the complete elliptic integral of first kind. Now,

$$\varphi\,(r, x_3) = 4G\widetilde{C}\int\limits_{r_p}^{r_a}\frac{r'\,dr'}{\sqrt{(r_a - r')\,(r' - r_p)}}\int\limits_0^{\pi/2}\frac{d\gamma}{\sqrt{(r' + r)^2 + x_3^2 - 4r'r\sin^2\gamma}}. \tag{9}$$

Since the algebraic equation

$$r'^2 + 2r'r\cos 2\gamma + r^2 + x_3^2 = 0 \tag{10}$$

has the complex roots

$$r' = -r\cos 2\gamma \pm i\sqrt{r^2\sin^2 2\gamma + x_3^2}, \tag{11}$$

it is possible to express (9) by the double integral

$$\varphi\,(r, x_3) = 4G\widetilde{C}\int\limits_0^{\pi/2}d\gamma\int\limits_{r_p}^{r_a}\frac{r'\,dr'}{\sqrt{(r_a - r')\,(r' - r_p)\,(r' - A - iB)\,(r' - A + iB)}}. \tag{12}$$

Here

$$A = -r\cos 2\gamma, \quad B = \sqrt{r^2\sin^2 2\gamma + x_3^2}. \tag{13}$$

By substitution

$$\tan^2\frac{\theta}{2} = \frac{\cos\theta_1}{\cos\theta_2}\frac{r_a - r_p}{r' - r_p}, \tag{14}$$

where

$$\tan\theta_1 = \frac{r_a - A}{B}, \quad \tan\theta_2 = \frac{r_p - A}{B}, \quad \tan\theta_1 \le \tan\theta_2, \quad \theta_1 \le \theta_2, \tag{15}$$

the internal integral in (12) can be reduced to the form

$$I = \mu \int_{\pi}^{0} \frac{r'(\theta)\, d\theta}{\sqrt{1 - \widetilde{k}^2 \sin^2\theta}} = \frac{\sqrt{\cos\theta_1 \cos\theta_2}}{B} \int_{0}^{\pi} \frac{r'(\theta)\, d\theta}{\sqrt{1 - \widetilde{k}^2 \sin^2\theta}}, \tag{16}$$

with

$$\widetilde{k}^2 = \sin^2 \frac{\theta_1 - \theta_2}{2}. \tag{17}$$

Besides, from (14) we find $r'$ as a function from variable $\theta$

$$r'(\theta) = b - 2(r_a - r_p) \frac{\cos\theta_1 \cos\theta_2}{(\cos\theta_1 - \cos\theta_2)^2} \frac{1}{a + \cos\theta}, \tag{18}$$

with

$$a = \frac{\cos\theta_1 + \cos\theta_2}{\cos\theta_1 - \cos\theta_2} < -1, \quad b = \frac{r_a \cos\theta_1 - r_p \cos\theta_2}{\cos\theta_1 - \cos\theta_2}. \tag{19}$$

As far as

$$a^2 - 1 = \frac{4\cos\theta_1 \cos\theta_2}{(\cos\theta 1 - \cos\theta_2)^2}, \tag{20}$$

we have instead of (18)

$$r'(\theta) = b - \frac{(r_a - r_p)}{2} \frac{a^2 - 1}{a + \cos\theta}. \tag{21}$$

With $r'(\theta)$ from (21) the integral (16) reduces to the form

$$\begin{aligned} I &= \frac{\sqrt{\cos\theta_1 \cos\theta_2}}{B} \left\{ 2b \int_0^{\pi/2} \frac{d\theta}{\sqrt{1 - \tilde{k}^2 \sin^2\theta}} - \frac{r_a - r_p}{2} (a^2 - 1) \times \right. \\ &\left. \left( \int_0^{\pi/2} \frac{d\theta}{(a + \cos\theta)\sqrt{1 - \tilde{k}^2 \sin^2\theta}} + \int_0^{\pi/2} \frac{d\theta}{(a - \cos\theta)\sqrt{1 - \tilde{k}^2 \sin^2\theta}} \right) \right\} \\ &= \frac{\sqrt{\cos\theta_1 \cos\theta_2}}{B} \left\{ 2b \int_0^{\pi/2} \frac{d\theta}{\sqrt{1 - \tilde{k}^2 \sin^2\theta}} - \right. \\ &\left. (r_a - r_p)(a^2 - 1)\, a \int_0^{\pi/2} \frac{d\theta}{(a^2 - \cos^2\theta)\sqrt{1 - \tilde{k}^2 \sin^2\theta}} \right\}. \end{aligned} \tag{22}$$

It is possible to express this $I$ through both the complete elliptic integrals of first $K\left(\widetilde{k}\right)$ and third $\Pi\left(\frac{\pi}{2}, -\frac{1}{a^2-1}, \widetilde{k}\right)$ kind

$$I = \frac{\sqrt{\cos\theta_1 \cos\theta_2}}{B}\left\{2bK\left(\widetilde{k}\right) - (r_a - r_p)\, a\, \Pi\right\}. \tag{23}$$

Thus, the ring potential (12) takes the form

$$\varphi(r, x_3) = \frac{4GM}{\pi^2 (r_a + r_p)} \int\limits_0^{\pi/2} \frac{\sqrt{\cos\theta_1 \cos\theta_2}}{\sqrt{r^2 \sin^2 2\gamma + x_3^2}} \left\{2bK\left(\widetilde{k}\right) - a(r_a - r_p)\,\Pi\right\} d\gamma, \tag{24}$$

where $a$ and $b$ from (19). There is the useful relation between $a$ and $b$

$$a + \nu = \frac{2b}{r_a - r_p}, \quad \nu = \frac{r_a + r_p}{r_a - r_p} > 1. \tag{25}$$

With account of (25) it is possible to exclude the value $2b$ from (24) and then

$$\varphi(r, x_3) = \frac{4GM}{\pi^2 \nu} \int\limits_0^{\pi/2} \frac{\sqrt{\cos\theta_1 \cos\theta_2}}{\sqrt{r^2 \sin^2 2\gamma + x_3^2}} \left\{a\left[K\left(\widetilde{k}\right) - \Pi\right] + \nu K\left(\widetilde{k}\right)\right\} d\gamma. \tag{26}$$

The expression (26) is equal to (24), but it more suitable for clearing up of some limiting cases.

Because of the relations (14) in general case the angles $\theta_1$ and $\theta_2$ depend on $\gamma$, this is valid for the function $\widetilde{k} = k(\gamma)$ also.

But if the sampling point is on the symmetry axis $Ox_3$, then $r = 0$ and

$$\begin{gathered}
A = 0, \quad B = x_3, \quad \tan\theta_1 = \frac{r_a}{x_3}, \quad \tan\theta_2 = \frac{r_p}{x_3}, \\
\cos\theta_1 = \frac{x_3}{\sqrt{r_a^2 + x_3^2}}, \quad \cos\theta_2 = \frac{x_3}{\sqrt{r_p^2 + x_3^2}}, \\
a = \frac{\sqrt{r_a^2 + x_3^2} + \sqrt{r_p^2 + x_3^2}}{\sqrt{r_a^2 + x_3^2} - \sqrt{r_p^2 + x_3^2}}, \quad \widetilde{k}^2 = \frac{1}{2}\left(1 - \frac{r_a r_p + x_3^2}{\sqrt{\left(r_a^2 + x_3^2\right)\left(r_p^2 + x_3^2\right)}}\right).
\end{gathered} \tag{27}$$

In this case the integrand function in (26) does not depend from $\gamma$ at all, and we have

$$\varphi(x_3) = \frac{2GM}{\pi\nu} \frac{1}{\left[\left(r_a^2 + x_3^2\right)\left(r_p^2 + x_3^2\right)\right]^{1/4}} \left\{a\left[K\left(\widetilde{k}\right) - \Pi\right] + \nu K\left(\widetilde{k}\right)\right\}. \tag{28}$$

In particular, under large $x_3$, the asymptotic is:

$$\widetilde{k} \to 0, \quad |a| \to \infty, \quad K\left(\widetilde{k}\right) \to \frac{\pi}{2}, \quad \Pi \to \frac{\pi}{2}; \tag{29}$$

then, the member in the square brackets in (28) vanishes and we have the desired result

$$\varphi(x_3) \approx \frac{GM}{x_3}. \tag{30}$$

The attraction force of the disk on symmetry axis has a maximum at the value $x_3/r_a$.

## References

Subbotin, M. F. 1968, Introduction to theoretical astronomy (Moscow: Nauka)

*Order and Chaos in Stellar and Planetary Systems*
*ASP Conference Series, Vol. 316, 2004*
*G. Byrd, K. Kholshevnikov, A. Mylläri, I. Nikiforov, and V. Orlov, eds.*

# Limited Mass Distribution Galaxy Model

S. A. Kutuzov

*Applied Math. & Control Processes Faculty, St. Petersburg State University, Universitetskij pr. 35, Staryj Peterhof, St. Petersburg 198504, Russia*

**Abstract.** A method of constructing limited mass distribution model is suggested. It is based on giving equipotentials and potential law separately (Kutuzov & Ossipkov 1980).

Unlimited mass distribution galaxy models are not physically correct for at least two reasons. First, when extending to infinity they do not allow existence of any other galaxy and pretend to be unique in the Universe. Second, if a mass density asymptotically diminishes as $r^{-4}$ or slower ($r$ is a distance from the model centre) the components of an inertia tensor get infinite values. Consecuently one can not use Lagrange–Jacobi equations of gross-dynamics.

A correctness of isolated limited galaxy models is so much the greater, the weaker is attraction of nearby galaxies. Two methods of constructing limited models are well known. First one can simply specify a limited mass distribution and find a potential function solving Poisson equation. Since a mass density outside of a boundary is equal to zero it is necessary that an external potential be a harmonic function. So one has to solve Laplace equation.

An equipotential (we use dimensionless quantities)

$$\varphi(R, z) = \varphi_* = \text{const} \tag{1}$$

could serve as a boundary (Kutuzov 1975). Here $R$, $z$ are cylindrical coordinates. We consider rotationally symmetric models. Inner equidensities need not be coincident with equipotentials (Kuzmin 1963). Unbounded spherical, flat and cylindrical models are an exclusion. One can cut unbounded mass distribution. Then internal potential of the cut shell vanishes, and resulting potential changes. Even in a spherical model an inner escape velocity decreases.

The second method is based on giving a potential function. Since we intend to study motion of galactic objects this would be more preferable. When giving the potential a mass density must be nonnegative everywhere and a total mass must be finite. Using the density boundary (1) we do not change an internal potential. External potential must be a harmonic function.

Here we consider a special mass distribution model of galaxy which consists of infinitesimal thin "sharp" disk embedded into a halo. Halo equidensities are lens-like surfaces formed by two families of concentric spheres. Their centres lie on the rotation axis symmetrically with respect to the disk plane and have $z = \mp\mu$. The equidensities have a sharp edge on the disk plane. They coincide with equipotentials of the system (Ossipkov & Kutuzov 1987). The distance $s_\mp$ of an arbitrary point $P(R, z)$ from the $\mp\mu$ attraction centre as well as a potential argument $\xi$ are expressed via cylindrical coordinates:

$$s_{\mp}^2 = R^2 + (z \pm \mu)^2, \qquad s_{\mp} \geq \mu, \qquad \xi^2 = s_{\mp}^2 - \mu^2. \tag{2}$$

Earlier we have adopted $\mu$ to be a constant. All the equipotentials had fixed centres. Now $\mu(\xi)$ varies, hence positions of the centres depend on equipotentials size and $\xi$ is non-explicit function of coordinates. Equipotentials do not intersect each other if the following condition is fulfilled:

$$\frac{d\,[\mu(\xi)]^2}{d\,\xi^2} > -1. \tag{3}$$

A general expression of halo density is found by Ossipkov & Kutuzov (1987) on the basis of Poisson equation. Disk surface density is (Kutuzov 1989):

$$\sigma_{disk}(R) = -2k(R)\,\frac{d\,\varphi(R)}{d\,R^2}, \qquad k(R) = 2\mu(R). \tag{4}$$

The limited halo model was constructed earlier (Kutuzov 1987; Ossipkov & Kutuzov 1987) on the basis of Genkin's (1966) model. In contradiction to Genkin's model the background density added to the halo is negative. Nevertheless the full halo density is positive everywhere.

We denote an equatorial radius of the halo as $\xi_*$. The following internal and external potential laws are used:

$$\begin{aligned} \varphi_{int}(\xi) &= \frac{\alpha^2}{\alpha^2 + \xi^2} + \kappa\xi^2, && 0 \leq \xi \leq \xi_*; \\ \varphi_{ext}(\xi) &= \frac{m}{s} = \frac{m}{\sqrt{\mu_a^2 + \xi^2}}, && \xi_* \leq \xi \leq a. \end{aligned} \tag{5}$$

Here a parameter $a$ may be equal to infinity, $\mu_a = \mu(a)$ is a constant.

The external potential coincides with potential of Kuzmin's (1953) disk and therefore the density outside disk is zero. So we have just cut the halo. There is a set of five positive parameters: $m$, $\alpha$, $\kappa$, $\mu_a$, $\xi_*$. The dimensionless mass $m$ is a scale parameter while the others are structural ones. Obviously the mass has an arbitrary value. To find remaining four parameters we use a continuity condition of the potential and its first and second derivatives on the boundary. This gives three algebraic equations. As a free parameter we choose the ratio $\beta$ which defines a ratio of boundary semiaxes:

$$\beta \equiv \frac{\mu_a}{\xi_*}, \qquad \frac{z_*}{\xi_*} = \sqrt{1 + \beta^2} - \beta. \tag{6}$$

Solving the system of equations we obtain three functions of $\beta$ (Ossipkov & Kutuzov 1987). Giving $\beta$ we find $\xi_*$, $\alpha$, $\kappa$ and finally $\mu_a = \beta\xi_*$. With the known parameters we can calculate the halo density as following:

$$\varrho_{\rm halo}(\xi) = \frac{8\alpha^2(\alpha^2 + \mu_a^2)}{(\alpha^2 + \xi^2)^3} - \frac{2\alpha^2}{(\alpha^2 + \xi^2)^2} - 6\kappa, \qquad 0 \leq \xi \leq \xi_*. \tag{7}$$

This function is positive everywhere inside the halo and tends to zero near the boundary as $\xi_* - \xi$. Halo equidensities coincide with equipotentials.

Finally let us cut the disk since with the external potential (5) it remains infinite. That is a corollary of broken edge of equipotentials. To annihilate disk surface density we must turn the coefficient $k(R)$ in eq. (4) into zero outside the radius $b$. So $\mu$ is the constant $\mu_a$ in the interval $[0, a]$ and diminishes to zero in the interval $[a, b]$. Then the attraction centres merge and equipotentials become spherical in the disk plane. The sharp edge as well as the outer ring of the disk disappear outside of the sphere with the radius $b$.

Now the external potential has been changed and it makes sense to see how the halo density changes. Owing to the symmetry we restrict ourselves to the case of positive $z$. Using the variables $s = s_-$, $Z = z + \mu(\xi)$ [see eq. (2)], and a potential law

$$\varphi(\xi) = \frac{m}{s} = \frac{m}{\sqrt{\mu^2(\xi) + \xi^2}}, \quad a \leq \xi \leq b, \quad \varphi(r) = \frac{m}{r}, \quad \xi \geq b, \tag{8}$$

we obtain from Poisson equation

$$\begin{aligned} \varrho_{\rm halo}(R, z) &= 2m \left\{ -\mu' pqZ + 2 \left[ (\mu')^2 q + \mu''(pz + q\mu) \right] s^2 \right\} / (qs)^3 , \\ a \leq \xi \leq b; &\qquad \varrho_{\rm halo}(R, z) = 0, \quad \xi > b. \end{aligned} \tag{9}$$

Here $p \equiv 1 + 2\mu\mu'$, $q \equiv 1 - 2\mu' z$, and the prime sign denotes a derivative with respect to $\xi^2$. The halo density (7) in the inner part does not change. A belt is added into the outer part. Here halo equidensities do not coincide with equipotentials. The boundary equidensity is a sphere. We have the following sufficient conditions of density non-negativity: $\mu' \leq 0$, $\mu'' \geq 0$ and eq. (3). We have to fulfill these conditions specifying the function $\mu(\xi)$. Since a mass is preserved, our potential transformation simply shifts the mass of the cut disk ring into the halo belt.

So we have constructed an analytical family of limited models consisting of a razor-thin disk and a halo. Both model components are bounded in space. The model parameters and the function of a mutual distance of the equipotential centres are free. It may be $\xi_* \leq a$. Hence one can get various kinds of model.

**Acknowledgments.** The author has a great pleasure to thank the reviewer for useful remarks.

## References

Genkin, I. L. 1966, Trudy Astrofiz. in-ta AN KazSSR (Alma-Ata), 7, 16 (in Russian)

Kutuzov, S. A. 1975, Vestnik Leningr.Univ., 7, 145 (in Russian)

Kutuzov, S. A. 1987 in Din. grav. sistem i metody analit. nebesn. mekhaniki (Alma-Ata), 31 (in Russian)

Kutuzov, S. A. 1989, Vestnik Leningr.Univ., 8 , 79 (in Russian)

Kutuzov, S. A., & Ossipkov, L. P. 1980, AZh, 57, 28 (in Russian)

Kuzmin, G. G. 1953, Izvestiya AN ESSR, 2, 368 [Tartu astron. observ. Contr. N 2] (in Russian)

Kuzmin, G. G. 1963, Publ. Tartu astron. observ., 34, 9 (in Russian)

Ossipkov, L. P., & Kutuzov, S. A. 1987, Astrofizika, 27, 523 (in Russian)

*Order and Chaos in Stellar and Planetary Systems*
*ASP Conference Series, Vol. 316, 2004*
*G. Byrd, K. Kholshevnikov, A. Mylläri, I. Nikiforov, and V. Orlov, eds.*

# The Galactic Bulge Modelling

N. V. Raspopova

*Department of Space Technologies & Applied Astrodynamics, St. Petersburg State University, Universitetskij pr. 28, Staryj Petergof, St. Petersburg 198504, Russia*

**Abstract.** Mass distribution models for the Milky Way bulge are briefly discussed. We use the special case $n = 4$ of Kuzmin's family of Stäckel potentials for bulge modelling. Equidensity surfaces of the model are similar spheroids. After truncating exterior part a finite size model is obtained. Unbounded model is also considered as it has no density jump on the border. In order to get concrete model in both cases parameters should be estimated. Kutuzov's interval method is suggested for this purpose.

## 1. Introduction

Mass distribution model for the Milky Way bulge is constructed. The purpose is to determine gross-parameters of the system such as mass, effective radius. The mass can be found by measuring the kinematics of various types of stars, from the Galactic rotation curve. Stellar surface brightness distribution can be converted into a mass distribution if a constant mass-to-light ratio is adopted.

The Milky Way bulge is usually modelled as a homogeneous or non-homogeneous spheroid of 0.6–0.8 axes ratio, with the effective radius of 1–1.5 kpc. Estimates of the mass vary from $1.1 \times 10^{10}$ to $1.9 \times 10^{10} M_\odot$ (Genzel & Townes 1987; Kent 1992; Minniti 1996; Petrovskaya & Ninković 1993).

## 2. Mass Distribution

As a rule mass modelling is approximation of the observational data, but theoretical approach is possible as well. Adopting a form of the gravitational potential we can find density distribution of the bulge. Steady model of galaxy allowing of the triaxial distribution of velocities was suggested by Kuzmin (1956). It has a gravitational field which allows, besides the integrals of energy and angular momentum, of a third integral of motion. We use the special case $n = 4$ of Kuzmin's family of potentials for bulge modelling.

Following Kuzmin's approach we take dimensionless ellipsoidal coordinates $\xi_1$, $\theta$, $\xi_2$:

$$R = z_0\sqrt{(\xi_1^2 - 1)(1 - \xi_2^2)}, \quad z = z_0\xi_1\xi_2,$$

where $R$, $\theta$, $z$ are cylindrical coordinates, $z_0$ is a constant.

For the case of $n = 4$ we obtain gravitational potential in the form:

$$\Phi = \frac{2GM}{z_0\pi}\left(\frac{\xi_1 \arctan\frac{\xi_1}{\xi_0} - \xi_2 \arctan\frac{\xi_2}{\xi_0}}{\xi_1^2 - \xi_2^2}\right), \tag{1}$$

where $\xi_0$ is a dimensionless parameter, $M$ is a mass of the bulge. The density $\rho$ is determined from the Poisson equation:

$$\rho = \rho_0\left(\frac{\zeta_0}{\zeta_1\zeta_2}\right)^4, \quad \rho_0 = \frac{M}{\pi^2 z_0^3 \xi_0(1+\xi_0^2)}, \tag{2}$$

$$\zeta_0^2 = \frac{\xi_0^2}{1+\xi_0^2}, \quad \zeta_1^2 = \frac{\xi_1^2+\xi_0^2}{1+\xi_0^2}, \quad \zeta_2^2 = \frac{\xi_2^2+\xi_0^2}{1+\xi_0^2}, \tag{3}$$

where $\zeta_1$, $\zeta_2$ are other ellipsoidal coordinates, $\rho_0$ is a central density. Let's show that equidensity surfaces are similar spheroids as it was declared by Kuzmin (1956). We transform (2) via the cylindrical coordinates. Then the density has the following form:

$$\rho = \frac{\rho_0}{(s^2/b^2+1)^2}, \quad s^2 = R^2 + z^2/\zeta_0^2, \quad b^2 = (1+\xi_0^2)z_0^2.$$

Density as a function of $s$ is monotone, it has a point of inflection. $\rho(s)$ is decreasing with the increase of the distance from the center.

In the case of ellipsoidal mass distribution the exterior layer doesn't attract the mass inside it. So we can truncate exterior part of the model (2). Neither equations of motion, nor condition of integrals of motion existence will change after truncating.

## 3. Conclusions

The potential of the bulge is constructed by composing two functions. The first function is potential (1), the other one is a harmonic function. Parameters $\rho_0$, $z_0$, $\xi_0$ should be estimated in order to get concrete model. Kutuzov's interval method (Kutuzov 1988) is suggested to be applied for this purpose. The method allows to obtain the estimates that will be in the best agreement with the observational data. We also need to take into account requirements of continuity of the potential and its first and second derivative on the border. As a result, a finite size mass model of the Galaxy bulge is obtained.

The case of unbounded model is of some interest too, as it has no density jump on the border.

## References

Blanco, V. M., & Terndrup, D. M. 1987, AJ, 98, 843
Genzel, R., & Townes, C. H. 1987, ARA&A, 25, 377
Kent, S. 1992, ApJ, 387, 181
Kutuzov, S. A. 1988, Kin.Fiz.Neb.Tel, 4, 39

Kuzmin, G. G. 1956, AZh, 33, 27

Minniti, D. 1996, ApJ, 459, 579

Petrovskaya, I. V., & Ninković, S. 1993, in IAU Symp. 153, Galactic Bulges, ed. H. Dejonghe & H. J. Habing (Ghent, Belgium), 353

*Order and Chaos in Stellar and Planetary Systems*
*ASP Conference Series, Vol. 316, 2004*
*G. Byrd, K. Kholshevnikov, A. Mylläri, I. Nikiforov, and V. Orlov, eds.*

# Three-Integral Models of the Milky Way Disk

B. Famaey and A. Jorissen

*Institut d'Astronomie et d'Astrophysique, Université Libre de Bruxelles, CP 226, Boulevard du Triomphe, B-1050 Bruxelles, Belgium*

H. Dejonghe

*Sterrenkundig Observatorium, Universiteit Gent, Krijgslaan 281, B-9000 Gent, Belgium*

S. Udry and M. Mayor

*Observatoire de Genève, Chemin des Maillettes 51, CH-1290 Sauverny, Switzerland*

**Abstract.** We briefly explain how kinematical data in the solar neighbourhood can be used in order to construct fully analytical dynamical models of the Galactic disk with three integrals of the motion. The Hipparcos mission gave rise to many dynamical studies but those studies lacked the third component of the space velocities, i.e, the radial velocities. Here, we combine Hipparcos parallaxes and proper motions with CORAVEL radial velocities.

## 1. Introduction

Even though the Milky Way is a spiral barred galaxy, axisymmetric models are a necessary starting point for perturbation analyses and are thus a prerequisite if one wants to understand the effects of the bar on a theoretical basis. In order to fully exploit kinematical stellar surveys, we should construct dynamical models based on Jeans theorem. This theorem states that the phase space distribution function of a stellar system in a steady state depends only on three isolating integrals of the motion: numerical experiments showed that most orbits in realistic galactic potentials admit three such integrals. The third integral, in addition to the binding energy and the vertical component of the angular momentum, is not analytic in a general potential, so we choose to construct models with an exact analytic third integral by using Stäckel potentials.

## 2. The Potentials

Axisymmetric Stäckel potentials are best expressed in spheroidal coordinates $(\lambda, \phi, \nu)$, with $\lambda$ and $\nu$ the roots for $\tau$ of

$$\frac{\varpi^2}{\tau - a^2} + \frac{z^2}{\tau - c^2} = 1, \quad a > c, \tag{1}$$

and $(\varpi, \phi, z)$ cylindrical coordinates.

We assume that the three mass components (thin disk, thick disk and halo) generate a Kuzmin-Kutuzov potential defined by

$$V(\lambda,\nu) = -\frac{GM}{\sqrt{\lambda}+\sqrt{\nu}}. \tag{2}$$

In order to remain in the Stäckel realm, the three mass components must have the same focal distance $\sqrt{a^2-c^2}$. Our potentials are described by five parameters (axis ratios of the three components and contributions of the disks to the total mass). In recent years, Hipparcos data have enabled an accurate determination of some fundamental galactic parameters in the solar neighbourhood: the mass density $\rho_\odot$ and the Oort constants. In a first step, we looked for the subset of potentials reproducing the parameters determined in these studies, namely, $0.06\ M_\odot\,\mathrm{pc}^{-3} < \rho_\odot < 0.12\ M_\odot\,\mathrm{pc}^{-3}$ and $-3.6\ \mathrm{km\,s^{-1}kpc^{-1}} < \frac{dv_c}{d\varpi} < -1.2\ \mathrm{km\,s^{-1}kpc^{-1}}$ in the solar neigbourhood. Many different forms of the potential are compatible with these fundamental galactic parameters, and five different valid potentials were selected (see Table 1 of Famaey & Dejonghe 2003).

## 3. The Distribution Functions

To describe a stellar disk, we also need a distribution function (DF, depending on the integrals of the motion according to Jeans theorem), that we choose to be a linear combination of basis functions depending on a few parameters. We define new component DF (modified Fricke components) with 9 parameters, that depend on three integrals, and that can represent realistic stellar disks when a judicious linear combination of them is chosen in a realistic galactic potential (Famaey, Van Caelenberg, & Dejonghe 2002).

## 4. The Data

The Hipparcos mission gave rise to many kinematical and dynamical studies of the solar neighbourhood. Nevertheless those studies lacked the third component of the space velocities, i.e, the radial velocities. Now, we are able to present the kinematics of K and M giants in the solar neighbourhood, based on Hipparcos data and on a radial velocity survey for 2775 K and M giants stars with relative error on the parallax smaller than 20% (large radial velocity survey performed with the CORAVEL spectrovelocimeter, see Udry et al. 1997). The striking conclusion of this survey is that some fine structure in the $UV$-plane (especially the Hercules stream in the neighbourhood of $U=-50\ \mathrm{km\,s^{-1}}$, $V=-50\ \mathrm{km\,s^{-1}}$) is clearly apparent. This is presumably due to the non-axisymmetric effects of a rotating central bar (Dehnen 2000; Fux 2001) and is responsible for a vertex deviation $l_v \equiv 1/2\arctan\left[2\sigma_{UV}^2/(\sigma_U^2-\sigma_V^2)\right] = 10.85° \pm 1.62°$.

For the mean motion, we obtain the classical values (cf. Dehnen & Binney 1998):

$$\begin{aligned}\langle U\rangle &= -10.24 \pm 0.66\ \mathrm{km\,s^{-1}} = -U_\odot,\\ \langle V\rangle &= -20.51 \pm 0.43\ \mathrm{km\,s^{-1}},\\ \langle W\rangle &= -7.77 \pm 0.34\ \mathrm{km\,s^{-1}} = -W_\odot.\end{aligned} \tag{3}$$

More details will be given on this data set in a paper in preparation.

## 5. Perspectives

The final goal is to choose one of the five Stäckel potentials listed in Table 1 of Famaey & Dejonghe (2003), and use the quadratic programming technique described by Dejonghe (1989) to determine, for the above data combined with radial velocity data of the inner Galaxy, a distribution function in the space of the integrals of the motion. So we shall have a completely analytical axisymmetric equilibrium phase space distribution function depending on three integrals of the motion, which gives ideal initial conditions for $N$-body simulations (Fux 1997) that can reproduce the non-axisymmetric features observed in the velocity distribution of our sample.

## References

Dehnen, W., Binney, J. J. 1998, MNRAS, 298, 387

Dehnen, W. 2000, AJ, 119, 800

Dejonghe, H. 1989, ApJ, 343, 113

Famaey, B., Van Caelenberg, K., & Dejonghe, H. 2002, MNRAS, 335, 201

Famaey, B., & Dejonghe, H. 2003, MNRAS, 340, 752

Fux, R. 1997, A&A, 327, 983

Fux, R. 2001, A&A, 373, 511

Udry, S., Mayor, M., Andersen, J., Crifo, F., Grenon, M., Imbert, M., Lindgren, H., Maurice, E., Nordström, B., Pernier, B., Prevot, L., Traversa, G., & Turon, C. 1997, in The Hipparcos Venice Symposium, ESA SP-402, ed. M. A. C. Perryman, 693

*Order and Chaos in Stellar and Planetary Systems*
*ASP Conference Series, Vol. 316, 2004*
*G. Byrd, K. Kholshevnikov, A. Mylläri, I. Nikiforov, and V. Orlov, eds.*

# The Gross-Dynamics of Star Systems

Leonid P. Ossipkov
*Department of Space Technologies & Applied Astrodynamics, St. Petersburg State University, Universitetskij pr. 28, Staryj Peterhof, St. Petersburg 198504, Russia*

**Abstract.** Some applications of the gross-dynamic methods for study the equilibrium and stability of stellar systems are briefly reviewed.

## 1. Introduction

According to Kuzmin (1965) the gross-dynamics is a branch of Stellar Dynamics dealing with parameters characterizing a stellar system as a whole, such as the energy, the angular momentum, the moment of inertia tensor. The virial theorem and the conservation laws for a closed system are its well known relations. The gross-dynamics supplements the hierarchy of various levels of description of many-body systems (e.g., Uhlenbeck 1957). It provides a natural tool for studying large-scale evolutionary processes in stellar systems. As examples we recall the classical gross-dynamic theories of star evaporation by Ambartsumian (1938), Gurevich & Levin (1950), and Agekian (1958). In this paper we will briefly discuss the gross-dynamics of the collisionless graviplasmas.

## 2. How to Study Non-Steady Gravitating Systems?

The final purpose of Galactic Dynamics is to construct theories of the dynamical evolution of stellar systems. Jeans (1928), Lindblad (e.g., 1933, 1958), Chandrasekhar (1942), Gurevich (1954) and others tried to develop such theories, but failed. The most elaborated one was the Chandrasekhar's non-stationary ellipsoidal dynamics. But it contains no physical ideas and results from the stationary theory by a space-time transformation (Schürer 1943). The Boltzmann's equation and the Poisson's equation conflict in Chandrasekhar's theory prescribing various laws of evolution (the Kurth's 1949 paradox; see also Kuzmin 1964[1]). Von der Pahlen (1947) suggested to study the evolution of stellar systems representing any description function $g(\mathbf{x}, \mathbf{v}, t)$ as a series $g = g_0 + g_1 t + g_2 t^2 + \ldots$ But only tendencies of the initial evolution can be revealed so.

In general we must solve the collisionless Boltzmann's equation

$$\frac{\partial f}{\partial t} + \mathbf{v}\frac{\partial f}{\partial \mathbf{x}} + \frac{\partial \Psi}{\partial \mathbf{x}}\frac{\partial f}{\partial \mathbf{v}} = 0 \quad (1)$$

[1]Genkin & Sadykhanov (1992) shown that the Kurth's paradox cannot be removed by taking into account the mass loss due to star evaporation (as it was proposed by Kuzmin 1964).

and (for self-gravitating systems) the Poisson's equation

$$\nabla^2\Psi = -4\pi G\varrho = -4\pi G\int f(\mathbf{x},\mathbf{v},t)\,d^3\mathbf{v}\,, \tag{2}$$

where $f(\mathbf{x},\mathbf{v},t)$ is the distribution function, $\varrho(\mathbf{x},t)$ is the mass density, $\Psi(\mathbf{x},t)$ is the gravitational potential signed such that it is a positive quantitative. Unfortunately, a rigorous analysis of Eqs. (1), (2) is out of possibilities of modern Mathematical Physics. The known non-steady solutions of these equations are, as a rule, homogeneous spheres, cylinders and disks with a sharp external boundary (e.g., Antonov & Nuritdinov 1975). But there is no sharp boundary for autonomous star systems (Kuzmin 1957; Ogorodnikov 1967), and it can be a result of an external tidal action only (Bottlinger 1931; Nezhinskij 1975). Munier et al. (1979) tried to solve Eqs. (1), (2) by separating the variables. The only self-similar non-stationary solution they succeeded to find was a homogeneous sphere.

The gross-dynamics allows to use for that purpose the known method of moments. When one is restricted to moments of low orders of the distribution function, only large-scale features of the structure and kinematics of a system can be taken into account. Their evolution is governed by ordinary differential equations. So, the problem of both analytical and numerical study is much simplified.

## 3. The Gross-Dynamic Equations

Let us multiply Eq. (1) by $\prod_{i=1}^{3} x_i^{k_i} v_i^{l_i}$ and integrate over all coordinates $\mathbf{x} = (x_i)$ and velocities $\mathbf{v} = (v_i)$ $(i = 1,\, 2,\, 3)$. Denote $s = \sum_{i=1}^{3} (k_i + l_i)$. If $s = 0$ and $s = 1$ the equations are trivial. One will find the law of the mass conservation ($s = 0$) and the equations of motion of the baricenter ($s = 1$). We will restrict ourselves to $s = 2$. Some of equations for $s = 3$ were derived by Chandrasekhar (1969).

Denote

$$\langle g(\mathbf{x},\mathbf{v},t)\rangle = \frac{1}{\mathcal{M}}\int\!\!\int g(\mathbf{x},\mathbf{v},t) f(\mathbf{x},\mathbf{v},t)\,d^3\mathbf{x}\,d^3\mathbf{v}\,,$$

where $\mathcal{M} = \int f(\mathbf{x},\mathbf{v},t)\,d^3\mathbf{x}\,d^3\mathbf{v}$ is the mass of a system. We introduce the following tensors: $\mathcal{I}_{ij} = \mathcal{M}\langle x_i x_j\rangle$ (the moment of inertia tensor), $\mathcal{K}_{ij} = \mathcal{M}\langle v_i v_j\rangle$ (the kinetic energy tensor[2]), $\mathcal{W}_{ij} = \mathcal{M}\langle x_i\,\partial\Psi/\partial x_j\rangle$ (the Chandrasekhar potential energy tensor; we recall that for self-gravitating systems it is symmetric), $\mathcal{V}_{ij} = \mathcal{M}\langle v_i x_j\rangle$, $\mathcal{W}_{ij} = \mathcal{M}\langle v_i\,\partial\Psi/\partial x_j\rangle$. The tensor $\mathcal{V}$ coincides with the "$H$-tensor" discussed by Som Sunder & Kochhar (1987). Now we are able to write down the following gross-dynamic equations:

$$\frac{d}{dt}\mathcal{I}_{ij} \;=\; \mathcal{L}_{ij} + \mathcal{L}_{ji}\,, \tag{3}$$

[2] It differs by factor 2 from the kinetic energy tensor as defined, e.g., by Binney & Tremaine (1987).

$$\frac{d}{dt}\mathcal{L}_{ij} = \mathcal{K}_{ij} + \mathcal{W}_{ij}\,, \tag{4}$$

$$\frac{d}{dt}\mathcal{K}_{ij} = \mathcal{V}_{ij} + \mathcal{V}_{ji}\,. \tag{5}$$

The Lagrange–Jacobi's equation follows immediately from Eqs. (3), (4) as their trace: if $\mathcal{I} = \mathrm{Spur}\,\mathcal{I}_{ij}$, $\mathcal{K} = {}^1\!/_2\mathrm{Spur}\,\mathcal{K}_{ij}$, $\mathcal{W} = \mathrm{Spur}\,\mathcal{W}_{ij}$, then ${}^1\!/_2\ddot{\mathcal{I}} = 2\mathcal{K} + \mathcal{W}$. The trace of Eqs. (5) gives the energy conservation.

## 4. Steady State Systems

If the system is in a steady state, then the left sides of Eqs. (3)–(5) are zero, and the gross-dynamic equations enable us to relate the gross kinematic and structure parameters of galaxies. We will consider some simple examples.

### 4.1. The Scalar Virial Theorem

Numerous applications of the virial theorem are well known (e.g., Binney & Tremaine 1987). We recall only one of them following Ogorodnikov (1957). We split the kinetic energy up into the contributions from the rotation and random motions: $\mathcal{K} = \mathcal{K}_{\rm rot} + \mathcal{Q}$. Then the following inequality is evident: $2\mathcal{K}_{\rm rot} < -\mathcal{W}$. If the systems rotates with a constant angular velocity $\omega$, then $\mathcal{K}_{\rm rot} = {}^1\!/_2\mathcal{J}_{\|}\omega^2$, where $\mathcal{J}_{\|}$ is the moment of inertia relative to a rotation axis. Then $\omega^2 < -\mathcal{W}/\mathcal{J}_{\|}$ .For homogeneous systems this inequality can be transformed into the known Poincaré's inequality of Theory of Classical Figures of Equilibrium.

### 4.2. Generalizations of Maclaurin Spheroids

Let us consider an axisymmetric system. Write down the equations of the tensor virial theorem in cylindrical coordinates $\varpi$, $\phi$, $\zeta$ (e.g., Binney & Tremaine 1987; Kondratjev 1989):

$$2\mathcal{K}_{\rm rot} + \mathcal{M}\left(\sigma_\varpi^2 + \sigma_\phi^2\right) = -\mathcal{W}_{\varpi\varpi}\,, \tag{6}$$

$$\mathcal{M}\sigma_\zeta^2 = -\mathcal{W}_{\zeta\zeta}\,. \tag{7}$$

Applications of these equations (under some additional assumptions) to elliptical galaxies brought direct evidences of the anisotropy of the velocity-dispersion tensors (Binney 1976, 1978; Kondratjev 1981). It follows immediately from Eq. (6) that $2\mathcal{K}_{\rm rot} < -\mathcal{W}_{\varpi}$. If the system rotates with a constant angular velocity $\omega$, the generalized Poincaré–Ogorodnikov inequality is valid, $\omega^2 < -\mathcal{W}_{\varpi\varpi}/\mathcal{J}_{\|}$. Generally, an upper limit for the angular momentum of the system can be found from Eq. (6) (Ossipkov 1999b).

Denote $\alpha = \left(\sigma_\varpi^2 + \sigma_\phi^2\right)/(2\sigma_\zeta^2)$, $\varepsilon^2 = 2\mathcal{W}_{\varpi\varpi}/\mathcal{W}_{\zeta\zeta}$. A dimensionless parameter $\alpha$ measures the anisotropy of the velocity-ellipsoid tensor, and $\varepsilon$ measures the flattening (for disks $\varepsilon = 0$, and for cylinders $\varepsilon = \infty$). Dividing Eq. (6) by Eq. (7), we obtain

$$\mathtt{t} = \frac{1 - \alpha\varepsilon^2}{2 + \varepsilon^2}\,, \tag{8}$$

where $\mathtt{t} = \mathcal{K}_{\rm rot}/(-\mathcal{W})$ is a parameter by Ostriker & Peebles (1973). For $\alpha \neq 1$ Eq. (8) determines generalizations of the Maclaurin's ellipsoids with the anisotropic velocity dispersions. We find that for a fixed $\alpha$ the maximal value $\varepsilon_{\max} = \alpha^{-2}$, so for $\alpha < 1$ the rotation around the large axis of the system is possible (Ossipkov 1999a).

### 4.3. The Influence of the Stationary Background

Let us consider the axisymmetric system embedded into the stationary homogeneous axisymmetric background (a halo of dark matter). Then the total gravitational potential $\Psi(\varpi,\zeta) = \Psi_g(\varpi,\zeta) + \Psi_b(\varpi,\zeta)$, where $\Psi_g$ is a potential of the system and $\Psi_b = \Psi_0 - {}^1\!/_2\left(\nu_\pi^2\pi^2 + \nu_\zeta^2\zeta^2\right)$ is the potential of the background. We define the following dimensionless parameters: $\lambda_b^2 = (\nu_\zeta/\nu_\varpi)^2$ (it measures the flattening of the background), $\varepsilon_*^2 = 2\mathcal{I}_{\zeta\zeta}/\mathcal{I}_{\varpi\varpi}$ (in general it differs from $\varepsilon$), $\lambda_\zeta^2 = \langle\zeta^2\rangle\nu_\zeta^2/\sigma_\zeta^2$ (it measures a relative density of the background). It is easily to find that the tensor virial theorem yields the following relation:

$$\mathtt{t} = \frac{\left(1-\lambda_\zeta^2\right) - \varepsilon^2\left[\alpha - \lambda_\zeta^2/(\varepsilon_*^2\lambda_b^2)\right]}{\left(1-\lambda_\zeta^2)(2+\varepsilon^2\right)}\,. \tag{9}$$

It determines the equilibrium of system in the external field. For instance, if $\alpha = 1$, $\varepsilon = \varepsilon_*$, $\lambda_b = 1$, then $\mathtt{t} = \left(1-\varepsilon^2\right)/\left[\left(1-\lambda_\zeta^2\right)\left(2+\varepsilon^2\right)\right]$.

## 5. Virial Oscillations

### 5.1. Nonlinear Radial Oscillations

Following Chandrasekhar & Elbert (1972) let us consider the simplest model of the gross-dynamic evolution. Combining the Lagrange–Jacobi equation with the energy conservation $\mathcal{K} + \mathcal{W} = \mathcal{H} = \text{const}$ we obtain that $\ddot{\mathcal{I}} = 2(2\mathcal{H} - \mathcal{W})$. We denote $s^2 = 2G^{-1}(-\mathcal{W})\mathcal{M}^{-5/2}\mathcal{I}^{1/2}$. The dimensionless parameter $s^2$ measures the mass concentration in the system. Calculations show, however, that it varies in very narrow intervals for wide clases of mass distribution. It is natural to suppose that $s^2 = \text{const}$ in the course of evolution. Then the Lagrange–Jacobi equation can be solved analytically, and qualitative considarations show that systems of negative energy ($\mathcal{H} < 0$) oscillate (Ossipkov 1985).

### 5.2. Axisymmetric Oscillations of Axisymmetric Systems

We have to solve the following equations:

$$ {}^1\!/_2\ddot{\mathcal{I}}_\parallel - \mathcal{K}_\parallel = \mathcal{W}_{\varpi\varpi}\,, \qquad \dot{\mathcal{K}}_\parallel = 2\mathcal{V}_{\varpi\varpi}\,, \tag{10}$$

$$ {}^1\!/_2\ddot{\mathcal{I}}_{\zeta\zeta} - \mathcal{K}_{\zeta\zeta} = \mathcal{W}_{\zeta\zeta}\,, \qquad \dot{\mathcal{K}}_{\zeta\zeta} = 2\mathcal{V}_{\zeta\zeta}\,. \tag{11}$$

For closing the system of equations (10), (11) we suppose (Ossipkov 2000a) that

$$\mathcal{W}_{\varpi\varpi} = -\frac{s^2 G\mathcal{M}^{5/2}\mathcal{I}_\parallel}{2\left(\mathcal{I}_\parallel + \mathcal{I}_{\zeta\zeta}\right)^{3/2}}\,, \qquad \mathcal{W}_{\zeta\zeta} = -\frac{s^2 G\mathcal{M}^{5/2}\mathcal{I}_{\zeta\zeta}}{2\left(\mathcal{I}_\parallel + \mathcal{J}_{\zeta\zeta}\right)^{3/2}}\,,$$

and

$$\mathcal{V}_{\varpi\varpi} = -\frac{s^2 G \mathcal{M}^{5/2} \dot{\mathcal{I}}_{\parallel}}{4\left(\mathcal{I}_{\parallel} + \mathcal{I}_{\zeta\zeta}\right)^{3/2}}, \qquad \mathcal{W}_{\zeta\zeta} = -\frac{s^2 G \mathcal{M}^{5/2} \dot{\mathcal{I}}_{\zeta\zeta}}{4\left(\mathcal{I}_{\parallel} + \mathcal{I}_{\zeta\zeta}\right)^{3/2}},$$

with $s^2 = \text{const}$. Then we restrict ourselves to small deviation from the equilibrium state. Eigenfrequences can be found (Ossipkov 2000b). They are imaginary, that is the equilibrium state is stable, and the evolution of the system can be represented as a superposition of oscillations changing its size and flattening.

## 6. Conclusions

The gross-dynamics allows to make preliminary conclusions on the large-scale structure, the equilibrium state, and the evolution of galaxies when more detailed kinetic or macroscopic studies are not possible. The non-linear evolution was analytically considered for the simplest one-dimensional problem only. We may expect that the non-linear evolution of axisymmetric systems is more complicated and stochasticity is possible. The latter may be of great significance for understanding early evolution of galaxies and their collisionless relaxation (Chernin et al. 2002). The gross-dynamics of star clusters in the Galactic tidal field is another application of methods described here. It was briefly described in another paper (Ossipkov 2001).

## References

Agekian, T. A. 1958, AZh, 35, 26 (in Russian)

Ambartsumian, V. A. 1938, Uchen.Zap.Leningr.Univ., 22, 19 (in Russian)

Antonov, V. A., & Nuritdinov, S. N. 1975, Vestnik Leningr.Univ., Ser. math.,mekh.,ast., 7, 133 (in Russian)

Binney, J. 1976, MNRAS, 177, 19

Binney, J. 1978, MNRAS, 183, 501

Binney, J., & Tremaine, S. 1987, Galactic Dynamics (Princeton: Princeton Univ. Press)

Bottlinger, K. F. 1931, Veröff.Universitätssternwarte (Berlin–Babelsberg), 8, H. 5

Chandrasekhar, S. 1942, Principles of Stellar Dynamics (Chicago: Chicago Univ. Press)

Chandrasekhar, S. 1969, Ellipsoidal Figures of Equilibrium (New Haven: Yale Univ. Press)

Chandrasekhar, S., & Elbert, D. 1972, MNRAS, 155, 435

Chernin, A. D., Valtonen, M., Zheng, J.-Q., & Ossipkov, L. P. 2001, in Stellar Dynamics: from Classic to Modern, ed. L. P. Ossipkov, I. I. Nikiforov (St. Petersburg: St. Petersburg Univ. Press), 431

Genkin, I. L., & Sadykhanov, D. A. 1992, Trudy Astrofiz.inst. (Alma-Ata), 50, 34 (in Russian)

Gurevich, L. E. 1954, Voprosy Cosmogonii, 2, 150 (in Russian)

Gurevich, L. E., & Levin, B. Yu. 1950, Dokl.AN SSSR, 70, 781 (in Russian)

Jeans, J. H. 1928, Astronomy and Cosmogony (Cambridge: Univ. Press)

Kondratjev, B. P. 1981, Pis'ma v AZh, 7, 83 (in Russian)

Kondratjev, B. P. 1989, The Dynamics of Ellipsoidal Gravitating Figures (Moscow: Nauka) (in Russian)

Kurth, R. 1949, ZAp, 26, 175

Kuzmin, G. G. 1957, Tartu Publ., 33, 75 (in Russian)

Kuzmin, G. G. 1964, Tartu Publ., 34, 457 (in Russian)

Kuzmin, G. G. 1965, Trudy Astrofiz.inst. (Alma-Ata), 5, 11 (in Russian)

Lindblad, B. 1933, in Handbuch der Astrophysik, Band V/2 (Berlin: Springer), 937

Lindblad, B. 1959, in Handbuch der Physik, Band LIII, ed. S. Flügge (Berlin: Springer), 21

Munier, A., Feix, M., Fijalkow, E., Burg, J. R., & Gutierrez, J. 1979, A&A, 78, 64

Nezhinskij, E. M. 1975, AZh, 52, 1007 (in Russian)

Ogorodnikov, K. F. 1957, Dokl.AN SSSR, 116, 38 (in Russian)

Ogorodnikov, K. F. 1967, AZh, 44, 390 (in Russian)

Ossipkov, L. P. 1981, Ast.Tsirk., 1181, 1 (in Russian)

Ossipkov, L. P. 1999a, Kin.Fiz.Neb.Tel, Suppl., no. 2, 22

Ossipkov, L. P. 1999b, Astrofizika, 42, 597 (in Russian)

Ossipkov, L. P. 2000a, Astrofizika, 43, 293 (in Russian)

Ossipkov, L. P. 2000b, Astrofizika, 43, 483 (in Russian)

Ossipkov, L. P. 2001, in ASP Conf. Ser., Vol. 228. Dynamics of Star Clusters and the Milky Way, ed. S. Dieters, B. Fuchs, A. Just, R. Spurzem & R. Wielen (San Francisco: ASP), 341

Ostriker, J. P., & Peebles, P. J. E. 1973, ApJ, 186, 467

Schürer, M. 1943, Ast.Nachr., 273, 230

Som Sunder, G., & Kochhar, R. K. 1987, MNRAS, 221, 553

Uhlenbeck, G. E. 1957, Supplement to M. Kac, Probability and Related Topics in Physical Sciences (London: Interscience Publ.)

von der Pahlen, E. 1947, ZAp, 24, 68

*Order and Chaos in Stellar and Planetary Systems*
*ASP Conference Series, Vol. 316, 2004*
*G. Byrd, K. Kholshevnikov, A. Mylläri, I. Nikiforov, and V. Orlov, eds.*

# On the Limiting Angular Velocity of the Rotation of the Stellar Systems

B. P. Kondratycv

*Mathematical Faculty, Udmurt State University, Universitetskaya 1, Izhevsk 426034, Russia*

**Abstract.** We proved: for any rotating stationary inhomogeneous gravitating stellar system of arbitrary form with continuous density distribution is always fulfilled the important inequality $\Omega^2 \leq 2\pi G\rho_{\max}$.

In 1900 H. Poincare (see, e.g., Poincare 2000) proved the existence of important limit of angular velocity for fluid gravitating figures

$$\Omega^2 < 2\pi G\,\bar{\rho}, \tag{1}$$

where $\bar{\rho}$ is the average density of configuration. Fulfilment of the inequality guarantees a direction of total gravity inside the rotating mass and nonnegativity of the hydrostatic pressure.

For stellar systems the similar inequality have been derived by K. F. Ogorodnikov (1958) and L. P. Osipkov (1999). They came from the virial theorem. In particular, the main Osipkov's inequality (20) includes three structured parameters and, in addition, flattening of model. Thereby, its result is in strong degrees model dependent. Here we consider this problem, leaning on equations of motion themselves a stars. Is it taken into account *possible anisotropy of velocity dispersion* in stellar systems.

We shall consider in the inertial cartesian system $Oxyz$ a configuration of $N$ gravitating stars with mass $m_i$. The internal potential of the system is well known

$$\varphi(\mathbf{r}) = G\sum_j \frac{m_j}{|r_j - r|}. \tag{2}$$

Let us define the function

$$S = \sum_i m_i(x_i\dot{x}_i + y_i\dot{y}_i), \tag{3}$$

which represents a moving of the system particles in planes $Oxy$. For stationary configuration

$$\frac{dS}{dt} = \sum_i m_i\left[\dot{x}_i^2 + \dot{y}_i^2 + x_i\ddot{x}_i + y_i\ddot{y}_i\right] = 0 \tag{4}$$

or, with account of the motion equations of star,

$$\frac{dS}{dt} = \sum_i m_i\left[\dot{x}_i^2 + \dot{y}_i^2 + x_i\frac{\partial\varphi}{\partial x_i} + y_i\frac{\partial\varphi}{\partial y_i}\right] = 0. \tag{5}$$

Now we introduce the angular velocity of the rotation of the configuration $\Omega$ and reject peculiar chaotic star velocities; then from (5) one has the following inequality

$$\Omega^2 \sum_i m_i \left(x_i^2 + y_i^2\right) + \sum_i m_i \left[x_i \frac{\partial \varphi}{\partial x_i} + y_i \frac{\partial \varphi}{\partial y_i}\right] \leq 0. \tag{6}$$

In the left part of the formula (6) we transform the second sum, using the expression (2) for the potential. Obviously, the force components on unit of the mass in testing point $r_i$ are

$$\begin{aligned} \frac{\partial \varphi}{\partial x_i} &= -G \sum_j \frac{m_j (x_i - x_j)}{|r_j - r_i|^3}, \\ \frac{\partial \varphi}{\partial y_i} &= -G \sum_j \frac{m_j (y_i - y_j)}{|r_j - r_i|^3}, \end{aligned} \tag{7}$$

so

$$\sum_{i=1}^{N} m_i \left[x_i \frac{\partial \varphi}{\partial x_i} + y_i \frac{\partial \varphi}{\partial y_i}\right] = -G \sum_i \sum_j \frac{m_i m_j \left[x_i \left(x_i - x_j\right) + y_i \left(y_i - y_j\right)\right]}{|r_i - r_j|^3}. \tag{8}$$

By virtue of symmetry, the indexes $i$ and $j$ here we may interchange

$$\sum_{i=1}^{N} m_i \left[x_i \frac{\partial \varphi}{\partial x_i} + y_i \frac{\partial \varphi}{\partial y_i}\right] = -G \sum_j \sum_i \frac{m_i m_j \left[x_j \left(x_j - x_i\right) + y_j \left(y_j - y_i\right)\right]}{|r_i - r_j|^3} \tag{9}$$

and take the half-sum of the expressions (8) and (9). Then

$$\sum_{i=1}^{N} m_i \left[x_i \frac{\partial \varphi}{\partial x_i} + y_i \frac{\partial \varphi}{\partial y_i}\right] = -\frac{1}{2} G \sum_i \sum_j \frac{m_i m_j \left[\left(x_i - x_j\right)^2 + \left(y_i - y_j\right)^2\right]}{|r_i - r_j|^3}. \tag{10}$$

As a result, we shall get the following inequality

$$\Omega^2 I \leq \frac{G}{2} \sum_i \sum_j \frac{m_i m_j \left[\left(x_i - x_j\right)^2 + \left(y_i - y_j\right)^2\right]}{|r_i - r_j|^3}, \tag{11}$$

where $I$ is the inertia moment of the system with respect to rotational axis.

Before this we had discrete masses. Under continuous distribution of the matter in stellar configuration instead of (11) we have

$$\Omega^2 I \leq \frac{G}{2} \int\limits_{V_1} \int\limits_{V} \frac{\rho \rho_1 \left[\left(x - x_1\right)^2 + \left(y - y_1\right)^2\right] dx\, dy\, dz\, dx_1\, dy_1\, dz_1}{\left[\left(x - x_1\right)^2 + \left(y - y_1\right)^2 + \left(z - z_1\right)^2\right]^{3/2}}. \tag{12}$$

Write here the notations $R_1 = \sqrt{x_1^2 + y_1^2}$, $R = \sqrt{x^2 + y^2}$ and take into consideration, that $R_1$ always less $R$—then each pair of material points is taken into account only once, consequently,

$$
\begin{aligned}
&\Omega^2 I \le G \int\limits_{R_1<R}\int \frac{\rho\rho_1[(x-x_1)^2+(y-y_1)^2]}{[(x-x_1)^2+(y-y_1)^2+(z-z_1)^2]^{3/2}}\, dV\, dV_1 < \\
&2G\rho_{\max}\int\limits_{-\infty}^{+\infty}\left(\int \frac{\rho[(x-x_1)^2+(y-y_1)^2]}{[(x-x_1)^2+(y-y_1)^2+(z-z_1)^2]^{3/2}}\, dV\, dx_1\, dy_1\right) dz_1 = \\
&2\pi G\rho_{\max}\int \rho R^2\, dV\, dx_1\, dy_1 = 2\pi G\rho_{\max} I.
\end{aligned} \tag{13}
$$

Thus, for any rotating stationary inhomogeneous gravitating stellar system with arbitrary figure is always fulfilled the important inequality

$$\Omega^2 \le 2\pi G\rho_{\max} \tag{14}$$

($\rho_{\max}$ is density maximum ), which extends the known Poincare's inequality (1) on equilibrium figures with possible anisotropy of velocity dispersion.

The Poincare's inequality (1) is realized only for figures with isotropic pressure and in right its part is average density. In the inequality (14) enters namely maximal density $\rho_{\max}$.

If system has an axial symmetry and differential rotation, it be possible to define some average angular velocity through the angular moment $L$

$$\bar{\Omega} = \frac{L}{I}, \tag{15}$$

and for it the similar criterion is

$$\bar{\Omega}^2 \le 2\pi G\rho_{\max}. \tag{16}$$

This inequality is valid for stellar systems with continuous density distribution.

As well known, for figures with isotropic pressure the Poincare's inequality was improved by Crudeli $\bar{\Omega}^2 < \pi G\bar{\rho}$. However, for stellar systems because of velocity dispersion anisotropy the Crudeli argument do not pass; so the inequality (14) it is impossible to improve, rejecting factor 2 in right its part. For instance, for the uniform collisionless Freeman ellipsoid (Kondratyev 2003), which has at most fast rotation $\Omega^2 = 2A_1$, in prolate spheroid limit will be fulfilled the relation $A_1 = \pi G\rho$, and here we have just that case of the formula (14) $\Omega^2 = 2\pi G\rho$, which fall outside Crudeli limit, but does not break the limit (14).

## References


Poincare, H. 2000, Equilibrium figures of fluid mass (Moscow–Izhevsk: Institute of Computer Science)

Ogorodnikov, K. F. 1958, Dynamics of stellar systems (Moskow: GIFML)

Osipkov, L. P. 1999, Astrofizika, 42, 597 (in Russian)

Kondratyev, B. P. 2003, Potential theory and equilibrium figures (Moscow–Izhevsk: Institute of Computer Science)

*Order and Chaos in Stellar and Planetary Systems*
*ASP Conference Series, Vol. 316, 2004*
*G. Byrd, K. Kholshevnikov, A. Mylläri, I. Nikiforov, and V. Orlov, eds.*

# Agekian's Factor and Relaxation Time

Slobodan Ninković
*Astronomical Observatory, Volgina 7, 11160 Beograd 74, Serbia and Montenegro; e-mail: sninkovic@aob.bg.ac.yu*

**Abstract.** Agekian's factor is proposed as a suitable measure of the mass scatter within a stellar system. Chandrasekhar's formula for the relaxation time is analysed from this point of view and an amendment of it is proposed, where another measure of the mass scatter is introduced, which has the same qualitative behaviour as Agekian's factor. A suggestion for the future work is to study in more details the relationship between these two measures of the mass scatter.

## 1. Introduction

Agekian's factor (the formula will be given in the next section) is a dimensionless quantity describing the mass scatter in stellar systems. It has already been the subject in some papers written by the present author (Ninković 1995a,b). As said in these papers and also emphasized in the original derivation of Agekian's criterion (e.g., Agekian 1962, p. 529), this criterion is aimed at estimating the role of the irregular forces in a stellar system. Its importance is in the fact that it indicates that the role of the irregular forces does not only depend on the total number of stars, but it is also affected by the mass scatter within a given stellar system. This property has been examined in more details by the present author (Ninković 1995a) who proposed for the term depending on the mass scatter to be named Agekian's factor.

Agekian studied the role of the irregular forces by estimating the fraction of the total volume of a stellar system within which the irregular forces prevail over the regular ones. However, as well known, the irregular forces have their own time scale, referred to as the relaxation time of a stellar system. A suitable criterion for estimating the role of the irregular forces can be formulated also on this basis. Clearly, the most reasonable solution is to use the crossing time of the system as a unit for its relaxation time. Then (e.g., Binney & Tremaine 1987, p. 489) the role of the irregular forces will be a function of the total number of stars in the system. This formula, clearly, is different from the first factor of Agekian's one, but, taking into account that such formulae only indicate the oder of magnitude, it may be said that the agreement is satisfactory (e.g., Ninković 1995c).

However, the main difference between the two formulae is in the presence of a term dependent of the mass scatter in Agekian's one (Agekian's factor), unlike the other formula containing the term dependent of the total number of stars only. In the present contribution it will be shown that the alternative criterion based on the time scales can be also amended by introducing a factor depending on the mass scatter.

## 2. Chandrasekhar's Formula and Mass Scatter

Agekian's factor, as defined in Ninković (1995a), is

$$A = \frac{\langle m^{3/2} \rangle}{\langle m \rangle^{3/2}}. \tag{1}$$

Here the signs $\langle\rangle$ means the mean value taken over the whole stellar system, whereas $m$ is the designation for the mass of an individual star. It should be emphasized that the quantity A is always $\geq 1$. The minimum ($A = 1$) is attained when all stars in the system have the same mass, otherwise $A > 1$. Clearly, the stronger the mass scatter in a stellar system is, the higher is the value of Agekian's factor. It is, thus, a measure of the mass scatter in a stellar system. With regard that the mass scatter, undoubtedly, affects the efficiency of the irregular forces, it is to expect that in the case of a stronger mass scatter the relaxation time becomes shorter, i.e., the irregular forces are more efficient.

As well known, Chandrasekhar (1960, p. 64) derived a formula for the relaxation time. This formula is assumed here and its form is

$$\tau_{rel} = \frac{V_2^3}{32\pi n G^2 m_1^2 f(x) \ln X};$$

the designations used are as follows: $\tau_{rel}$ is the relaxation time, $V_2$ is the velocity of star No 2 taking part in the encounter with respect to the system centre, $n$ is the number density, $G$ is the gravitation constant, $m_1$ is the mass of star No 1, $f(x)$ is a dimensionless function, $X$ is a dimensionless quantity to be defined below, whereas the other quantities are universal constants. The dimensionless quantity $X$ is

$$X = \frac{DV_2^2}{Gm_p},$$

$D$ is the impact parameter, $G$ was already defined, $m_p$ is the mass of the pair (stars No 1 and No 2).

In Chandrasekhar's further derivation the velocity $V_2$ and the mass $m_1$ were referred to the average star of the system. Here, however, the pair is considered as consisting of "perturbed" star (No 2) and the "perturber" star (No 1). In what follows star No 2 is treated as the average star of the system, but one also takes into account the mass ratio within the pair, $\chi = m_1/m_2$. In this way the term $m_1^2$ in the denominator is replaced by the product $(\chi m_2)^2$. Since star No 2 is now "the average star" of the system, all the quantities referred to it are included in the calculation of the quantities averaged over the whole of the system, exactly as done by Chandrasekhar. Taking into account that the product of the mean mass of a single star ("mass of average star") and the number density is equal to the (mass) density, $\rho$, and that the crossing time is approximately

$$\tau_{cr} \approx \frac{1}{(G\rho)^{1/2}},$$

after some self-evident transformations, also present in Chandrasekhar's derivation, one can write

$$\frac{\tau_{rel}}{\tau_{cr}} \sim \frac{0.1N}{\ln N} \frac{1}{\langle \chi^2 \rangle}. \tag{2}$$

This is the same formula as that given in Binney & Tremaine (1987, p. 489). The only difference is the presence of the factor $1/\langle\chi^2\rangle$. This factor is analogous to Agekian's one. In formula (2) one gives its mean value taken over the whole of the system, just as in the case of Agekian's factor. For each among the possible star pairs in a stellar system there are two solutions for the mass ratio, depending on which star is considered as "perturbed", i.e., "perturbing". However after averaging these two possibilities for each pair over the whole of the system one obtains a dimensionless quantity which is $\geq 1$, where equality corresponds to the case of equal masses of all stars. Thus the qualitative behaviour is exactly the same as in the case of Agekian's factor. As could be expected, the mass scatter affects the relaxation time by shortening it.

## 3. Conclusion

Formula (2) appears as an improvement of its earlier version. The second factor contains the improvement since it indicates the importance of the mass scatter in evaluating the role of the irregular forces. It is similar to Agekian's factor and, as a consequence, the two criteria of effectiveness of the irregular forces can be compared. As said above, the qualitative behaviour of both Agekian's factor and the quantity $\langle\chi\rangle$ is the same. Therefore, both can be used as measures of the mass scatter in a stellar system. However, from the practical point of view, Agekian's factor is more suitable because it is easier for calculation provided that the mass distribution in the stellar system is known. As a task for a future work it is foreseen to study in more details the relationship of the two mass-scatter measures. It should be, certainly, also said that there are many dimensionless quantities which can be used as mass-scatter measures. For instance, the ratio of the dispersion (square root) to the mean mass, but then, as easily seen, in the case of equal masses, one obtains zero. Such a value is not practical for formulating criteria like those given above since the multiplying by zero yields zero, whereas the role of the irregular forces cannot be evaluated as zero, especially not if the total number of stars is sufficiently small. For this reason quantities for which the minimum is equal to one are very suitable.

**Acknowledgments.** This work is a part of the project "Structure, Kinematics and Dynamics of the Milky Way", No 1468, supported by the Ministry of Science, Technology and Development of Serbia.

## References

Agekian, T. A. 1962, in Kurs astrofiziki i zvezdnoj astronomii, Vol. II, ed. A. A. Mikhajlov (Moskva: gos. izd. fiz.-mat. lit.), 528

Binney, J., & Tremaine, S. 1987, Galactic Dynamics (Princeton, New Jersey: Princeton Univ. Press)

Chandrasekhar, S. 1960, Principles of Stellar Dynamics (New York: Dover Publications, Inc.)

Ninković, S. 1995a, Bull.Ast.Belgrade, 151, 1

Ninković, S. 1995b, A&A Trans., 7, 195

Ninković, S. 1995c, Publ.Obs.Ast.Belgrade, 48, 87

*Order and Chaos in Stellar and Planetary Systems*
*ASP Conference Series, Vol. 316, 2004*
*G. Byrd, K. Kholshevnikov, A. Mylläri, I. Nikiforov, and V. Orlov, eds.*

# Numerical Integration of the Landau Kinetic Equation

Evgeny Griv

*Department of Physics, Ben-Gurion University of the Negev, P.O. Box 653, Beer-Sheva 84105, Israel*

**Abstract.** The problem of the relaxation to a Maxwellian of the Landau kinetic equation is solved numerically to examine in detail the nature of the approach of the distribution function of stars in the Galaxy to equilibrium. The problem is solved as an initial value problem for stars situated in a uniformly smeared out background of a few thousand giant molecular clouds. Comparison of the exact numerical solutions with analytical estimations based on the simple Chandrasekhar (1960) "molecular-kinetic" theory shows that in some cases the difference between them is not negligibly small, and a probable reason for this is advanced.

## 1. Introduction

One can learn much about the properties of stellar systems of disk-shaped galaxies by applying mathematical methods developed previously in considerations of hot magnetized plasmas. A formal analogy between the dynamics of galaxies containing a large number of mutually gravitating stars and one-component plasmas has been pointed out repeatedly (e.g., Lynden-Bell 1967; Lin & Bertin 1984). Similarities between self-gravitating systems and ordinary plasmas arise from the common long-range nature of the basic forces, whereas differences arise from the opposite signs of these forces. Thus a connection between several plasma physics phenomena and the dynamics of disk galaxies was established. (In many respects, however, "gravitational plasmas" differ strongly from laboratory ones. For instance, in a system of $N \gg 1$ gravitationally interacting particles Debye screening, as distinct from plasmas, is absent.) Such a connection deepens our understanding of the nature and broadens the reader audience.

The heating of the stellar subsystem by giant molecular clouds was already investigated by numerically or semianalytically integrating the space-independent Fokker–Planck equation by Binney & Lacey (1988) and Jenkins & Binney (1990).[1] Important conclusions were obtained about the heating of a stellar disk under the combined influence of stochastic spiral structure and giant molecular clouds.[2] Because of the nature of the gravitational force, however, gravitating systems are always spatially *inhomogeneous*. In this work unlike

[1] The relevant equation has been derived first by Landau (1936) and then fully independently by Kuzmin (1957) for gravitational systems and Rosenbluth, MacDonald, & Judd (1957) for an ionized gas. See Trubnikov (1965) and Binney & Tremaine (1987, §8.3 therein) for a discussion.

[2] Griv, Gedalin, & Yuan (1997) have used the Landau collision integral in galactic dynamics.

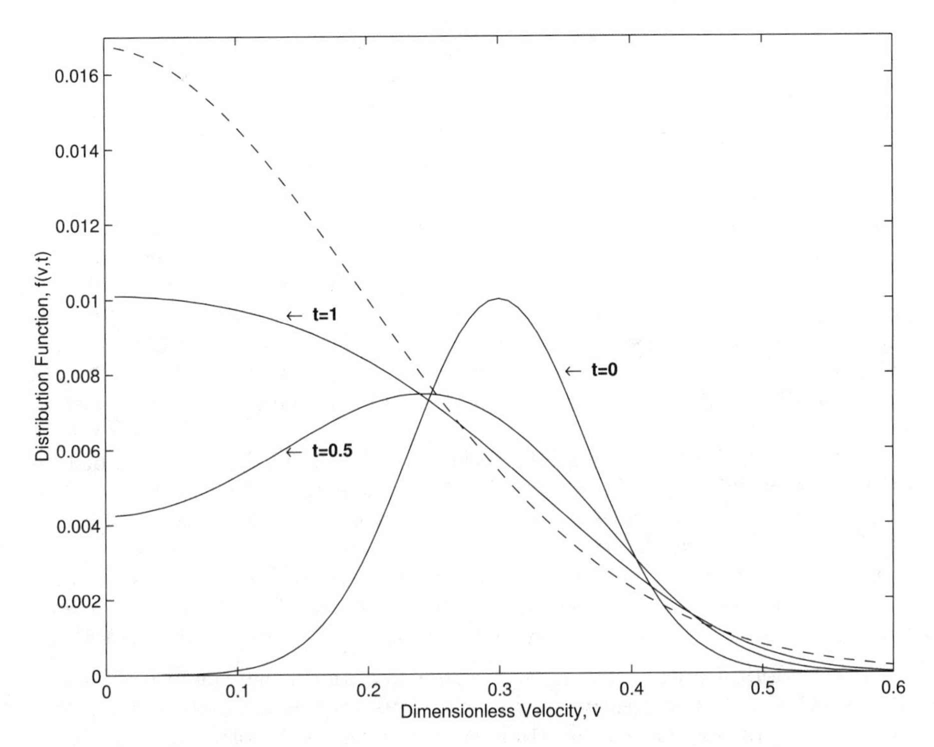


Figure 1. The relaxation of a system of stars with a mass $M_\odot$ each to the equilibrium Maxwellian distribution. The dashed line shows a Maxwellian of the same density and energy. The time is normalized so that $t = 1$ corresponds to a Chandrasekhar star–cloud relaxation time.

Binney & Lacey (1988) and Jenkins & Binney (1990), a much more realistic case is considered: the spatially inhomogeneous three-dimensional system.

## 2. Numerical Integration

The local distribution function of stars $f(\mathbf{r}, \mathbf{v}, t)$, where $\int\int f\, d\mathbf{v}\, d\mathbf{r} = N$, must satisfy the Boltzmann kinetic equation [$\Phi(\mathbf{r}, t)$ is the gravitational potential]

$$\partial f/\partial t + \mathbf{v} \cdot (\partial f/\partial \mathbf{r}) - \nabla\Phi \cdot (\partial f/\partial \mathbf{v}) = (\partial f/\partial t)^{\beta}_{\mathrm{St}}\,, \tag{1}$$

where $\mathbf{r}$ and $\mathbf{v}$ are the coordinates and velocities of stars, respectively, and the term on the right-hand side describes gravitational collisions (encounters) of stars with particles of a kind $\beta$. Landau (1936) collision integrals take the form of divergences of the particle currents in velocity space:

$$\left(\frac{\partial f}{\partial t}\right)^{\beta}_{\mathrm{St}} = 2\pi G^2 \Lambda \sum_{\beta} m_{\beta}^2 \frac{\partial}{\partial v_i} \int d^3\mathbf{v}_{\beta} I_{ij} \left[ f_{\beta}(\mathbf{v}_{\beta}) \frac{\partial f(\mathbf{v})}{\partial \mathbf{v}_j} - f(\mathbf{v}) \frac{\partial f_{\beta}(\mathbf{v}_{\beta})}{\partial \mathbf{v}_j} \right], \tag{2}$$

where $f_\beta$ is the distribution function of the particles $\beta$, $\mathbf{v}$ and $\mathbf{v}_\beta$ are the velocities, $I_{ij} = (w^2\delta_{ij} - w_i w_j)/w^3$, $\mathbf{w} = \mathbf{v} - \mathbf{v}_\beta$ is the relative velocity of the colliding particles, and $\Lambda$ is the so-called Coulomb logarithm (summation is carried out with respect to all kinds of particles including stars). The solution of Eq. (1) for a steady state is the Maxwellian distribution. In plasma physics, Eq. (1) with the collision integral (2) is called the Landau (LD) kinetic equation (Trubnikov 1965; Alexandrov, Bogdankevich, & Rukhadze 1984).

As yet the work is preliminary and has not been published, but I believe it shows promise of yielding greater understanding of the physics of galaxies. As an example, let us consider the simplest case of a weakly inhomogeneous system with the isotropic distribution function of stars in velocity space. In accord with observations, in our model the number of stars $N_s = 10^{11}$, the number of clouds $N_{cl} = 2\,000$, and the mass of each cloud $5 \times 10^5\, M_\odot$. I solved the LD equation by numerical integration, subject to the initial condition that at time $t = 0$, the initial distribution is $f(v, 0) \propto \exp[-(v - 0.3)^2]$. The resulting solution for $f(v, t)$ is plotted in Figure 1 for various values of time $t$. In fair agreement with the simple Chandrasekhar (1960) theory, at $t \approx 1$ the distribution function of stars becomes almost the Maxwellian. As calculations show, near $v = 0.25$ and $v = 0.4$ the distribution function departs slightly from the equilibrium Maxwellian. However, at $t = 1$ the lower energy portion of the spectrum is somewhat underpopulated, while the higher energy part is overpopulated. Thus, the higher- and lower-energy parts of the distribution are filled in at a much later time (at a time $t \sim 20$) as one may expect from the Chandrasekhar theory.

**Acknowledgments.** The author wishes to acknowledge David Eichler and Michael Gedalin for suggesting this study. This work was carried out under a joint Israeli Ministry of Science–Ministry of Immigrant Absorption program.

## References

Alexandrov, A. F., Bogdankevich, L. S., & Rukhadze, A. A. 1984, Principles of Plasma Electrodynamics (Berlin: Springer)

Binney, J., & Lacey, C. G. 1988, MNRAS, 230, 597

Binney, J., & Tremaine, S. 1987, Galactic Dynamics (Princeton, NJ: Princeton Univ. Press)

Chandrasekhar, S. 1960, Principles of Stellar Dynamics (New York: Dover)

Griv, E., Gedalin, M., & Yuan, C. 1997, A&A, 328, 531

Jenkins, A., & Binney, J. 1990, MNRAS, 245, 305

Kuzmin, G. G. 1957, Publ.Tartu Obs., 33, 75

Landau, L. D. 1936, Phys.Z.Sowjetunion, 10, 154; 1937, Zhur.Eksptl.Teoret.Fiz., 7, 203

Lin, C. C., & Bertin, G. 1984, Adv.Appl.Mech., 24, 155

Lynden-Bell, D. 1967, in Relativity Theory and Astrophysics, Vol. 2, Galactic Structure, ed. J. Ehlers (Providence, RI: AMS), 131

Rosenbluth, M. N., MacDonald, W. M., & Judd, D. L. 1957, Phys.Rev., 107, 1

Trubnikov, B. 1965, in Reviews of Plasma Physics, Vol. 1 (New York: Consultants Bureau), 105

# Part VIII

# Dynamics of Non-Stationary Systems

*Order and Chaos in Stellar and Planetary Systems*
*ASP Conference Series, Vol. 316, 2004*
*G. Byrd, K. Kholshevnikov, A. Mylläri, I. Nikiforov, and V. Orlov, eds.*

# Non-Linear Dynamics of Galaxies

V. A. Antonov

*Central Astronomical Observatory at Pulkovo, Pulkovskoe sh. 65/1, St. Petersburg 196140, Russia*

**Abstract.** The linear dynamics of star systems is the dynamics of small oscillations. However one has to take into account often some specific features of nonlinear processes. (1) The period-amplitude dependence. (2) The energy transfer from one to another modes. The throb phenomena are possible. Besides, the stability can be lost because of interacting modes with different energy signs. (3) To take into account the nonlinearity is necessary when the evolution path passes through critical points. The most typical examples correspond to the theory of equilibrium figures. (4) The mixing processes at their sufficiently developed stage are also nonlinear.

The non-linear dynamics of galaxies covers a rather broad domain, therefore we are interested in certain problems only. The non-linearity can be manifested either in small corrections or in a qualitative change of results. The latter is a more tempting object for an investigation.

In particular, such qualitative problems arise when the system achieves a critical state and the instability begins beyond it—in the future will a relatively sharp transition or an evolution take place? Will the system go smoothly along a different branch? In the solid body physics, the similar problem is known as one of distinguishing between the phase transitions of the first and second kind. The analogous interchange of the stability is well known in the theory of equilibrium figures. The analysis of members of the order higher than two in the expansion of energy functional in powers of the disturbance amplitude is the classical method of this distinguishing.

Let us attempt to apply this procedure to the ring-like instability of the stellar disk in the interior and exterior (predominant) gravitational fields. The exact solution appears to be very complicated and we use, besides the usual quasi-classical approximation, an additional assumption about the invariability of epicycle motions when the density waves pass. Really, the stability depends on the struggle of two factors: the "gyroscopic" effect and the self-gravitation of the disk (Toomre 1964). The presence of epicyclic motions modifies (i.e., smooths) the self-gravitation. As the result, the instability induced by waves of small length disappears. We approximately take into account this circumstance substituting individual stars by a conditional hard elliptic disklet whose density is proportional to the frequency of given stellar peculiar velocities with a fixed angular momentum.

When calculating the energy excess, we take into account the following values:

1. The excess of individual energy of the "hard disklet", that is the rotational energy plus the gravitational one in a common force field of the galaxy (the interior kinetic energy must be assumed as constant);

2. The gravitational energy of the density disturbance with regard to the smoothing factor.

The condition of vanishing the second order members in the energy excess at special critical values of $k$ gives the relation between the system parameters on the stability boundary.

For a greater length, assume that $\sigma$ is the surface density of the galactic disk, $a$ and $b$ are the transversal and radial semi-axes of the average epicyclic ellipse, $\bar{v}_r$ is the average radial velocity; then $b = \bar{v}_r/\kappa$, where $\kappa$ is the epicyclic frequency. The surface density of stars in a disklet with unit mass is assumed as

$$\mu(x, y) = \frac{1}{2\pi ab} \exp\left[-\frac{1}{2}\left(\frac{x^2}{a^2} + \frac{y^2}{b^2}\right)\right].$$

Here $x$ and $y$ are the deviations from the circular orbit along the transversal and radial directions. On the other hand, $\xi$ and $\eta$ are the shifts of the "hard disklet" center along the same directions. We next obtain the perturbation of the "disklets" density in a sliding point

$$\Delta\sigma = -\sigma\frac{\partial\eta}{\partial y}.$$

We use the Fourier expansion

$$\eta = \alpha_1 \sin ky + \alpha_2 \sin 2ky + \ldots$$

The linear approximation being used, stability thresholds on the $k$-scale are determined through

$$\kappa^2 = 2\pi Gk\sigma \exp\left(-\frac{k^2b^2}{2}\right).$$

This relation is similar to that derived on the base of exact phase density (Lin et al. 1969; Lynden-Bell & Kalnajs 1972).

The non-linear approach enforces us to distinguish $\Delta\sigma$ and the density perturbation in a fixed (non-moving) point $\delta\sigma$:

$$\delta\sigma = \beta_1 \cos ky + \beta_2 \cos 2ky + \ldots,$$

$$\beta_1 = -k\sigma\alpha_1 - \frac{3k^3\sigma}{8}\alpha_1^3 - \frac{k^2\sigma}{2}\alpha_1\alpha_2 + \ldots,$$

$$\beta_2 = -2k\sigma\alpha_2 + k^2\sigma\alpha_1^2 + \ldots$$

Then the doubled total energy of the perturbation per unit area is

$$2T = \sigma\kappa^2(\alpha_1^2 + \alpha_2^2 + \ldots) - \frac{2\pi G}{k}\left[\beta_1^2 \exp\left(-\frac{k^2b^2}{2}\right) + \frac{\beta_2^2}{2}\exp(-2k^2b^2) + \ldots\right].$$

Here we substitute the above expressions of $\beta_1$, $\beta_2$ and search for $\min T$ with respect to $\alpha_2$. For the system in the critical state, when the instability zone on the $k$-scale falls out, i.e., $kb = 1$ and $\kappa^2 = (2\pi G\sigma)/(b\sqrt{e})$, we have

$$\min T = 0 \cdot \alpha_1^2 + 0 \cdot \alpha_1^4 + O(\alpha_1^6).$$

The members of the fourth order have vanished, but this was not obvious beforehand.

The conclusion is rather disappointing: the method appears to be insufficiently exact to judge about the consequent evolution in such non-linear cases. Simultaneously, this conclusion provokes the interest to the problem, since it is clear that varied very subtle properties of the phase density have actually to influence on the type of evolution. It is required a very careful analysis to obtain a correct result (which is probably not single-valued). Also, it is a risk if a scientist extends some results of numerical experiments obtained in certain specific conditions on the general case.

A. M. Fridman has noted that he and his colleagues (Mikhailovskij et al. 1979) deduced a similar result for a gaseous disk without stars.

This is evidence of a profound general law.

Qualitatively, new properties of non-linear processes can appear for the systems consisting of parts with fully different parameters in presence of the resonance interaction between a large-scale oscillation with great energy related to the stellar system as a whole, on the one hand, and the local waves in the disk, on the other hand.

The typical features of such interaction are more pronounced for the asymmetric oscillations. So far we considered a linear disk response to the azimuthal harmonics of the second order. In the general form, the problem has been solved for a cold cylinder or disk (see, e.g., Fridman & Polyachenko 1976). We keep only main members in the density disturbance. Those must be either inversely proportional to the "untuning" of the resonance, or directly proportional to the highest (second) derivative of the self-gravitational potential. The problem is assigned as self-consisted: the reciprocal gravitational effect of a density disturbance is taken into account. It is assumed that an exterior disturbance $\varphi$ and all its consequences are restricted to the ring zone whose width is small compared with the radius of whole disk. Such an assumption is introduced for formal convenience, however it is partially justified because the dependence of the zone width (parameter $\alpha^{-1}$) falls out in final calculations. Once more we introduce the averaging according to the "hard disklets" principle.

We use the following designations:

- $r$ and $\theta$ are the polar coordinates (the disturbances include the factor of $e^{2i\theta}$);
- $\delta\sigma$ is the disturbance of the surface density $\sigma$;
- $\Omega(r)$ is the disk angular velocity;

- $\Omega'(r)$ is its derivative;
- $q$ and $\Omega_0$ are the increment and angular velocity of exterior disturbance ($q \ll \Omega_0$);
- $r_0$ is the middle radius of the zone $[\Omega(r_0) = \Omega_0]$;
- $y = r - r_0$ is the local coordinate.

The referred main members for $|y| \ll r_0$ give

$$\delta\sigma = \frac{i\beta(\varphi + \varphi_1)}{q + 2i\Omega' y} + \gamma \frac{d^2\varphi_1}{dy^2},$$

where

$$\beta = \frac{4}{r}\frac{d}{dr}\left(\frac{\Omega\sigma}{\kappa^2}\right), \quad \gamma = \frac{\sigma}{\kappa^2}.$$

Further, the coefficients of this equation can be assumed as constant. Then the Fourier expansion gives

$$\varphi(y) = \int_{-\infty}^{+\infty} e^{iky} h(k)\, dk,$$

$$s(y) \equiv \frac{\varphi + \varphi_1}{y - i\varepsilon} = \int_{-\infty}^{+\infty} \tilde{s}(k) e^{iky}\, dk \ \left(\varepsilon = \frac{q}{2\Omega'}\right)$$

etc.

We derive the following differential equation

$$\frac{d\tilde{s}}{dk} + \left(-\varepsilon + \frac{\frac{i\beta}{2\Omega'}}{\gamma k^2 - \frac{|k|}{2\pi G} e^{\frac{b^2k^2}{2}}}\right)\tilde{s} + ih(k) = 0.$$

A non-trivial solution may be found for the semi-axis $k > 0$ only (since the harmonics drift on the $k$-scale in the positive direction). Taking into account a similar drift from $k = 0$, we have approximately

$$\tilde{s} = -ik^{\frac{i\pi\beta G}{\Omega'}} \int_0^k h(k) k^{-\frac{i\pi\beta G}{\Omega'}}\, dk \ (0 < k \ll b^{-1}).$$

With the concrete function

$$h(k) = \frac{\alpha\varphi_0}{2} e^{-\alpha|k|} \ (\alpha \gg b),$$

we have at $\alpha^{-1} \ll k \ll b^{-1}$

$$\tilde{s} \approx -\frac{i\varphi_0}{2} \alpha^{\frac{i\pi\beta G}{\Omega'}} \Gamma\left(1 - \frac{i\pi\beta G}{\Omega'}\right)$$

and

$$|\tilde{s}| \approx \frac{\varphi_0}{2}\sqrt{\frac{\pi\mu}{\sinh \pi\mu}} \ \left(\mu = \frac{\pi\beta G}{\Omega'}\right).$$

To determine the subsequent behaviour of the function $\tilde{s}(k)$ and, on this base, the transfer of moment from the exterior field to the disk or *vice versa*

$$M = -\pi r_0 \, \Im \int_{-\infty}^{+\infty} \delta\sigma\varphi_\star \, dy$$

(in units of time) is a matter of some difficulty. Anyway, generally $M \neq 0$. A more accurate calculation of $M$ should permit to determine a law for changing the wave amplitude.

Let us pay attention to a contradiction with the "anti-spiral" theorem, according to which the disturbance in the steady-state disk must have a meridional symmetry plane. This circumstance excepts the systematic moment transfer, e.g., from the center to the periphery or *vice versa.*

However, actually two factors make the process to be non-stationary from the formal point of view:

1. The irreversible and, generally speaking, non-linear accumulation of the resonance consequences in a very narrow zone. The absolutely analogous phenomenon is observed when considering the Landau damping in plasma. We take into account this circumstance introducing a small positive increment $q$.

2. The irreversible divergence of density waves from the resonance zone in true sense. Moreover, in our consideration they have no any clearly fixed length. Also they are hardly capable to be concentrated again and to be reflected from any boundary to create a correct stationary situation.

The "exterior" field in typical cases is connected with the galaxy itself, namely with its spheroidal components, but not with the disk. The source can be either the mutual rotation of two separated components, the triaxial oscillation of the bulk spheroidal mass, or combinations of an intermediate nature. For the distant interaction with the disk, certain invariants are conserved: the mass, phase volume, and additional phase invariants. However, the total moment (and the energy) changes. If we know all possible states of the source for given invariants and the moment balance, we can calculate the change of the oscillation amplitude $A$. Actually, corresponding relations are hardly studied; it is only clear that there is a significant variety of possible cases. In principle, during disk interaction with something else, its oscillations can damp or increase. From general considerations, it follows only that the state of the system as a whole has to approach the thermodynamic equilibrium, i.e., the rigid rotation. However, according to all indications, normal galaxies have no time for such long-term evolution.

Do the above representations contribute to the development of the density waves theory as the base for spiral structure understanding? Yes, partially. Indeed, such considerable obstacle as the anti-spiral theorem becomes non-essential. The independence, to a considerable extent, of the resonance zone response from the exterior fields of any nature appears to be a favourable circumstance—always in actual galaxies, a chance remains to adjust a suitable exciting mode.

However, some fundamental obstacles remain as before:

1. The connection, in many cases, of spiral arms with bridges between galaxies.
2. The predominance of two-arm structures.
3. The uniformity of gaseous subsystem reaction—star formation—in spite of inevitable differences of the wave amplitudes in many times.

Toomre (1981) achieved a similar opinion.

## References

Fridman, A. M., & Polyachenko V. L. 1976, Equilibrium and stability of gravitating systems (Moscow: Nauka)

Lin, C. C., Yuan, C., & Shu, F. H. 1969, ApJ, 155, 721

Lynden-Bell, D., & Kalnajs, A. J. 1972, MNRAS, 157, 1

Mikhailovskij, A. B., Petviashvili, V. I., & Fridman, A. M. 1979, AZh, 56, 279

Toomre, A. 1964, ApJ, 139, 1217

Toomre, A. 1981, in The structure and evolution of normal galaxies, ed. S. M. Fall & D. Lynden-Bell (Cambridge), 111

*Order and Chaos in Stellar and Planetary Systems*
*ASP Conference Series, Vol. 316, 2004*
*G. Byrd, K. Kholshevnikov, A. Mylläri, I. Nikiforov, and V. Orlov, eds.*

# Towards Theory of Compulsive Phase Mixing for Non-Stationary Stellar Systems

K. T. Mirtadjieva, I. I. Kirbijekova, and S. N. Nuritdinov

*National University of Uzbekistan, Tashkent; e-mail: nurit@astrin.uzsci.net*

## 1. Introduction

The observed regular structures of a number of collisionless star systems confirm the existence of phase mixing processes at their non-stationary stage. In this work we study the strongly non-stationary case at which the compulsive phase mixing takes place (Antonov et al. 1973). In this case any small phase volume cannot evolve independently and evolve under action of the general field. Then the gravitational field at any point of phase space can stochastically vary with the time. As a result a continuous change of phase volume occurs which also has random character. It is possible to observe that phase volume is exposed to the series of random pushes of various intensity playing a deforming role. Hence the application of the statistical approach is necessary for describing and studying phase evolution of collisionless star systems.

## 2. Modelling the Compulsive Mixing

We are interested in studying the behavior of selected phase volume in the process of compulsive phase mixing. Obviously any such volume is stretched with the time in complicated way. We study here the stretching of the types sphere–ellipsoid and ellipsoid–ellipsoid. Any two states can be connected by affine transformation in phase space. Therefore the deformation of selected phase volume can be described by multiplication of random matrices. Then the power and the speed of stretching of phase volume can be characterized by eigenvalues of a resulting matrix. This method for the first time was suggested by Nuritdinov (1992).

Assume that at the moment of time $t_n$ the push effect is described by a matrix $M_n$. The choice of the matrix $M_n$ from given set of matrices is defined in random way independently of all previous process. Any two matrices should satisfy to the following conditions

$$AB \neq BA, \quad A\vec{r} \neq B\vec{r} \neq \vec{r}, \quad \det A = \det B = 1. \tag{1}$$

The meaning of first two conditions is obvious and last one means the conservation of the phase volume. Then the resulting effect at the time $t_n$ can be presented as product $M_n M_{n-1} \ldots M_1 \equiv M$ and the power and speed of the stretching at the moment $t_n$ are characterized by formulas

$$\gamma_{nj} = \lambda_{nj}/\lambda_{nj+1}, \quad \mu_{nj} = n^{-1} \ln \gamma_{nj} \quad (j = \overline{1, 2n-1}, \ \mid \lambda_{n1} \mid > \mid \lambda_{n2} \mid > \ldots), \tag{2}$$

where $\lambda_{nj}$ are eigenvalues of matrix $M$. Obviously the mixing rate will depend on concrete form of $M_n$. Hereafter we assume that $M_n = A$ with the probability $p$ (A is the matrix of the volume deformation) and $M_n = B$ with the probability $(1-p)$ (B is the matrix of turn of phase volume). Parameters $\gamma$, $\mu$ and characteristic time have to be evaluated separately for the system with the different geometry.

## 3. Two-Dimensional Case

First we consider a two-dimensional matrices in the following form

$$A(k_1, k_2) = \left\| \begin{matrix} k_1 & k_2 \\ 0 & k_1^{-1} \end{matrix} \right\|, \qquad B(\alpha) = \left\| \begin{matrix} \cos\alpha & -\sin\alpha \\ \sin\alpha & \cos\alpha \end{matrix} \right\|. \tag{3}$$

Here $\alpha$ is an angle of turn with random distribution in interval $[0, 2\pi]$, $k_1$ is the parameter of a stretching and compression on the main axes ($a_1 < k_1 < b_1$) and $k_2$ is deformation parameter, which is taken in the interval $[a_2, b_2]$. More wide intervals correspond to the more strong non-stationary power of the general field. We consider two types of deformations: 1) very strong and 2) moderate. In some special cases we can analytically calculate the step $n$ appropriate to asymptotical behavior of system. It is known the law of large numbers begins to occur on a step $n_{\min}$ satisfying the condition $n_{\min} \geq 9\mathrm{D}x/\mathrm{M}x$ (Tutubalin 1972), where M$x$ and D$x$ are the mathematical expectation and dispersion of random number $x$. Let's fix a certain random vector $\vec{r}$ in phase volume. Then $r_{n+1} = r_n\sqrt{(k\cos\theta)^2 + (k^{-1}\sin\theta)^2}$, where $\theta$ is the random angle of the interval $(0, 2\pi)$ (angle between $\vec{r}$ and the main axis of stretching at the moment of time $t_n$). The estimations of mathematical expectation and asymptotically stretching dispersion $\ln(r_n/r_0)$ are more complicated. It is possible to consider that $\mu$ has random character. Then the value $\mu_n$ statistically averaged on several realizations and corresponding to it dispersion are equal to

$$\mu_n = m^{-1}\sum_{i=1}^{m}\mu_{ni}, \quad \sigma_\mu(n) = \left[m^{-1}\sum_{i=1}^{m}\mu_{ni}^2 - \left(m^{-1}\sum_{i=1}^{m}\mu_{ni}\right)^2\right]^{1/2}, \tag{4}$$

where $m$ is number of realizations. If $\mu$ has the defined value then the ratio of eigenvalues $\lambda$ on step $n$ can be presented as $\ln\gamma_{ni} = n\mu + \xi_i$. Linearizing the sum of squares of deviations $(\ln\gamma_{ni} - n\mu + \xi_i)^2$ we have

$$\begin{aligned} \mu &= \frac{12}{m(N-\nu)(N-\nu)^2 - 1}\sum_{n=n_0}^{N}\sum_{i=1}^{m} n\ln\gamma_{ni}\frac{N+\nu+1}{2}\sum_{n=n_0}^{N}\sum_{i=1}^{m}\ln\gamma_{ni}, \\ \xi_i &= \frac{1}{N-\nu}\sum_{n=n_0}^{N}\ln\gamma_{ni} - \mu\frac{N+\nu+1}{2}, \end{aligned} \tag{5}$$

where $\nu = n_0 - 1$. First $n_0$ steps are expelled (here $n_0 = 5$) for exclusion of strong fluctuations. For random numbers the dispersion in asymptotic subjects to the law $\sigma(n) = \sigma_0(n)^{-\alpha}$, where $\alpha = 0.5$ according to the central limit theorem of the probability theory. It is interesting to find out whether asymptotical multiplication of random matrices follows to the same law or not. With this purpose and for the better clarification we present the graphic dependence for $\sigma_\mu^2 n$. In most cases the diagrams go out on "plateau" starting from $n = 25 \div 28$.

## 4. Four-Dimensional Case

Now we concentrate on the models with the higher dimension. Here we consider 4-D case for the estimation of the stretching power in a plane $(x, y)$. 4-D matrices satisfying conditions (1) should be symplectic (Tutubalin 1972). Therefore representing the matrix of effect as $M = \begin{Vmatrix} A & B \\ C & D \end{Vmatrix}$, where A, B, C, D are 2-D matrices, we find from (1) the following conditions $\det M = 1$ and $M^*LM = L$, where $L = \begin{Vmatrix} 0 & -E \\ E & 0 \end{Vmatrix}$, the mark * means the transposition, $E$ is identity matrix, and 0 is zero one. Writing by the line the last expression we get $-C^*A + A^*C = 0$, $-C^*B + A^*D = E$, $-D^*B + B^*D = 0$. We can now make the selection of matrices of a suitable kind. We are limited by four transformations matrixes: by two deformation matrixes and by two turn matrixes. As in 2-D case the numerical calculations were carried out here under two schemes: assuming random character of a stretching speeds when $\mu_1 = \mu_3 = (1/n)\ln\lambda_2/\lambda_1 = (1/n)\ln\lambda_4/\lambda_3$ and for the determined case. In selected cases the going out on the "plateau" starts from step $n = 40 \div 50$. Due to lack of memory space we reduced the number of realizations $m$ and steps $N$ up to 50.

## 5. Conclustions

The basic results of this work are following: i) the process of compulsive phase mixing by the method of product of random matrices is simulated both in 2-D and 4-D phase spaces, ii) the selection of suitable matrices describing the effects of various intensity powers is made, iii) the numerical calculations for strongly non-stationary systems are made, iv) the estimations of characteristic time of compulsive mixing are made whereas the time of mixing has been increased twice during the transition from 2-D to 4-D matrices.

At present the research of 6-D model and comparative analysis of results are carried out.

## References

Antonov, V. A., Nuritdinov, S. N., & Osipkov, L. P. 1973, in Dinamika galaktik i zvezdnih skopleniy, ed. T. B. Omarov (Alma-Ata: Nauka), 55

Nuritdinov, S. N. 1993, in Proc. IAU Coll. 132, Instability, Chaos and Predictability in Celestial Mechanics and Stellar Dynamics, ed. K. B. Bhatnagar (Nova Science Publishers), 39

Tutubalin, V. N. 1972, Teoriya veroyatnostej (Moskva: MGU), 17, 266

*Order and Chaos in Stellar and Planetary Systems*
*ASP Conference Series, Vol. 316, 2004*
*G. Byrd, K. Kholshevnikov, A. Mylläri, I. Nikiforov, and V. Orlov, eds.*

# On the Dynamics of Non-Stationary Binary Stellar Systems

A. A. Bekov

*Fesenkov Astrophysical Institute, Almaty 480020, Kazakhstan;*
*e-mail: bekov@mail.ru*

**Abstract.** The motion of a test body (gas, dust particle, star) in the external gravitational field of the binary stars and galaxy systems with slowly variable some physical parameters of radiating components is considered on the base of restricted non-stationary photogravitational three and two bodies problems. The families of polar and coplanar solutions are obtained. These solutions give the possibility of the dynamical and structure interpretations of the binary young evolving stars and galaxies.

## 1. Introduction

At present time the non-stationary dynamical problems of astronomy are intensively investigated (Omarov 2002; Bekov & Omarov 2003). The variations of some physical parameters of massive celestial bodies allow the experimental definition. Therefore, the formulation and investigation of the celestial mechanics problems are necessary, taking into account the variation with the time of some these physical parameters. Consequently, we can take into account the variations of the gravitational effect in the motion of the test body. The motion of test body (gas, dust particle, star) in the external gravitational field of the binary stars and galaxy systems with slowly variable physical parameters: mass, sizes and form, is considered as the dynamical model. In addition we take into account the variable of reduction parameters for radiating and gravitating bodies in the photogravitational formulation of problem. The motion of particles is investigated in the frame of restricted non-stationary photogravitational three and two bodies problems. The obtained results allow the possibility to carry out the quantitative and qualitative analysis of effects of variable gravitation in the motion of celestial bodies.

## 2. The Motion in the Neighbourhood of Binary Stellar System

Let's consider the motion of test body (gas, dust, star) in gravitational field of the binary star or galaxy on the base of the restricted non-stationary photogravitational three-body problem. Particular solutions of stationary case of the problem are considered in work by Radzievskij (1953). For the non-stationary case we can point out analogous particular solutions of the problem (Bekov & Ristigulova 2002). The equations of motion for passive gravitating material point in rotating barycentric system of coordinates $Oxyz$, the plane $xy$ of that coincides with the plane of motion of the main bodies, and $x$ axis always passes through these points, have the following form:

$$\begin{aligned} \ddot{x} - 2\omega\dot{y} &= \omega^2 x + \dot{\omega} y - \mu_1 \frac{x - x_1}{r_1^3} - \mu_2 \frac{x - x_2}{r_2^3}, \\ \ddot{y} + 2\omega\dot{x} &= \omega^2 y - \dot{\omega} x - \mu_1 \frac{y}{r_1^3} - \mu_2 \frac{y}{r_2^3}, \\ \ddot{z} &= -\mu_1 \frac{z}{r_1^3} - \mu_2 \frac{z}{r_2^3}. \end{aligned} \tag{1}$$

Here $r_1$, $r_2$ are the distances of a test body from the main bodies, $\omega$ is their angular velocity of motion, and

$$\mu_i = G q_i M_i \quad (i = 1, 2), \tag{2}$$

where $G$ is the gravitational constant, $M_i$ are masses of the main bodies, $q_i$ is reduction parameters, which are functions of time.

Let's consider the case of changing of parameters $q_i$ in the interval of the real scale for the planet systems:

$$0 < q_i \leq 1. \tag{3}$$

Then, as in the case of the restricted three-body problem with the variable masses (Bekov 1993), we can point out particular solutions of considering problem at the law of variable parameters $q_i$:

$$\mu(t) = \mu_1 + \mu_2 = \frac{\mu_0}{\sqrt{\alpha t^2 + 2\beta t + \gamma}}. \tag{4}$$

The equations (1), via the transformation

$$r(x, y, z) = \frac{\mu_0}{\mu} \rho(\xi, \eta, \zeta), \quad \tau = \left(\frac{\mu}{\mu_0}\right)^2 dt, \quad \omega = \left(\frac{\mu}{\mu_0}\right)^2 \omega_0, \tag{5}$$

are taken to autonomous form

$$\begin{aligned} \xi'' - 2\omega_0 \eta' &= \frac{\partial U}{\partial \xi}, \\ \eta'' + 2\omega_0 \xi' &= \frac{\partial U}{\partial \eta}, \\ \zeta'' &= \frac{\partial U}{\partial \zeta}, \end{aligned} \tag{6}$$

where

$$\begin{aligned} U &= \frac{\chi\omega_0^2}{2}\left(\xi^2 + \eta^2 + \zeta^2\right) - \frac{\omega_0^2 \zeta^2}{2} + \frac{\mu_{01}}{\rho_1} + \frac{\mu_{02}}{\rho_2}, \\ \rho_i^2 &= (\xi - \xi_i)^2 + \eta^2 + \zeta^2 \quad (i = 1, 2), \\ \xi_1 &= -\frac{\mu_{02}}{\mu_0}\rho_{12}, \quad \xi_2 = \frac{\mu_{01}}{\mu_0}\rho_{12}, \end{aligned} \tag{7}$$

$\rho_{12}$ and $\chi$ are the constants:

$$\rho_{12}\mu = \chi C^2 \quad (\chi > 0)\,. \tag{8}$$

Here $\rho_{12}$ is the distance between main bodies, $C$ is the constant of the area integral.

The particular solutions of equations (6) are defined by systems of equations:

$$\frac{\partial U}{\partial \xi} = 0, \quad \frac{\partial U}{\partial \eta} = 0, \quad \frac{\partial U}{\partial \zeta} = 0. \tag{9}$$

There are rectilinear solutions $L_i$ $(i = 1,\, 2,\, 3)$

$$\xi_L = \alpha_i, \quad \eta = 0, \quad \zeta = 0 \quad (i = 1,\; 2,\; 3), \tag{10}$$

triangular solutions $L_4$, $L_5$ that are defined by a condition

$$\rho_1^3 = \rho_2^3 = \rho_{12}^3 = \frac{\chi^3 C^6}{\mu_0^3}. \tag{11}$$

If we suppose $\zeta \neq 0$ in equations (9), then there are the coplanar solutions $L_6$, $L_7(\xi, 0, \zeta)$, that may be defined, as well as in the case of the restricted three-body problem with the variable masses (Bekov 1993; Luk'yanov 1989), from the equation

$$2\,(\xi + \mu_{20}) - 1 - \left[\frac{\mu_{10}\chi}{\xi + \mu_{10}\,(\chi - 1)}\right]^{\frac{2}{3}} + \left[\frac{\mu_{20}\chi}{-\xi + \mu_{20}\,(\chi - 1)}\right]^{\frac{2}{3}} = 0. \tag{12}$$

Thus, the considered photogravitational problem has the seven particular solutions in the region of parameter's varying (3), analogous to solutions of the restricted variable mass three-body problem.

## 3. The Motion in Neighbourhood of the Massive Radiating Component of Star and Galaxy

Let's now investigate the motion near one component of binary stellar system (star or galaxy), supposing the gravitational influence from secondary is small, and considering this influence as a perturbation, we can neglect this, in comparison with the influence of the main component. Or we consider the case, when the mass of the secondary component is infinitesimal in comparison with the mass of the main component of stellar system. Then, as the dynamical model, we consider the motion on the base of the restricted non-stationary photogravitational two-body problem. Particular solutions of the stationary problem are considered in works by Batrakov (1957) and Zhuravlev (1990). In our case we additionally take into account the variability of mass, sizes and form of the main component, that is taken as the triaxial radiating and gravitating ellipsoid (Bekov 1992).

Let's consider the motion of passive gravitating point in the external gravitational field of rotating with angular velocity $\Omega$ radiating triaxial ellipsoid with the mass $M(t)$ slowly varying with time, the reduction parameter $q$ $(0 < q \leq 1)$, sizes and form. We suppose, that the slow varying of ellipsoid's physical parameters don't lead to displacement of its center of mass. Semiaxes of ellipsoid $a, b, c$ in common are the functions of time, and let's, as in the stationary case, take the ellipsoid as having a small difference from a homogeneous sphere with radius $R$ and with a volume equal to the volume of this sphere. Then

$$a^2 = R^2 + \alpha', \quad b^2 = R^2 + \beta', \quad c^2 = R^2 + \sigma', \tag{13}$$

where $\alpha'$, $\beta'$, $\sigma'$ are the small quantities in comparison with $R^2$ which in consequence of equality of the volumes of ellipsoid and sphere satisfy, with the accuracy until to small quantities of higher order, the relation $\alpha' + \beta' + \sigma' = 0$.

The equations of motion for a material point in a rotating Cartesian system of coordinates $Oxyz$ with the origin in the center of mass $O$ of ellipsoid, and axes $Ox$, $Oy$, $Oz$, coinciding with the main central inertia axes of ellipsoid, and with the direction of angle velocity $\Omega$ of ellipsoid rotation coinciding with $Oz$ axis direction, may be transformed

$$\vec{r}\,(x, y, z) = l(t)\vec{\rho}\,(\xi, \eta, \varsigma)\,, \quad d\tau = \Omega\, dt, \tag{14}$$

where $l^3\Omega^2\kappa = \mu\,(t)$, $\kappa = \text{const}$, $\mu(t) = GqM(t)$, $G$ is the gravitational constant, in autonomous form:

$$\xi'' - 2\eta' = \frac{\partial V}{\partial \xi}, \quad \eta'' + 2\xi' = \frac{\partial V}{\partial \eta}, \quad \varsigma'' = \frac{\partial V}{\partial \varsigma}, \tag{15}$$

where

$$V = \kappa U, \quad U = \frac{\rho^2}{2} - \frac{1}{\kappa}\frac{\varsigma^2}{2} + \frac{1}{\rho} + \varepsilon\frac{\alpha\xi^2 + \beta\eta^2 + \sigma\varsigma^2}{\rho^5} + \ldots, \quad \rho^2 = \xi^2 + \eta^2 + \varsigma^2. \tag{16}$$

Here, $\varepsilon$ is the parameter $(0 < \varepsilon \ll 1)$; $\alpha$, $\beta$, $\sigma$ are the constants that are defined by relations

$$\frac{3}{10}\frac{\alpha'(t)}{l^2(t)} = \varepsilon\alpha, \quad \frac{3}{10}\frac{\beta'(t)}{l^2(t)} = \varepsilon\beta, \quad \frac{3}{10}\frac{\sigma'(t)}{l^2(t)} = \varepsilon\sigma. \tag{17}$$

The functions $l(t)$ and $\Omega(t)$, which determine the transformation (14), are found from relations

$$l^2\Omega = l_0^2\Omega_0 = C_0, \quad \ddot{l} + (\kappa - 1)\Omega^2 l = 0, \tag{18}$$

and due to adiabatic invariant

$$l\mu = \kappa C_0^2 = \text{const} \tag{19}$$

we find

$$\mu = \frac{\mu_0}{\sqrt{At^2 + 2Bt + C}}. \tag{20}$$

The system (15) has particular solutions in the form

$$\xi = \text{const}, \quad \eta = \text{const}, \quad \varsigma = \text{const}, \tag{21}$$

analogous to equatorial and polar solutions of the stationary problem.

Equatorial solutions $P_i$ are determined from expressions:

$$P_1(P_3):\quad \xi = \pm 1 \pm \varepsilon\alpha + \ldots,\quad \eta = 0,\quad \varsigma = 0, \tag{22}$$

$$P_2(P_4):\quad \xi = 0,\quad \eta = \pm 1 \pm \varepsilon\beta + \ldots,\quad \varsigma = 0. \tag{23}$$

The polar solutions are determinated in the form

$$P_5(P_6):\quad \xi = 0,\quad \eta = 0,\quad \varsigma = \pm\left(\frac{\kappa}{\kappa-1}\right)^{\frac{1}{3}} \pm \varepsilon\left(\frac{\kappa-1}{\kappa}\right)^{\frac{1}{3}}\sigma + \ldots \tag{24}$$

Besides this solutions there is other class of polar solutions—$z$-solutions in the neighbourhood of gravitating and radiating ellipsoid, that disposed along the rotation axies of ellipsoid (Bekov 1992).

## 4. Conclusion

The results of investigation of the dynamics of binary stellar systems on the base of the photogravitational three-body and two-body problems with variable mass and radiation pressure of system's components are presented as important, because we can investigate the new properties of conforming homographic solutions and build the analogous of Hill surfaces for following quantitative and qualitative analysis of the dynamical problem (Bekov 2001). The obtained particular solutions may be used in difference problems of stellar dynamics, for example in investigation of the motion of the gas, dust particles in neighbourhood of binary or single forming variable star or the motion of stars and the gas, dust particles in external gravitational field of binary galaxy with slowly changing physical parameters of galaxy's nucleus, and also in the astrophysical supplements for possible interpretation of originated transient structure peculiarities in neighbourhood of such evolving stars and galaxies.

## References

Batrakov, Yu. V. 1957, Bull.ITA AS USSR, 6, no.8, 524

Bekov, A. A. 1992, Trudy Astrofiz.inst. (Kazakhstan), 50, 45

Bekov, A. A. 1993, Problems of physics of stars and extragalactic astronomy (Almaty, Kazakhstan), 91

Bekov, A. A. 2001, Transactions of Kazakh-American University, no.2. 23

Bekov, A. A., & Omarov, T. B. 2003, A&A Trans., 22(2), 145

Bekov, A. A., & Ristigulova, V. B. 2002, Isv.Akad.Nauk R.K., no.4, 47

Luk'yanov, L. G. 1989, AZh, 66, 180

Omarov, T. B., ed. 2002, Non-Stationary Dynamical Problems in Astronomy (New-York: Nova Science Publishers Inc.), 248

Radzievskij, V. V. 1953, AZh, 30, 265

Zhuravlev, S. G. 1990, Questions of celestial mechanics and stellar dynamics (Alma-Ata), 23

*Order and Chaos in Stellar and Planetary Systems*
*ASP Conference Series, Vol. 316, 2004*
*G. Byrd, K. Kholshevnikov, A. Mylläri, I. Nikiforov, and V. Orlov, eds.*

# Chaotic Behavior of Oscillations of Self-Gravitating Spheroid

Ch. T. Omarov and E. A. Malkov

*Fesenkov Astrophysical Institute, 480020 Almaty, Kazakhstan*

**Abstract.** Non-linear oscillations of self-gravitating spheroid is studied in this work. It is shown that at the certain values of governing parameter the system becomes unstable and evolve to stochastic state passing through a sequence of period doubling bifurcation. Poincare's sections illustrating such evolution are presented.

## 1. Introduction

One of the important problems of stellar dynamics is construction and investigation of ellipsoidal models of collisionless gravitating systems. In spite of their idealization, e.g., such as homogeneous spatial density, the study is of interest to understand formation and evolution of large-scale gravitating systems. The most fruitful progress in theory of homogeneous collisionless ellipsoids was done in 70th, 80th years (Antonov, Nuritdinov, Kondratiev, Malkov, Vietri). A basic result in the field is derivation of system of ordinary differential equations of 18th order, describing arbitrary oscillations of collisionless ellipsoidal configurations (see Kondratiev & Malkov 1987). However, only linearized form of these equations has been yet studied. Therefore, as a basis for further investigations of dynamics of ellipsoids, we carried out investigation of nonlinear oscillations for relatively simple model, namely, for non-rotating spheroid. So, in our case, the system is defined by the following Lagrangian:

$$L = \frac{1}{2}\left(\frac{da}{dt}\right)^2 + \left(\frac{dc}{dt}\right)^2 - \frac{\alpha^2}{2a^2} - \frac{\alpha^2}{c^2} + \frac{3}{2}\int_0^\infty \frac{dc}{(c^2+s)\sqrt{\alpha^2+s}}, \tag{1}$$

where $a$, $c$ are the spheroid semi-axes, the parameter $\alpha$ characterizes a part of heat energy in a total one ($\alpha \in [0,1]$). It is convenient to introduce a new variables $R = \frac{a+2c}{3}$, $\delta = a - c$. It is obvious $R$ and $\delta$ define average radius and deformation of spheroid respectively. Accordingly from (1) we have equations of motion:

$$\frac{d^2R}{dt^2} = \frac{1}{3}\left(\frac{\alpha^2}{\left(R+\frac{1}{3}\delta\right)^3} + \frac{2\alpha^2}{\left(R-\frac{1}{3}\delta\right)^3}\right) - \frac{1}{3}\frac{\partial\Phi}{\partial R}, \tag{2}$$

$$\frac{d^2\delta}{dt^2} = \frac{\alpha^2}{\left(R-\frac{2}{3}\delta\right)^3} - \frac{\alpha^2}{\left(R-\frac{1}{3}\delta\right)^3} - \frac{3}{2}\frac{\partial\Phi}{\partial\delta}, \tag{3}$$

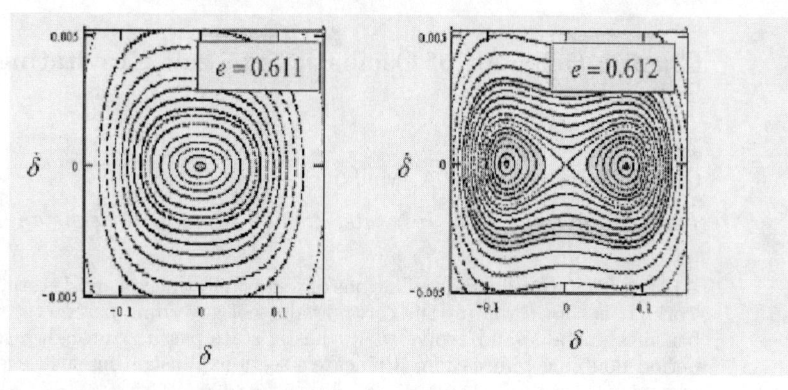


Figure 1. Illustrates that in passing to parameter $e = 0.612$ a new periodic solution gains stability.

where

$$\Phi = -\frac{3}{2}\int_0^\infty \frac{ds}{\left(\left(\frac{3R-\delta}{3}\right)^2 + s\right)\sqrt{\left(\frac{3R+2\delta}{3}\right)^2 + s}}.$$

From now on we will use parameter $e$ instead of $\alpha$ by means of the relation $e = \sqrt{1-\alpha^2}$. It is known that evolution of cold system when $e^2 = 1$ leads to the flat singularity and case when $e^2 = 0$ corresponds to stationary system. In general case, when parameter is ranged from 0 to 1, but in linear approximation the problem was resolved. Moreover, the critical values of $e$ parameter when the system becomes unstable due to the parametric resonance were found (Malkov 1986), namely, when $e \in [0.612, 0.88] \cup [0.92, 1]$, the system is unstable. Unfortunately, linear analysis is not able to give a comprehensive information and to trace down evolution of the system in region of linear instability. In the present paper, we implemented non-linear analysis of spheroid oscillations and in doing so basic properties of non-linear dynamical instability have been detected.

## 2. Poincare's Section

Investigation of nonlinear oscillations of the system of above equations has been curried out numerically. Poincare's section at various $e$ values and various initial conditions (total energy $E$ is conserved: $E = -1.5$ according to our normalization) has been built. Namely, points on the phase plane $(\delta, \dot{\delta})$, $\dot{R} = 0$ are fixed when $\dot{R}$ changes a sign from "+" to "−". Thus, location of the points on this plane defines topology of the phase-space. As a result, we found a sequence of values of the governing parameter $e$ when bifurcation is occurred. In fact, the Figure 1 demonstrates phenomenon of period doubling bifurcation of primordial $2\pi$-periodic solution as a result of parametric resonance with the frequency ratio $2:1$ in passing from value $e = 0.611$ to $e = 0.612$. As seen, the new $4\pi$-periodic solution gains (or inherits) stability. Then, with the further increasing of $e$ parameter the solution is subjected to successive period doubling bifurcation. After all, $8\pi$-, $16\pi$- and $32\pi$-periodic solutions are detected at values $e = 0.915$,

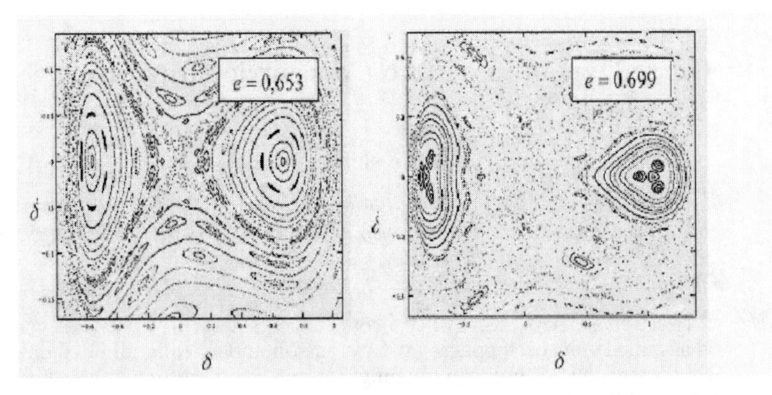


Figure 2. Illustrates a growth of stochastic region in vicinity of $2\pi$-periodic solution and presence of nonlinear resonance in vicinity of $4\pi$-periodic solution with frequency ratio $3:1$ (right figure).

0.9184 and 0.91872 respectively. One can see (Figure 2) that, with increasing of parameter $e$ a growth of stochastic region in vicinity of $2\pi$ solution takes place. Meanwhile, a presence of nonlinear resonance, in particular, with frequency ratio $3:1$ in vicinity of $2\pi$ solution is revealed.

Thus, obtained numerical results give useful information about nonlinear oscillations of the self-gravitating spheroid and will serve us as a base for comparative analysis in finding of an adequate analytical mapping.

## References

Chandrasekhar, S. 1973, Ellipsoid figures of equilibrium (Moscow)

Malkov, E. A. 1986, in Dynamics of stationary and non-stationary gravitating systems (Almaty), 78

Vietri, M. 1990, MNRAS, 245, 40

*Order and Chaos in Stellar and Planetary Systems*
*ASP Conference Series, Vol. 316, 2004*
*G. Byrd, K. Kholshevnikov, A. Mylläri, I. Nikiforov, and V. Orlov, eds.*

# Gross-Dynamics of Two Concentric Gravitating Spheres

L. P. Ossipkov and A. G. Shoshin

*Department of Space Technologies & Applied Astrodynamics, St. Petersburg State University, Universitetskij pr. 28, Staryj Peterhof, St. Petersburg 198504, Russia*

**Abstract.** A problem of two concentric gravitating spheres is considered. It is assumed that each sphere evolves quasi-homologically and expands according to a linear law. Equilibrium configurations are found, and their linear stability is analyzed.

## 1. Introduction

Many authors used the *method of quasi-particles* for studying collective processes in graviplasmas. For instance, spherical gravitating spheres were modelled by a set of spherical concentric shells (Campbell 1962; Hénon 1964; Bisnovatyi-Kogan & Yangurazova 1984; Bisnovatyi-Kogan 2002; Barkov, Belinski, & Bisnovatyi-Kogan 2002). It is interesting that vibrations of shells were found to be stochastic under some conditions (Barkov et al. 2002). The latter can be significant for understanding violent relaxation at early evolutionary stages of galaxies (e.g., Chernin et al. 2002). Following Danilov (1983) we will discuss here another model. A spherical system is considered as an ensemble of concentric spheres. Generally, spheres can be non-uniform, but it is supposed that each sphere evolves *quasi-homologically*, i.e., without forming a halo-core structure. We will restrict ourselves to studying the most general (or integral) parameters of the model, such as a size or energy of a sphere. That means we will use the gross-dynamic level of description (Kuzmin 1965).

## 2. Gross-Dynamic Equations of the Problem

We consider a system consisting of two concentric spheres (labeled as "1" and "2"). Let $r$ be a spherical radius, $a_i(t)$ the radius of the $i$-th sphere, $\varrho_i(r,t)$ its mass density, $V_{r_i}(r,t)$ its velocity of expansion (or contraction) at distance $r$, $M_i(r,t)$ be its mass inside a sphere of radius $r$, $\sigma_i^2(r,t)$ be a dispersion of total (not only residual!) velocities of stars belonging to the $i$th sphere at distance $r$. Then $\mathcal{M}_i = M_i[a_i(t),t] = \text{const}$ is a mass of the $i$-th sphere. For an arbitrary function $g(r,t)$ we define its average value $\langle g\rangle_i(t)$ inside the $i$-th sphere as

$$\langle g\rangle_i(t) = \frac{1}{\mathcal{M}_i}\, 4\pi \int\limits_0^{a_i(t)} \varrho_i(r,t)\, g(r,t)\, r^2\, dr\,. \tag{1}$$

It can be easily found (Ossipkov 1984) that

$$\begin{aligned} {}^1\!/_2\, \frac{d^2}{dt^2}\, \mathcal{M}_i \, \langle r^2 \rangle_i - \mathcal{M}_i \langle \sigma_i^2 \rangle_i + G\mathcal{M}_i \langle \frac{M_1 + M_2}{r} \rangle_i &= 0\,, \\ \frac{d}{dt} \mathcal{M}_i \, \langle \sigma_i^2 \rangle_i + 2G\mathcal{M}_i \langle \frac{(M_1 + M_2)V_{r_i}}{r^2} \rangle_i &= 0\,. \end{aligned} \tag{2}$$

The first of these equations is a generalization of the known Lagrange–Jacobi equation, and a sum of second ones yields an energy conservation.

We put a moment of inertia of the $i$-th sphere (relative to a barycenter) $\mathcal{M}_i \langle r^2 \rangle_i$ equal to $l_i \mathcal{M}_i a_i^2$, where $l_i$ is a dimensionless parameter depending on mass distribution, $l_i = {}^3\!/_5$ for a uniform sphere. Also we put $G\mathcal{M}_i \, \langle M_j / r \rangle_j = k_{ij} G \mathcal{M}_i \mathcal{M}_j / a_j$. Dimensionless factors $k_{11} = k_{22} = {}^3\!/_5$ for a uniform sphere. Note that $k_{12} \neq k_{21}$. If both spheres are uniform then $k_{12} = {}^3\!/_5 \, (a_1/a_2)^3$ if $a_1 \leq a_2$ and $k_{12} = {}^3\!/_5 \, (a_2/a_1)^2 + {}^3\!/_2 \left[ 1 - (a_2/a_1)^2 \right]$ if $a_1 \geq a_2$. It is natural to suppose that the expansion of each sphere follows a linear law. Then $V_{r_i}(r) = r \, \dot{a}_i(t)/a_i(t)$, and $G \, \mathcal{M}_i \, \langle \mathcal{M}_j \, V_{r_i}/r^2 \rangle_i = (\dot{a}_i/a_i) \; k_{ij} \, G \, \mathcal{M}_i \mathcal{M}_j / a_i$. Supposing that each sphere evolves quasi-homologically we consider $l_i$, $k_{ii}$ (but not $k_{12}$, $k_{21}$!) as constants.

## 3. A Dimensionless Form of Equations

Now we transform our equations into a dimensionless form. We choose a total mass $\mathcal{M} = \mathcal{M}_1 + \mathcal{M}_2$ as a unit of mass and denote $m_i = \mathcal{M}_i / \mathcal{M}$. Also we choose an initial size of the first sphere $a_i(t = 0) = a_0$ as a unit of length and denote $\delta_i = a_i / a_0$. Of course, $t_0 = \left( a_0^3 / G\mathcal{M} \right)^{1/2}$ is a natural unit of time. Denote $\tau = t/t_0$ and $' = d/d\tau$. A dimensionless kinetic energy of the $i$-th sphere $æ_i = {}^1\!/_2 \, a_0 \, \langle \sigma_i^2 \rangle_i \, / \, (G\mathcal{M})$. Then Eqs. (2) can be rewritten as follows:

$$l_i m_i (\delta_i^2)'' - 4\, æ_i + 2\, k_{i1} \, \frac{m_1 m_i}{\delta_i} + 2\, k_{i2} \, \frac{m_2 m_i}{\delta_i} = 0\,, \tag{3}$$

$$(æ_i)' + \frac{\delta_i'}{\delta_i} \left[ +k_{i1} \, \frac{m_1 m_i}{\delta_i} + k_{i2} \, \frac{m_2 m_i}{\delta_i} \right] = 0\,. \tag{4}$$

If both spheres are uniform, then Eqs. (3), (4) can be simplified:

$$3\, R_i'' - 20\, \varsigma_i + 6\, C_i / R_i^{1/2} = 0\,, \tag{5}$$

$$3\, \varsigma_i' + C_i \, R_i' / R_i = 0\,. \tag{6}$$

Here $R_i = \delta_i^2$, $\varsigma_i = æ_i / m_i$, $C_1 = m_1 + k_{12} m_2$, $C_2 = k_{21} m_1 + m_2$. Note that $C_i$ are functions of $R_1 / R_2$.

## 4. Equilibrium and Linear Stability

Denote equilibrium values of variables $R_1$, $R_2$, $\varsigma_1$, $\varsigma_2$, $C_1$, $C_2$ as $r_1 (= 1)$, $r_2 (< r_1)$, $s_1$, $s_2$, $c_1$, $c_2$. It is evident that $s_i = {}^3\!/_{10} \, c_i r_i^{-1/2}$. Now we put $R_1 = r_1 + x$,

$R_2 = r_2 + y$, $\varsigma_1 = s_1 + \xi$, $\varsigma_2 = s_2 + \eta$, $C_i = c_i + \gamma_i$ and will consider $x$, $y$, $\xi$, $\eta$, $\gamma_i$ as small quantities. In linear approximation $c_1 = m_1 + {}^3/_2\, m_2\, (1 - {}^3/_5\, r_2/r_1)$, $c_2 = {}^3/_5\, m_1\, (r_2/r_1)^{3/2}$, and also one finds $\gamma_1 = {}^9/_{10}\, m_2\, (x/r_1 - y/r_2)\, (r_2/r_1)$, $\gamma_2 = {}^9/_{10}\, m_1\, (y/r_2 - x/r_1)\, (r_2/r_1)$. Linearization of Eqs. (5) yields

$$x'' = \frac{20}{3}\,\xi + \alpha_1\, x + \beta_1\, y\,, \qquad y'' = \frac{20}{3}\,\eta + \alpha_2\, x + \beta_2\, y\,, \tag{7}$$

where coefficients $\alpha_i$, $\beta_i$ can be expressed in terms of $m_i$, $r_i$. It follows from Eqs. (6) that $\xi' \propto x'$, $\eta' \propto y'$. Then Eqs. (7) can be rewritten in the following form:

$$x'' = A_1\, x + B_1\, y + k_1\,, \qquad y'' = A_2\, x + B_2\, y + k_2\,, \tag{8}$$

where $A_1 = \alpha_1 - {}^{10}/_3\, (c_1/r_1)$, $B_1 = \beta_1$, $A_2 = \alpha_2$, $B_2 = \beta_2 - {}^{10}/_3\, (c_2/r_2)$. Now we have to find eigenfrequencies $\lambda_i$ of Eqs. (8) and to solve a characteristic equation. The latter is reduced to a biquadratic equation

$$p^2 - (A_1 + B_2)\, p + (A_1\, B_2 - A_2\, B_1) = 0$$

with $p = \lambda^2$. Then $p_{1,2} = {}^1/_2\, (A_1 + B_2) \pm {}^1/_2\, \left[(A_1 - B_2)^2 + 4\, A_2\, B_1\right]^{1/2}$. Substituting expressions for $A_i$, $B_i$ we can calculate eigenfrequencies for any set of $m_1$, $r_2$ (remember that $m_2 = 1 - m_1$, $r_1 = 1$). We restrict ourselves here with a case of $m_2 \ll m_1$, $r_2 \ll 1$, so the second sphere is a nucleus of small size and small mass. Then $\beta_1 \ll 1$ and $p_1 \approx A_1 \approx -{}^7/_3\, m_1$, $p_2 \approx B_2 \approx -{}^9/_5\, m_1 r_2^{-1/2}$. Both $p_i$ are negative, i.e., all $\lambda_i$ are imaginary. The latter means stability of the equilibrium state. The system exerts undamping oscillations.

## References

Barkov, M. V., Belinski, V. A., & Bysnovatyi-Kogan, G. S. 2002, MNRAS, 334, 338

Bysnovatyi-Kogan, G. S. 2002, Space Sci.Rev., 102, 9

Bisnovatyi-Kogan, G. S., & Yangurazova L. R. 1984, Ap&SS, 100, 319

Campbell, P. M. 1962, Proc.Nat.Acad.Sci., 48, 1993

Chernin, A. D., Valtonen, M., Ossipkov, L. P., Zheng, J. Q., & Wiren, S. 2002, Astrofizika, 45, 296 (in Russian)

Danilov, V. M. 1983, in Star Clusters and Problems of Stellar Evolution, ed. K. A. Barkhatova (Sverdlovsk: Urals Univ. Press), 39 (in Russian)

Hénon, M. 1964, Ann.d'Astrophys., 27, 83

Kuzmin, G. G. 1965, Trudy Astrofiz.inst. (Alma-Ata), 5, 11 (in Russian)

Ossipkov, L. P. 1984, in Problems of Astrophysics, ed. K. A. Barkhatova (Saransk: Mordovia Univ. Press), 55 (in Russian)

*Order and Chaos in Stellar and Planetary Systems*
*ASP Conference Series, Vol. 316, 2004*
*G. Byrd, K. Kholshevnikov, A. Mylläri, I. Nikiforov, and V. Orlov, eds.*

# Modes of High Degrees for Collapsing Galaxies: Formation of Globular Cluster Systems

S. N. Nuritdinov, I. U. Tadjibaev, and K. T. Mirtadjieva

*National University of Uzbekistan and Astronomical Institute of Uzbek Academy of Sciences, Tashkent; nurit@astrin.uzsci.net*

## 1. Introduction

Up to now the problems of an origin and early evolution stage of galaxies and their subsystems are basically studied by numerical experiments. Theoretical research of the formation problem of the globular cluster systems (GCSs) and nonlinearly nonstationary stages of their evolution requires, first of all, construction of exact analytically solvable models of early evolution stages of galaxies and revealing the instabilities on a background of their non-equilibrium states. In this work we look for instability conditions of formation of GCS at an early stage of collapsing protogalaxy relatively to the oscillation modes of the high degrees which correspond to rather small-scale perturbations of density of collapsing system. Thus the mode degree defines in average the number of clusters in the GCS. The formation process of the GCS covers a long period of time as one begins from a dark matter state at least. So far it is difficult to simulate formation of galaxy and its GCS *ab initio*. That is why it is important to find initial conditions for formation of the GCS. Here the preliminary results of the analysis of the problem of modes of high degrees are given.

## 2. Problem Statement

Earlier (Nuritdinov et al. 2000) we had first suggested a formation model of the GCSs in the collapsing protogalaxies, while not specifying the type of a galaxy. There two possible ways of evolution were specified: i) due to gravitational instability of radial motions the collapse of the protogalaxy forms small clouds (limiting case); these clouds move radially and easily stick together and then a formation of enough massive clouds takes place; ii) at certain conditions enough massive clouds can be formed at the begining.

To analyse these phenomena we investigate the non-stationary dispersion equation (NDE) of a stability problem of collapse of self-gravitating systems (Nuritdinov 1993). To understand the more realistic evolution, we need to study as a limiting case where to find the asymptotic of the NDE and concrete discrete modes of high degrees. Moreover we have to note that results of our analysis could be applied for two states: i) state of collisionless dark matter, ii) state of gas matter. In the last case the applicability of results demands that the ratio of heat capacities was equal 5/3.

For the NDE for spherical nonlinear collapsing model being non-stationary the generalization of an Einstein's equilibrium sphere has next form (Nuritdinov 1993)

$$\frac{a_0(\psi)\Pi^3}{3(N-n-2)!!(N+n-1)!!} =$$

$$\sum_{s=0}^{N-1} \sum_{\substack{k=N-1-s \\ \text{(even } k\text{'s)}}}^{N-1} \frac{2^{-k} s! V_{ks} Y_k [N(s+1)+s-1-(n-1)(n+2)]}{(k/2)![(N-s-1)/2]!(s-n+1)!!(s+n+2)!!} + im\mu \times \tag{1}$$

$$\sum_{s=0}^{N-2} \sum_{\substack{k=N-1-s \\ \text{(odd } k\text{'s)}}}^{N-1} \frac{[(k-1)/2]! s! V_{ks} Y_k [N(s+1)+2(s-1)-(n-2)(n+3)]}{(k+1)![(N-s-2)/2]!(s-n+2)!!(s+n+3)!!},$$

where

$$V_{ks} = \frac{i^{s+k-N+1}}{(N-1-k)![(s+k-N+1)/2]!},$$

$$Y_k = \int_{-\infty}^{\psi} (W \cos e)^{N-1-k} (W \sin e)^k E \, d\psi_1,$$

$\mu$ is a parameter of rotation ($0 \leq \mu \leq 1$), $S(\psi, \psi_1)$ is an analogue of the Green function and $\lambda$ is the pulsation amplitude which is defined as $\lambda = 1 - (2T/|U|)_0$ through virial ratio (other notations were given, for example, in the paper of Gaynullina & Nuritdinov 2000). Up to now the NDE (1) has been investigated for modes of low and moderate degrees (Gaynullina & Nuritdinov 2000). The cases of modes of high degrees result in the NDE with very cumbersome differential equations (Tadjibaev 2003). The NDE (1) is convenient for analysing modes of high degrees, and for analysis of the limiting case we have studied asymptotic behaviour of other NDE (Nuritdinov 1993)

$$\frac{N(N+1)}{6} a(\psi)\Pi^3 = \int_{-\infty}^{\psi} E \frac{dP_n(\theta)}{d\theta} Sa(\psi_1) \, d\psi_1, \tag{2}$$

where $P_n(\theta)$ is the Legendre polynomial of $n$ order.

## 3. The Analysis of NDE

We study the instability of two modes of oscillations by numerically solving the NDE (1) separately for each mode. For the small-scale modes of oscillations we find the appropriate dependences of the growth rates of instability on the virial ratio at the beginning moment of the collapse. The results of calculations are given in Figure 1 which shows that with growth of the mode degree the area of stability is increased. In each case the growth rates are also increased with growth of values of azimuth wave number and the rotation parameter. The comparisons of the growth rates for different instability modes are given in Figure 2 which shows that among all instabilities the mode (4,2) are more strong [even than the mode (2,2)]. It is interesting that the maximum of the growth rate of the mode (9,7) is very close to the one of mode (2,2).

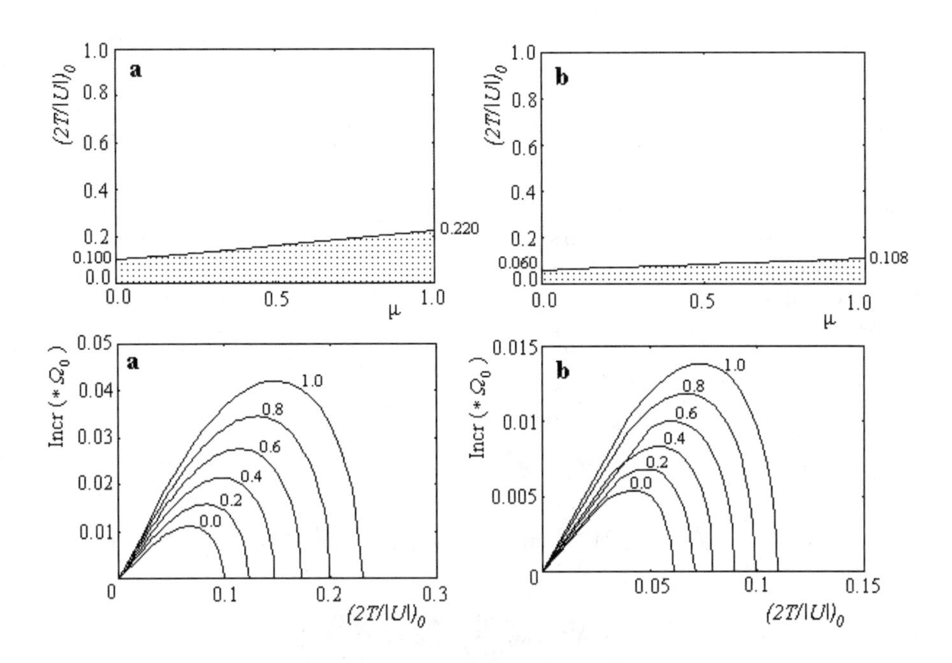


Figure 1. Critical dependences of initial virial ratio on the rotation parameter and critical dependences of the growth rate on the virial ratio in cases: a) $N = 11$, $n = 9$; b) $N = 18$, $n = 16$.

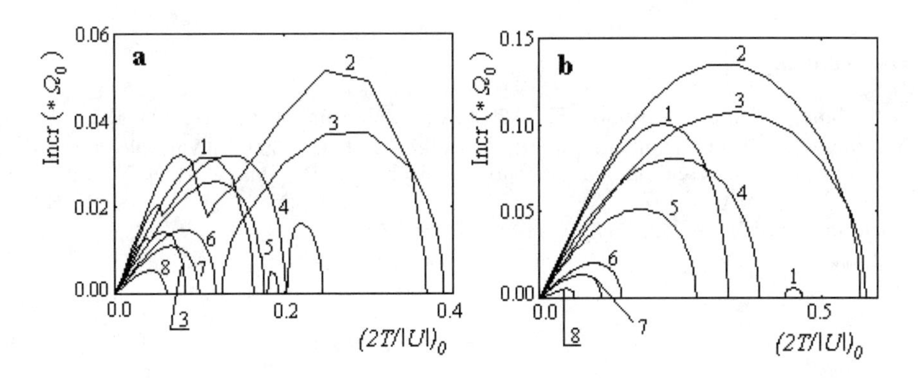


Figure 2. Comparison of maximal growth rates of instabilities: a) $\mu = 0.0$; b) $\mu = 1.0$: 1) $N = 3$, $n = 1$; 2) $N = 4$, $n = 2$; 3) $N = 2$, $n = 2$; 4) $N = 5$, $n = 3$; 5) $N = 6$, $n = 4$; 6) $N = 9$, $n = 7$; 7) $N = 11$, $n = 9$; 8) $N = 18$, $n = 16$.

In the limiting case the formation criteria of the GCS demands on the exact asymptotic solution of above indicated NDE. For the beginning we consider the case of non-rotating model. We use next asymptotics for Legandre polynomial at $N \gg 1$

$$P_n(\cos e) = J_0(\eta) + \sin^2\frac{e}{2}\left[\frac{1}{2\eta}J_1(\eta) - J_2(\eta) + \frac{\eta}{6}J_3(\eta)\right] + O\left(\sin^4\frac{e}{2}\right), \quad (3)$$

where $\eta = (2N+1)\sin\frac{e}{2}$, $J_n(\eta)$ are the Bessel functions.

Further we are limited by first member in (3). Here $\eta$ depends on $\psi$ and $\psi_1$. Finding an asymptotics demands a separation of functions of $\psi$ and $\psi_1$. It is very difficult problem for (3). That's why at a number of concrete conditions we have found the following NDE:

$$\Lambda F_k(\psi) = \frac{3\sqrt{1-\lambda^2}\sin^{2-k}\psi(\lambda+\cos\psi)^{k-1}}{2(1+\lambda\cos\psi)}[(\lambda+\cos\psi)F_1 - \sin\psi F_2], \quad (4)$$

where

$$F_k(\psi) = \int_{-\infty}^{\psi} E(\psi,\psi_1)W^{-1}\sin^{2-k}\psi_1(\lambda+\cos\psi_1)^{k-1}\,d\psi_1 \quad (k=1,2),$$

$$\Lambda = (1+\lambda\cos\psi)\frac{d^2}{d\psi^2} + \lambda\sin\psi\frac{d}{d\psi} + 1.$$

The numerical calculations show that at $(2T/|U|)_0 \ll 2\times10^{-5}$ we have the instability of radial orbits which can bring on formation of the GCS.

We are grateful to reviewer for useful remarks.

## References

Gaynullina, E. R., & Nuritdinov, S. N. 2000, Uzbek Journal of Physics, 2, 3, 185

Nuritdinov, S. N., Orazimbetov, J. R., & Tadjibaev, I. U. 2000, in Variable stars as key to understanding of a structure and evolution of a Galaxy, ed. N. N. Samus' (Nijniy Arhyz), 197

Nuritdinov, S. N. 1993, Dr. Sci. Thesis (St. Petersburg)

Tadjibaev, I. U. 2003, Uzbek Journal of Physics, 5, 1, 1

*Order and Chaos in Stellar and Planetary Systems*
*ASP Conference Series, Vol. 316, 2004*
*G. Byrd, K. Kholshevnikov, A. Mylläri, I. Nikiforov, and V. Orlov, eds.*

# Lin–Shu's Quasi-Stationary Spiral Structure Hypothesis and Critical Behaviour of Stellar Disks in a Marginal State

M. N. Maksumov
*Institute of Astrophysics, 22 Bukhoro Street, Dushanbe, Tajikistan 734042*

**Abstract.** The possibility of the quasi-stationary spiral structure origination due to nonequilibrium process of the fluctuating type in the critical state is discussed. It is proposed to consider spiral structure as one influenced by fluctuations of disc matter, which correlation statistically and continually maintains ordered and concerted behaviour on the long range of the disk.

## 1. Introduction

The maximum in Toomre's stability diagram (Toomre 1964), which separates gravitationally stable and unstable states with respect to axisymmetric perturbations, can be treated as a critical point in the sense of a phase transition of the second kind, because the disk symmetry gets broken and an equilibrium or dynamical structure can arise. Taken by itself, a globally stabilized galactic disk has a strong self-action (the susceptibility) at this point. In the broad sense the question is critical phenomena (Ivanov 1992) because fluctuations, displacement and shift effects of disk substance can take place also. Author considers a stellar disk in the collisionless hydrodynamic approach (Rohlfs 1977; Marochnik & Suchkov 1984) for simplicity, but with good reasons.

Spiral structure is an evolutionary phenomenon. It does not exist from the outset. It is important to know how it could arise in an initially marginal stellar disk. It would be true to include the possible critical behaviour of a marginally stable stellar disc like a phase transition of the second kind, when the disk changes its symmetry and turnes into nonaxisymmetric state. And what is more, criticality is most suitable to connect excitation and lasting existance of spiral density waves with collective nature of stellar systems. First of all, because of its lasting nature the density wave must be considered as a nonaxsymmetric formation coexisting with axisymmetrical background. Therefore a situation must be realized, which is typical for nonequilibrium second-order phase transition. I.e., for such property as breaking of a symmetry, a global nature of the process and its fastness must be realized. Thus taking into account the criticality is the step in development of Lin–Shu's quasi-stationary spiral structure (QSSS) hypothesis to include the variety of patterns. Especially as no power resources are needed.

Are marginally stable stellar disks possible things? Those can exist quite enough. As

$$\overline{Q} = \frac{c_r}{c_T} = \frac{c_r}{(\pi G\sigma)/\kappa} = \frac{c_r^2 \kappa r}{c_r \pi G\sigma r} = \frac{c_r^2}{\pi G\sigma r}\frac{\kappa r}{c_r},$$

$$\frac{c_r^2}{\pi G\sigma r} = \frac{c_r^2}{\Omega^2 r^2}\frac{\Omega^2 r^2}{\pi G\sigma r}, \qquad \frac{\Omega^2 r^2}{\pi G\sigma r} = \frac{GM_t}{\pi G\sigma r^2} = \frac{M_t}{M_d},$$

then $\overline{Q} = (c_r/\Omega r)(M_t/M_d)(\kappa/\Omega)$, which can be $\approx$1 for many galactic disks because first multiplier is approximately a few or ten per cent, second one is 2–5, and third one is $\leq$2 for bright spirals. Here, $c_r$ is the radial velocity dispersion; $c_T = \pi G\sigma/\kappa$ is Toomre's characteristic velocity; $G$ is Newton's gravitational constant; $\kappa$ the epicyclic frequency; $r$ is spanwise radial distance; $\Omega$ is the angular velocity of the disc rotation; $M_t$ and $M_d$ are total galaxy mass and disk mass respectively.

Previously equilibrium properties of stellar disks near their critical point were studied and critical exponents were calculated by Ivannikova & Maksumov (2001). They considered according to Rohlfs (1977, §5.7) a nonaxisymmetric perturbation of the stellar surface density $\sigma_m$ (the order parameter), whose background value is $\sigma$, induced by small perturbations of gravitational potential $\psi_1 = \psi_{1,\text{int}} + \psi_{1,\text{ext}}$ (positive definite), where the potential is represented as a sum of the internal, self-consistent field and external field. The perturbed quantities (including $\psi_1$) were proportional to $\exp[i(\omega t - m\varphi + kr)]$ where $\omega$ is the angular frequency of the perturbations, and $m/r$ and $k^*$ are the azimuthal and total wave numbers, respectively). Besides the linear mode analysis they used the general equilibrium relations Landau & Lifshitz (1976, §25) in order to describe incomplete equilibrium in the marginal state. They found three critical exponents, $\gamma = 1$ [by definition $\sigma_m = \chi\psi_{1,\text{ext}}$ the susceptibility $\chi$ and the definition $\gamma$ as $\chi \sim (c_r^2 - c_{\text{T}}^2)^{-\gamma}$], $\beta = 1/2$ [by definition a spontaneous value of the order parameter $\sigma_m/\sigma \sim (c_T^2 - c_r^2)^{\beta}$], and $\delta = 3$ (which evaluated at $c_r = c_{\text{T}}$ after $\sigma_m \sim \psi_{1,\text{ext}}^{1/\delta}$).

The condition of conversion to zero of the sound speed of Saslaw (1987) for gravithermodynamic instability relates in a certain sense to the condition of marginal stability (when $\omega - m\Omega = 0$). For all this, the physics of gravithermodynamic instability and criticality of the stellar disc are quite different things.

Criticality of the marginal state gives birth to the initial, weakly marked, almost melted into the background large scale structure like ripples, as it was shown by Ivannikova & Maksumov (2001). Because of ripple inhomogeneity chaotic fluctuational motions of the drift type (shear modes) can be exited which bunch, "gather up" disk substance in small "heaps". "Heaps" correlate and form a large scale structure.

After the analogy of the mathematical description of the critical behaviour of a marginally stable collisionless stellar disk (see below) it is possible to draw an analogy of the nature of such behaviour with phase transitions of the second kind. The singular nature of solutions in the marginal state ($c_r = c_{\text{T}}$) and its small vicinity in the post-critical state ($c_r < c_{\text{T}}$, $|c_r - c_{\text{T}}| \ll c_{\text{T}}$) resembles that of mentioned transitions. We speak mainly just about the critical behavior.

## 2. The Response Function, Coherence Length, Susceptibility, Dynamical Scale Invariance and Spiral Structure

In terms of $(\omega - m\Omega)/\kappa = \widetilde{\omega}$, using $\psi_{1,\mathrm{int}} = 2\pi G\sigma_m/|k^*|$, the Rohlfs (1977, §5.7) density response can be written as

$$\left(1 - \widetilde{\omega}^2 + \frac{c_r^2 k^{*2}}{\kappa^2} - \frac{2\pi G\sigma|k^*|}{\kappa^2}\right)\frac{\sigma_m}{\sigma} = \frac{\psi_{1,\mathrm{ext}} k^{*2}}{\kappa^2}\,. \tag{1}$$

Completing the square in $k^*$ on the left-hand side:

$$\begin{aligned} & 1 + (c_r^2 k^{*2})/\kappa^2 - (2\pi G\sigma)/(c_r\kappa)(c_r|k^*|)/\kappa + \\ & (\pi^2 G^2\sigma^2)/(c_r^2\kappa^2) - (\pi^2 G^2\sigma^2)/(c_r^2\kappa^2) \\ = \; & 1 + [(c_r k^*)/\kappa - (\pi G\sigma)/(c_r\kappa)]^2 - c_\mathrm{T}^2/c_r^2 \\ = \; & 1 + (c_r^2/\kappa^2)[k^* - (\pi G\sigma)/c_r^2]^2 - c_\mathrm{T}^2/c_r^2 \\ = \; & (c_r^2 - c_\mathrm{T}^2)/c_r^2 + (c_r^2 k^2)/\kappa^2, \end{aligned}$$

adopting $k^* - \pi G\sigma/c_r^2 = k$, and a characteristic length (analogous to the correlation length) quantifying the spatial coherence of the perturbations in the uniform self-consistent field $L_\mathrm{cohr}^2 = r_\mathrm{c}^2 c_r^2/(c_r^2 - c_\mathrm{T}^2)$ ($r_\mathrm{c}^2 = c_r^2/\kappa^2$) the density response (1) reduces to

$$\frac{\sigma_m}{\sigma} = \frac{\psi_{1,\mathrm{ext}}}{c_r^2}\frac{k^{*2}}{L_\mathrm{cohr}^{-2} + k^2 - r_\mathrm{c}^{-2}\widetilde{\omega}^2}. \tag{2}$$

Having accepted $L_\mathrm{cohr}$ as the correlation length, we see that the critical exponent $\nu$ is equal to 1/2, as it should be in Landau's theory (Landau & Lifshitz 1976, §148) by definition $\nu$ as $L_\mathrm{corr} \sim (c_r^2 - c_\mathrm{T}^2)^{-\nu}$. See also Ivannikova et al. (1997).

Now consider the response in the vicinity of the top point in Toomre's diagram.

Eq. (2) yields the ratio controlling the susceptibility $\partial\sigma_m/\partial\psi_{1,\mathrm{ext}}$:

$$\begin{aligned} \frac{\partial\sigma_m}{\partial\psi_{1,\mathrm{ext}}} = \chi = \frac{\sigma}{c_r^2 L_{cohr}^{-2}}\frac{k^{*2}}{1 + k^2 L_\mathrm{cohr}^2 - r_\mathrm{c}^{-2} L_\mathrm{cohr}^2\widetilde{\omega}^2} = \\ = |\, c_r^2 - c_T^2 \,|^{-1}\frac{\sigma k^{*2} r_c^2}{1 + k^2 L_\mathrm{cohr}^2 - r_c^{-2} L_\mathrm{cohr}^2\widetilde{\omega}^2}, \end{aligned} \tag{3}$$

i.e., just of the form which is in the agreement with requirements of the dynamical scale invariance for the phase transition of the second kind (Halperin & Hohenberg 1969; Lifshitz & Pitayevsky 1979): $\chi = |\, c_r^2 - c_T^2 \,|^{-\gamma} f(k^* r_c, kL_\mathrm{cohr}, r_c^{-1} L_\mathrm{cohr}\widetilde{\omega})$. And therefore, the susceptibility in a marginal state is such that the disk could change its symmetry and turn into nonaxisymmetric state.

Assuming $k = 0$, $\widetilde{\omega} = 0$ in Eq. (1), i.e., $k^* = k_\mathrm{cr} = \pi G\sigma/c_r^2$, $\omega - m\Omega = 0$ (cf. Saslaw 1987, ch. 31), we obtain for the amplitude of a mode whose scale is close to the critical value $k_\mathrm{cr}^{-1}$:

$$\frac{\sigma_m}{\sigma} = \frac{\psi_{1,\mathrm{ext}}}{c_r^2}\frac{k^{*2}}{L_\mathrm{corr}^{-2}} = \frac{\psi_{1,\mathrm{ext}}}{c_r^2}\frac{c_\mathrm{T}^2}{c_r^2 - c_\mathrm{T}^2}\,. \tag{4}$$

This yields the critical exponent as $\gamma = 1$ for $c_r \simeq c_{\rm T}$ by definition $\sigma_m = \chi\psi_{1,\rm ext}$ the susceptibility $\chi$ and the definition $\gamma$ as $\chi \sim (c_r^2 - c_{\rm T}^2)^{-\gamma}$.

As it was be expected, the coherence (correlation) length is related to the susceptibility (Ma 1976; Patashinsky & Pokrovsky 1975). The spontaneous value of the order parameter $\sigma_m/\sigma$ in the post-critical state also must be related to $L_{\rm cohr}$ (Patashinsky & Pokrovsky 1975).

Equating potential gravitational energy of ordering $\sigma_m\psi_m \approx \sigma_m\pi G\sigma_m L_{\rm cohr}$ to the energy of compressive strain of the disk surface $\sigma_m c_T^2$, one can derive

$$L_{\rm cohr} = \frac{\sigma_m c_T^2}{\pi G\sigma_m^2} = \frac{c_T^2}{\pi G\sigma \mid \sigma_m/\sigma \mid}.$$

So we derive again the formula for the coherence (correlation) length of this paper when $\sigma_m$ is given by $\sigma_m/\sigma \sim (c_T^2 - c_r^2)^\beta$, $\beta = 1/2$.

This relation has also another view. We can write $\mid \sigma_m/\sigma \mid = c_T^2/(\pi G\sigma L_{\rm cohr})$. And since $\pi G\sigma = [(\pi G\sigma)/\kappa]\kappa = c_T\kappa$, $\mid \sigma_m/\sigma \mid = (c_T/(\kappa L_{\rm cohr}) = (c_T/\kappa)/L_{\rm cohr}$.

Controlling the critical behaviour of the marginally stable stellar disk the chatacteristic length $L_{\rm cohr}$ is evidently non-Jeansean. If $L_{\rm cohr}^2 = r_{\rm c}^2 c_r^2/(c_r^2 - c_{\rm T}^2) \gg \lambda_J^2 = r_{\rm c}^2$, it is possible to derive the inequality when one can prefer criticality to Jeans instability $(c_r^2 - c_{\rm T}^2) \ll c_r^2$, i.e., $Q^2 - 1 \ll Q^2$. The inequality may be concretized if $L_{\rm cohr}$ is compared with $\lambda_T = 4\pi^2 G\sigma/\kappa^2 = 4\pi c_T/\kappa$.

So the marginally stable stellar disk can be found in incomplete equilibrium (critical) state and is prepared to fluctuate (Ivannikova et al. 1994). Unlike molecules of liquids which force to displace Brownian particles, fluctuations "gather up" "heaps" of disk substance. "Heaps" correlate and form a large scale structure.

"Heaps" dynamics can be understood as follows.

Time variations of the velocity of the part of the disk matter $\Delta v$ can be written as $\Delta v \sim (\varepsilon t)^{1/2}$ from the dimension reason, if $\varepsilon$ is the energy lost per second by mass unity and $t$ is time.

Equating the order of the Coriolis and gravitational forces $\Omega\Delta v \sim \pi G\sigma_m$, one can connect time variations $\Delta v$ and $\sigma_m$. So $\sigma_m \sim t^{1/2}$, as if the changing of stars pozitions $\overline{(\Delta r)^2}$ would be "Brownian" one because on the other side one may consider that $\sigma_m$ at a given radial distance $r$ $\sigma_m \sim r\sqrt{\overline{(\Delta r)^2}}$ and therefore $\overline{(\Delta r)^2} \sim t$.

Since $\sigma_m \sim t^a, a < 1$, origination process is in some sense "avalanche"-type, $\partial\sigma_m/\partial t$ is large in the beginning.

## 3. Conclusions

In the spirit of the generic nature of critical phenomena, we have shown the possibility of critical behaviour of galactic stellar discs. The disc can undergo a critical transformation (like phase transition) when the stellar velocity dispersion $c_r$ is equal to Toomre's critical value, $c_{\rm T}$. It appears that a fundamental spiral structure can originate from the criticality. In any case, multiarmes and flocculent spiral structures requie a special approach. Besides, scale invariance due to self-similarity of the underlying structure is a frequent and common thing (McKane et al. 1995).

The heuristic reasons given above are confirmed by numerical experiments of Hohl (1975a,b) which demonstrate that a stellar disc embedded in a bulky halo appears to be rather changeable at $Q = c_r/c_T = Q_{\rm cr} = 1$. Processes different from the bar instability "thermalize" the disc quickly (practically instantaneously), heating it up to $Q \approx (2 \div 3)Q_{\rm cr}$. At the initial stage of the process, the disc is populated by a set of stellar splinters and crescents, and a few large-scale spiral branches develop later.

Anomalous growth of density fluctuations in marginal state and its lack far from that was convincingly demonstrated by (Ivanov 1989, 1992) in numerical experiment.

## References

Halperin, B. I., & Hohenberg P. C. 1969, Phys.Rev., 177, 952

Hohl, F. 1975a, in La dynamique des galaxies spirales, Coll. Intern. du CNRS, ed. L. Weliachew (Paris), 55

Hohl, F. 1975b, in IAU Symp. 69, Dynamics of Stellar Systems, ed. A. Hayli (Dordrecht: Kluwer), 349

Ivannikova, E. I., Ivanov, A. V., & Maksumov, M. N. 1994, in Astrophysics and Cosmology after Gamov, Abstract Book (Moscow: Kozmozinform), 18

Ivannikova, E. I., Ivanov, A. V., & Maksumov, M. N. 1997, in Structure and Evolution of Stellar Systems, ed. T. A. Agekian, A. A. Müllări & V. V. Orlov (St. Petersburg: St. Petersburg Univ. Press), 271

Ivannikova, E. I., & Maksumov, M. N. 2001, in Stellar Dynamics: from Classic to Modern, ed. L. P. Osipkov & I. I. Nikiforov (Saint Petersburg: St. Petersburg Univ. Press), 367

Ivanov, A. V. 1989, Ap&SS, 159, 47

Ivanov, A. V. 1992, MNRAS, 259, 576

Landau, L. D., & Lifshitz, E. M. 1976, Statistical physics, Pt. 1 (Moscow: Nauka) (in Russian)

Lifshitz, E. M., & Pitayevsky, L. P., 1979, Physical kinetics (Landau, L. D., & Lifshitz, E. M., Theoretical Physics, Vol. 10) (Moscow: Nauka) (in Russian)

Lin, C. C., & Shu, F. H. 1964, ApJ, 140, 646

Ma, Sh.-K. 1976, Modern Theory of Critical Phenomena (London: Benjamin)

Marochnik, L. S., & Suchkov, A. A. 1984, Our Galaxy (Moscow: Nauka) (in Russian)

McKane, A., Droz, M., Vannimenus, J., & Wolf, D., ed. 1995, Scale Invariance, Interfaces, and Non-Equilibrium Dynamics. NATO ASI Series, Ser. B: Physics, Vol. 344 (New York and London: Plenum Press)

Patashinsky, A. Z., & Pokrovsky, V. L. 1975, Fluctuational Theory of Phase Transitions, §4 of Ch. 1 and §1 of Ch. 2 (Moscow: Nauka) (in Russian)

Rohlfs, K. 1977, Lectures on Density Wave Theory (Berlin: Springer)

Saslaw, W. C. 1987, Gravitational Physics of Stellar and Galactic Systems (Cambridge: Cambridge Univ. Press)

Toomre, A. 1964, ApJ, 139, 1217

*Order and Chaos in Stellar and Planetary Systems*
*ASP Conference Series, Vol. 316, 2004*
*G. Byrd, K. Kholshevnikov, A. Mylläri, I. Nikiforov, and V. Orlov, eds.*

# Scattering of Stars by Jeans-Unstable Density Waves

Evgeny Griv, Michael Gedalin, and David Eichler

*Department of Physics, Ben-Gurion University, Beer-Sheva, Israel*

**Abstract.** For the first time in galactic dynamics, the stellar disk of the Galaxy is treated by using the quasilinear (or weakly nonlinear) theory. It is shown that the nonresonant interaction of stars with Jeans-unstable aperiodic density waves in the disk of the Galaxy accounts for the observed increase in the residual (chaotic) stellar plane velocities with age.

## 1. Introduction

Many observations indicate ongoing dynamical relaxation on the timescale of less than 10 rotation periods ($< 2 \times 10^9$ yr) in the disk of the Milky Way (Binney & Tremaine 1987; Binney 2001). It was observed that in the solar neighborhood the residual (chaotic) velocity distribution function of stars with an age $t \gtrsim 10^8$ yr is close to the Schwarzschild distribution—a set of Gaussian distributions along each coordinate in velocity space, i.e., close to equilibrium along each coordinate. Also, older stellar populations are observed to have a higher velocity dispersion than younger ones. Yet a simple calculation of the relaxation time of the local disk of the Milky Way due to pairwise star–star encounters brings the standard value of $\sim 10^{14}$ yr (Binney & Tremaine 1987), which greatly exceeds the lifetime of the universe. The latter indicates a significant irregular gravitational field in the Galaxy's disk (Wielen 1977). The irregular field causes a diffusion of stellar orbits in velocity (and positional) space. Various mechanisms for the relaxation have been proposed (e.g., Grivnev & Fridman 1990). We elaborate upon the idea of the collective *collisionless* relaxation: gravitationally unstable perturbations affect the averaged velocity distribution of stars. Accordingly, the increase in chaotic velocities may be explained naturally by "collisions" of stars with the Jeans-unstable (almost) aperiodic density waves (see Griv, Gedalin, & Eichler 2001 and Griv, Gedalin, & Yuan 2002 for a detailed explanation).

## 2. Equilibrium

In the rotating frame of a disk galaxy, the collisionless motion of an ensemble of identical stars in the plane of the system can be described by the collisionless Boltzmann equation for the distribution function $f(\mathbf{r}, \mathbf{v}, t)$:

$$\frac{\partial f}{\partial t} + v_r \frac{\partial f}{\partial r} + \left(\Omega + \frac{v_\varphi}{r}\right)\frac{\partial f}{\partial \varphi} + \left(2\Omega v_\varphi + \frac{v_\varphi^2}{r} + \Omega^2 r \frac{\partial \Phi}{\partial r}\right)\frac{\partial f}{\partial v_r}$$
$$- \left(\frac{\kappa^2}{2\Omega} v_r + \frac{v_r v_\varphi}{r} + \frac{1}{r}\frac{\partial \Phi}{\partial \varphi}\right)\frac{\partial f}{\partial v_\varphi} = 0, \quad (1)$$

where the total azimuthal velocity of the stars is represented as a sum of the chaotic $v_\varphi$ and the basic rotation velocity $V_{\rm rot} = r\Omega$, $v_r$ is the velocity in the radial direction, the epicyclic frequency $\kappa(r)$ is defined by $\kappa = 2\Omega[1 + (r/2\Omega)(d\Omega/dr)]^{1/2}$, and $r$, $\varphi$, $z$ are the galactocentric cylindrical coordinates. The quantity $\Omega(r)$ denotes the angular velocity of galactic rotation at the distance $r$ from the center, and $\kappa \sim \Omega$. Random velocities are usually small compared with $V_{\rm rot}$. Collisions are neglected here because the collision frequency is much smaller than the cyclic frequency $\Omega$. In the kinetic equation (1), $\Phi(\mathbf{r}, t)$ is the total gravitational potential determined self-consistently from the Poisson equation $\nabla^2\Phi = 4\pi G \int f\, d\mathbf{v}$. Equation (1) and the Poisson equation give a complete self-consistent description of the problem for disk oscillation modes.

The equilibrium state is assumed, for simplicity, to be an axisymmetric and spatially homogeneous stellar disk. Second, in our simplified model, the perturbation is propagating in the plane of the disk. This approximation of an infinitesimally thin disk is a valid approximation if one considers perturbations with a radial wavelength that is greater than the typical disk thickness $h$. We assume here that the stars move in the disk plane so that $v_z = 0$. This allows us to use the two-dimensional distribution function $f = f(v_r, v_\varphi, t)\delta(z)$ such that $\int f\, dv_r\, dv_\varphi\, dz = \sigma$, where $\sigma$ is the surface density. From Eq. (1), the disk in the equlibrium is described by the following equation: $\nabla\Phi(\partial f/\partial \mathbf{v}) = 0$, where the angular velocity of rotation is such that the necessary centrifugal acceleration is exactly provided by the central gravitational force. The later equation does not determine the equilibrium distribution $f_{\rm e}$ uniquely. We choose $f_{\rm e}$ in the form of the Schwarzschild distribution (Griv & Peter 1996): $f_{\rm e} = (2\Omega/\kappa)(\sigma_{\rm e}/2\pi c_r^2)\exp\left[-(v_\perp^2/2c_r^2)\right]$. The Schwarzschild distribution function is a function of the two epicyclic constants of motion $\mathcal{E} = v_\perp^2/2$ and $r_0^2\Omega(r_0)$, where $r_0 = r + (2\Omega/\kappa^2)v_\varphi$. These constants of motion are related to the unperturbed epicyclic star orbits.[1] The distribution function $f_{\rm e}$ has been normalized according to $\int_{-\infty}^{\infty}\int_{-\infty}^{\infty} f_{\rm e}\, dv_r\, dv_\varphi = 2\pi(\kappa/2\Omega)\int_0^\infty v_\perp\, dv_\perp f_{\rm e} = \sigma_{\rm e}$, where $\sigma_{\rm e}$ is the equilibrium surface density. Such a distribution function for the unperturbed system is particularly important because it provides a fit to observations.

## 3. Diffusion Equation

We proceed by applying the procedure of the weakly nonlinear (or quasilinear) approach and decompose the time dependent distribution function $f = f_0(\mathbf{v}, t) + f_1(\mathbf{r}, \mathbf{v}, t)$ and the gravitational potential $\Phi = \Phi_0(r, t) + \Phi_1(\mathbf{r}, t)$ with

[1] Here $v_\perp^2 = v_r^2 + (2\Omega/\kappa)^2 v_\varphi^2$, $v_\perp/\kappa r_0 \sim \rho/r_0 \ll 1$, $\rho$ is the mean epicycle radius, and we follow Lin, Yuan, & Shu (1969), Shu (1970), and Morozov (1980) making use of Lindblad's expressions for the unperturbed epicyclic trajectories of stars in the equilibrium central field $\Phi_{\rm e}(r)$.

$|f_1/f_0| \ll 1$ and $|\Phi_1/\Phi_0| \ll 1$ for all $\mathbf{r}$ and $t$. The functions $f_1$ and $\Phi_1$ are oscillating rapidly in space and time, while the functions $f_0$ and $\Phi_0$ describe the slowly (and monotonically) developing "background" against which small perturbations develop; $f_0(t=0) \equiv f_{\rm e}$ and $\Phi_0(t=0) \equiv \Phi_{\rm e}$. Linearizing Eq. (1) and separating fast and slow varying variables one obtains

$$\frac{df_1}{dt} = \frac{\partial \Phi_1}{\partial r}\frac{\partial f_0}{\partial v_r} + \frac{1}{r}\frac{\partial \Phi_1}{\partial \varphi}\frac{\partial f_0}{\partial v_\varphi}, \tag{2}$$

$$\frac{\partial f_0}{\partial t} = \left\langle \frac{\partial \Phi_1}{\partial r}\frac{\partial f_1}{\partial v_r} + \frac{1}{r}\frac{\partial \Phi_1}{\partial \varphi}\frac{\partial f_1}{\partial v_\varphi} \right\rangle, \tag{3}$$

where $d/dt$ means the total derivative along the star orbit and $\langle \cdots \rangle$ denotes the time average over the fast oscillations, $f_0 = \langle f \rangle$ and $\langle f_1 \rangle = \langle \Phi_1 \rangle = 0$. To emphasize it again, we are concerned with the growth or decay of small perturbations from an equilibrium state. In the epicyclic approximation, the partial derivatives in these equations are transformed as follows (Shu 1970; Morozov 1980; Griv & Peter 1996; Griv et al. 2001, 2002; Griv, Gedalin, & Yuan 2003):

$$\partial/\partial v_r = v_r(\partial/\partial \mathcal{E}), \quad \partial/\partial v_\varphi = (2\Omega/\kappa)^2\, v_\varphi(\partial/\partial \mathcal{E}). \tag{4}$$

To determine oscillation spectra, let us consider the stability problem in the lowest (local) WKB approximation: the perturbation scale is sufficiently small for the disk to be regarded as spatially homogeneous. This is accurate for short wave perturbations only, but qualitatively correct even for perturbations with a longer wavelength, of the order of the disk radius $R \approx 20$ kpc. In the local WKB approximation in Eqs (2) and (3), assuming the weakly inhomogeneous disk, the perturbation is selected in the form of a plane wave (in the rotating frame): $\aleph_1 = \aleph_{\mathbf{k}}\left(e^{ik_r r + im\varphi - i\omega_* t} + \text{c.c.}\right)$, where $\aleph_{\mathbf{k}}$ is an amplitude that is a constant in space and time, $m$ is the nonnegative azimuthal mode number (= number of spiral arms), $\omega_* = \omega - m\Omega$ is the Doppler-shifted complex wavefrequency, suffixes $\mathbf{k}$ denote the $\mathbf{k}$-th Fourier component, c. c. means the complex conjugate, and $|k_r|R \gg 1$. The solution in such a form represents a spiral wave with $m$ arms. With $\varphi$ increasing in the rotation direction, we have $k_r > 0$ for trailing spiral patters, which are the most frequently observed among spiral galaxies. With $m = 0$, we have the density waves in the form of concentric rings.

In Eq. (2), using the transformation of the partial derivatives $\partial/\partial v_r$ and $\partial/\partial v_\varphi$ given by Eqs (4), one obtains

$$f_1 = \int_{-\infty}^{t} dt'\, \mathbf{v}_\perp \cdot \frac{\partial \Phi_1}{\partial \mathbf{r}}\frac{\partial f_0}{\partial \mathcal{E}}, \tag{5}$$

where $f_1(t' \to -\infty) = 0$, i.e., only growing perturbations are considered. In Eq. (5) making use of the time dependence of perturbations in the form of WKB wave and the unperturbed epicyclic trajectories of stars given, e.g., by Griv & Peter (1996) in the exponential factor, it is straightforward to show that

$$f_1 = -\kappa \Phi_1 \frac{\partial f_0}{\partial \mathcal{E}} \sum_{l=-\infty}^{\infty} \sum_{n=-\infty}^{\infty} l \frac{e^{i(n-l)(\phi_0 - \zeta)} J_l(\chi) J_n(\chi)}{\omega_* - l\kappa} + \text{c.c.}, \tag{6}$$

where $J_l(\chi)$ is the Bessel function of the first kind of order $l$, $\chi = k_* v_\perp/\kappa \sim k_* \rho$, $k_* = k\{1 + [(2\Omega/\kappa)^2 - 1]\sin^2\psi\}^{1/2}$ is the effective wavenumber, and $\tan\psi = k_\varphi/k_r = m/rk_r$. In the equation above the denominator vanishes when $\omega_* - l\kappa \to 0$. This occurs near corotation ($l = 0$) and other resonances ($l = \pm 1, \pm 2, \ldots$). The location of these resonances depends on the rotation curve and the spiral pattern speed (Lin et al. 1969). Only the main part of the disk is studied which lies sufficiently far from the resonances: below in all equations $\omega_* - l\kappa \neq 0$.

Next, we substitute the solution (6) into Eq. (3). Taking into account that the terms $l \neq n$ vanish for axially symmetric functions $f_0$, we obtain

$$\frac{\partial f_0}{\partial t} = i\frac{\pi}{2}\sum_{\mathbf{k}}\sum_{l=-\infty}^{\infty} |\Phi_{\mathbf{k}}|^2 \frac{\partial}{\partial v_\perp}\frac{k_*\kappa}{v_\perp \chi}\left[\frac{l^2 J_l^2(\chi)}{\omega_* - l\kappa} - \frac{l^2 J_l^2(\chi)}{\omega_*^* - l\kappa}\right]\frac{\partial f_0}{\partial v_\perp}, \tag{7}$$

where $\omega_*^*$ is the complex conjugate wavefrequency. As it is seen, the initial distribution of stars will change upon time only under the action of growing, that is, Jeans-unstable, perturbations ($\Im\omega_*$ and $\Im\omega_*^* > 0$).

As usual in the weakly nonlinear theory, in order to close the system one must engage an equation for $\Phi_{\mathbf{k}}$. Averaging over the fast oscillations, we have

$$(\partial/\partial t)|\Phi_{\mathbf{k}}|^2 = 2\Im\omega_*|\Phi_{\mathbf{k}}|^2. \tag{8}$$

Equations (7) and (8) form the closed system of quasilinear equations for Jeans-unstable oscillations of the rotating homogeneous disk of stars, and describe a diffusion in velocity space. The spectrum of oscillations and their growth rate are obtained from Eq. (2) and the Poisson equation (Morozov 1980; Griv & Peter 1996; Griv et al. 2001, 2002):

$$\frac{k^2 c_r^2}{2\pi G\sigma_0|k|} = -\kappa\sum_{l=-\infty}^{\infty} l\frac{e^{-x}I_l(x)}{\omega_* - l\kappa}, \tag{9}$$

and $\Im\omega_* \approx \sqrt{2\pi G\sigma_0|k|F(x)}$, respectively. Here, $F(x) \approx 2\kappa^2 e^{-x}I_1(x)/k^2c_r^2$ is the so-called reduction factor, which takes into account the fact that the wave field only weakly affects the stars with high chaotic velocities (cf. Lin et al. 1969; Shu 1970). Also, $I_l(x)$ is a Bessel function of an imaginary argument with its argument $x \approx k_*^2\rho^2$ and $\rho = c_r/\kappa$ is now the mean epicyclic radius. An important feature of the instability under consideration is the fact that it is aperiodic (the real part of $\omega_*$ vanishes). A further simplification results from restricting the frequency range of the waves examined by taking the low frequency limit ($|\omega_*|$ less than the epicyclic frequency $\kappa$ of any disk star). See Griv & Peter (1996) for an explanation. One concludes that the terms in series of Eqs (7) and (9) for which $|l| \geq 2$ may be neglected, and consideration will be limited to the transparency region between the inner and outer Lindblad resonances. In this case, in Eq. (9) the function $\Lambda(x) = \exp(-x)I_1(x)$ starts from $\Lambda(0) = 0$, reaches a maximum $\Lambda_{\max} < 1$ at $x \approx 0.5$, and then decreases. Hence, the growth rate has a maximum at $x \approx 0.5 < 1$ (Griv et al. 2002, their Figure 1).

In general, the growth rate of the Jeans instability is high and is comparable to the epicyclic period $|\Im\omega_*| \sim \Omega$; perturbations with wavelength $\lambda_{\text{crit}} \approx (4\pi\Omega/\kappa)c_r/\kappa \approx 3\pi\rho$ have the fastest growth rate (Griv & Peter 1996). In the

Solar vicinity of the Galaxy for the subsystem of young, low-dispersion stars with $c_r \sim 10$ km s$^{-1}$, $\Omega \approx 3 \times 10^{-8}$ yr$^{-1}$ and $3\pi\rho \approx 2$ kpc. Interestingly, from observations in the Galaxy, the distance between spiral arms $\lambda = 2$–$3$ kpc also. So, the Jeans length is small compared to the disk radius $R$ and the radial scale of the perturbations is sufficiently small, $\lambda_{\text{crit}} \ll R$ but $\lambda_{\text{crit}} \gg h \sim 100$ pc.

## 4. Astronomical Implications

As an application of the theory we investigate the relaxation of low-frequency and Jeans-unstable ($|\omega_*| \lesssim \kappa$ and $\omega_*^2 < 0$, respectively) oscillations in the homogeneous galactic disk. Evidently, the unstable Jeans oscillations must influence the distribution function of the main, nonresonant part of stars in such a way as to hinder the wave excitation, i.e., to increase the velocity dispersion. This is because the Jeans instability, being essentially a gravitational one, tends to be stabilized by chaotic motions (Griv & Peter 1996). Therefore, along with the growth of the oscillation amplitude, chaotic velocities must increase at the expense of circular motion, and eventually in the disk there can be established a quasi-stationary distribution so that the Jeans stability sets in.

In the following, we restrict ourselves to the fastest growing long-wavelength perturbations, $\chi^2$ and $x^2 < 1$ [see the explanation after Eq. (9)]. Then in Eqs (7) and (9) one can use the expansions $J_1^2(\chi) \approx \chi^2/4$ and $e^{-x}I_1(x) \approx (1/2)x - (1/2)x^2 + (5/16)x^3$. As a result, Eq. (7) takes the simple form

$$(\partial/\partial t)f_0 = \mathcal{D}_v(\partial^2/\partial v_\perp^2)f_0, \tag{10}$$

where $\mathcal{D}_v \approx (\pi/2\kappa^2)\sum_{\mathbf{k}} k_*^2 \Im\omega_* |\Phi_{\mathbf{k}}|^2$, and $\Im\omega_*$ is a function of $t$. Equation (10) shows that the velocity diffusion coefficient for nonresonant stars $\mathcal{D}_v$ is independent of $\mathbf{v}_\perp$ (to lowest order). This is a qualitative result of the nonresonant character of the star's interaction with collective aggregates. Whereas Binney & Lacey (1988) and Jenkins & Binney (1990) attributed the effect to some unspecified time behavior of the spiral density wave, we attribute the effect to the nonresonant term resulting from wave growth of Jeans-unstable perturbations. Hence in the lowest approximation, we obtain a velocity diffusion coefficient $\mathcal{D}_v$ that is independent of velocity $\mathbf{v}_\perp$, without any further assumptions.

By introducing the definitions $d\tau/dt = \mathcal{D}_v(t)$ and $d/dt = (d\tau/dt)(d/d\tau)$, Eqs (8) and (10) are rewritten as follows: $(\partial/\partial\tau)f_0 - (\partial^2/\partial v_\perp^2)f_0 = 0$ and $(\partial/\partial\tau)\mathcal{D}_v = 2\Im\omega_*$, which have the particular solution

$$f_0(v_\perp, \tau) = \frac{\text{const}}{\sqrt{\tau + c_r^2/2}} \exp\left[-\frac{v_\perp^2}{4(\tau + c_r^2/2)}\right]. \tag{11}$$

Accordingly, during the development of the Jeans instability on the timescale $\sim(\Im\omega_*)^{-1} \sim \Omega^{-1}$ (that is, on the timescale of the epicyclic period), the Schwarzschild distribution of chaotic velocities (i.e., Gaussian spread along $v_r, v_\varphi$ coordinates in velocity space) is established. As the perturbation energy increases, the initial distribution spreads ($f_0$ becomes less peaked), and the effective temperature $T$ grows with time (a Gaussian spread increases): $T = 2\tau \propto \int \mathcal{D}_v(t)\,dt \propto \sum_{\mathbf{k}} k_*^2|\Phi_{\mathbf{k}}|^2$ (or $c_r^2 \propto t$, respectively). The energy of the oscillation field $\propto \sum_{\mathbf{k}} k_*^2 |\Phi_{\mathbf{k}}|^2$ plays the role of a "temperature" in the nonresonant-particle distribution. The well-known Lindblad relation for the velocity ellipsoid in a rotating frame $c_r/c_\varphi \approx 2\Omega/\kappa$ is clearly conserved.

This mechanism increases the plane velocity dispersion of stars in the Galaxy after they are born. The collisionless relaxation mimics thermal relaxation, leading to Schwarzschild-type velocity distributions with an effective temperature that increases with time. The diffusion of nonresonant stars takes place because they gain wave energy as the instability develops. The velocity diffusion, however, presumably tapers off as Jeans stability is approached: $c_r$ becomes greater than $c_{\mathrm{crit}} \approx (2\Omega/\kappa)c_{\mathrm{T}}$, where $c_{\mathrm{T}} \approx 3.4G\sigma_0/\kappa$ is the well-known Toomre's (1964) velocity dispersion to suppress the instability of only axisymmetric perturbations (Griv & Peter 1996; Griv, Rosenstein, & Gedalin 1999; Griv et al. 2002).

Let us finally evaluate the heating $\Delta v$ for a realistic model of the Galaxy. In accord with observations, we adopt the ratio $|\Phi_1/\Phi_0| \approx 0.01$, $\Phi_0 \approx 0.5r_0^2\Omega^2$, $r_0 \approx 8.5$ kpc, $k_* \approx 2$ kpc$^{-1}$, $\kappa \approx 1.5\Omega$, and $\Im\omega_* \approx \Omega$. Then from Eq. (10), one defines $\Delta v \approx \sqrt{\mathcal{D}_v t} \approx 27\frac{100\Phi_1}{\Phi_0}$ km s$^{-1}$ if $t \sim 10^9$ yr. This velocity spread is in agreement with estimates (Wielen 1977; Binney & Tremaine 1987; Grivnev & Fridman 1990; Binney 2001) based on the observed stellar velocities.

We propose this collisionless relaxation as an explanation of the observed Schwarzschild distribution of the Galaxy's disk stars. The actual timescale for relaxation in the Galaxy may be much shorter than its standard value of $\sim 10^{14}$ yr for Chandrasekhar's collisional star–star relaxation; it may be of the order of $(\Im\omega_*)^{-1} \lesssim \Omega^{-1} \lesssim 3 \times 10^8$ yr, i.e., comparable with 1–2 periods of the Milky Way rotation. The above time is in fair agreement with both observations (Wielen 1977; Binney & Tremaine 1987; Dehnen & Binney 1998; Binney 2001) and $N$-body computer simulations (Sellwood & Carlberg 1984; Griv & Chiueh 1998).

**Acknowledgments.** This work was supported in part by the Israel Science Foundation and the Israeli Ministry of Immigrant Absorption.

## References

Binney, J. 2001, in ASP Conf. Proc., 230, Galaxy Disks and Disk Galaxies, ed. J. G. Funes & E. M. Corsini (San Francisco: ASP), 63

Binney, J., & Lacey, C. 1988, MNRAS, 230, 597

Binney, J., & Tremaine, S. 1987, Galactic Dynamics (Princeton, NJ: Princeton Univ. Press)

Dehnen, W., & Binney, J. 1998, MNRAS, 298, 387

Griv, E., & Chiueh, T. 1998, ApJ, 503, 186

Griv, E., Gedalin, M., & Eichler, D. 2001, ApJ, 555, L29

Griv, E., Gedalin, M., & Yuan, C. 2002, A&A, 383, 338

Griv, E., Gedalin, M., & Yuan, C. 2003, MNRAS, 342, 1102

Griv, E., & Peter, W. 1996, ApJ, 469, 84

Griv, E., Rosenstein, B., & Gedalin, M. 1999, A&A, 347, 821

Grivnev, E. M., & Fridman, A. M. 1990, Soviet Ast., 34, 10

Jenkins, A., & Binney, J. 1990, MNRAS, 245, 305

Lin, C. C., Yuan, C., & Shu, F. H. 1969, ApJ, 155, 721
Morozov, A. G. 1980, Soviet Ast., 24, 391
Sellwood, J. A., & Carlberg, R. G. 1984, ApJ, 282, 61
Shu, F. H. 1970, ApJ, 160, 99
Toomre, A. 1964, ApJ, 139, 1217
Wielen, R. 1977, A&A, 60, 263

# Author Index

**THE FOLLOWING IS A LISTING OF THE VOLUMES**

Published
by

ASTRONOMICAL SOCIETY OF THE PACIFIC
(ASP)

**An international, nonprofit, scientific and educational organization founded in 1889**

All book orders or inquiries concerning

ASTRONOMICAL SOCIETY OF THE PACIFIC CONFERENCE SERIES (ASP - CS)

and

INTERNATIONAL ASTRONOMICAL UNION VOLUMES (IAU)

should be directed to the:

**Astronomical Society of the Pacific Conference Series**
**390 Ashton Avenue**
**San Francisco CA 94112-1722 USA**

**Phone: 800-335-2624 (within USA)**
**Phone: 415-337-2126**
**Fax: 415-337-5205**

**E-mail: service@astrosociety.org**
**Web Site: http://www.astrosociety.org**

Complete lists of proceedings of past IAU Meetings are maintained at the IAU Web site at the URL: http://www.iau.org/publicat.html

Volumes 32 - 189 in the IAU Symposia Series may be ordered from:

Kluwer Academic Publishers
P. O. Box 117
NL 3300 AA Dordrecht
The Netherlands

Kluwer@wKap.com

# ASP CONFERENCE SERIES VOLUMES

Published by the Astronomical Society of the Pacific

**PUBLISHED: 1988 (* asterisk means OUT OF PRINT)**

Vol. CS -1 — PROGRESS AND OPPORTUNITIES IN SOUTHERN HEMISPHERE OPTICAL ASTRONOMY: CTIO 25$^{TH}$ Anniversary Symposium
eds. V. M. Blanco and M. M. Phillips
ISBN 0-937707-18-X

Vol. CS-2 — PROCEEDINGS OF A WORKSHOP ON OPTICAL SURVEYS FOR QUASARS
eds. Patrick S. Osmer, Alain C. Porter, Richard F. Green, and Craig B. Foltz
ISBN 0-937707-19-8

Vol. CS-3 — FIBER OPTICS IN ASTRONOMY
ed. Samuel C. Barden
ISBN 0-937707-20-1

Vol. CS-4 — THE EXTRAGALACTIC DISTANCE SCALE:
Proceedings of the ASP 100$^{th}$ Anniversary Symposium
eds. Sidney van den Bergh and Christopher J. Pritchet
ISBN 0-937707-21-X

Vol. CS-5 — THE MINNESOTA LECTURES ON CLUSTERS OF GALAXIES AND LARGE-SCALE STRUCTURE
ed. John M. Dickey
ISBN 0-937707-22-8

**PUBLISHED: 1989**

Vol. CS-6 * — SYNTHESIS IMAGING IN RADIO ASTRONOMY: A Collection of Lectures from the Third NRAO Synthesis Imaging Summer School
eds. Richard A. Perley, Frederic R. Schwab, and Alan H. Bridle
ISBN 0-937707-23-6

**PUBLISHED: 1990**

Vol. CS-7 — PROPERTIES OF HOT LUMINOUS STARS: Boulder-Munich Workshop
ed. Catharine D. Garmany
ISBN 0-937707-24-4

Vol. CS-8 * — CCDs IN ASTRONOMY
ed. George H. Jacoby
ISBN 0-937707-25-2

Vol. CS-9 — COOL STARS, STELLAR SYSTEMS, AND THE SUN:
Sixth Cambridge Workshop
ed. George Wallerstein
ISBN 0-937707-27-9

Vol. CS-10 — EVOLUTION OF THE UNIVERSE OF GALAXIES:
Edwin Hubble Centennial Symposium
ed. Richard G. Kron
ISBN 0-937707-28-7

Vol. CS-11 — CONFRONTATION BETWEEN STELLAR PULSATION AND EVOLUTION
eds. Carla Cacciari and Gisella Clementini
ISBN 0-937707-30-9

## ASP CONFERENCE SERIES VOLUMES

Published by the Astronomical Society of the Pacific

---

**PUBLISHED: 1991 (* asterisk means OUT OF PRINT)**

Vol. CS-12 THE EVOLUTION OF THE INTERSTELLAR MEDIUM
ed. Leo Blitz
ISBN 0-937707-31-7

Vol. CS-13 THE FORMATION AND EVOLUTION OF STAR CLUSTERS
ed. Kenneth Janes
ISBN 0-937707-32-5

Vol. CS-14 ASTROPHYSICS WITH INFRARED ARRAYS
ed. Richard Elston
ISBN 0-937707-33-3

Vol. CS-15 LARGE-SCALE STRUCTURES AND PECULIAR MOTIONS IN THE UNIVERSE
eds. David W. Latham and L. A. Nicolaci da Costa
ISBN 0-937707-34-1

Vol. CS-16 Proceedings of the 3rd Haystack Observatory Conference on ATOMS, IONS, AND MOLECULES: NEW RESULTS IN SPECTRAL LINE ASTROPHYSICS
eds. Aubrey D. Haschick and Paul T. P. Ho
ISBN 0-937707-35-X

Vol. CS-17 LIGHT POLLUTION, RADIO INTERFERENCE, AND SPACE DEBRIS
ed. David L. Crawford
ISBN 0-937707-36-8

Vol. CS-18 THE INTERPRETATION OF MODERN SYNTHESIS OBSERVATIONS OF SPIRAL GALAXIES
eds. Nebojsa Duric and Patrick C. Crane
ISBN 0-937707-37-6

Vol. CS-19 RADIO INTERFEROMETRY: THEORY, TECHNIQUES, AND APPLICATIONS, IAU Colloquium 131
eds. T. J. Cornwell and R. A. Perley
ISBN 0-937707-38-4

Vol. CS-20 FRONTIERS OF STELLAR EVOLUTION:
50th Anniversary McDonald Observatory (1939-1989)
ed. David L. Lambert
ISBN 0-937707-39-2

Vol. CS-21 THE SPACE DISTRIBUTION OF QUASARS
ed . David Crampton
ISBN 0-937707-40-6

**PUBLISHED: 1992**

Vol. CS-22 NONISOTROPIC AND VARIABLE OUTFLOWS FROM STARS
eds. Laurent Drissen, Claus Leitherer, and Antonella Nota
ISBN 0-937707-41-4

Vol CS-23 * ASTRONOMICAL CCD OBSERVING AND REDUCTION TECHNIQUES
ed. Steve B. Howell
ISBN 0-937707-42-4

Vol. CS-24 COSMOLOGY AND LARGE-SCALE STRUCTURE IN THE UNIVERSE
ed. Reinaldo R. de Carvalho
ISBN 0-937707-43-0

Vol. CS-25 ASTRONOMICAL DATA ANALYSIS, SOFTWARE AND SYSTEMS (ADASS) I
eds. Diana M. Worrall, Chris Biemesderfer, and Jeannette Barnes
ISBN 0-937707-44-9

## ASP CONFERENCE SERIES VOLUMES

Published by the Astronomical Society of the Pacific

---

**PUBLISHED: 1992 (* asterisk means OUT OF PRINT)**

Vol. CS-26 COOL STARS, STELLAR SYSTEMS, AND THE SUN:
Seventh Cambridge Workshop
eds. Mark S. Giampapa and Jay A. Bookbinder
ISBN 0-937707-45-7

Vol. CS-27 THE SOLAR CYCLE: Proceedings of the National Solar Observatory/Sacramento Peak 12th Summer Workshop
ed. Karen L. Harvey
ISBN 0-937707-46-5

Vol. CS-28 AUTOMATED TELESCOPES FOR PHOTOMETRY AND IMAGING
eds. Saul J. Adelman, Robert J. Dukes, Jr., and Carol J. Adelman
ISBN 0-937707-47-3

Vol. CS-29 Viña del Mar Workshop on CATACLYSMIC VARIABLE STARS
ed. Nikolaus Vogt
ISBN 0-937707-48-1

Vol. CS-30 VARIABLE STARS AND GALAXIES
ed. Brian Warner
ISBN 0-937707-49-X

Vol. CS-31 RELATIONSHIPS BETWEEN ACTIVE GALACTIC NUCLEI AND STARBURST GALAXIES
ed. Alexei V. Filippenko
ISBN 0-937707-50-3

Vol. CS-32 COMPLEMENTARY APPROACHES TO DOUBLE AND MULTIPLE STAR RESEARCH, IAU Colloquium 135
eds. Harold A. McAlister and William I. Hartkopf
ISBN 0-937707-51-1

Vol. CS-33 RESEARCH AMATEUR ASTRONOMY
ed. Stephen J. Edberg
ISBN 0-937707-52-X

Vol. CS-34 ROBOTIC TELESCOPES IN THE 1990's
ed. Alexei V. Filippenko
ISBN 0-937707-53-8

**PUBLISHED: 1993**

Vol. CS-35 * MASSIVE STARS: THEIR LIVES IN THE INTERSTELLAR MEDIUM
eds. Joseph P. Cassinelli and Edward B. Churchwell
ISBN 0-937707-54-6

Vol. CS-36 PLANETS AROUND PULSARS
ed. J. A. Phillips, S. E. Thorsett, and S. R. Kulkarni
ISBN 0-937707-55-4

Vol. CS-37 FIBER OPTICS IN ASTRONOMY II
ed. Peter M. Gray
ISBN 0-937707-56-2

Vol. CS-38 NEW FRONTIERS IN BINARY STAR RESEARCH: Pacific Rim Colloquium
eds. K. C. Leung and I.-S. Nha
ISBN 0-937707-57-0

# ASP CONFERENCE SERIES VOLUMES

Published by the Astronomical Society of the Pacific

**PUBLISHED: 1993 (* asterisk means OUT OF PRINT)**

Vol. CS-39 THE MINNESOTA LECTURES ON THE STRUCTURE AND DYNAMICS OF THE MILKY WAY
ed. Roberta M. Humphreys
ISBN 0-937707-58-9

Vol. CS-40 INSIDE THE STARS, IAU Colloquium 137
eds. Werner W. Weiss and Annie Baglin
ISBN 0-937707-59-7

Vol. CS-41 ASTRONOMICAL INFRARED SPECTROSCOPY: FUTURE OBSERVATIONAL DIRECTIONS
ed. Sun Kwok
ISBN 0-937707-60-0

Vol. CS-42 GONG 1992: SEISMIC INVESTIGATION OF THE SUN AND STARS
ed. Timothy M. Brown
ISBN 0-937707-61-9

Vol. CS-43 SKY SURVEYS: PROTOSTARS TO PROTOGALAXIES
ed. B. T. Soifer
ISBN 0-937707-62-7

Vol. CS-44 PECULIAR VERSUS NORMAL PHENOMENA IN A-TYPE AND RELATED STARS, IAU Colloquium 138
eds. M. M. Dworetsky, F. Castelli, and R. Faraggiana
ISBN 0-937707-63-5

Vol. CS-45 LUMINOUS HIGH-LATITUDE STARS
ed. Dimitar D. Sasselov
ISBN 0-937707-64-3

Vol. CS-46 THE MAGNETIC AND VELOCITY FIELDS OF SOLAR ACTIVE REGIONS, IAU Colloquium 141
eds. Harold Zirin, Guoxiang Ai, and Haimin Wang
ISBN 0-937707-65-1

Vol. CS-47 THIRD DECENNIAL US-USSR CONFERENCE ON SETI -- Santa Cruz, California, USA
ed. G. Seth Shostak
ISBN 0-937707-66-X

Vol. CS-48 THE GLOBULAR CLUSTER-GALAXY CONNECTION
eds. Graeme H. Smith and Jean P. Brodie
ISBN 0-937707-67-8

Vol. CS-49 GALAXY EVOLUTION: THE MILKY WAY PERSPECTIVE
ed. Steven R. Majewski
ISBN 0-937707-68-6

Vol. CS-50 STRUCTURE AND DYNAMICS OF GLOBULAR CLUSTERS
eds. S. G. Djorgovski and G. Meylan
ISBN 0-937707-69-4

Vol. CS-51 OBSERVATIONAL COSMOLOGY
eds. Guido Chincarini, Angela Iovino, Tommaso Maccacaro, and Dario Maccagni
ISBN 0-937707-70-8

## ASP CONFERENCE SERIES VOLUMES

Published by the Astronomical Society of the Pacific

**PUBLISHED: 1994 (* asterisk means OUT OF PRINT)**

Vol. CS-52 ASTRONOMICAL DATA ANALYSIS SOFTWARE AND SYSTEMS (ADASS) II
eds. R. J. Hanisch, R. J. V. Brissenden, and Jeannette Barnes
ISBN 0-937707-71-6

Vol. CS-53 BLUE STRAGGLERS
ed. Rex A. Saffer
ISBN 0-937707-72-4

Vol. CS-54 THE FIRST STROMLO SYMPOSIUM: THE PHYSICS OF ACTIVE GALAXIES
eds. Geoffrey V. Bicknell, Michael A. Dopita, and Peter J. Quinn
ISBN 0-937707-73-2

Vol. CS-55 OPTICAL ASTRONOMY FROM THE EARTH AND MOON
eds. Diane M. Pyper and Ronald J. Angione
ISBN 0-937707-74-0

Vol. CS-56 * INTERACTING BINARY STARS
ed. Allen W. Shafter
ISBN 0-937707-75-9

Vol. CS-57 STELLAR AND CIRCUMSTELLAR ASTROPHYSICS
eds. George Wallerstein and Alberto Noriega-Crespo
ISBN 0-937707-76-7

Vol. CS-58 THE FIRST SYMPOSIUM ON THE INFRARED CIRRUS AND DIFFUSE INTERSTELLAR CLOUDS
eds. Roc M. Cutri and William B. Latter
ISBN 0-937707-77-5

Vol. CS-59 ASTRONOMY WITH MILLIMETER AND SUBMILLIMETER WAVE INTERFEROMETRY, IAU Colloquium 140
eds. M. Ishiguro and Wm. J. Welch
ISBN 0-937707-78-3

Vol. CS-60 THE MK PROCESS AT 50 YEARS: A POWERFUL TOOL FOR ASTRO-PHYSICAL INSIGHT, A Workshop of the Vatican Observatory -- Tucson, Arizona, USA
eds. C. J. Corbally, R. O. Gray, and R. F. Garrison
ISBN 0-937707-79-1

Vol. CS-61 ASTRONOMICAL DATA ANALYSIS SOFTWARE AND SYSTEMS (ADASS) III
eds. Dennis R. Crabtree, R. J. Hanisch, and Jeannette Barnes
ISBN 0-937707-80-5

Vol. CS-62 THE NATURE AND EVOLUTIONARY STATUS OF HERBIG Ae/Be STARS
eds. Pik Sin Thé, Mario R. Pérez, and Ed P. J. van den Heuvel
ISBN 0-9837707-81-3

Vol. CS-63 SEVENTY-FIVE YEARS OF HIRAYAMA ASTEROID FAMILIES: THE ROLE OF COLLISIONS IN THE SOLAR SYSTEM HISTORY
eds. Yoshihide Kozai, Richard P. Binzel, and Tomohiro Hirayama
ISBN 0-937707-82-1

Vol. CS-64 * COOL STARS, STELLAR SYSTEMS, AND THE SUN: Eighth Cambridge Workshop
ed. Jean-Pierre Caillault
ISBN 0-937707-83-X

# ASP CONFERENCE SERIES VOLUMES

Published by the Astronomical Society of the Pacific

**PUBLISHED: 1994 (* asterisk means OUT OF PRINT)**

Vol. CS-65 * CLOUDS, CORES, AND LOW MASS STARS:
The Fourth Haystack Observatory Conference
eds. Dan P. Clemens and Richard Barvainis
ISBN 0-937707-81-8

Vol. CS-66 * PHYSICS OF THE GASEOUS AND STELLAR DISKS OF THE GALAXY
ed. Ivan R. King
ISBN 0-937707-85-6

Vol. CS-67 UNVEILING LARGE-SCALE STRUCTURES BEHIND THE MILKY WAY
eds. C. Balkowski and R. C. Kraan-Korteweg
ISBN 0-937707-86-4

Vol. CS-68 SOLAR ACTIVE REGION EVOLUTION: COMPARING MODELS WITH OBSERVATIONS
eds. K. S. Balasubramaniam and George W. Simon
ISBN 0-937707-87-2

Vol. CS-69 REVERBERATION MAPPING OF THE BROAD-LINE REGION IN ACTIVE GALACTIC NUCLEI
eds. P. M. Gondhalekar, K. Horne, and B. M. Peterson
ISBN 0-937707-88-0

Vol. CS-70 * GROUPS OF GALAXIES
eds. Otto-G. Richter and Kirk Borne
ISBN 0-937707-89-9

**PUBLISHED: 1995**

Vol. CS-71 TRIDIMENSIONAL OPTICAL SPECTROSCOPIC METHODS IN ASTROPHYSICS, IAU Colloquium 149
eds. Georges Comte and Michel Marcelin
ISBN 0-937707-90-2

Vol. CS-72 MILLISECOND PULSARS: A DECADE OF SURPRISE
eds. A. S. Fruchter, M. Tavani, and D. C. Backer
ISBN 0-937707-91-0

Vol. CS-73 AIRBORNE ASTRONOMY SYMPOSIUM ON THE GALACTIC ECOSYSTEM: FROM GAS TO STARS TO DUST
eds. Michael R. Haas, Jacqueline A. Davidson, and Edwin F. Erickson
ISBN 0-937707-92-9

Vol. CS-74 PROGRESS IN THE SEARCH FOR EXTRATERRESTRIAL LIFE:
1993 Bioastronomy Symposium
ed. G. Seth Shostak
ISBN 0-937707-93-7

Vol. CS-75 MULTI-FEED SYSTEMS FOR RADIO TELESCOPES
eds. Darrel T. Emerson and John M. Payne
ISBN 0-937707-94-5

Vol. CS-76 GONG '94: HELIO- AND ASTERO-SEISMOLOGY FROM THE EARTH AND SPACE
eds. Roger K. Ulrich, Edward J. Rhodes, Jr., and Werner Däppen
ISBN 0-937707-95-3

## ASP CONFERENCE SERIES VOLUMES

Published by the Astronomical Society of the Pacific

**PUBLISHED: 1995 (* asterisk means OUT OF PRINT)**

Vol. CS-77 ASTRONOMICAL DATA ANALYSIS SOFTWARE AND SYSTEMS (ADASS) IV
eds. R. A. Shaw, H. E. Payne, and J. J. E. Hayes
ISBN 0-937707-96-1

Vol. CS-78 ASTROPHYSICAL APPLICATIONS OF POWERFUL NEW DATABASES:
Joint Discussion No. 16 of the 22nd General Assembly of the IAU
eds. S. J. Adelman and W. L. Wiese
ISBN 0-937707-97-X

Vol. CS-79 ROBOTIC TELESCOPES: CURRENT CAPABILITIES, PRESENT DEVELOPMENTS, AND FUTURE PROSPECTS FOR AUTOMATED ASTRONOMY
eds. Gregory W. Henry and Joel A. Eaton
ISBN 0-937707-98-8

Vol. CS-80 * THE PHYSICS OF THE INTERSTELLAR MEDIUM AND INTERGALACTIC MEDIUM
eds. A. Ferrara, C. F. McKee, C. Heiles, and P. R. Shapiro
ISBN 0-937707-99-6

Vol. CS-81 LABORATORY AND ASTRONOMICAL HIGH RESOLUTION SPECTRA
eds. A. J. Sauval, R. Blomme, and N. Grevesse
ISBN 1-886733-01-5

Vol. CS-82 * VERY LONG BASELINE INTERFEROMETRY AND THE VLBA
eds. J. A. Zensus, P. J. Diamond, and P. J. Napier
ISBN 1-886733-02-3

Vol. CS-83 * ASTROPHYSICAL APPLICATIONS OF STELLAR PULSATION,
IAU Colloquium 155
eds. R. S. Stobie and P. A. Whitelock
ISBN 1-886733-03-1

ATLAS INFRARED ATLAS OF THE ARCTURUS SPECTRUM, 0.9–5.3 μm
eds. Kenneth Hinkle, Lloyd Wallace, and William Livingston
ISBN: 1-886733-04-X

Vol. CS-84 THE FUTURE UTILIZATION OF SCHMIDT TELESCOPES, IAU Colloquium 148
eds. Jessica Chapman, Russell Cannon, Sandra Harrison, and Bambang Hidayat
ISBN 1-886733-05-8

Vol. CS-85 CAPE WORKSHOP ON MAGNETIC CATACLYSMIC VARIABLES
eds. D. A. H. Buckley and B. Warner
ISBN 1-886733-06-6

Vol. CS-86 FRESH VIEWS OF ELLIPTICAL GALAXIES
eds. Alberto Buzzoni, Alvio Renzini, and Alfonso Serrano
ISBN 1-886733-07-4

**PUBLISHED: 1996**

Vol. CS-87 NEW OBSERVING MODES FOR THE NEXT CENTURY
eds. Todd Boroson, John Davies, and Ian Robson
ISBN 1-886733-08-2

## ASP CONFERENCE SERIES VOLUMES

Published by the Astronomical Society of the Pacific

**PUBLISHED: 1996 (* asterisk means OUT OF PRINT)**

Vol. CS-88 * CLUSTERS, LENSING, AND THE FUTURE OF THE UNIVERSE
eds. Virginia Trimble and Andreas Reisenegger
ISBN 1-886733-09-0

Vol. CS-89 ASTRONOMY EDUCATION: CURRENT DEVELOPMENTS, FUTURE COORDINATION
ed. John R. Percy
ISBN 1-886733-10-4

Vol. CS-90 THE ORIGINS, EVOLUTION, AND DESTINIES OF BINARY STARS IN CLUSTERS
eds. E. F. Milone and J.-C. Mermilliod
ISBN 1-886733-11-2

Vol. CS-91 BARRED GALAXIES, IAU Colloquium 157
eds. R. Buta, D. A. Crocker, and B. G. Elmegreen
ISBN 1-886733-12-0

Vol. CS-92 * FORMATION OF THE GALACTIC HALO .... INSIDE AND OUT
eds. Heather L. Morrison and Ata Sarajedini
ISBN 1-886733-13-9

Vol. CS-93 RADIO EMISSION FROM THE STARS AND THE SUN
eds. A. R. Taylor and J. M. Paredes
ISBN 1-886733-14-7

Vol. CS-94 MAPPING, MEASURING, AND MODELING THE UNIVERSE
eds. Peter Coles, Vicent J. Martinez, and Maria-Jesus Pons-Borderia
ISBN 1-886733-15-5

Vol. CS-95 SOLAR DRIVERS OF INTERPLANETARY AND TERRESTRIAL DISTURBANCES: Proceedings of 16th International Workshop National Solar Observatory/Sacramento Peak
eds. K. S. Balasubramaniam, Stephen L. Keil, and Raymond N. Smartt
ISBN 1-886733-16-3

Vol. CS-96 HYDROGEN-DEFICIENT STARS
eds. C. S. Jeffery and U. Heber
ISBN 1-886733-17-1

Vol. CS-97 POLARIMETRY OF THE INTERSTELLAR MEDIUM
eds. W. G. Roberge and D. C. B. Whittet
ISBN 1-886733-18-X

Vol. CS-98 FROM STARS TO GALAXIES: THE IMPACT OF STELLAR PHYSICS ON GALAXY EVOLUTION
eds. Claus Leitherer, Uta Fritze-von Alvensleben, and John Huchra
ISBN 1-886733-19-8

Vol. CS-99 COSMIC ABUNDANCES:
Proceedings of the 6th Annual October Astrophysics Conference
eds. Stephen S. Holt and George Sonneborn
ISBN 1-886733-20-1

Vol. CS-100 ENERGY TRANSPORT IN RADIO GALAXIES AND QUASARS
eds. P. E. Hardee, A. H. Bridle, and J. A. Zensus
ISBN 1-886733-21-X

## ASP CONFERENCE SERIES VOLUMES

Published by the Astronomical Society of the Pacific

---

**PUBLISHED: 1996 (* asterisk means OUT OF PRINT)**

Vol. CS-101 ASTRONOMICAL DATA ANALYSIS SOFTWARE AND SYSTEMS (ADASS) V
eds. George H. Jacoby and Jeannette Barnes
ISBN 1080-7926

Vol. CS-102 THE GALACTIC CENTER, 4th ESO/CTIO Workshop
ed. Roland Gredel
ISBN 1-886733-22-8

Vol. CS-103 THE PHYSICS OF LINERS IN VIEW OF RECENT OBSERVATIONS
eds. M. Eracleous, A. Koratkar, C. Leitherer, and L. Ho
ISBN 1-886733-23-6

Vol. CS-104* PHYSICS, CHEMISTRY, AND DYNAMICS OF INTERPLANETARY DUST, IAU Colloquium 150
eds. Bo Å. S. Gustafson and Martha S. Hanner
ISBN 1-886733-24-4

Vol. CS-105 PULSARS: PROBLEMS AND PROGRESS, IAU Colloquium 160
ed. S. Johnston, M. A. Walker, and M. Bailes
ISBN 1-886733-25-2

Vol. CS-106 THE MINNESOTA LECTURES ON EXTRAGALACTIC NEUTRAL HYDROGEN
ed. Evan D. Skillman
ISBN 1-886733-26-0

Vol. CS-107 COMPLETING THE INVENTORY OF THE SOLAR SYSTEM:
A Symposium held in conjunction with the 106th Annual Meeting of the ASP
eds. Terrence W. Rettig and Joseph M. Hahn
ISBN 1-886733-27-9

Vol. CS-108 M.A.S.S. -- MODEL ATMOSPHERES AND SPECTRUM SYNTHESIS:
5th Vienna - Workshop
eds. Saul J. Adelman, Friedrich Kupka, and Werner W. Weiss
ISBN 1-886733-28-7

Vol. CS-109 COOL STARS, STELLAR SYSTEMS, AND THE SUN:
Ninth Cambridge Workshop
eds. Roberto Pallavicini and Andrea K. Dupree
ISBN 1-886733-29-5

Vol. CS-110 BLAZAR CONTINUUM VARIABILITY
eds. H. R. Miller, J. R. Webb, and J. C. Noble
ISBN 1-886733-30-9

Vol. CS-111 MAGNETIC RECONNECTION IN THE SOLAR ATMOSPHERE:
Proceedings of a Yohkoh Conference
eds. R. D. Bentley and J. T. Mariska
ISBN 1-886733-31-7

Vol. CS-112 THE HISTORY OF THE MILKY WAY AND ITS SATELLITE SYSTEM
eds. Andreas Burkert, Dieter H. Hartmann, and Steven R. Majewski
ISBN 1-886733-32-5

## ASP CONFERENCE SERIES VOLUMES

Published by the Astronomical Society of the Pacific

---

**PUBLISHED: 1997 (* asterisk means OUT OF PRINT)**

| | |
|---|---|
| Vol. CS-113 | EMISSION LINES IN ACTIVE GALAXIES: NEW METHODS AND TECHNIQUES, IAU Colloquium 159<br>eds. B. M. Peterson, F.-Z. Cheng, and A. S. Wilson<br>ISBN 1-886733-33-3 |
| Vol. CS-114 | YOUNG GALAXIES AND QSO ABSORPTION-LINE SYSTEMS<br>eds. Sueli M. Viegas, Ruth Gruenwald, and Reinaldo R. de Carvalho<br>ISBN 1-886733-34-1 |
| Vol. CS-115 | GALACTIC CLUSTER COOLING FLOWS<br>ed. Noam Soker<br>ISBN 1-886733-35-X |
| Vol. CS-116 | THE SECOND STROMLO SYMPOSIUM: THE NATURE OF ELLIPTICAL GALAXIES<br>eds. M. Arnaboldi, G. S. Da Costa, and P. Saha<br>ISBN 1-886733-36-8 |
| Vol. CS-117 | DARK AND VISIBLE MATTER IN GALAXIES<br>eds. Massimo Persic and Paolo Salucci<br>ISBN-1-886733-37-6 |
| Vol. CS-118 | FIRST ADVANCES IN SOLAR PHYSICS EUROCONFERENCE: ADVANCES IN THE PHYSICS OF SUNSPOTS<br>eds. B. Schmieder. J. C. del Toro Iniesta, and M. Vázquez<br>ISBN 1-886733-38-4 |
| Vol. CS-119 | PLANETS BEYOND THE SOLAR SYSTEM AND THE NEXT GENERATION OF SPACE MISSIONS<br>ed. David R. Soderblom<br>ISBN 1-886733-39-2 |
| Vol. CS-120 | LUMINOUS BLUE VARIABLES: MASSIVE STARS IN TRANSITION<br>eds. Antonella Nota and Henny J. G. L. M. Lamers<br>ISBN 1-886733-40-6 |
| Vol. CS-121 | ACCRETION PHENOMENA AND RELATED OUTFLOWS, IAU Colloquium 163<br>eds. D. T. Wickramasinghe, G. V. Bicknell, and L. Ferrario<br>ISBN 1-886733-41-4 |
| Vol. CS-122 | FROM STARDUST TO PLANETESIMALS:<br>Symposium held as part of the 108th Annual Meeting of the ASP<br>eds. Yvonne J. Pendleton and A. G. G. M. Tielens<br>ISBN 1-886733-42-2 |
| Vol. CS-123 | THE 12th 'KINGSTON MEETING': COMPUTATIONAL ASTROPHYSICS<br>eds. David A. Clarke and Michael J. West<br>ISBN 1-886733-43-0 |
| Vol. CS-124 | DIFFUSE INFRARED RADIATION AND THE IRTS<br>eds. Haruyuki Okuda, Toshio Matsumoto, and Thomas Roellig<br>ISBN 1-886733-44-9 |
| Vol. CS-125 | ASTRONOMICAL DATA ANALYSIS SOFTWARE AND SYSTEMS (ADASS) VI<br>eds. Gareth Hunt and H. E. Payne<br>ISBN 1-886733-45-7 |

## ASP CONFERENCE SERIES VOLUMES

Published by the Astronomical Society of the Pacific

**PUBLISHED: 1997 (* asterisk means OUT OF PRINT)**

Vol. CS-126 FROM QUANTUM FLUCTUATIONS TO COSMOLOGICAL STRUCTURES
eds. David Valls-Gabaud, Martin A. Hendry, Paolo Molaro, and Khalil Chamcham
ISBN 1-886733-46-5

Vol. CS-127 PROPER MOTIONS AND GALACTIC ASTRONOMY
ed. Roberta M. Humphreys
ISBN 1-886733-47-3

Vol. CS-128 MASS EJECTION FROM AGN (Active Galactic Nuclei)
eds. N. Arav, I. Shlosman, and R. J. Weymann
ISBN 1-886733-48-1

Vol. CS-129 THE GEORGE GAMOW SYMPOSIUM
eds. E. Harper, W. C. Parke, and G. D. Anderson
ISBN 1-886733-49-**X**

Vol. CS-130 THE THIRD PACIFIC RIM CONFERENCE ON RECENT DEVELOPMENT ON BINARY STAR RESEARCH
eds. Kam-Ching Leung
ISBN 1-886733-50-3

**PUBLISHED: 1998**

Vol. CS-131 BOULDER-MUNICH II: PROPERTIES OF HOT, LUMINOUS STARS
ed. Ian D. Howarth
ISBN 1-886733-51-1

Vol. CS-132 STAR FORMATION WITH THE INFRARED SPACE OBSERVATORY (ISO)
eds. João L. Yun and René Liseau
ISBN 1-886733-52-X

Vol. CS-133 SCIENCE WITH THE NGST (Next Generation Space Telescope)
eds. Eric P. Smith and Anuradha Koratkar
ISBN 1-886733-53-8

Vol. CS-134 * BROWN DWARFS AND EXTRASOLAR PLANETS
eds. Rafael Rebolo, Eduardo L. Martin, and Maria Rosa Zapatero Osorio
ISBN 1-886733-54-6

Vol. CS-135 A HALF CENTURY OF STELLAR PULSATION INTERPRETATIONS: A TRIBUTE TO ARTHUR N. COX
eds. P. A. Bradley and J. A. Guzik
ISBN 1-886733-55-4

Vol. CS-136 GALACTIC HALOS: A UC SANTA CRUZ WORKSHOP
ed. Dennis Zaritsky
ISBN 1-886733-56-2

Vol. CS-137 WILD STARS IN THE OLD WEST: PROCEEDINGS OF THE 13th NORTH AMERICAN WORKSHOP ON CATACLYSMIC VARIABLES AND RELATED OBJECTS
eds. S. Howell, E. Kuulkers, and C. Woodward
ISBN 1-886733-57-0

Vol. CS-138 1997 PACIFIC RIM CONFERENCE ON STELLAR ASTROPHYSICS
eds. Kwing Lam Chan, K. S. Cheng, and H. P. Singh
ISBN 1-886733-58-9

## ASP CONFERENCE SERIES VOLUMES

Published by the Astronomical Society of the Pacific

---

**PUBLISHED: 1998 (* asterisk means OUT OF PRINT)**

Vol. CS-139 PRESERVING THE ASTRONOMICAL WINDOWS:
Proceedings of Joint Discussion No. 5 of the 23rd General Assembly of the IAU
eds. Syuzo Isobe and Tomohiro Hirayama
ISBN 1-886733-59-7

Vol. CS-140 SYNOPTIC SOLAR PHYSICS --18th NSO/Sacramento Peak Summer Workshop
eds. K. S. Balasubramaniam, J. W. Harvey, and D. M. Rabin
ISBN 1-886733-60-0

Vol. CS-141 ASTROPHYSICS FROM ANTARCTICA:
A Symposium held as a part of the 109th Annual Meeting of the ASP
eds. Giles Novak and Randall H. Landsberg
ISBN 1-886733-61-9

Vol. CS-142 THE STELLAR INITIAL MASS FUNCTION: 38th Herstmonceux Conference
eds. Gerry Gilmore and Debbie Howell
ISBN 1-886733-62-7

Vol. CS-143 * THE SCIENTIFIC IMPACT OF THE GODDARD HIGH RESOLUTION SPECTROGRAPH (GHRS)
eds. John C. Brandt, Thomas B. Ake III, and Carolyn Collins Petersen
ISBN 1-886733-63-5

Vol. CS-144 RADIO EMISSION FROM GALACTIC AND EXTRAGALACTIC COMPACT SOURCES, IAU Colloquium 164
eds. J. Anton Zensus, G. B. Taylor, and J. M. Wrobel
ISBN 1-886733-64-3

Vol. CS-145 ASTRONOMICAL DATA ANALYSIS SOFTWARE AND SYSTEMS (ADASS) VII
eds. Rudolf Albrecht, Richard N. Hook, and Howard A. Bushouse
ISBN 1-886733-65-1

Vol. CS-146 THE YOUNG UNIVERSE GALAXY FORMATION AND EVOLUTION AT INTERMEDIATE AND HIGH REDSHIFT
eds. S. D'Odorico, A. Fontana, and E. Giallongo
ISBN 1-886733-66-X

Vol. CS-147 ABUNDANCE PROFILES: DIAGNOSTIC TOOLS FOR GALAXY HISTORY
eds. Daniel Friedli, Mike Edmunds, Carmelle Robert, and Laurent Drissen
ISBN 1-886733-67-8

Vol. CS-148 ORIGINS
eds. Charles E. Woodward, J. Michael Shull, and Harley A. Thronson, Jr.
ISBN 1-886733-68-6

Vol. CS-149 SOLAR SYSTEM FORMATION AND EVOLUTION
eds. D. Lazzaro, R. Vieira Martins, S. Ferraz-Mello, J. Fernández, and C. Beaugé
ISBN 1-886733-69-4

Vol. CS-150 NEW PERSPECTIVES ON SOLAR PROMINENCES, IAU Colloquium 167
eds. David Webb, David Rust, and Brigitte Schmieder
ISBN 1-886733-70-8

## ASP CONFERENCE SERIES VOLUMES

Published by the Astronomical Society of the Pacific

---

**PUBLISHED: 1998 (* asterisk means OUT OF PRINT)**

Vol. CS-151 COSMIC MICROWAVE BACKGROUND AND LARGE SCALE STRUCTURES OF THE UNIVERSE
eds. Yong-Ik Byun and Kin-Wang Ng
ISBN 1-886733-71-6

Vol. CS-152 FIBER OPTICS IN ASTRONOMY III
eds. S. Arribas, E. Mediavilla, and F. Watson
ISBN 1-886733-72-4

Vol. CS-153 LIBRARY AND INFORMATION SERVICES IN ASTRONOMY III -- (LISA III)
eds. Uta Grothkopf, Heinz Andernach, Sarah Stevens-Rayburn, and Monique Gomez
ISBN 1-886733-73-2

**PUBLISHED: 1999**

Vol. CS-154 COOL STARS, STELLAR SYSTEMS AND THE SUN: Tenth Cambridge Workshop
eds. Robert A. Donahue and Jay A. Bookbinder
ISBN 1-886733-74-0

Vol. CS-155 SECOND ADVANCES IN SOLAR PHYSICS EUROCONFERENCE: THREE-DIMENSIONAL STRUCTURE OF SOLAR ACTIVE REGIONS
eds. Costas E. Alissandrakis and Brigitte Schmieder
ISBN 1-886733-75-9

Vol. CS-156 HIGHLY REDSHIFTED RADIO LINES
eds. C. L. Carilli, S. J. E. Radford, K. M. Menten, and G. I. Langston
ISBN 1-886733-76-7

Vol. CS-157 ANNAPOLIS WORKSHOP ON MAGNETIC CATACLYSMIC VARIABLES
eds. Coel Hellier and Koji Mukai
ISBN 1-886733-77-5

Vol. CS-158 SOLAR AND STELLAR ACTIVITY: SIMILARITIES AND DIFFERENCES
eds. C. J. Butler and J. G. Doyle
ISBN 1-886733-78-3

Vol. CS-159 BL LAC PHENOMENON
eds. Leo O. Takalo and Aimo Sillanpää
ISBN 1-886733-79-1

Vol. CS-160 ASTROPHYSICAL DISCS: An EC Summer School
eds. J. A. Sellwood and Jeremy Goodman
ISBN 1-886733-80-5

Vol. CS-161 HIGH ENERGY PROCESSES IN ACCRETING BLACK HOLES
eds. Juri Poutanen and Roland Svensson
ISBN 1-886733-81-3

Vol. CS-162 QUASARS AND COSMOLOGY
eds. Gary Ferland and Jack Baldwin
ISBN 1-886733-83-X

Vol. CS-163 STAR FORMATION IN EARLY-TYPE GALAXIES
eds. Jordi Cepa and Patricia Carral
ISBN 1-886733-84-8

## ASP CONFERENCE SERIES VOLUMES

Published by the Astronomical Society of the Pacific

**PUBLISHED: 1999 (* asterisk means OUT OF PRINT)**

Vol. CS-164 ULTRAVIOLET–OPTICAL SPACE ASTRONOMY BEYOND HST
eds. Jon A. Morse, J. Michael Shull, and Anne L. Kinney
ISBN 1-886733-85-6

Vol. CS-165 THE THIRD STROMLO SYMPOSIUM: THE GALACTIC HALO
eds. Brad K. Gibson, Tim S. Axelrod, and Mary E. Putman
ISBN 1-886733-86-4

Vol. CS-166 STROMLO WORKSHOP ON HIGH-VELOCITY CLOUDS
eds. Brad K. Gibson and Mary E. Putman
ISBN 1-886733-87-2

Vol. CS-167 HARMONIZING COSMIC DISTANCE SCALES IN A POST-HIPPARCOS ERA
eds. Daniel Egret and André Heck
ISBN 1-886733-88-0

Vol. CS-168 NEW PERSPECTIVES ON THE INTERSTELLAR MEDIUM
eds. A. R. Taylor, T. L. Landecker, and G. Joncas
ISBN 1-886733-89-9

Vol. CS-169 11th EUROPEAN WORKSHOP ON WHITE DWARFS
eds. J.-E. Solheim and E. G. Meištas
ISBN 1-886733-91-0

Vol. CS-170 THE LOW SURFACE BRIGHTNESS UNIVERSE, IAU Colloquium 171
eds. J. I. Davies, C. Impey, and S. Phillipps
ISBN 1-886733-92-9

Vol. CS-171 LiBeB, COSMIC RAYS, AND RELATED X- AND GAMMA-RAYS
eds. Reuven Ramaty, Elisabeth Vangioni-Flam, Michel Cassé, and Keith Olive
ISBN 1-886733-93-7

Vol. CS-172 ASTRONOMICAL DATA ANALYSIS SOFTWARE AND SYSTEMS (ADASS) VIII
eds. David M. Mehringer, Raymond L. Plante, and Douglas A. Roberts
ISBN 1-886733-94-5

Vol. CS-173 THEORY AND TESTS OF CONVECTION IN STELLAR STRUCTURE:
First Granada Workshop
ed. Álvaro Giménez, Edward F. Guinan, and Benjamín Montesinos
ISBN 1-886733-95-3

Vol. CS-174 CATCHING THE PERFECT WAVE: ADAPTIVE OPTICS AND INTERFEROMETRY IN THE 21st CENTURY,
A Symposium held as a part of the 110th Annual Meeting of the ASP
eds. Sergio R. Restaino, William Junor, and Nebojsa Duric
ISBN 1-886733-96-1

Vol. CS-175 STRUCTURE AND KINEMATICS OF QUASAR BROAD LINE REGIONS
eds. C. M. Gaskell, W. N. Brandt, M. Dietrich, D. Dultzin-Hacyan, and M. Eracleous
ISBN 1-886733-97-X

Vol. CS-176 OBSERVATIONAL COSMOLOGY: THE DEVELOPMENT OF GALAXY SYSTEMS
eds. Giuliano Giuricin, Marino Mezzetti, and Paolo Salucci
ISBN 1-58381-000-5

## ASP CONFERENCE SERIES VOLUMES

Published by the Astronomical Society of the Pacific

**PUBLISHED: 1999 (* asterisk means OUT OF PRINT)**

Vol. CS-177 ASTROPHYSICS WITH INFRARED SURVEYS: A Prelude to SIRTF
eds. Michael D. Bicay, Chas A. Beichman, Roc M. Cutri, and Barry F. Madore
ISBN 1-58381-001-3

Vol. CS-178 STELLAR DYNAMOS: NONLINEARITY AND CHAOTIC FLOWS
eds. Manuel Núñez and Antonio Ferriz-Mas
ISBN 1-58381-002-1

Vol. CS-179 ETA CARINAE AT THE MILLENNIUM
eds. Jon A. Morse, Roberta M. Humphreys, and Augusto Damineli
ISBN 1-58381-003-X

Vol. CS-180 SYNTHESIS IMAGING IN RADIO ASTRONOMY II
eds. G. B. Taylor, C. L. Carilli, and R. A. Perley
ISBN 1-58381-005-6

Vol. CS-181 MICROWAVE FOREGROUNDS
eds. Angelica de Oliveira-Costa and Max Tegmark
ISBN 1-58381-006-4

Vol. CS-182 GALAXY DYNAMICS: A Rutgers Symposium
eds. David Merritt, J. A. Sellwood, and Monica Valluri
ISBN 1-58381-007-2

Vol. CS-183 HIGH RESOLUTION SOLAR PHYSICS: THEORY, OBSERVATIONS, AND TECHNIQUES
eds. T. R. Rimmele, K. S. Balasubramaniam, and R. R. Radick
ISBN 1-58381-009-9

Vol. CS-184 THIRD ADVANCES IN SOLAR PHYSICS EUROCONFERENCE: MAGNETIC FIELDS AND OSCILLATIONS
eds. B. Schmieder, A. Hofmann, and J. Staude
ISBN 1-58381-010-2

Vol. CS-185 PRECISE STELLAR RADIAL VELOCITIES, IAU Colloquium 170
eds. J. B. Hearnshaw and C. D. Scarfe
ISBN 1-58381-011-0

Vol. CS-186 THE CENTRAL PARSECS OF THE GALAXY
eds. Heino Falcke, Angela Cotera, Wolfgang J. Duschl, Fulvio Melia, and Marcia J. Rieke
ISBN 1-58381-012-9

Vol. CS-187 THE EVOLUTION OF GALAXIES ON COSMOLOGICAL TIMESCALES
eds. J. E. Beckman and T. J. Mahoney
ISBN 1-58381-013-7

Vol. CS-188 OPTICAL AND INFRARED SPECTROSCOPY OF CIRCUMSTELLAR MATTER
eds. Eike W. Guenther, Bringfried Stecklum, and Sylvio Klose
ISBN 1-58381-014-5

Vol. CS-189 CCD PRECISION PHOTOMETRY WORKSHOP
eds. Eric R. Craine, Roy A. Tucker, and Jeannette Barnes
ISBN 1-58381-015-3

# ASP CONFERENCE SERIES VOLUMES

Published by the Astronomical Society of the Pacific

**PUBLISHED: 1999 (* asterisk means OUT OF PRINT)**

Vol. CS-190 GAMMA-RAY BURSTS: THE FIRST THREE MINUTES
eds. Juri Poutanen and Roland Svensson
ISBN 1-58381-016-1

Vol. CS-191 PHOTOMETRIC REDSHIFTS AND HIGH REDSHIFT GALAXIES
eds. Ray J. Weymann, Lisa J. Storrie-Lombardi, Marcin Sawicki, and Robert J. Brunner
ISBN 1-58381-017-X

Vol. CS-192 SPECTROPHOTOMETRIC DATING OF STARS AND GALAXIES
eds. I. Hubeny, S. R. Heap, and R. H. Cornett
ISBN 1-58381-018-8

Vol. CS-193 THE HY-REDSHIFT UNIVERSE:
GALAXY FORMATION AND EVOLUTION AT HIGH REDSHIFT
eds. Andrew J. Bunker and Wil J. M. van Breugel
ISBN 1-58381-019-6

Vol. CS-194 WORKING ON THE FRINGE:
OPTICAL AND IR INTERFEROMETRY FROM GROUND AND SPACE
eds. Stephen Unwin and Robert Stachnik
ISBN 1-58381-020-X

**PUBLISHED: 2000**

Vol. CS-195 IMAGING THE UNIVERSE IN THREE DIMENSIONS:
Astrophysics with Advanced Multi-Wavelength Imaging Devices
eds. W. van Breugel and J. Bland-Hawthorn
ISBN 1-58381-022-6

Vol. CS-196 THERMAL EMISSION SPECTROSCOPY AND ANALYSIS OF DUST, DISKS, AND REGOLITHS
eds. Michael L. Sitko, Ann L. Sprague, and David K. Lynch
ISBN: 1-58381-023-4

Vol. CS-197 XV$^{th}$ IAP MEETING DYNAMICS OF GALAXIES:
FROM THE EARLY UNIVERSE TO THE PRESENT
eds. F. Combes, G. A. Mamon, and V. Charmandaris
ISBN: 1-58381-24-2

Vol. CS-198 EUROCONFERENCE ON "STELLAR CLUSTERS AND ASSOCIATIONS: CONVECTION, ROTATION, AND DYNAMOS"
eds. R. Pallavicini, G. Micela, and S. Sciortino
ISBN: 1-58381-25-0

Vol. CS-199 ASYMMETRICAL PLANETARY NEBULAE II:
FROM ORIGINS TO MICROSTRUCTURES
eds. J. H. Kastner, N. Soker, and S. Rappaport
ISBN: 1-58381-026-9

Vol. CS-200 CLUSTERING AT HIGH REDSHIFT
eds. A. Mazure, O. Le Fèvre, and V. Le Brun
ISBN: 1-58381-027-7

Vol. CS-201 COSMIC FLOWS 1999: TOWARDS AN UNDERSTANDING OF LARGE-SCALE STRUCTURES
eds. Stéphane Courteau, Michael A. Strauss, and Jeffrey A. Willick
ISBN: 1-58381-028-5

## ASP CONFERENCE SERIES VOLUMES

Published by the Astronomical Society of the Pacific

**PUBLISHED: 2000 (* asterisk means OUT OF PRINT)**

Vol. CS-202 * PULSAR ASTRONOMY – 2000 AND BEYOND, IAU Colloquium 177
eds. M. Kramer, N. Wex, and R. Wielebinski
ISBN: 1-58381-029-3

Vol. CS-203 THE IMPACT OF LARGE-SCALE SURVEYS ON PULSATING STAR RESEARCH, IAU Colloquium 176
eds. L. Szabados and D. W. Kurtz
ISBN: 1-58381-030-7

Vol. CS-204 THERMAL AND IONIZATION ASPECTS OF FLOWS FROM HOT STARS: OBSERVATIONS AND THEORY
eds. Henny J. G. L. M. Lamers and Arved Sapar
ISBN: 1-58381-031-5

Vol. CS-205 THE LAST TOTAL SOLAR ECLIPSE OF THE MILLENNIUM IN TURKEY
eds. W. C. Livingston and A. Özgüç
ISBN: 1-58381-032-3

Vol. CS-206 HIGH ENERGY SOLAR PHYSICS – *ANTICIPATING HESSI*
eds. Reuven Ramaty and Natalie Mandzhavidze
ISBN: 1-58381-033-1

Vol. CS-207 NGST SCIENCE AND TECHNOLOGY EXPOSITION
eds. Eric P. Smith and Knox S. Long
ISBN: 1-58381-036-6

ATLAS VISIBLE AND NEAR INFRARED ATLAS OF THE ARCTURUS SPECTRUM 3727–9300 Å
eds. Kenneth Hinkle, Lloyd Wallace, Jeff Valenti, and Dianne Harmer
ISBN: 1-58381-037-4

Vol. CS-208 POLAR MOTION: HISTORICAL AND SCIENTIFIC PROBLEMS, IAU Colloquium 178
eds. Steven Dick, Dennis McCarthy, and Brian Luzum
ISBN: 1-58381-039-0

Vol. CS-209 SMALL GALAXY GROUPS, IAU Colloquium 174
eds. Mauri J. Valtonen and Chris Flynn
ISBN: 1-58381-040-4

Vol. CS-210 DELTA SCUTI AND RELATED STARS: Reference Handbook and Proceedings of the 6th Vienna Workshop in Astrophysics
eds. Michel Breger and Michael Houston Montgomery
ISBN: 1-58381-043-9

Vol. CS-211 MASSIVE STELLAR CLUSTERS
eds. Ariane Lançon and Christian M. Boily
ISBN: 1-58381-042-0

Vol. CS-212 FROM GIANT PLANETS TO COOL STARS
eds. Caitlin A. Griffith and Mark S. Marley
ISBN: 1-58381-041-2

Vol. CS-213 BIOASTRONOMY `99: A NEW ERA IN BIOASTRONOMY
eds. Guillermo A. Lemarchand and Karen J. Meech
ISBN: 1-58381-044-7

# ASP CONFERENCE SERIES VOLUMES

Published by the Astronomical Society of the Pacific

**PUBLISHED: 2000 (* asterisk means OUT OF PRINT)**

Vol. CS-214 THE Be PHENOMENON IN EARLY-TYPE STARS, IAU Colloquium 175
eds. Myron A. Smith, Huib F. Henrichs and Juan Fabregat
ISBN: 1-58381-045-5

Vol. CS-215 COSMIC EVOLUTION AND GALAXY FORMATION: STRUCTURE, INTERACTIONS AND FEEDBACK
The 3rd Guillermo Haro Astrophysics Conference
eds. José Franco, Elena Terlevich, Omar López-Cruz, and Itziar Aretxaga
ISBN: 1-58381-046-3

Vol. CS-216 ASTRONOMICAL DATA ANALYSIS SOFTWARE AND SYSTEMS (ADASS) IX
eds. Nadine Manset, Christian Veillet, and Dennis Crabtree
ISBN: 1-58381-047-1 ISSN: 1080-7926

Vol. CS-217 IMAGING AT RADIO THROUGH SUBMILLIMETER WAVELENGTHS
eds. Jeffrey G. Mangum and Simon J. E. Radford
ISBN: 1-58381-049-8

Vol. CS-218 MAPPING THE HIDDEN UNIVERSE: THE UNIVERSE BEHIND THE MILKY WAY – THE UNIVERSE IN HI
eds. Renée C. Kraan-Korteweg, Patricia A. Henning, and Heinz Andernach
ISBN: 1-58381-050-1

Vol. CS-219 DISKS, PLANETESIMALS, AND PLANETS
eds. F. Garzón, C. Eiroa, D. de Winter, and T. J. Mahoney
ISBN: 1-58381-051-X

Vol. CS-220 AMATEUR – PROFESSIONAL PARTNERSHIPS IN ASTRONOMY:
The 111th Annual Meeting of the ASP
eds. John R. Percy and Joseph B. Wilson
ISBN: 1-58381-052-8

Vol. CS-221 STARS, GAS AND DUST IN GALAXIES: EXPLORING THE LINKS
eds. Danielle Alloin, Knut Olsen, and Gaspar Galaz
ISBN: 1-58381-053-6

**PUBLISHED: 2001**

Vol. CS-222 THE PHYSICS OF GALAXY FORMATION
eds. M. Umemura and H. Susa
ISBN: 1-58381-054-4

Vol. CS-223 COOL STARS, STELLAR SYSTEMS AND THE SUN:
Eleventh Cambridge Workshop
eds. Ramón J. García López, Rafael Rebolo, and María Zapatero Osorio
ISBN: 1-58381-056-0

Vol. CS-224 PROBING THE PHYSICS OF ACTIVE GALACTIC NUCLEI BY MULTIWAVELENGTH MONITORING
eds. Bradley M. Peterson, Ronald S. Polidan, and Richard W. Pogge
ISBN: 1-58381-055-2

Vol. CS-225 VIRTUAL OBSERVATORIES OF THE FUTURE
eds. Robert J. Brunner, S. George Djorgovski, and Alex S. Szalay
ISBN: 1-58381-057-9

# ASP CONFERENCE SERIES VOLUMES

Published by the Astronomical Society of the Pacific

**PUBLISHED: 2001 (* asterisk means OUT OF PRINT)**

Vol. CS-226 12th EUROPEAN CONFERENCE ON WHITE DWARFS
eds. J. L. Provencal, H. L. Shipman, J. MacDonald, and S. Goodchild
ISBN: 1-58381-058-7

Vol. CS-227 BLAZAR DEMOGRAPHICS AND PHYSICS
eds. Paolo Padovani and C. Megan Urry
ISBN: 1-58381-059-5

Vol. CS-228 DYNAMICS OF STAR CLUSTERS AND THE MILKY WAY
eds. S. Deiters, B. Fuchs, A. Just, R. Spurzem, and R. Wielen
ISBN: 1-58381-060-9

Vol. CS-229 EVOLUTION OF BINARY AND MULTIPLE STAR SYSTEMS
A Meeting in Celebration of Peter Eggleton's 60th Birthday
eds. Ph. Podsiadlowski, S. Rappaport, A. R. King, F. D'Antona, and L. Burderi
IBSN: 1-58381-061-7

Vol. CS-230 GALAXY DISKS AND DISK GALAXIES
eds. Jose G. Funes, S. J. and Enrico Maria Corsini
ISBN: 1-58381-063-3

Vol. CS-231 TETONS 4: GALACTIC STRUCTURE, STARS, AND THE INTERSTELLAR MEDIUM
eds. Charles E. Woodward, Michael D. Bicay, and J. Michael Shull
ISBN: 1-58381-064-1

Vol. CS-232 THE NEW ERA OF WIDE FIELD ASTRONOMY
eds. Roger Clowes, Andrew Adamson, and Gordon Bromage
ISBN: 1-58381-065-X

Vol. CS-233 P CYGNI 2000: 400 YEARS OF PROGRESS
eds. Mart de Groot and Christiaan Sterken
ISBN: 1-58381-070-6

Vol. CS-234 X-RAY ASTRONOMY 2000
eds. R. Giacconi, S. Serio, and L. Stella
ISBN: 1-58381-071-4

Vol. CS-235 SCIENCE WITH THE ATACAMA LARGE MILLIMETER ARRAY (ALMA)
ed. Alwyn Wootten
ISBN: 1-58381-072-2

Vol. CS-236 ADVANCED SOLAR POLARIMETRY: THEORY, OBSERVATION, AND INSTRUMENTATION, The 20th Sacramento Peak Summer Workshop
ed. M. Sigwarth
ISBN: 1-58381-073-0

Vol. CS-237 GRAVITATIONAL LENSING: RECENT PROGRESS AND FUTURE GOALS
eds. Tereasa G. Brainerd and Christopher S. Kochanek
ISBN: 1-58381-074-9

Vol. CS-238 ASTRONOMICAL DATA ANALYSIS SOFTWARE AND SYSTEMS (ADASS) X
eds. F. R. Harnden, Jr., Francis A. Primini, and Harry E. Payne
ISBN: 1-58381-075-7

## ASP CONFERENCE SERIES VOLUMES

Published by the Astronomical Society of the Pacific

**PUBLISHED: 2001 (* asterisk means OUT OF PRINT)**

Vol. CS-239 MICROLENSING 2000: A NEW ERA OF MICROLENSING ASTROPHYSICS
eds. John Menzies and Penny D. Sackett
ISBN: 1-58381-076-5

Vol. CS-240 GAS AND GALAXY EVOLUTION,
A Conference in Honor of the 20th Anniversary of the VLA
eds. J. E. Hibbard, M. P. Rupen, and J. H. van Gorkom
ISBN: 1-58381-077-3

Vol. CS-241 THE 7TH TAIPEI ASTROPHYSICS WORKSHOP ON
COSMIC RAYS IN THE UNIVERSE
ed. Chung-Ming Ko
ISBN: 1-58381-079-X

Vol. CS-242 ETA CARINAE AND OTHER MYSTERIOUS STARS:
THE HIDDEN OPPORTUNITIES OF EMISSION SPECTROSCOPY
eds. Theodore R. Gull, Sveneric Johannson, and Kris Davidson
ISBN: 1-58381-080-3

Vol. CS-243 FROM DARKNESS TO LIGHT:
ORIGIN AND EVOLUTION OF YOUNG STELLAR CLUSTERS
eds. Thierry Montmerle and Philippe André
ISBN: 1-58381-081-1

Vol. CS-244 YOUNG STARS NEAR EARTH: PROGRESS AND PROSPECTS
eds. Ray Jayawardhana and Thomas P. Greene
ISBN: 1-58381-082-X

Vol. CS-245 ASTROPHYSICAL AGES AND TIME SCALES
eds. Ted von Hippel, Chris Simpson, and Nadine Manset
ISBN: 1-58381-083-8

Vol. CS-246 SMALL TELESCOPE ASTRONOMY ON GLOBAL SCALES,
IAU Colloquium 183
eds. Wen-Ping Chen, Claudia Lemme, and Bohdan Paczyński
ISBN: 1-58381-084-6

Vol. CS-247 SPECTROSCOPIC CHALLENGES OF PHOTOIONIZED PLASMAS
eds. Gary Ferland and Daniel Wolf Savin
ISBN: 1-58381-085-4

Vol. CS-248 MAGNETIC FIELDS ACROSS THE HERTZSPRUNG-RUSSELL DIAGRAM
eds. G. Mathys, S. K. Solanki, and D. T. Wickramasinghe
ISBN: 1-58381-088-9

Vol. CS-249 THE CENTRAL KILOPARSEC OF STARBURSTS AND AGN:
THE LA PALMA CONNECTION
eds. J. H. Knapen, J. E. Beckman, I. Shlosman, and T. J. Mahoney
ISBN: 1-58381-089-7

Vol. CS-250 PARTICLES AND FIELDS IN RADIO GALAXIES CONFERENCE
eds. Robert A. Laing and Katherine M. Blundell
ISBN: 1-58381-090-0

Vol. CS-251 NEW CENTURY OF X-RAY ASTRONOMY
eds. H. Inoue and H. Kunieda
ISBN: 1-58381-091-9

## ASP CONFERENCE SERIES VOLUMES

Published by the Astronomical Society of the Pacific

---

**PUBLISHED: 2001 (* asterisk means OUT OF PRINT)**

Vol. CS-252 HISTORICAL DEVELOPMENT OF MODERN COSMOLOGY
eds. Vicent J. Martínez, Virginia Trimble, and María Jesús Pons-Bordería
ISBN: 1-58381-092-7

**PUBLISHED: 2002**

Vol. CS-253 CHEMICAL ENRICHMENT OF INTRACLUSTER AND INTERGALACTIC MEDIUM
eds. Roberto Fusco-Femiano and Francesca Matteucci
ISBN: 1-58381-093-5

Vol. CS-254 EXTRAGALACTIC GAS AT LOW REDSHIFT
eds. John S. Mulchaey and John T. Stocke
ISBN: 1-58381-094-3

Vol. CS-255 MASS OUTFLOW IN ACTIVE GALACTIC NUCLEI: NEW PERSPECTIVES
eds. D. M. Crenshaw, S. B. Kraemer, and I. M. George
ISBN: 1-58381-095-1

Vol. CS-256 OBSERVATIONAL ASPECTS OF PULSATING B AND A STARS
eds. Christiaan Sterken and Donald W. Kurtz
ISBN: 1-58381-096-X

Vol. CS-257 AMiBA 2001: HIGH-*Z* CLUSTERS, MISSING BARYONS, AND CMB POLARIZATION
eds. Lin-Wen Chen, Chung-Pei Ma, Kin-Wang Ng, and Ue-Li Pen
ISBN: 1-58381-097-8

Vol. CS-258 ISSUES IN UNIFICATION OF ACTIVE GALACTIC NUCLEI
eds. Roberto Maiolino, Alessandro Marconi, and Neil Nagar
ISBN: 1-58381-098-6

Vol. CS-259 RADIAL AND NONRADIAL PULSATIONS AS PROBES OF STELLAR PHYSICS, IAU Colloquium 185
eds. Conny Aerts, Timothy R. Bedding, and Jørgen Christensen-Dalsgaard
ISBN: 1-58381-099-4

Vol. CS-260 INTERACTING WINDS FROM MASSIVE STARS
eds. Anthony F. J. Moffat and Nicole St-Louis
ISBN: 1-58381-100-1

Vol. CS-261 THE PHYSICS OF CATACLYSMIC VARIABLES AND RELATED OBJECTS
eds. B. T. Gänsicke, K. Beuermann, and K. Reinsch
ISBN: 1-58381-101-X

Vol. CS-262 THE HIGH ENERGY UNIVERSE AT SHARP FOCUS: CHANDRA SCIENCE, held in conjunction with the 113th Annual Meeting of the ASP
eds. Eric M. Schlegel and Saeqa Dil Vrtilek
ISBN: 1-58381-102-8

Vol. CS-263 STELLAR COLLISIONS, MERGERS AND THEIR CONSEQUENCES
ed. Michael M. Shara
ISBN: 1-58381-103-6

## ASP CONFERENCE SERIES VOLUMES

Published by the Astronomical Society of the Pacific

**PUBLISHED: 2002 (* asterisk means OUT OF PRINT)**

Vol. CS-264 CONTINUING THE CHALLENGE OF EUV ASTRONOMY: CURRENT ANALYSIS AND PROSPECTS FOR THE FUTURE
eds. Steve B. Howell, Jean Dupuis, Daniel Golombek, Frederick M. Walter, and Jennifer Cullison
ISBN: 1-58381-104-4

Vol. CS-265 ω CENTAURI, A UNIQUE WINDOW INTO ASTROPHYSICS
eds. Floor van Leeuwen, Joanne D. Hughes, and Giampaolo Piotto
ISBN: 1-58381-105-2

Vol. CS-266 ASTRONOMICAL SITE EVALUATION IN THE VISIBLE AND RADIO RANGE, IAU Technical Workshop
eds. J. Vernin, Z. Benkhaldoun, and C. Muñoz-Tuñón
ISBN: 1-58381-106-0

Vol. CS-267* HOT STAR WORKSHOP III: THE EARLIEST STAGES OF MASSIVE STAR BIRTH
ed. Paul A. Crowther
ISBN: 1-58381-107-9

Vol. CS-268 TRACING COSMIC EVOLUTION WITH GALAXY CLUSTERS
eds. Stefano Borgani, Marino Mezzetti, and Riccardo Valdarnini
ISBN: 1-58381-108-7

Vol. CS-269 THE EVOLVING SUN AND ITS INFLUENCE ON PLANETARY ENVIRONMENTS
eds. Benjamín Montesinos, Álvaro Giménez, and Edward F. Guinan
ISBN: 1-58381-109-5

Vol. CS-270 ASTRONOMICAL INSTRUMENTATION AND THE BIRTH AND GROWTH OF ASTROPHYSICS: A Symposium held in honor of Robert G. Tull
eds. Frank N. Bash and Christopher Sneden
ISBN: 1-58381-110-9

Vol. CS-271 NEUTRON STARS IN SUPERNOVA REMNANTS
eds. Patrick O. Slane and Bryan M. Gaensler
ISBN: 1-58381-111-7

Vol. CS-272 THE FUTURE OF SOLAR SYSTEM EXPLORATION, 2003–2013
Community Contributions to the NRC Solar System Exploration Decadal Survey
ed. Mark V. Sykes
ISBN: 1-58381-113-3

Vol. CS-273 THE DYNAMICS, STRUCTURE AND HISTORY OF GALAXIES
eds. G. S. Da Costa and H. Jerjen
ISBN: 1-58381-114-1

Vol. CS-274 OBSERVED HR DIAGRAMS AND STELLAR EVOLUTION
eds. Thibault Lejeune and João Fernandes
ISBN: 1-58381-116-8

Vol. CS-275 DISKS OF GALAXIES: KINEMATICS, DYNAMICS AND PERTURBATIONS
eds. E. Athanassoula, A. Bosma, and R. Mujica
ISBN: 1-58381-117-6

# ASP CONFERENCE SERIES VOLUMES

Published by the Astronomical Society of the Pacific

**PUBLISHED: 2002 (* asterisk means OUT OF PRINT)**

Vol. CS-276 SEEING THROUGH THE DUST:
THE DETECTION OF HI AND THE EXPLORATION OF THE ISM IN GALAXIES
eds. A. R. Taylor, T. L. Landecker, and A. G. Willis
ISBN: 1-58381-118-4

Vol. CS 277 STELLAR CORONAE IN THE CHANDRA AND XMM-NEWTON ERA
eds. Fabio Favata and Jeremy J. Drake
ISBN: 1-58381-119-2

Vol. CS 278 NAIC–NRAO SCHOOL ON SINGLE-DISH ASTRONOMY:
TECHNIQUES AND APPLICATIONS
eds. Snezana Stanimirovic, Daniel Altschuler, Paul Goldsmith, and Chris Salter
ISBN: 1-58381-120-6

Vol. CS 279 EXOTIC STARS AS CHALLENGES TO EVOLUTION, IAU Colloquium 187
eds. Christopher A. Tout and Walter Van Hamme
ISBN: 1-58381-122-2

Vol. CS 280 NEXT GENERATION WIDE-FIELD MULTI-OBJECT SPECTROSCOPY
eds. Michael J. I. Brown and Arjun Dey
ISBN: 1-58381-123-0

Vol. CS 281 ASTRONOMICAL DATA ANALYSIS SOFTWARE AND SYSTEM (ADASS) XI
eds. David A. Bohlender, Daniel Durand, and Thomas H. Handley
ISBN: 1-58381-124-9 ISSN: 1080-7926

Vol. CS 282 GALAXIES: THE THIRD DIMENSION
eds. Margarita Rosado, Luc Binette, and Lorena Arias
ISBN: 1-58381-125-7

Vol. CS 283 A NEW ERA IN COSMOLOGY
eds. Nigel Metcalfe and Tom Shanks
ISBN: 1-58381-126-5

Vol. CS 284 AGN SURVEYS
eds. R. F. Green, E. Ye. Khachikian, and D. B. Sanders
ISBN: 1-58381-127-3

Vol. CS 285 MODES OF STAR FORMATION AND THE ORIGIN OF FIELD POPULATIONS
eds. Eva K. Grebel and Walfgang Brandner
ISBN: 1-58381-128-1

**PUBLISHED: 2003**

Vol. CS 286 CURRENT THEORETICAL MODESL AND HIGH RESOLUTION SOLAR OBSERVATIONS: PREPARING FOR ATST
eds. Alexei A. Pevtsov and Han Uitenbroek
ISBN: 1-58381-129-X

Vol. CS 287 GALACTIC STAR FORMATION ACROSS THE STELLAR MASS SPECTRUM
eds. J. M. De Buizer and N. S. van der Bliek
ISBN:1-58381-130-3

Vol. CS 288 STELLAR ATMOSPHERE MODELING
eds. I. Hubeny, D. Mihalas and K. Werner
ISBN: 1-58381-131-1

## ASP CONFERENCE SERIES VOLUMES

Published by the Astronomical Society of the Pacific

---

**PUBLISHED: 2003 (* asterisk means OUT OF PRINT)**

Vol. CS 289 THE PROCEEDINGS OF THE IAU 8TH ASIAN-PACIFIC REGIONAL MEETING, VOLUME 1
eds. Satoru Ikeuchi, John Hearnshaw and Tomoyuki Hanawa
ISBN: 1-58381-134-6

Vol. CS 290 ACTIVE GALACTIC NUCLEI: FROM CENTRAL ENGINE TO HOST GALAXY
eds. S. Collin, F. Combes and I. Shlosman
ISBN: 1-58381-135-4

Vol. CS-291 HUBBLE'S SCIENCE LEGACY:
FUTURE OPTICAL/ULTRAVIOLET ASTRONOMY FROM SPACE
eds. Kenneth R. Sembach, J. Chris Blades, Garth D. Illingworth and Robert C. Kennicutt, Jr.
ISBN: 1-58381-136-2

Vol. CS-292 INTERPLAY OF PERIODIC, CYCLIC AND STOCHASTIC VARIABILITY IN SELECTED AREAS OF THE H-R DIAGRAM
ed. Christiaan Sterken
ISBN: 1-58381-138-9

Vol. CS-293 3D STELLAR EVOLUTION
eds. S. Turcotte, S. C. Keller and R. M. Cavallo
ISBN: 1-58381-140-0

Vol. CS-294 SCIENTIFIC FRONTIERS IN RESEARCH ON EXTRASOLAR PLANETS
eds. Drake Deming and Sara Seager
ISBN: 1-58381-141-9

Vol. CS-295 ASTRONOMICAL DATA ANALYSIS SOFTWARE AND SYSTEMS (ADASS) XII
eds. Harry E. Payne, Robert I. Jedrzejewski and Richard N. Hook
ISBN: 1-58381-142-7

Vol. CS-296 NEW HORIZONS IN GLOBULAR CLUSTER ASTRONOMY
eds. Giampaolo Piotto, Georges Meylan, S. George Djorgovski and Marco Riello
ISBN: 1-58381-143-5

Vol. CS-297 STAR FORMATION THROUGH TIME, A Conference to Honour Robert J. Terlevich
eds. Enrique Pérez, Rosa M. González Delgado and Guillermo Tenorio-Tagle
ISBN: 1-58381-144-3

Vol. CS-298 GAIA SPECTROSCOPY: SCIENCE AND TECHNOLOGY
ed. Ulisse Munari
ISBN: 1-58381-145-1

Vol. CS-299 HIGH ENERGY BLAZAR ASTRONOMY, An International Conference held to Celebrate the 50th Anniversary of Tuorla Observatory
eds. Leo O. Takalo and Esko Valtaoja
ISBN: 1-58381-146-X

Vol. CS-300 RADIO ASTRONOMY AT THE FRINGE, A Conference held in honor of Kenneth I. Kellermann, on the occasion of his 65th Birthday
eds. J. Anton Zensus, Marshall H. Cohen and Eduardo Ros
ISBN: 1-58381-147-8

## ASP CONFERENCE SERIES VOLUMES

Published by the Astronomical Society of the Pacific

**PUBLISHED: 2003 (* asterisk means OUT OF PRINT)**

Vol. CS-301 MATTER AND ENERGY IN CLUSTERS OF GALAXIES
eds. Stuart Bowyer and Chorng-Yuan Hwang
ISBN: 1-58381-149-4

Vol. CS-302 RADIO PULSARS, In celebration of the contributions of Andrew Lyne, Dick Manchester and Joe Taylor – A Festschrift honoring their 60th Birthdays
eds. Matthew Bailes, David J. Nice and Stephen E. Thorsett
ISBN: 1-58381-151-6

Vol. CS-303 SYMBIOTIC STARS PROBING STELLAR EVOLUTION
eds. R. L. M. Corradi, J. Mikołajewska and T. J. Mahoney
ISBN: 1-58381-152-4

Vol. CS-304 CNO IN THE UNIVERSE
eds. Corinne Charbonnel, Daniel Schaerer and Georges Meynet
ISBN: 1-58381-153-2

Vol. CS-305 International Conference on MAGNETIC FIELDS IN O, B AND A STARS: ORIGIN AND CONNECTION TO PULSATION, ROTATION AND MASS LOSS
eds. Luis A. Balona, Huib F. Henrichs and Rodney Medupe
ISBN: 1-58381-154-0

Vol. CS-306 NEW TECHNOLOGIES IN VLBI
ed. Y. C. Minh
ISBN: 1-58381-155-9

Vol. CS-307 SOLAR POLARIZATION 3
eds. Javier Trujillo Bueno and Jorge Sanchez Almeida
ISBN: 1-58381-156-7

Vol. CS-308 FROM X-RAY BINARIES TO GAMMA-RAY BURSTS
eds. Edward P. J. van den Heuvel, Lex Kaper, Evert Rol and Ralph A. M. J. Wijers
ISBN: 1-58381-158-3

**PUBLISHED: 2004**

Vol. CS-309 ASTROPHYSICS OF DUST
eds. Adolf N. Witt, Geoffrey C. Clayton and Bruce T. Draine
ISBN: 1-58381-159-1

Vol. CS-310 VARIABLE STARS IN THE LOCAL GROUP, IAU Colloquium 193
eds. Donald W. Kurtz and Karen R. Pollard
ISBN: 1-58381-162-1

Vol. CS-311 AGN PHYSICS WITH THE SLOAN DIGITAL SKY SURVEY
eds. Gordon T. Richards and Patrick B. Hall
ISBN: 1-58381-164-8

Vol. CS-312 Third Rome Workshop on GAMMA-RAY BURSTS IN THE AFTERGLOW ERA
eds. Marco Feroci, Filippo Frontera, Nicola Masetti and Luigi Piro
ISBN: 1-58381-165-6

Vol. CS-313 ASYMMETRICAL PLANETARY NEBULAE III: WINDS, STRUCTURE AND THE THUNDERBIRD
eds. Margaret Meixner, Joel H. Kastner, Bruce Balick and Noam Soker
ISBN: 1-58381-168-0

## ASP CONFERENCE SERIES VOLUMES

Published by the Astronomical Society of the Pacific

---

**PUBLISHED: 2004 (* asterisk means OUT OF PRINT)**

Vol. CS 314 — ASTRONOMICAL DATA ANALYSIS SOFTWARE AND SYSTEMS (ADASS) XIII
eds. Francois Ochsenbein, Mark G. Allen and Daniel Egret
ISBN: 1-58381-169-9 ISSN: 1080-7926

Vol. CS 315 — MAGNETIC CATACLYSMIC VARIABLES, IAU Colloquium 190
eds. Sonja Vrielmann and Mark Cropper
ISBN: 1-58381-170-2

Vol. CS 316 — ORDER AND CHAOS IN STELLAR AND PLANETARY SYSTEMS
eds. Gene G. Byrd, Konstantin V. Kholshevnikov, Aleksandr A. Mylläri, Igor' I. Nikiforov, and Victor V. Orlov
ISBN: 1-58381-172-9

A list of the IAU Volumes published by the ASP follows on the next page.

## INTERNATIONAL ASTRONOMICAL UNION (IAU) VOLUMES

Published by the Astronomical Society of the Pacific

**PUBLISHED: 1999 (* asterisk means OUT OF STOCK)**

Vol. No. 190 — NEW VIEWS OF THE MAGELLANIC CLOUDS
eds. You-Hua Chu, Nicholas B. Suntzeff, James E. Hesser, and David A. Bohlender
ISBN: 1-58381-021-8

Vol. No. 191 — ASYMPTOTIC GIANT BRANCH STARS
eds. T. Le Bertre, A. Lèbre, and C. Waelkens
ISBN: 1-886733-90-2

Vol. No. 192 — THE STELLAR CONTENT OF LOCAL GROUP GALAXIES
eds. Patricia Whitelock and Russell Cannon
ISBN: 1-886733-82-1

Vol. No. 193 — WOLF-RAYET PHENOMENA IN MASSIVE STARS AND STARBURST GALAXIES
eds. Karel A. van der Hucht, Gloria Koenigsberger, and Philippe R. J. Eenens
ISBN: 1-58381-004-8

Vol. No. 194 — ACTIVE GALACTIC NUCLEI AND RELATED PHENOMENA
eds. Yervant Terzian, Daniel Weedman, and Edward Khachikian
ISBN: 1-58381-008-0

**PUBLISHED: 2000**

Vol. XXIVA — TRANSACTIONS OF THE INTERNATIONAL ASTRONOMICAL UNION REPORTS ON ASTRONOMY 1996–1999
ed. Johannes Andersen
ISBN: 1-58381-035-8

Vol. No. 195 — HIGHLY ENERGETIC PHYSICAL PROCESSES AND MECHANISMS FOR EMISSION FROM ASTROPHYSICAL PLASMAS
eds. P. C. H. Martens, S. Tsuruta, and M. A. Weber
ISBN: 1-58381-038-2

Vol. No. 197 * — ASTROCHEMISTRY: FROM MOLECULAR CLOUDS TO PLANETARY SYSTEMS
eds. Y. C. Minh and E. F. van Dishoeck
ISBN: 1-58381-034-X

Vol. No. 198 — THE LIGHT ELEMENTS AND THEIR EVOLUTION
eds. L. da Silva, M. Spite, and J. R. de Medeiros
ISBN: 1-58381-048-X

**PUBLISHED: 2001**

IAU SPS — ASTRONOMY FOR DEVELOPING COUNTRIES
Special Session of the XXIV General Assembly of the IAU
ed. Alan H. Batten
ISBN: 1-58381-067-6

Vol. No. 196 — PRESERVING THE ASTRONOMICAL SKY
eds. R. J. Cohen and W. T. Sullivan, III
ISBN: 1-58381-078-1

Vol. No. 200 * — THE FORMATION OF BINARY STARS
eds. Hans Zinnecker and Robert D. Mathieu
ISBN: 1-58381-068-4

## INTERNATIONAL ASTRONOMICAL UNION (IAU) VOLUMES

Published by the Astronomical Society of the Pacific

---

**PUBLISHED: 2001 (* asterisk means OUT OF STOCK)**

Vol. No. 203 RECENT INSIGHTS INTO THE PHYSICS OF THE SUN AND HELIOSPHERE: HIGHLIGHTS FROM SOHO AND OTHER SPACE MISSIONS
eds. Pål Brekke, Bernhard Fleck, and Joseph B. Gurman
ISBN: 1-58381-069-2

Vol. No. 204 THE EXTRAGALACTIC INFRARED BACKGROUND AND ITS COSMOLOGICAL IMPLICATIONS
eds. Martin Harwit and Michael G. Hauser
ISBN: 1-58381-062-5

Vol. No. 205 GALAXIES AND THEIR CONSTITUENTS AT THE HIGHEST ANGULAR RESOLUTIONS
eds. Richard T. Schilizzi, Stuart N. Vogel, Francesco Paresce, and Martin S. Elvis
ISBN: 1-58381-066-8

Vol. XXIVB TRANSACTIONS OF THE INTERNATIONAL ASTRONOMICAL UNION REPORTS ON ASTRONOMY
ed. Hans Rickman
ISBN: 1-58381-087-0

**PUBLISHED: 2002**

Vol. No. 12 HIGHLIGHTS OF ASTRONOMY
ed. Hans Rickman
ISBN: 1-58381-086-2

Vol. No. 199 THE UNIVERSE AT LOW RADIO FREQUENCIES
eds. A. Pramesh Rao, G. Swarup, and Gopal-Krishna
ISBN: 58381-121-4

Vol. No. 206 COSMIC MASERS: FROM PROTOSTARS TO BLACKHOLES
eds. Victor Migenes and Mark J. Reid
ISBN: 1-58381-112-5

Vol. No. 207 EXTRAGALACTIC STAR CLUSTERS
eds. Doug Geisler, Eva K. Grebel, and Dante Minniti
ISBN: 1-58381-115-X

**PUBLISHED: 2003**

Vol. XXVA TRANSACTIONS OF THE INTERNATIONAL ASTRONOMICAL UNION REPORTS ON ASTRONOMY 1999–2002
ed. Hans Rickman
ISBN: 1-58381-137-0

Vol. No. 208 ASTROPHYSICAL SUPERCOMPUTING USING PARTICLE SIMULATIONS
eds. Junichiro Makino and Piet Hut
ISBN: 1-58381-139-7

Vol. No. 209 PLANETARY NEBULAE: THEIR EVOLUTION AND ROLE IN THE UNIVERSE
eds. Sun Kwok, Michael Dopita and Ralph Sutherland
ISBN: 1-58381-148-6

Vol. No. 210 MODELLING OF STELLAR ATMOSPHERES
eds. N. Piskunov, W. W. Weiss and D. F. Gray
ISBN: 1-58381-160-5

# INTERNATIONAL ASTRONOMICAL UNION (IAU) VOLUMES

Published by the Astronomical Society of the Pacific

---

**PUBLISHED: 2003 (* asterisk means OUT OF STOCK)**

Vol. No. 211 BROWN DWARFS
ed. Eduardo Martín
ISBN: 1-58381-132-X

Vol. No. 212 A MASSIVE STAR ODYSSEY: FROM MAIN SEQUENCE TO SUPERNOVA
eds. Karel A. van der Hucht, Artemio Herrero and César Esteban
ISBN: 1-58381-133-8

Vol. No. 214 HIGH ENERGY PROCESSES AND PHENOMENA IN ASTROPHYSICS
eds. X. D. Li, V. Trimble and Z. R. Wang
ISBN: 1-58381-157-5

**PUBLISHED: 2004**

Vol. No. 202 PLANETARY SYSTEMS IN THE UNIVERSE: OBSERVATION, FORMATION AND EVOLUTION
eds. Alan Penny, Pawel Artymowicz, Anne-Marie LaGrange and Sara Russell
ISBN: 1-58381-176-1

Vol. No. 213 BIOASTRONOMY 2002: LIFE AMONG THE STARS
eds. Ray P. Norris and Frank H. Stootman
ISBN: 1-58381-171-0

Vol. Nol 215 STELLAR ROTATION
eds. André Maeder and Philippe Eenens
ISBN: 1-58381-180-X

Vol. No. 217 RECYCLING INTERGALACTIC AND INTERSTELLAR MATTER
eds. Pierre-Alain Duc, Jonathan Braine and Elias Brinks
ISBN: 1-58381-166-4

Vol. No. 218 YOUNG NEUTRON STARS AND THEIR ENVIRONMENTS
eds. Fernando Camilo and Bryan M. Gaensler
ISBN: 1-58381-178-8

Vol. No. 219 STARS AS SUNS: ACTIVITY, EVOLUTION AND PLANETS
eds. A. K. Dupree and A. O. Benz
ISBN: 1-58381-163-X

Vol. No. 220 DARK MATTER IN GALAXIES
eds. S. D. Ryder, D. J. Pisano, M. A. Walker and K. C. Freeman
ISBN: 1-58381-167-2

Vol. No. 221 STAR FORMATION AT HIGH ANGULAR RESOLUTION
eds. Michael Burton, Ray Jayawardhana and Tyler Bourke
ISBN: 1-58381-161-3

Ordering information is available at the beginning of the listing